D1691686

Ultrathin Magnetic Structures I

J.A.C. Bland · B. Heinrich (Eds.)

Ultrathin Magnetic Structures I

An Introduction to the Electronic, Magnetic and Structural Properties

With 130 Figures

Springer

J. Anthony C. Bland
The Cavendish Laboratory
Department of Physics
University of Cambridge
Madingley Road
CB3 0HE Cambridge
United Kingdom
e-mail: jacb1@phy.cam.ac.uk

Bretislav Heinrich
Physics Department
Simon Fraser University
Burnaby, BC, V5A 1S6
Canada
e-mail: bheinric@sfu.ca

Library of Congress Control Number: 2004104844

ISBN 3-540-21955-2 Second Printing Springer Berlin Heidelberg New York
ISBN 3-540-57407-7 First Printing Springer Berlin Heidelberg New York

This work is subject to copyright. All rights are reserved, whether the whole or part of the material is concerned, specifically the rights of translation, reprinting, reuse of illustrations, recitation, broadcasting, reproduction on microfilm or in any other way, and storage in data banks. Duplication of this publication or parts thereof is permitted only under the provisions of the German Copyright Law of September 9, 1965, in its current version, and permission for use must always be obtained from Springer. Violations are liable for prosecution under the German Copyright Law.

Springer is a part of Springer Science+Business Media

springeronline.com

© Springer-Verlag Berlin Heidelberg 1994, 2005
Printed in Germany

The use of general descriptive names, registered names, trademarks, etc. in this publication does not imply, even in the absence of a specific statement, that such names are exempt from the relevant protective laws and regulations and therefore free for general use.

Production: LE-TeX Jelonek, Schmidt & Vöckler GbR, Leipzig
Cover production: Erich Kirchner, Heidelberg

Printed on acid-free paper   57/3141/YL - 5 4 3 2 1 0

Preface

The field of magnetic metallic structures is a truly multidisciplinary field. Scientists from various specialized fields, such as magnetism, molecular beam epitaxy (MBE) growth of III/V compounds, materials science, surface science, physics and chemistry, all strongly contribute to the research on ultrathin magnetic metallic films. Their interests are usually concentrated around topics that are closely related to their previous expertise. However, a full understanding of the systems studied can be achieved only if all results are considered together and understood in their entirety. There is therefore an enormous need for a comprehensive description of the various physical concepts used in this very diversified field. The purpose of this book is to present a wide but at the same time very detailed account of the most important phenomena and their underlying principles which govern the behavior of ultrathin metallic magnetic films. An immense range of techniques has been employed. We have emphasized the inclusion and description of the physical principles of those experimental techniques which in our view have played a crucial role in the development of metallic ultrathin structures.

The treatment is organized in two volumes, each divided into several chapters. Volume I includes chapters on: (1) Introduction and units, (2) The ground state electronic structure of ultrathin films and magnetic anisotropies, (3) The thermodynamic behavior of ultrathin films, (4) Spin polarized electron spectroscopy as a probe of ultrathin magnetic films, (5) Structural studies of MBE grown ultrathin films, and (6) Magnetic studies using spin-polarized neutrons.

Volume II includes: (1) MBE structures grown on III/V compound substrates and their magnetic properties, (2) Exchange coupling and magnetoresistance, (3) RF techniques applied to ultrathin structures, (4) Magnetic measurements of ultrathin films by surface magneto-optic Kerr effect (SMOKE), and (5) Mössbauer spectroscopy.

Most of the above chapters describe a wide range of physical aspects and techniques which in our view would be very difficult for only one author to cover properly. For that reason we decided to divide some of the chapters into several sections which are covered by different contributors. The chapters and their sections cover the most important aspects of metallic magnetic ultrathin structures.

The prime purpose of this treatment is to provide an accessible, self-contained monograph which in our view does not exist at present. Moreover, the study of metallic magnetic ultrathin structures has reached a point of such sophistication and complexity that a proper overview is now highly desirable. Due to the diversity of the various theoretical and experimental approaches adopted in the literature, it is our view that much can be learned from a single treatment in which the emphasis is put on physical concepts and their mutual relationships. This book therefore intends to present material that the listener and the reader miss at meeting presentations and from reading the existing literature. Given the present proliferation of scientific ideas, the editors and contributors agree that this is a particularly worthwhile and useful task. We also hope, by presenting in a single unified survey, the key ideas, developments and techniques in this field, that the result will be in itself stimulating from a research viewpoint.

We have been very fortunate in being able to enlist an excellent panel of first class scientists working in the field of ultrathin magnetic metallic structures. All contributors agree that the primary goal of this book is educational. Their contributions are not written in the absence of direct contacts between individual contributors. All contributors have worked for many years in this field and their work represents the main stream of ultrathin magnetic metallic structures. They have shared their ideas and in some cases fruitfully collaborated during the development of this field. All contributors understand perfectly well that none of them possess infinite wisdom in this field and they believe that the sharing of new ideas and concepts is essential to the healthy development of the field of magnetic ultrathin structures.

We would like to express our thanks to all participating authors for their willingness to put aside an appreciable amount of time to write their chapters and to cross-correlate their writings with other contributors. We greatly appreciate the authors' attempt to share their experience and expertise which allowed them to contribute so successfully to magnetic structures.

The authors and editors of this book hope that their contributions will be useful to experts already working in the field of magnetic metallic ultrathin structures as well as to those who are thinking about entering this field. We hope that the reader will find this book a pleasure to read and that the material presented will enrich the reader's understanding of this truly fascinating and rapidly developing field.

Cambridge, UK J.A.C. Bland
Burnaby, Canada B. Heinrich
February 1994

List of Acronyms and Abbreviations

AGM	Alternating Gradient Magnetometry
ASA	Atomic Sphere Approximation
BLS	Brillouin Light Scattering
FLAPW	Full Potential Linear Augmented Plane Wave
FMR	Ferromagnetic Resonance
KKR	Korringa Kohn Rostoker
LDA	Local Density Approximation
LEED	Low Energy Electron Diffraction
LMTO	Linear Muffin Tin Orbital
LSDA	Local-Spin-Density Approximation
MAE	Magnetic Anisotropy Energy
MBE	Molecular Beam Epitaxy
MOKE	Magneto-Optic Kerr Effect
NMR	Nuclear Magnetic Resonance
PMA	Perpendicular Magnetic Anisotropy
RHEED	Reflection High Energy Electron Diffraction
RLP	Reciprocal Lattice Point
RPA	Random Phase Approximation
RS	Rotating Sample
SCLO	Self-Consistent Local Orbital
SMOKE	Surface Magneto-Optic Kerr Effect
SQUID	Superconducting Quantum Interference Device
TEM	Transmission Electron Microscopy
TOF	Time of Flight
TOM	Torsion Oscillating Magnetometry
UMS	Ultraviolet Magnetic Structures
UPS	Ultraviolet Photoemission Spectroscopy
VSM	Vibrating Sample Magnetometry
XPS	X-ray Photoemission Spectroscopy

Contents

1. Introduction
1.1 Overview
   B. Heinrich and J.A.C. Bland 1
1.2 Magnetism in SI Units and Gaussian Units
   A.S. Arrott 7
   1.2.1 Equations of Electricity and Magnetism 9
   1.2.2 Translation Keys 17
References 19

2. Magnetic Anisotropy, Magnetization and Band Structure
2.1 Electronic Structure of Magnetic Thin Films
   J.G. Gay and R. Richter (With 3 Figures) 21
   2.1.1 Underlying Theory 22
   2.1.2 Calculation of the Magnetic Properties of Bulk Fe, Co and Ni 27
   2.1.3 Calculation of the Magnetic Properties of Thin Films 31
   2.1.4 Conclusions 39
2.2 Magnetic Anisotropy from First Principles
   G.H.O. Daalderop, P.J. Kelly, and M.F.H. Schuurmans
   (With 19 Figures) 40
   2.2.1 Method 41
   2.2.2 Spin-Orbit Coupling Matrix 42
   2.2.3 Co Monolayer 42
   2.2.4 Co/Pd, Co/Ag and Co/Cu Multilayers 51
   2.2.5 Co/Ni Multilayers 57
   2.2.6 Analysis of the Anisotropy Energy of Co_1Pd_2 Multilayers . . 61
   2.2.7 Conclusions 64
2.3 Experimental Investigations of Magnetic Anisotropy
   W.J.M. de Jonge, P.J.H. Bloemen, and F.J.A. den Broeder
   (With 9 Figures) 65
   2.3.1 Origin of the Magnetic Anisotropy in Thin Films 65
   2.3.2 Experimental Methods 73
   2.3.3 Experimental Results 78
   2.3.4 Concluding Remarks 85
References 86

3. **Thermodynamic Properties of Ultrathin Ferromagnetic Films**
   D.L. Mills (With 1 Figure) 91
   3.1 Introduction 91
   3.2 Interactions Between Spins: A Basic Spin Hamiltonian 93
   3.3 Properties of Ultrathin Ferromagnetic Films at Low Temperatures:
       The Ground State and the Spin Wave Regime 97
       3.3.1 The Ground State 97
       3.3.2 The Nature of Spin Waves in Ultrathin Films; Low
             Temperature Thermodynamic Properties 100
   3.4 Beyond Spin Wave Theory: The Intermediate Temperature Regime . . 110
   3.5 The Transition Temperature of Ultrathin Films 114
   3.6 Concluding Remarks 115
   References 121

4. **Spin-Polarized Spectroscopies**
   4.1 Spin-Polarized Electron Spectroscopies
       H. Hopster (With 19 Figures) 123
       4.1.1 Introduction 123
       4.1.2 Instrumentation 128
       4.1.3 Secondary Electrons (SPSEES) 132
       4.1.4 Elastic Scattering (Spin-Polarized Low-Energy Electron
             Diffraction: SPLEED) 135
       4.1.5 Inelastic Scattering (Spin-Polarized Electron Energy-Loss
             Spectroscopy: SPEELS) 139
       4.1.6 Photoemission Techniques 145
       4.1.7 Concluding Remarks 152
   4.2 Probing Magnetic Properties with Spin-Polarized Electrons
       H.C. Siegmann and E. Kay (With 7 Figures) 152
       4.2.1 Magnetic Information from Measurement of Spin Polarization
             or Spin Asymmetry 153
       4.2.2 Unique Features of Magnetometry with Spin Polarized
             Electrons 155
       4.2.3 Field Dependence of the Magnetization 157
       4.2.4 Temperature Dependence of the Magnetization 162
       4.2.5 Magnetism away from Equilibrium 169
   References 171

5. **Epitaxial Growth of Metallic Structures**
   5.1 Introduction to Reflection High Energy Electron
       Diffraction (RHEED)
       A.S. Arrott (With 20 Figures) 177
       5.1.1 Real Space and k-space 177
       5.1.2 The Surface as a Diffraction Grating 182
       5.1.3 Waves Inside a Slab 200
       5.1.4 Applications of RHEED to the Study of Growth 215

5.2 X-Ray Photoelectron and Auger Electron Forward Scattering:
    A Structural Diagnostic for Epitaxial Thin Films
    W.F. Egelhoff Jr. (With 25 Figures). 220
    5.2.1 Introduction. 221
    5.2.2 The Basics of Electron-Atom Scattering 223
    5.2.3 Experimental Problems of Current Interest 250
    5.2.4 Conclusions 263
5.3 X-Ray Studies of Ultrathin Magnetic Structures
    R. Clarke and F.J. Lamelas (With 15 Figures) 264
    5.3.1 Introduction. 264
    5.3.2 Overview of the Problem 265
    5.3.3 X-Ray Diffuse Scattering 266
    5.3.4 Modeling Ultrathin Layered Structures 269
    5.3.5 Future Directions 284
References. 285

6. Polarized Neutron Reflection
J.A.C. Bland (With 12 Figures) 305
6.1 Introduction 305
6.2 Theory of Polarized Neutron Reflection. 306
    6.2.1 The Optical Potential for a Magnetized Medium 307
    6.2.2 Transfer Matrix Methods and the Polarization
          Dependent Reflectivity 311
    6.2.3 PNR Magnetometry of Single Magnetic Films 315
    6.2.4 The Diffraction Limit 320
    6.2.5 Rough Interfaces and Wave Coherence 322
6.3 Experimental Methods 329
    6.3.1 Time of Flight Methods. 329
    6.3.2 Fixed Wavelength Methods 331
6.4 Experimental Results for Fe Films 332
    6.4.1 Magnetic Moments in Ultrathin Fe Films 332
    6.4.2 Comparison of the Experimentally Determined Moment
          with Theory. 338
    6.4.3 Conclusions 341
References. 342

Subject Index 345

Contributors

A.S. Arrott
Physics Department, Simon Fraser University, Burnaby, BC, V5A 1S6, Canada

J.A.C. Bland
Department of Physics, Cavendish Laboratory, University of Cambridge,
Madingley Road, Cambridge, CB3 0HE, UK

P.J.H. Bloemen
Department of Physics, Eindhoven University of Technology (EUT),
5600 MB Eindhoven, The Netherlands

F.J.A. den Broeder
Philips Research Laboratories, PO Box 80.000, 5600 JA Eindhoven,
The Netherlands

R. Clarke
Department of Physics, Randall Laboratory, The University of Michigan,
Ann Arbor, MI 48109, USA

G.H.O. Daalderop
Philips Research Laboratories, PO Box 80.000, 5600 JA Eindhoven,
The Netherlands

W.F. Egelhoff, Jr
Surface and Microanalysis Science Division, National Institute of Standards
and Technology, Gaithersburg, MD 20899, USA

J.G. Gay
Physics Department, General Motors Research Laboratories, Warren,
MI 48090-9055, USA

H. Hopster
Department of Physics and Institute for Surface and Interface Science,
University of California, Irvine, CA 92717, USA

W.J.M. de Jonge
Department of Physics, Eindhoven University of Technology (EUT),
5600 MB Eindhoven, The Netherlands

E. Kay
IBM Almaden Research Center, 650 Harry Road, San Jose, CA 95120, USA

P.J. Kelly
Philips Research Laboratories, PO Box 80.000, 5600 JA Eindhoven,
The Netherlands

F.J. Lamelas
Department of Physics, Marquette University, Milwaukee,
WI 53233, USA

D.L. Mills
Department of Physics and Institute for Surface and Interface Science,
University of California, Irvine, CA 92717, USA

R. Richter
Physics Department, General Motors Research Laboratories, Warren,
MI 48090-9055, USA

M.F.H. Schuurmans
Philips Research Laboratories, PO Box 80.000, 5600 JA Eindhoven,
The Netherlands

H.C. Siegmann
Swiss Federal Institute of Technology, CH-8093 Zurich, Switzerland

1. Introduction

B. HEINRICH and J.A.C. BLAND

1.1 Overview

Progress in modern science is driven by intellectual curiosity which is often triggered by the availability of new techniques. Unbounded optimism and stubbornness in pursuing scientific investigations is often richly rewarded by the discovery of phenomena which were neither expected nor even envisioned and by the development of entirely new ideas and concepts. The field of magnetic ultrathin metallic structures has evolved in exactly this way and is rapidly becoming one of the most active and exciting areas of current solid state research, offering the promise of unravelling problems which have lain at the heart of magnetism for at least half a century while the prospect of valuable applications based on novel magnetic phenomena is already in sight.

Some of the underlying ideas in the field of magnetic ultrathin metallic structures considerably predate the current research activity. For example, when *F. Bloch* introduced the concept of spin waves in 1930 he already recognized that a two-dimensional spin system with Heisenberg interactions cannot be ferromagnetic at finite temperatures. However the possibility of synthesizing a two-dimensional metallic film was well beyond the reach of experimental physics at that time. Moreover no techniques were available within a sensitivity which permitted the investigation of the magnetic properties of a single film of atomic thickness. Research on thin (i.e. nm lengthscale) metallic films was already a center of activity in magnetism in the late 1950s, motivated by the prospect of discovering novel magnetic phenomena with potential for applications. It was anticipated from theoretical considerations that the reduced thickness and the presence of the interface should modify such properties as the magnetic domain structure, spin wave spectrum and magnetic anisotropy, for example. Studies of spin wave resonances well illustrate the successes and difficulties of that time. The prediction of spin waves in thin films by *C. Kittel* and their subsequent observation in Permalloy films were significant advances. However, most of the work done at that time was hindered by insufficiently high vacuum and a lack of probes for structural and spectroscopic characterization, while the necessary control of crystalline structure, chemical purity, interface quality and homogeneity during growth were not yet available. However, many interesting topics were raised already at that time and the fact that the available experimental techniques did not give clear answers did not make these topics irrelevant. They waited to be readdressed when the availability of ultra high vacuum technology and the advances in surface science made research in magnetic films again very attractive.

It was indeed the success of molecular beam epitaxy (MBE) of III–V compounds which turned attention once again to magnetic metallic films. The ability to grow ultrathin semiconducting structures which exhibited unique properties created an opportunity which could not be missed by experimentalists working in magnetism. Approximately 10–12 years ago several research groups were willing to take up this challenge and set up MBE systems with the intention of growing magnetic metallic epitaxial films. The past ten years have shown that this decision was both prudent and timely. Tools for the study of MBE growth of semiconductors such as reflection high-energy electron diffraction (RHEED) were also applied with great success to the study of the epitaxial growth of metals despite the significantly different details of film growth occurring in semiconductors and metals – see for example Chap. 5 of Volume I. The availability of powerful supercomputers in the last decade made it possible to perform sophisticated spin resolved electronic structure and total energy calculations with high accuracy. The predictions of first principles calculations of modified magnetic moments in 2D structures in the mid-eighties came to serve as an ever present reminder that the magnetism of ultrathin structures was extremely attractive for both those who wanted to understand the magnetic moment formation in systems with reduced dimensionality and those who wanted to create a new class of magnetic materials based on superlattice structures in which the magnetic properties could be controllably modified. Concurrently, significant progress in the development of probes of surface magnetism occurred. For example, advances in techniques based on polarized electron spectroscopies were made possible by the development of novel electron spin analyzers in the late seventies and novel spin-polarized electron sources, such as the optically pumped GaAs source – for example see Volume I Chap. 4 and Volume II Chap. 2. Such techniques are particularly valuable in providing a means of studying the surface magnetic properties in-situ because of their high surface sensitivity. The availability of MBE metals growth also stimulated advances in the use of conventional probes of magnetism which permitted the study of magnetic properties of ultrathin magnetic layers in the form of sandwich or multilayer structures – for example Volume I Chap. 6 on neutron scattering techniques, Volume II Chap. 3 on RF techniques, Volume II Chap. 4 on magneto-optical techniques and Volume II Chap. 5 on Mossbauer spectroscopy techniques.

The development of this field reflects the diversity of modern sciences. The techniques of surface science became a part of research in magnetism and many researchers in magnetism began to attend conferences which were devoted to surface science. This was a crucial step which resulted not only in a better understanding of the structures grown, but at the same time encouraged the creation of fruitful links between the scientists working in traditional areas of magnetism and those who were either proponents of surface science or who were involved in MBE growth of semiconducting structures. The successes of MBE of magnetic structures is based on the ability to epitaxially grow a wide variety of systems: for example a range of high quality metastable phases can be synthesized as can bulk phases with modified lattice parameters. It is not our intention here to provide a comprehensive account of work done in magnetic metallic epitaxy. In fact this book will not cover the very successful field of rare-earth superlattices, but will be only devoted to 3d transition metal ferromagnetic films and superlattices.

Metallic magnetic epitaxial films have been grown on both semiconducting and metallic substrates. The strain induced by epitaxial growth on a single crystal substrate provides a way of stabilizing crystallographic structures at room temperature which correspond to high pressure or high temperature phases in the bulk material phase diagram. Such artificially occurring metastable structures often exhibit unique magnetic properties and provide an opportunity to perform stringent tests of computations of magnetic and structural peoperties. In addition bulk-like phases can be prepared as ultrathin (i.e. few atoms thick) layers in order to study the effects of film thickness and the presence of the interface in determining magnetic properties. The list below shows the great diversity available in the epitaxy of magnetic metals. III/V and II/VI compound substrates provided very good templates for the growth of 3d transition metals. Bcc Fe and metastable bcc Co were grown on GaAs and ZnSe substrates. Seeded epitaxy in which a buffer layer of Co, Pt or Fe is first prepared to stabilize a single crystal non-magnetic layer allowed one to grow for example [Fe^{bcc}, Cr], [Fe^{bcc}, Ag], [Co^{fcc}, Pt] and [Co^{fcc}, Cu] superlattices on oriented semiconductor substrates. [Co^{hcp}, Au], [Co^{hcp}, Cr], [$Co^{fcc, hcp}$/Pd] superlattices have also been successfully grown on oriented semiconductor substrates. Many epitaxial systems have been grown on metallic substrates such as $Mn^{hcp,\ Laves\ structures}$ on Ru, Fe^{bcc} on W and Ni^{fcc} on Re, Fe^{bcc} on Ag, Fe^{fcc} on Cu, Co^{fcc} on Cu, Co^{hcp} on Au, Ni^{bcc} on Fe and Fe^{bcc} on Pd. The above list of stable and metastable epitaxial structures is far from being complete; more detailed accounts can be found in the individual chapters. The requirements for epitaxial growth are demanding and sometimes conflicting. In-plane lattice matching of the epitaxial phase to the substrate is not alone sufficient – in addition strain along the growth direction, surface energies, diffusion constants and bulk phase behavior are important considerations. In transition metals the energies associated with changes in crystalline structure are often of the same magnitude as those associated with a change in magnetic structure, resulting in a very sensitive dependence of magnetic properties on film structure and growth conditions. The goal of ultrathin film magnetism research is to systematically identify and understand how magnetic properties depend on film structure and quality. It is for this reason that the study of epitaxial phases plays such an important role in the subject. The presence of surface and material scientists in this field has been very beneficial since they have contributed greatly to the use and appreciation of various structural techniques which then helped to improve the characterization of the ultrathin films used for magnetic studies. In this book we describe some of the techniques essential for MBE but do not attempt to describe particular MBE systems.

The expressions "ultrathin" and "superlattice" are used frequently throughout this book and therefore a word of explanation of their physical meaning is appropriate. What is meant by "thin" depends upon the physical quantity under consideration. We begin by considering the lengthscale over which the magnetization is uniform across the film since this can be used to define "ultrathin". The concept of ultrathin films and structures is also closely associated with the presence of interface (surface) magnetic anisotropies (i.e. a dependence of the total energy of a single magnetic domain on the orientation of the magnetization direction with respect to the surface normal) since such anisotropies impose boundary conditions on the magnetization. It was shown by *L. Néel* in the 1950s that the broken symmetry at interfaces results in a surface magnetic anisotropy. Their presence was clearly identified already in the 1960s in ultrathin films of NiFe(1 1 1) grown epitaxially on Cu(1 1 1) for example. Interest in the surface anisotropies has significantly increased in the last six to seven years. Spin-polarized local density band calculations which included relativistic spin-orbit contributions in Fe(0 0 1) showed that the broken symmetry at the interfaces can significantly alter the magnetic anisotropies of interface 3d valence band electrons, and could result in giant uniaxial anisotropies perpendicular to the film surface. It is interesting to note that in a pure bcc structure the Néel surface anisotropy is absent in (0 0 1) surfaces for the nearest neighbor interaction.

The lack of spin-polarized photoemission from Fe(0 0 1)/Ag(0 0 1) ultrathin films suggested that the surface anisotropies in these samples were sufficiently strong to overcome the demagnetizing field causing the saturation magnetization to be oriented perpendicularly to the film surface. Shortly after that the surface anisotropies in Fe(0 0 1) ultrathin films and superlattices were quantitatively determined in several experiments. It was also shown that the surface anisotropies in Fe(0 0 1) ultrathin films are very close to those found in bulk Fe(0 0 1). The presence of strong uniaxial anisotropies in Fe(0 0 1) raised the question of the nature of the magnetic configuration around the interfaces in the presence of an interface uniaxial anisotropy with its easy axis perpendicular to the surface. Similar questions had already been discussed in the sixties in an attempt to explain the then reported observation of decreased magnetic moments at the interfaces of magnetic films, although it is now known that the decrease was in this case due to the surface chemical contamination. The orientation θ of magnetic moments with respect to the surface normal (z axis) near the interface is determined by the competition between exchange energy, proportional to $d^2\theta/dz^2$ and the relevant anisotropy energies (for example, the demagnetizing energy is $\frac{1}{2}\mu_0 M_s^2 \cos^2\theta$, while an additional term of the form $K_s \cos^2\theta$ can also appear due to the existence of a surface uniaxial magnetic anisotropy). In bulk samples the saturation magnetization remains parallel with the sample surface until the uniaxial anisotropy reaches a critical value ($\sim 6 \times 10^{-3}$ J m^{-2} in Fe), then the surface saturation magnetization becomes oriented perpendicular to the surface and the bulk saturation magnetization gradually rotates back to its parallel configuration.

Denoting the coefficient of the exchange interaction by A and the value of the saturation magnetization by M_S, the length scale over which the magnetization returns back to its parallel configuration is given by the coefficient $(2A/\mu_0 M_S^2)^{-1/2}$ (30ML in Fe), and it is usually referred to as the exchange length. In films thinner than the exchange length the magnetic moments across the film thickness remain parallel. Films in which the saturation magnetization is uniform across the film thickness can be referred to as "ultrathin". The total magnetic moment is given by an algebraic sum of all atomic magnetic moments across the film thickness. Due to a strong exchange interaction, the magnetic torques acting on the individual spins are shared equally among all the spins across the film thickness. Consequently the surface torques act like bulk torques. Since surface torques are shared equally across the sample thickness their contribution scales inversely with the film thickness, $1/t$, and in that way they can be distinguished from those effective fields which have their origin in the bulk. Caution is needed as the above simplified description is inevitably partly a fiction. To proceed further one has to introduce a more general definition of the exchange length – see Volume II Chap. 3, for example. In fact, atomic moments are not exactly parallel even in ultrathin films. However the deviation from parallel alignment is very small and has very little effect on the overall behavior of ultrathin films. Ultrathin films exhibit magnetic properties associated with the effective fields which are an admixture of the interface (inversely proportional to the film thickness) and true bulk effective fields.

Ultrathin films behave like giant magnetic molecules and it is indeed remarkable that well grown ultrathin films follow this simplified concept. However first-principles band calculations for ideal, sharp interfaces indicate that the electron charge densities and the majority and minority spin densities of the valence electrons very quickly reach bulk-like values just a few atomic layers away from the interface, a result which is far from obvious. The lengthscale over which this happens is typically around 3ML for Fe for example. The admixture of surface and bulk magnetic properties in ultrathin films allows one to engineer new magnetic properties which are not obtainable in bulk materials and lends a true meaning to the concept of giant magnetic molecules. Unique magnetic materials have been grown: for example [Co/Pt] superlattices can be engineered with perpendicular anisotropies far surpassing the perpendicular demagnetizing fields (resulting in a remanent magnetization along the surface normal) and can be employed in magneto-optical recording.

Giant magnetic molecules are often incorporated into a superlattice structure in order to create layers which are sufficiently thick to be technologically useful. In these structures the magnetic layers can be coupled through non-magnetic interlayers. In metallic magnetic structures the word "superlattice" should be taken with caution since we need to distinguish between the true superlattice effects associated with the artificial periodicity and effects which merely correspond to the addition of single layer properties. Also some authors use the term "superlattice" to refer to an epitaxially grown structure in distinction to the term "multilayer" which is not epitaxially grown, although this usage does not have universal acceptance. The concept of superlattices was developed in the context of semiconductor MBE. In semiconductors the superlattice potential can strongly affect the quantum levels of the valence electrons. In this case the superlattice potential can be responsible for unique properties which are otherwise absent in the individual building block. In metallic magnetic superlattices, as we understand them so far, the situation is different. The basic magnetic properties such as magnetic anisotropies are mostly determined by the magnetic properties of the individual unit blocks (giant magnetic molecules) and by the strength of the exchange coupling through the non-magnetic interlayers. However, the existence of "magnetic" quantum well states in thin magnetic layers and multilayers has been reported recently and it is known that spin-split interface states exist at interfaces between magnetic and non-magnetic layers in superlattices. The importance of quantum well (resonance) states in the interlayer exchange coupling has recently been well established experimentally and theoretically. Zone folding effects have also been reported for magnetic excitations in magnetic superlattices, and the diffraction of X-rays or electrons into super-reflections defined by the multiple layering is a clearly a superlattice effect.

Studies of ultrathin magnetic structures are not limited to epitaxial structures only. In fact an essential part of this field is concerned with polycrystalline ultrathin structures with sharp interfaces grown by sputtering since a comparison of these systems yields insight into the morphology dependence of specific magnetic properties. Such structures can also exhibit properties which are comparable to those of structures grown epitaxially and in some cases the properties of sputtered structures are even enhanced with respect to those of the MBE grown structures – see for example Volume II Chap. 2 describing the magnetoresistivity of sputtered multilayer structures and Volume I Chap. 2 on the magnetic anisotropy behavior of sputtered films. Furthermore, from an application-oriented viewpoint, sputtered films are of strong interest because of the viability of producing samples commercially.

Finally, a word about units. This is a difficulty since many magneticians tend to use Gaussian units, partly because a large body of literature now exists which is written in these units, whereas many Europeans tend to use SI units automatically (or in some cases because they are required to). This issue is a particular concern for those entering the field. In writing this book it was first thought that it would be best to use one system of units only. But since it is by no means clear which units to use, it was decided that it would prove more educational if the book were to make use of both units and to include a conversion table between the two systems. For this reason, some sections are written in Gaussian units and others in SI, according to the authors' preferences. The reader is referred to the next section in which both systems of units are discussed and the conversion between them is given. While at first sight the reader may find it inconvenient to have to convert between units, we hope that after using this book he or she will agree that it is indeed necessary to do this and that anyone wishing to seriously read the literature in magnetism must be fully conversant with both systems.

1.2 Magnetism in SI Units and Gaussian Units

A.S. ARROTT

Equations, units, dimensions and conversion factors are provided here as an aid to translation between the two languages of magnetism.

The late William Fuller Brown, Jr., founder of the field of micromagnetics, wrote a *Tutorial Paper on Dimensions and Units* [1.1], in which he said:

> "*If this seems a bit artibrary and confusing, bear in mind two principles: first, dimensions are the invention of man, and man is at liberty to assign them in any way he pleases, as long as he is consistent throughout any one interrelated set of calculations. Second, international committees arrive at their decisions by the same irrational procedures as do various IEEE committees that you have served on.*"

To writers on the subject, Brown advises, "At all costs avoid tables: with them, you never know whether to multiply or divide." To the reader he consoles, "This terminology is quite arbitrary; don't take it too seriously, or you may get yourself into philosophical dilemmas."

The field of magnetism has been recently enriched by attachment to the world of surface science. This has created some demand for those in the field of magnetism to convert from the Gaussian system to the International System (SI). Workers in magnetism have been reluctant to use SI units for reasons that become apparent when the two systems are compared. Both systems have been said to be "a bit arbitrary and confusing." The arbitrariness adds to the confusion when translation is attempted. Both systems are self-consistent, but the translations exhibit apparent contradictions. Sympathy must be expressed for those surface scientists new to the field of magnetism. The following equations, units, dimensions and conversion factors are provided as an aid to translation between the two languages of magnetism.

The SI is based on the meter, kilogram, second and ampere, mksa. The Gaussian system is based on the centimeter, gram and second, cgs. It uses the electrostatic cgs units for electrical quantities and the electromagnetic cgs units for magnetic properties. In the Gaussian system, the velocity of light c appears in equations that have both magnetic and electrical quantities. In Maxwell's equations in Gaussian units the velocity of light multiplies each occurrence of the variable t whereever it appears, as in the expression $(1/c)\partial E/\partial t$, or wherever it is implied by a hidden time derivative, as in the combination $(4\pi/c)I$. In the Gaussian system each source term is multiplied by 4π. In the SI system 4π is banished from the fundamental equations, but reappears when fields are calculated from sources.

Tables for converting between the two systems appear in many texts. It is harder to find direct comparisons of the equations of electricity and magnetism side by side as given here along with a discussion of units and dimensions. The presentation starts with the Lorentz force equation to introduce $\boldsymbol{E}$ and $\boldsymbol{B}$. This is followed by the electrical and then magnetic quantities leading to Maxwell's equations and the equation of motion for magnetism. The latter is a torque equation with the dimensions of energy density. As it is the same for both systems of units it appears in the center of the line:[1]

$$\frac{1}{\gamma}\frac{\partial \boldsymbol{M}}{\partial t} = \boldsymbol{M} \times \boldsymbol{B}_{\text{eff}}, \tag{1.45}$$

where effective magnetic induction $\boldsymbol{B}_{\text{eff}}$ is the vector gradient of the energy density $w(M)$ with respect to the components of the magnetization $\boldsymbol{M}$:

$$\boldsymbol{B}_{\text{eff}} = -\nabla_M w(M) + \lambda \boldsymbol{M}. \tag{1.46}$$

When the equation of motion is presented in terms of the magnetic field $\boldsymbol{H}$, the equations are different. The SI equations are put on the left and the Gaussian equations on the right.

In SI units:

$$\frac{1}{\gamma}\frac{\partial \boldsymbol{M}}{\partial t} = \boldsymbol{M} \times \mu_0 \boldsymbol{H}_{\text{eff}}, \tag{1.62}$$

where

$$\boldsymbol{H}_{\text{eff}} = -\frac{\nabla_M w(M)}{\mu_0} + \lambda \boldsymbol{M}. \tag{1.63}$$

In Gaussian units:

$$\frac{1}{\gamma}\frac{\partial \boldsymbol{M}}{\partial t} = \boldsymbol{M} \times \boldsymbol{H}_{\text{eff}}, \tag{1.64}$$

where

$$\boldsymbol{H}_{\text{eff}} = -\nabla_M w(M) + \lambda \boldsymbol{M}. \tag{1.65}$$

The equations are explained as they are introduced. The units and dimensions of the quantities are listed on the left and right. The conversions between quantities in the two systems are given as equations in the center of the line. For example, some of the principle magnetic quantities are:

SI units	conversion	Gaussian units
m is in $A\,m^2$ = joule/tesla	10^{-3} J/T = 1 erg/G	m is in erg/G, emu
B, $\mu_0 M$, and $\mu_0 H$ are in teslas, (T)	10^{-4} T = 1 G	B and $4\pi M$ are in gausses, G
M is in $A\,m^{-1}$	1000 $A\,m^{-1}$ = 1 erg ($G^{-1}\,cm^{-3}$)	M is in emu cm^{-3}
γ is in $T^{-1}s^{-1}$	$10^4 (T\,s)^{-1}$ = 1 $(G\,s)^{-1}$	γ is in $G^{-1}\,s^{-1}$
H is in ampere(turns) m^{-1}	$1000/4\pi\ A\,m^{-1}$ = 1 Oe	H is in oersteds, Oe
Φ_B is in webers = $T\,m^2$	10^{-8} Wb = 1 maxwell	Φ_B is in maxwells = $G\,cm^2$

---

[1] These equations appear again later in this section and the numbering refers to the sequence in which they are presented there.

1.2 Magnetism in SI Units and Gaussian Units

In the Gaussian system B, H and M have the same dimensions, but not the same units. This may seem strange to a visitor from SI space to Gaussian space. Indeed it can be a source of confusion, but note that torque and energy have the same dimensions, but are quoted in different units, e.g. N m and joule.

A general method of translating equations between the two systems is given in the final section. There are eight separate forms for the translation keys. The application of these to one particular equation is shown. The number of required steps illustrates why one might rather have all the pertinent equations side by side.

1.2.1 Equations of Electricity and Magnetism

in SI Units: and in Gaussian Units:
{always on the left} {always on the right}

The Lorentz force equation defines the electric field E and the magnetic induction B in terms of the force F on a charge q and its velocity v with respect to the frame of reference:

$$F = qE + qv \times B, \quad (1.1) \qquad F_{cgs} = q_{esu}\left(E_{esu} + \frac{v_{cgs} \times B_{emu}}{c_{cgs}}\right). \quad (1.2)$$

F is in newton, N	10^{-5} N = 1 dyn	F_{cgs} is in dyn
q is in coulomb, C	1/2997924580 C = 1 statcoulomb	q_{esu} is in statcoulomb
v is in m s^{-1}	0.01 m s^{-1} = 1 cm s^{-1}	v is in cm s^{-1}
qv is in ampere meter, A m		qv is in statampere cm
I is in ampere, A	$3.33564095 \times 10^{-10}$ A = 1 statampere	I_{esu} is in statampere
E is in N C^{-1}	29979.2458 N C^{-1} = 1 dyn statcoulomb^{-1}	E_{esu} is in dyn statcoulomb^{-1}
1 N C^{-1} = 1 volt meter^{-1} = 1 V m^{-1}	299.792458 v = 1 statvolt	1 dyn/statcoulomb = 1 statvolt cm^{-1}
B is in teslas, T		B_{emu} is in gauss, G
1 T = 1 N A^{-1} m^{-1} = J A^{-1} m^{-2}	10^{-4} T = 1 G	1 G = 1 dyn statcoulomb^{-1}

The dimensions [ ] of E are the same in the two systems but the dimensions of

B are not:

$$[E] = [\text{energy}] [\text{current}]^{-1} \times [\text{distance}]^{-1} [\text{time}]^{-1}$$

$$[E] = [\text{energy}] [\text{current}]^{-1} \times [\text{distance}]^{-1} [\text{time}]^{-1}$$

$$[B] = [\text{energy}] [\text{current}]^{-1} \times [\text{distance}]^{-2}$$

$$[B] = [\text{energy}] [\text{current}]^{-1} \times [\text{distance}]^{-1} [\text{time}]^{-1}$$

Electrical quantities

Charge is a source of the electric field. In differential form the field has the total charge density ρ_T as a source:

$$\nabla \cdot \boldsymbol{E} = 4\pi k_e \rho_T, \tag{1.3}$$

$$\rho_T = \rho + \rho_P = \rho - \nabla \cdot \boldsymbol{P}_{\text{dip}}, \tag{1.4}$$

where $\boldsymbol{P}_{\text{dip}}$ is the dipole moment per unit volume. The subscript "dip" is dropped in the remainder of this chapter.

In SI:

$$k_e = \frac{1}{4\pi\varepsilon_0} = Kc^2, \tag{1.5}$$

$$K \equiv 10^{-7} \, \text{N A}^{-2}, \tag{1.6}$$

$$\frac{1}{4\pi\varepsilon_0} = 8.98755179 \times 10^9 \, \text{V m C}^{-1}, \tag{1.7}$$

$$c \equiv 299792458 \, \text{m s}^{-1}, \tag{1.8}$$

$$\varepsilon_0 \nabla \cdot \boldsymbol{E} = \rho_T. \tag{1.9}$$

In esu and Gaussian units:

$$k_e = 1 \frac{\text{statvolt cm}}{\text{statcoulomb}}. \tag{1.10}$$

In the Gaussian system

$$1 \, \text{statvolt} = 1 \, \text{statcoulomb cm}^{-1},$$

so that k_e is usually considered to be dimensionless, but there are certain advantages in dimensional analysis if the dimensions of k_e are carried along.

$$\nabla \cdot \boldsymbol{E} = 4\pi \rho_T. \tag{1.11}$$

The dimensions of $\boldsymbol{P}$ are the same in the two systems:

$$[P] = [\text{current}] [\text{time}] [\text{length}]^{-2}.$$

In both systems of units the electric dipole is $\boldsymbol{p} = q\boldsymbol{l}$ and the polarization $\boldsymbol{P}$ is the dipole moment per unit volume. The torque on an electric dipole $\boldsymbol{p}_{\text{dip}}$ in a field $\boldsymbol{E}$ is

$$\boldsymbol{T} = \boldsymbol{p} \times \boldsymbol{E}. \tag{1.12}$$

$\boldsymbol{T}$ is in joule rad^{-1}
J m m^{-1}

10^{-7} J m m^{-1}
$= 1$ erg cm cm^{-1}

$\boldsymbol{T}$ is in erg rad^{-1}

1.2 Magnetism in SI Units and Gaussian Units

The energy W of a permanent electric dipole p in a field E is

$$W = -\boldsymbol{p}\cdot\boldsymbol{E}. \tag{1.13}$$

W is in J	10^{-7} J = 1 erg	W is in erg
ρ is in C m^{-3}		ρ is in statcoulomb cm^{-3}
$\boldsymbol{p}$ is in C m	$3.33564095 \times 10^{-12}$ C m = 1 statcoulomb cm	$\boldsymbol{p}$ is in statcoulomb cm
$\boldsymbol{P}$ is in C m^{-2}	$3.33564095 \times 10^{-6}$ C m^{-2} = 1 statcoulomb cm^{-2}	$\boldsymbol{P}$ is in statcoulomb cm^{-2}
$\boldsymbol{E}$ is in V m^{-1}		$\boldsymbol{E}$ is in statvolt cm^{-1}
$4\pi k_e \boldsymbol{P}$ is in V m^{-1}		$4\pi k_e \boldsymbol{P}$ is in statvolt cm^{-1}

$[p]$ = [current] [time] [length]	$[p]$ = [current] [time] [length]
$[P]$ = [current] [time] [length]$^{-2}$	$[P]$ = [current] [time] [length]$^{-2}$
$[E]$ = [energy] [current]$^{-1}$ × [time]$^{-1}$ [length]$^{-1}$	$[E]$ = [energy] [current]$^{-1}$ × [time]$^{-1}$ [length]$^{-1}$

$4\pi k_e \boldsymbol{P}$ appears as a source of $\boldsymbol{E}$. The units and dimensions of $4\pi k_e \boldsymbol{P}$ and $\boldsymbol{E}$ are the same as each other in both systems. Unfortunately, the definition of the electric susceptibility is different in the two systems.

$$4\pi k_e \boldsymbol{P} = \chi_e \boldsymbol{E}, \tag{1.14}$$
$$\boldsymbol{P}_{esu} = \chi_e \boldsymbol{E}_{esu}. \tag{1.15}$$

The SI electric susceptibility is dimensionless. The Gaussian electric susceptibility is in statcoulombs/statvolt·cm, which is actually dimensionless.

$$\boxed{(\chi_e)_{SI} = 4\pi(\chi_e)_{Gaussian}.}$$

The translation between the systems is further confused by the contrary definitions of the electric flux density $\boldsymbol{D}$.

$$4\pi k_e \boldsymbol{D} = \boldsymbol{E} + 4\pi k_e \boldsymbol{P}, \tag{1.16}$$
$$\boldsymbol{D} \equiv \varepsilon_0 \boldsymbol{E} + \boldsymbol{P}, \tag{1.17}$$
$$\nabla \cdot \boldsymbol{D} = \rho, \tag{1.18}$$

$$\boldsymbol{D} \equiv \boldsymbol{E} + 4\pi k_e \boldsymbol{P}, \tag{1.19}$$
$$\boldsymbol{D} \equiv \boldsymbol{E} + 4\pi \boldsymbol{P}, \tag{1.20}$$
$$\nabla \cdot \boldsymbol{D} = 4\pi k_e \rho, \tag{1.21}$$
$$\nabla \cdot \boldsymbol{D} = 4\pi \rho. \tag{1.22}$$

| D is in C m^{-2} | 3.33564095 × 10^{-6} C m^{-2} = 1 statvolt cm^{-1} | D is in statvolt cm^{-1} |

| $[D]$ = [current] [time] [length]$^{-2}$ | $[D]$ = [energy] [current]$^{-1}$ × [time]$^{-1}$ [length]$^{-1}$ |

Magnetic quantities

The magnetic induction B has no divergence. E and B are related by Faraday's laws of induction:

$$\nabla \cdot B = 0, \quad (1.23) \qquad \nabla \cdot B = 0, \quad (1.25)$$

$$\nabla \times E + \frac{\partial B}{\partial t} = 0, \quad (1.24) \qquad \nabla \times E + \frac{1}{c}\frac{\partial B}{\partial t} = 0. \quad (1.26)$$

In Gaussian units the c appears explicitly in Faraday's Law. Maxwell introduced the analogous relation between curl B and the time derivative of E valid in free space:

$$\nabla \times B - \frac{1}{c^2}\frac{\partial E}{\partial t} = 0, \quad (1.27) \qquad \nabla \times B - \frac{1}{c}\frac{\partial E}{\partial t} = 0. \quad (1.28)$$

In the presence of current density j these two equations are:

$$\nabla \times B - \frac{1}{c^2}\frac{\partial E}{\partial t} = 4\pi k_m j, \quad (1.29) \qquad \nabla \times B - \frac{1}{c}\frac{\partial E}{\partial t} = 4\pi k_m j, \quad (1.32)$$

where   where

$$k_m = \frac{\mu_0}{4\pi} = 10^{-17} \text{ N A}^{-2}, \quad (1.30) \qquad k_m = 1/c, \quad (1.33)$$

$$\mu_0 = 4\pi \times 10^{-7} \text{ N A}^{-2}$$
$$= 4\pi \times 10^{-7} \text{ henry m}^{-1} \quad (1.31) \qquad c = 2.99792458 \times 10^{10} \text{ cm s}^{-1}. \quad (1.34)$$

In SI units the c's are hidden in the expression for the sources of curl B, which in differential form are current densities including the polarization currents and the amperian currents as well as the true current density j:

$$\nabla \times B - \varepsilon_0 \mu_0 \frac{\partial E}{\partial t} \qquad \qquad \nabla \times E - \frac{1}{c}\frac{\partial E}{\partial t}$$

$$= \mu_0 \left(j + \frac{\partial P}{\partial t} + \nabla \times M \right), \quad (1.35) \qquad = \frac{4\pi}{c}\left(j + \frac{\partial P}{\partial t} + c\nabla \times M \right). \quad (1.36)$$

There is a $1/c^2$ in the SI equation from the combination $\mu_0 \varepsilon_0$. The c appears in front of the curl M term in Gaussian units to cancel the $1/c$ outside the bracket, because this term is measured in emu. The magnetization M in both systems is the magnetic moment per unit volume. In both cases the magnetic moment m is defined in terms of a current I in a loop with a finite area A and normal $\hat{n}$:

$$m = IA\hat{n}, \qquad (1.37) \qquad m = \frac{IA}{c}\hat{n}. \qquad (1.38)$$

The c appears in Gaussian units because I is measured in esu and m in emu. The magnetic moment of the electron is the Bohr magneton with a quantum electrodynamic correction. The Bohr magneton is:

$$\mu_B = \frac{eh}{4\pi m_e}, \qquad (1.39) \qquad \mu_B = \frac{eh}{4\pi c m_e}. \qquad (1.40)$$

Some fundamental constants in the two systems are:

Bohr magneton:
$\mu_B = 9.2740154 \times 10^{-24}$ J T^{-1}

electron moment:
$\mu_e = 9.2750909 \times 10^{-24}$ J T^{-1}

Planck's constant:
$h = 6.6260755 \times 10^{-34}$ J s

electron mass:
$m_e = 9.1093897 \times 10^{-31}$ kg

Bohr magneton:
$\mu_B = 9.2740154 \times 10^{-21}$ erg G^{-1}

electron moment:
$\mu_e = 9.2750909 \times 10^{-21}$ erg G^{-1}

Planck's constant:
$h = 6.6260755 \times 10^{-27}$ erg s

electron mass:
$m_e = 9.1093897 \times 10^{-28}$ g

The magnetic induction produces a torque on a magnetic moment, which follows from the Lorentz force equations (1.1, 2). In both cases:

$$T = m \times B. \qquad (1.41)$$

The energy W_m of a magnetic moment follows from the torque equation if the zero of energy is taken from the position where m is perpendicular to B:

$$W_m = -m \cdot B. \qquad (1.42)$$

The dynamic response of a magnetic moment to a torque is determined by the ratio of the magnetic moment to the angular momentum, called the gyromagnetic ratio γ. Again the equations in the two systems differ by the factor c:

$$\gamma = -g\frac{|e|}{2m_e}, \qquad (1.43) \qquad \gamma = -g\frac{|e|}{2cm_e}. \qquad (1.44)$$

The dynamic equation for a magnetic moment m in the presence of a magnetic induction B in both systems is:

$$\frac{1}{\gamma}\frac{\partial m}{\partial t} = m \times B. \tag{1.45}$$

There are also nonmagnetic sources of torque on a magnetic moment, e.g. exchange torques and spin-orbit interactions. These are included in the equation of motion using an effective magnetic induction B_{eff}, which can include any arbitrary vector parallel to m. When expressed in terms of M, the equation of motion becomes:

$$\frac{1}{\gamma}\frac{\partial M}{\partial t} = M \times B_{\text{eff}}, \tag{1.46}$$

where B_{eff} is the vector gradient of the energy density $w(M)$ with respect to the magnetization components:

$$B_{\text{eff}} = -\nabla_M w(M) + \lambda M. \tag{1.47}$$

Because $M \times M = 0$, one can include any λM into (1.46), where λ is chosen for convenience and can vary with position. Much confusion results in comparing various authors, because B_{eff} is not uniquely defined. This carries over when the magnetic field H is introduced into Maxwell's equations. The magnetic field H is defined as that part of B which is not directly the contribution of M to B. The contribution of M to B is different in the two systems. It is $\mu_0 M$ in SI and $4\pi M$ in the Gaussian system. To keep the units of H and M the same in the SI system, H is defined with B divided by μ_0. The distinction between H and B is not so reinforced in Gaussian units, where they have different names for the units, but the two quantities have the same dimensions, in fact the same dimensions as M and $4\pi M$. The definitions of H are:

$$H \equiv \frac{B}{\mu_0} - M. \quad (1.48) \qquad H \equiv B - 4\pi M. \tag{1.49}$$

B, $\mu_0 M$ and $\mu_0 H$ are in teslas, T	10^{-4} T = 1 G	B is in gausses, G
H is in A (turns) m^{-1}	$1000/4\pi$ A m^{-1} = 1 Oe	H is in oersteds, Oe {1 Oe = 1 G}
M is in A m^{-1}	1000 A m^{-1} = 1 emu cm^{-3}	M is in ergs (G^{-3} cm^{-3}) or emu cm^{-3} {1 emu cm^{-3} = 1 erg (G^{-3} cm^{-3}) = 1 Oe = 1 G}

1.2 Magnetism in SI Units and Gaussian Units

$$[H] = [M] = [\text{current}] [\text{distance}]^{-1}$$

$$[B] = [H] = [M] = [\text{energy}] \times [\text{current}]^{-1} [\text{time}]^{-1} [\text{distance}]^{-1}$$

> If both H and M are measured in amperes/meters in SI and both H and M are measured in oersteds in Gaussian units, then the conversion from amperes/meter to oersteds depends on whether it is H or M that is being converted! This confusion is avoided by not using Oe for M. If M is quoted in Oe, first convert to emu cm^{-3} and then convert to A m^{-1}.

Some confusion can be avoided also if one uses the magnetic polarization, either $\mu_0 M$ in tesla or $4\pi M$ in gauss, to report data. Similarly the magnetic field can be reported as $\mu_0 H$ in tesla, particularly for the applied field.

The introduction of H is particularly useful in ferromagnetism. It plays an important role in magnetostatics, where div M provides a source of H. The magnetic charge density is defined as proportional to $-$ div M, but the proportionality factor is different in the two systems:

$$\nabla \cdot H = -\nabla \cdot M = \rho_m, \quad (1.50) \qquad \nabla \cdot H = -4\pi \nabla \cdot M = \rho_m. \quad (1.51)$$

The factor of 4π enters SI in the expressions for the magnetic field and magnetic induction arising from a magnetic moment in the absence of material media, expressed in terms of the magnetic moment m and the unit vector $\hat{r}$:

$$\frac{1}{\mu_0} B = H = \frac{3(m \cdot \hat{r})\hat{r} - m}{4\pi r^3}, \quad (1.52) \qquad B = H = \frac{3(m \cdot \hat{r})\hat{r} - m}{r^3}. \quad (1.53)$$

Maxwell's equations are often written in terms of E, D, B, and H, suppressing the roles of P and M. These are particularly simple in SI units, containing neither of the constants ε_0 or μ_0 and none of the 4π's or c's that appear in the Gaussian version of Maxwell's equations:

$$\nabla \times H - \frac{\partial D}{\partial t} = j, \quad (1.54) \qquad \nabla \times H - \frac{1}{c}\frac{\partial D}{\partial t} = \frac{4\pi}{c} j, \quad (1.58)$$

$$\nabla \times E + \frac{\partial B}{\partial t} = 0, \quad (1.55) \qquad \nabla \times E + \frac{1}{c}\frac{\partial B}{\partial t} = 0, \quad (1.59)$$

$$\nabla \cdot D = \rho, \quad (1.56) \qquad \nabla \cdot D = 4\pi \rho, \quad (1.60)$$

$$\nabla \cdot B = 0, \quad (1.57) \qquad \nabla \cdot B = 0. \quad (1.61)$$

The equations of motion (1.45, 46) in terms of H include a factor of μ_0 in the SI units. As above, in (1.43, 44) and [44], the gyromagnetic ratio is defined differently in the two systems.

$$\frac{1}{\gamma}\frac{\partial M}{\partial t} = M \times \mu_0 H_{\text{eff}}, \quad (1.62) \qquad \frac{1}{\gamma}\frac{\partial M}{\partial t} = M \times H_{\text{eff}}, \quad (1.64)$$

where

where

$$H_{\text{eff}} = -\frac{\nabla_M w(M)}{\mu_0} + \lambda M. \quad (1.63) \qquad H_{\text{eff}} = -\nabla_M w(M) + \lambda M. \quad (1.65)$$

Because the magnetic field differs from the magnetic induction only by a vector along M, it does not matter whether one talks about an effective field or an effective induction, but in SI units it is necessary to remember that the torque equation contains the factor μ_0. Derivatives of energy densities with respect to M produce magnetic inductions. In Gaussian units these also have the dimensions of magnetic field. In SI units the derivatives of energies with respect to M must be divided by μ_0 to make them effective fields rather than effective inductions.

The magnetic contribution to the energy density w is:

$$w = \frac{B \cdot H}{2}. \quad (1.66) \qquad w = \frac{B \cdot H}{8\pi}. \quad (1.67)$$

The magnetostatic self energy of an aggregation of magnetic moments, excluding the self energy of the individual moments, is:

$$W = -\frac{\mu_0}{2}\sum_i m_i H'_i = -\frac{1}{2}\sum_i m_i B'_i, \quad (1.68) \qquad W = -\frac{1}{2}\sum_i m_i H'_i, \quad (1.70)$$

where H'_i is the demagnetizing field from all the other moments:

where H'_i is the demagnetizing field from all the other moments:

$$H'_i = \sum_{j \neq i} \frac{3(m_j \cdot \hat{r}_{ij})\hat{r}_{ij} - m_j}{4\pi r_{ij}^3} = \frac{B'_i}{\mu_0}. \quad (1.69) \qquad H'_i = \sum_{j \neq i} \frac{3(m_j \cdot \hat{r}_{ij})\hat{r}_{ij} - m_j}{r_{ij}^3} = B'_i. \quad (1.71)$$

For a uniformly magnetized ellipsoid the demagnetizing field is uniform and proportional to the magnetic moment per unit volume. For a uniformly magnetized sphere, the demagnetizing field H_d is:

$$H_d = -M/3. \quad (1.72) \qquad H_d = -4\pi M/3. \quad (1.73)$$

For a slab magnetized perpendicularly to the surface, the demagnetizing field H_d is:

$$H_d = -M. \quad (1.74) \qquad H_d = -4\pi M. \quad (1.75)$$

The factor of 4π also appears in the conversion of magnetic susceptibility between the two systems. In both systems $M = \chi H$ defines the volume suscepti-

1.2 Magnetism in SI Units and Gaussian Units

bility. This is a dimensionless ratio if M and H are in the same units, but

$$\boxed{\chi_{\mathrm{SI}} = 4\pi\chi_{\mathrm{Gaussian}}}. \tag{1.76}$$

One could say that the units of χ_{SI} are turn^{-1}, and the units of χ_{Gaussian} are erg/(G Oe cm^3).

The magnetic flux is defined in the same way in each system:

$$\Phi_B = \int_{\text{area}} \boldsymbol{B} \cdot \hat{\boldsymbol{n}} \, dA, \tag{1.77}$$

$\boxed{\Phi_B \text{ is in webers,} \\ 1 \text{ Wb} = 1 \text{ T m}^2}$ $\boxed{10^{-8} \text{ Wb} = 1 \text{ maxwell.}}$ $\boxed{\Phi_B \text{ is in maxwells} \\ = \text{G cm}^2.}$

Faraday's law of induction in integral form is:

$$\oint_{\text{loop}} \boldsymbol{E} \cdot \hat{\boldsymbol{s}} \, ds = -\frac{\partial}{\partial t}\Phi_B. \tag{1.78} \qquad \oint_{\text{loop}} \boldsymbol{E} \cdot \hat{\boldsymbol{s}} \, ds = -\frac{1}{c}\frac{\partial}{\partial t}\Phi_B. \tag{1.79}$$

1.2.2 Translation Keys

To translate from **equations** in the Gaussian system to **equations** in SI, the following table of translation equations provides the keys [1.2]. Solve the appropriate equation for the starred variable, substitute for each quantity in the Gaussian system, and then clean up using the relation $\varepsilon_0 \mu_0 c^2 = 1$ to remove the c's which are hidden in SI. To translate from **equations** in SI to **equations** in the Gaussian system, solve for the unstarred variable, substitute for each quantity in SI, and then remove the ε_0 and μ_0 factors using $\varepsilon_0 \mu_0 c^2 = 1$. The fact that eight different translation equations are needed accounts for much of the confusion that attends the subject.

$$\boxed{\begin{aligned} \sqrt{4\pi\varepsilon_0} &= \frac{E^*}{E} = \frac{V^*}{V} = \frac{Q}{Q^*} = \frac{\rho}{\rho^*} = \frac{j}{j^*} = \frac{I}{I^*} = \frac{P}{P^*}; & (1.80) \\ \sqrt{4\pi\mu_0} &= \frac{H^*}{H}; \quad \sqrt{\frac{4\pi}{\varepsilon_0}} = \frac{D^*}{D}; \quad \sqrt{\frac{4\pi}{\mu_0}} = \frac{B^*}{B} = \frac{M}{M^*} = \frac{\gamma}{\gamma^*} & (1.81) \\ 1 &= \frac{x^*}{x} = \frac{t^*}{t} = \frac{m^*}{m} = \frac{c^*}{c} = \frac{F^*}{F} = \frac{W^*}{W}; \quad 4\pi = \frac{\chi_e}{\chi_e^*} = \frac{\chi_m}{\chi_m^*}; \quad \varepsilon_0\mu_0 c^2 = 1. \end{aligned}} \tag{1.82}$$

For example:
start with

$$\gamma = -g\frac{|e|}{2m_e},\qquad(1.83)$$

look in the table for the transfer relation of γ and γ^*:

$$\gamma = \sqrt{\frac{4\pi}{\mu_0}}\gamma^*\qquad(1.84)$$

and the transfer relation of charge:

$$|e| = |e^*|\sqrt{4\pi\varepsilon_0}.\qquad(1.85)$$

Substitute these, $g = g^*$ and $m = m^*$ in (1.83) to obtain:

$$\sqrt{\frac{4\pi}{\mu_0}}\gamma^* = -g^*\frac{|e^*|\sqrt{4\pi\varepsilon_0}}{2m_e^*},$$
$$(1.86)$$

which using $c^2\mu_0\varepsilon_0 = 1$ and $c = c^*$ simplifies to:

$$\gamma^* = -g^*\frac{|e^*|}{2c^*m_e^*}.\qquad(1.87)$$

or start with

$$\gamma^* = -g^*\frac{|e^*|}{2c^*m_e^*},\qquad(1.88)$$

look in the table for the transfer relation of γ and γ^*:

$$\gamma^* = \sqrt{\frac{\mu_0}{4\pi}}\gamma\qquad(1.89)$$

and the transfer relation of charge:

$$|e^*| = \frac{1}{\sqrt{4\pi\varepsilon_0}}|e|\qquad(1.90)$$

Substitute these, $g = g^*$, $c^* = c$ and $m^* = m$ in (1.88) to obtain:

$$\sqrt{\frac{\mu_0}{4\pi}}\gamma = -g\frac{|e|}{2cm_e\sqrt{4\pi\varepsilon_0}},$$
$$(1.91)$$

which using $c^2\mu_0\varepsilon_0 = 1$ simplifies to:

$$\gamma = -g\frac{|e|}{2m_e}.\qquad(1.92)$$

Acknowledgement. Helpful discussions with Dr. R.B. Goldfarh of the National Institute of Standards and Technology in the preparation of this document are gratefully acknowledged.

References

Section 1.2

1.1 William Fuller Brown, Jr.: IEEE Trans. Mag. MAG-20, 112–117 (1984)
1.2 *Symbols, Units, Nomenclature and Fundamental Constants in Physics*, 1987 Revision, Prepared by E. Richard Cohen and Pierre Giacomo for the International Union of Pure and Applied Physics: Physica **146A**, 1–68 (1987)

2. Magnetic Anisotropy, Magnetization and Band Structure

The fundamental magnetic parameters of magnetization and magnetic anisotropy energy (MAE) are related to the ground state electronic structure. These static magnetic properties provide a basic description of any magnetic system, and knowledge of them is clearly needed before the dynamic properties and finite temperature behavior can be described in ultrathin films, as discussed in later chapters (e.g., see *Mills* in this volume and the chapters by *Heinrich*, *Cochran* and also by *Hillebrands* and *Guntherodt* in Volume II). Our aim is to discuss the physical origin of these parameters and to gain an understanding of why they may differ in ultrathin films in comparison with the bulk. We have included a treatment of both theoretical and experimental aspects in this chapter so that the interplay between experiment and theory may be better appreciated. We therefore wish to evaluate the status of the theory in describing the current experimental situation, particularly with regard to the magnetic anisotropy. We should stress at the outset that the accuracy and reliability of the theoretical predictions remain a contentious issue, as discussed in this chapter, and the sensitivity of the anisotropies to the interface structure and quality also further complicates the comparison between experiment and theory. In an important advance, the computational accuracy of MAE calculations was recently significantly improved by the research groups of Freeman and Victora. In the first section, *Gay* and *Richter* discuss density functional theoretical calculations of the magnetic moment and anisotropy for the transition metal ferromagnets, focusing on a direct comparison between the bulk properties and those of single ultrathin films. In the second section, *Daalderop*, *Kelly* and *Schuurman's* describe how the magnetic anisotropy energy of overlayer and multilayer systems is related to the detailed electronic band structure and how high symmetry wave functions play an important role in giving rise to magnetic anisotropy. In the final section, *de Jonge*, *den Broeder* and *Bloemen* describe recent experimental investigations of the magnetic anisotropy in magnetic films and multilayers.

2.1 Electronic Structure of Magnetic Thin Films

J.G. Gay and R. Richter

The experimentalist working in the field of thin-film magnetism is confronted with two problems. The first is sample size. One is dealing with a million times fewer atoms than are found in bulk samples. This makes for difficult experiments requiring sensitive detection equipment. The second is characterization: what is the structure and composition of the film one is studying? This problem is especially acute if, as is often the case, the film which the experimentalist wishes to prepare and study is thermodynamically unstable.

Because of these problems, especially the problem of characterization, we believe it is safe to say that there is no completely unequivocal experimental determination, for example, of the magnetic properties of a monolayer of Fe on Cu {100}. It would therefore be a great help to the experimentalist if the theorist, who has no difficulty in making structurally perfect thin films even when they are thermodynamically unstable, could reliably predict the physical and magnetic properties of these perfect films.

It is the intent of this first part of the chapter to assess the ability of first-principles theory to calculate the ground state electronic structures of magnetic thin films composed of transition and noble metal atoms. Since there is no competing theory, this amounts to an assessment of density functional theory within the local density approximation.

In the next section we will outline the basic theory that underlies that approach beginning with the Dirac equation and ending with the relativistic *Kohn-Sham* equations. Since the magnetic properties of the bulk elemental ferromagnets are well known, we will preface the assessment for magnetic films with a comparison of the prdictions of theory with the experimental results for the bulk elemental ferromagnets.

2.1.1 Underlying Theory

The phenomenon of ferromagnetism is a consequence of the fact that electrons have spin and obey the exclusion principle. These properties are relativistic in origin but can be regarded as empirical facts and incorporated into non-relativistic quantum mechanics by assigning a spin quantum number to the electrons and requiring that the wave function be antisymmetric under interchange of electrons. This non-relativistic theory, in which the electrical interactions of the electrons do not depend on their spins, gives a satisfactory account of some, but not all, aspects of ferromagnetism. One aspect that is not accounted for at all is the magnetocrystalline anisotropy. In this theory the energy of a ferromagnet is independent of the direction of spin quantization, so there is no easy[1] magnetization axis.

A fully satisfactory theory of ferromagnetism has to be based on relativistic electron dynamics, i.e., on the Dirac equation and its many-electron extension. The Dirac equation for an electron in an external electromagnetic field is [2.1]

$$(E + mc^2)\psi = [-\boldsymbol{\alpha}\cdot(c\boldsymbol{p} + e\boldsymbol{A}) - \boldsymbol{\beta}mc^2 - e\phi]\psi. \tag{2.1}$$

Symbols have their usual meanings. ψ is a four component spinor, $\alpha_x, \alpha_y, \alpha_z$, and

---

[1] The easy axis is that along which the magnetic moments prefer to align in the absence of an applied field.

2.1 Electronic Structure of Magnetic Thin Films

β are the Dirac matrices. The equation is in Gaussian units and e is the magnitude of the electronic charge.

When the electrostatic potential is weak in the sense that $e\phi \ll mc^2$ the Dirac equation simplifies to the Pauli equation [2.2, 3]

$$\left\{ \frac{p^2}{2m} - e\phi + \frac{e\hbar}{2mc} \mathbf{V} \times \mathbf{A} \cdot \boldsymbol{\sigma} + \frac{e}{mc} \mathbf{A} \cdot \mathbf{p} + \frac{e\hbar^2}{4m^2c^2} \mathbf{V}\phi \cdot \mathbf{V} \right.$$
$$\left. - \frac{p^4}{8m^3c^2} - \frac{e\hbar}{4m^2c^2} \boldsymbol{\sigma} \cdot (\mathbf{V}\phi \times \mathbf{p}) \right\} \psi = E\psi. \tag{2.2}$$

ψ is now a two component spinor and $\boldsymbol{\sigma}$ is the Pauli spin matrix vector.

When written in Hartree atomic units (2.2) becomes

$$\left\{ \frac{1}{2}p^2 - \phi + \frac{1}{2} \mathbf{V} \times \mathbf{A} \cdot \boldsymbol{\sigma} + \mathbf{A} \cdot \mathbf{p} + \frac{\alpha^2}{4} \left[\mathbf{V}\phi \cdot \mathbf{V} - \frac{p^4}{2} - \frac{1}{2}\boldsymbol{\sigma} \cdot (\mathbf{V}\phi \times \mathbf{p}) \right] \right\} \psi = E\psi, \tag{2.3}$$

where α is the fine structure constant. The vector potential has units such that the magnetic energy of a spin is $(1/2)\mathbf{V} \times \mathbf{A} \cdot \boldsymbol{\sigma}$ in Hartrees. Except for the terms proportional to α^2 this is the Schrödinger equation of an electron with spin. The terms proportional to α^2 are respectively the Darwin, mass-velocity and spin-orbit corrections. For hydrogen-like ions of nuclear charge Z, an energy level of (2.2) corresponding to an energy level E of (2.1) is in error by an amount of order $(Z\alpha)^4 E$ [2.2].

There exists no exact relativistic equation for a many-electron system [2.2] but equations giving the same relative accuracy as the Pauli equation (2.2) can be derived from quantum electrodynamics [2.2] or by the following heuristic argument of *Slater* [2.3]. Slater assumes in the Pauli equation (2.2) for electron i that the electromagnetic potentials of another electron j can be treated as external potentials acting on electron i. The potentials at electron i due to electron j are

$$A_{ij} = -\frac{e\hbar}{m} \left[\frac{\mathbf{p}_j}{r_{ij}} + \frac{1}{2} \frac{\boldsymbol{\sigma}_j \times (\mathbf{r}_i - \mathbf{r}_j)}{r_{ij}^3} \right],$$

$$\phi_{ij} = -\frac{e}{r_{ij}}.$$

When these are used in (2.2) they yield, after conversion to atomic units, the electron–electron interaction

$$g_{ij} = \frac{1}{r_{ij}} + \alpha^2 \left\{ -\frac{\mathbf{p}_i \cdot \mathbf{p}_j}{2r_{ij}} - \frac{1}{2r_{ij}^3}[(\mathbf{r}_i - \mathbf{r}_j) \cdot \mathbf{p}_i][(\mathbf{r}_i - \mathbf{r}_j) \cdot \mathbf{p}_j] \right.$$
$$+ \frac{\boldsymbol{\sigma}_i \cdot [(\mathbf{r}_j - \mathbf{r}_i) \times \mathbf{p}_j] + \boldsymbol{\sigma}_j \cdot [(\mathbf{r}_i - \mathbf{r}_j) \times \mathbf{p}_i]}{4r_{ij}^3} + \frac{\boldsymbol{\sigma}_i \cdot \boldsymbol{\sigma}_j}{4r_{ij}^3}$$
$$- 3\frac{[\boldsymbol{\sigma}_i \cdot (\mathbf{r}_j - \mathbf{r}_i)][\boldsymbol{\sigma}_j \cdot (\mathbf{r}_j - \mathbf{r}_i)]}{4r_{ij}^5} - \frac{\boldsymbol{\sigma}_i \cdot [(\mathbf{r}_i - \mathbf{r}_j) \times \mathbf{p}_i] + \boldsymbol{\sigma}_j \cdot [(\mathbf{r}_j - \mathbf{r}_i) \times \mathbf{p}_j]}{4r_{ij}^3}$$
$$\left. + \frac{(\mathbf{r}_i - \mathbf{r}_j) \cdot \mathbf{V}_i + (\mathbf{r}_i - \mathbf{r}_j) : \mathbf{V}_j}{4r_{ij}^3} \right\}. \tag{2.4}$$

A N-electron Hamiltonian will then consist of a one-electron Hamiltonian h_i [the quantity in braces in (2.3)] for each electron plus a sum of the pairwise interactions (2.4) i.e.,

$$H = \sum_i^N h_i + \sum_{j<i} g_{ij} \,. \tag{2.5}$$

The terms in this Hamiltonian which couple the spatial part of a wave function to its spin part are the spin–orbit and the spin–other–orbit interaction. The spin–orbit interaction comes from the field of the nuclei [the last term in (2.3)] and from the field of the other electrons [the next to last term in the braces in (2.4)]. The spin–other–orbit interaction is the third term in the braces in (2.4). These are the interactions that produce multiplet splittings in atoms and the magnetocrystalline anisotropy in ferromagnets[2]. There is much evidence that the Hamiltonian (2.5) correctly predicts atomic multiplet structure [2.3].

We note for future reference that the spin–other–orbit interaction is not negligible. *Blume* and *Watson* [2.4] have calculated multiplet splittings for a number of atoms and ions using the full interaction (2.4) and then omitting the spin–other–orbit interaction. They find that the first calculation generally gives good agreement with experiment but the second gives splitting that are too large: by 50% for light elements and by 10% for 3d elements.

Obtaining a solution to the wave equation posed by (2.5) is which the electrons are correlated is possible only when N is small. Thus solving that equation even approximately for such a wave function is only possible for atoms and small molecules. For condensed matter systems like ferromagnets we must resort to a one-electron approximation in which exchange and correlation are handled by a one-electron exchange-correlation potential. In non-relativistic quantum mechanics such an approximation is justified by density functional theory [2.5] and made practical by the local density approximation (LDA) [2.6]. An excellent review of non-relativistic density functional theory and the LDA are contained in the first three chapters of the book of *Mahan* and *Subbaswamy* [2.7].

The density functional formalism generalizes to the case that the electron dynamics is relativistic [2.8–10]. The result is a Dirac equation [2.10],

$$(\varepsilon_i + mc^2)\psi_i = [-\boldsymbol{\alpha}\cdot(c\boldsymbol{p} + e\boldsymbol{A}_{\text{eff}}) - e\boldsymbol{\sigma}'\cdot\boldsymbol{B}_{\text{eff}} - \beta mc^2 - e\phi_{\text{eff}}]\psi_i \,, \tag{2.6}$$

which is the relativistic *Kohn–Sham* equation for orbital i with eigenvalue ε_i. This is of the form of (2.1) except for the addition of the $\boldsymbol{\sigma}'\cdot\boldsymbol{B}_{\text{eff}}$ term. $\boldsymbol{\sigma}'$ is the vector of 'diagonal' Dirac spin matrices [2.1]. ϕ_{eff} is an effective one-electron

---

[2] The fifth term of (2.4) which is the dipole–dipole interaction of the spin magnetic moments leads to another anisotropy. Because of its long range it will lead to an energy that depends on the shape of the ferromagnet and thus to the shape anisotropy.

2.1 Electronic Structure of Magnetic Thin Films

electrostatic potential,

$$\phi_{\text{eff}} = \phi_{\text{ext}} + e \int \frac{\rho(r')}{|r - r'|} d^3 r' + \frac{1}{e} \frac{\delta E_{\text{xc}}}{\delta \rho(r)},$$

comprising an external, a Hartree and an exchange-correlation potential. A_{eff} is an effective vector potential,

$$A_{\text{eff}} = A_{\text{ext}} + \int \frac{J(r')}{|r - r'|} d^3 r' + \frac{\delta E_{\text{xc}}}{\delta J(r)},$$

comprising corresponding terms. B_{eff} is an effective induction,

$$B_{\text{eff}} = \frac{\delta E_{\text{xc}}}{\delta m(r)},$$

that only interacts with the electron spin. The relativistic exchange-correlation energy E_{xc} is a functional of the electron density $\rho(r)$, the spin moment $m(r)$ and the current $J(r)$. The latter cannot be assumed to vanish. To convert the formalism into a practical calculational technique we make the LDA in which E_{xc} is approximated by the exchange-correlation energy of a uniform relativistic electron gas at the same density, moment and current.

As above when $e\phi_{\text{eff}}$ is weak (for atoms with $Z \ll \alpha$), (2.6) reduces to a Pauli equation

$$\left\{ \frac{1}{2} p^2 - \phi_{\text{eff}} + \frac{1}{2}(B_{\text{eff}} \cdot \sigma) + \frac{\alpha^2}{4} \left[(\nabla \phi_{\text{eff}} \cdot \nabla) - \frac{p^4}{2} \right] \right.$$

$$- \frac{\alpha^2}{4} \left[\sigma \cdot (\nabla \phi_{\text{eff}} \times p) \right]$$

$$\left. + \frac{1}{2} (\nabla \times A_{\text{eff}} \cdot \sigma) + (A_{\text{eff}} \cdot p) \right\} \psi_i = \varepsilon_i \psi_i \quad (2.7)$$

for the *Kohn–Sham* orbital i.

The terms in the first row of (2.7) constitute the scalar-relativistic approximation. The first three terms are just those of the spin polarized non-relativistic Kohn–Sham equation[3] except that E_{xc} contains relativistic corrections. The terms proportional to α^2 are the Darwin and mass–velocity corrections. In this approximation a net magnetic moment, i.e. ferromagnetism, can occur because of the $B_{\text{eff}} \cdot \sigma$ term. However B_{eff} is not a true induction but the 'exchange' field

---

[3] The connection with the usual form of the spin polarized non-relativistic equation which expresses the exchange-correlation energy as a functional of spin-up and spin-down densities, $\rho_\uparrow$ and $\rho_\downarrow$, is not entirely transparent. When operating on eigenfunctions of spin the three terms lead to the equations

$[\frac{1}{2} p^2 + \phi_{\text{eff}} + \frac{1}{2} B_{\text{eff}}] \psi_\uparrow = \varepsilon_\uparrow \psi_\uparrow,$

$[\frac{1}{2} p^2 + \phi_{\text{eff}} - \frac{1}{2} B_{\text{eff}}] \psi_\downarrow = \varepsilon_\downarrow \psi_\downarrow.$

But $\rho = \rho_\uparrow + \rho_\downarrow$ and $m = \frac{1}{2}(\rho_\uparrow - \rho_\downarrow)$ so that the exchange-correlation contribution to the two equations is respectively $\delta E_{\text{xc}}/\delta \rho_\uparrow$ and $\delta E_{\text{xc}}/\delta \rho_\downarrow$.

that couples just to the electron spins and not to any orbital moments. L and S are still good quantum numbers and the energy is independent of the direction of spin quantization. There is no magnetocrystalline anisotropy.

The second row of (2.7) is the spin–orbit interaction. When it is included L and S are no longer separately conserved and the energy of the system will depend on the direction of quantization of the spin. The spin–orbit interaction is the primary source of magnetocrystalline anisotropy.

The third row contains terms associated with true magnetic fields. The A_{ext} part of A_{eff} arises not only from any external applied field but also from the dipole field to the spin moments induced by the $B_{eff} \cdot \sigma$ term. The latter field is typically 10^{-4} of B_{eff} and will produce negligible anisotropy in a spherical sample [2.11]. In subsequent discussion we assume that A_{ext} is zero and thereby avoid the complication of Landau levels induced by the second term of the third row of (2.7). Shape anisotropy arising from the long range dipole field of the spin moments acting through the first term of the row can be handled independently.

In a scalar-relativistic calculation with A_{ext} zero the wave functions can be chosen to be real. The expectation value of the orbital angular momentum is then imaginary and hence vanishes [2.12]. The orbital momentum is quenched, there is no current and A_{eff} vanishes. However, when the spin-orbit interaction is introduced orbital currents are inevitably induced. The resulting A_{eff} acting through the first term of row there is an additional source of magnetocrystalline anisotropy. This is how the spin–other–orbit interactions enter in density functional theory. Note that A_{eff} is implicitly proportional to α^2.

First-principles calculations of the ground state properties of ferromagnets solve the Dirac equation (2.6) or the Pauli equation (2.7) self-consistently at some level of approximation for the one-electron orbitals ψ_i and eigenvalues ε_i of the ferromagnet. Presumably the two equations agree to order $(Z\alpha)^4$. The magnetic moment and magnetocrystalline anisotropy are then calculated in approximations that employ only these one-electron quantities. The calculations are justifiably labeled first-principles since they are parameter free in general. The process of iterating to self-consistency removes any influence of the ad hoc potentials used to start the calculations, and precise problem-independent recipes are used to construct the LDA exchange-correlation protential. The purpose of this chapter is to review such first-principles calculations for magnetic thin films, assess their accuracy and, where possible, compare their predictions with experiment. We do not review the generally earlier literature of calculations which are not first-principles in the sense just described.

There have been a number of first-principles calculations of the magnetic properties of bulk ferromagnetic Fe, Co and Ni. These calculations provide a more conclusive test of theory than the thin film calculations because the experimental values for the magnetic moment and magnetocrystalline anisotropy are well known. For this reason and because there is every reason to believe that a method must succeed in the bulk to succeed in a thin film, we will preface our review of the thin film calculations with a discussion of these bulk calculations in the following section.

2.1.2 Calculation of the Magnetic Properties of Bulk Fe, Co and Ni

Calculations of magnetic properties of ferromagnets can be placed in two groups. The first group consists of scalar-relativistic calculations, i.e., calculations in which only the first row of (2.7) is retained[4]. These calculations can predict the occurrence of ferromagnetism [2.13], the size of the spin moment $\langle S \rangle$ and non-magnetic properties such as cohesive energies, atomic volumes and bulk moduli [2.13]. The second group we will call full-relativistic calculations by which we mean that at least the first two rows of (2.7) are retained. These calculations can additionally determine the magnetocrystalline anisotropy, the induced orbital moment, $\langle L_\parallel \rangle$, parallel to $\langle S \rangle$ and thereby the gyromagnetic ratio. Results of a number of calculations for magnetic quantities of Fe, Co and Ni have been collected in Tables 2.1(a)–(c) along with experimental values. Unless noted to the contrary, the calculations employ no disposable parameters. Scalar- and full-relativistic entries in the tables are discussed separately.

Table 2.1a. Calculated magnetic properties of Fe compared with experiment. Magnetic moments are in μ_B/atom. The magnetocrystalline anisotropy energy ΔE is the difference $E_{001} - E_{111}$ in μeV/atom. Experimental quantities are those quoted in [2.11]

	$\langle S \rangle$	$\langle L_\parallel \rangle$	ΔE
Experiment	2.13	0.09	−1.4
Connolly [2.14]	2.55		
Tawil and *Callaway* [2.16]	2.26		
Moruzzi et al. [2.13]	2.15		
Fritsche et al. [2.19]	2.13	0.039	7.4
Daalderop et al. [2.11]	2.16	0.048	−0.5
Strange et al. [2.22]	2.13	0.06	−9.6

Table 2.1b. Calculated magnetic properties of Co compared with experiment. Magnetic moments are in μ_B/atom. The magnetocrystalline anisotropy energy ΔE is the difference $E_{0001} - E_{10\bar{1}0}$ in μeV/atom. Experimental quantities are those quoted in [2.11]

	$\langle S \rangle$	$\langle L_\parallel \rangle$	ΔE
Experiment	1.59	0.16	−65
Connolly [2.14]	1.68		
Moruzzi et al. [2.13]	1.56		
Daalderop et al. [2.11]	1.57	0.079	16

---

[4] We include in this group non-relativistic calculations for which $\alpha = 0$ in the first row of (2.7).

Table 2.1c. Calculated magnetic properties of Ni compared with experiment. Magnetic moments are in μ_B/atom. The magnetocrystalline anisotropy energy ΔE is the difference $E_{001} - E_{111}$ in μeV/atom. Experimental quantities are those quoted in [2.11]

	$\langle S \rangle$	$\langle L_\parallel \rangle$	ΔE
Experiment	0.56	0.05	2.7
Connolly [2.14]	0.62		
Callaway and *Wang* [2.15]	0.58		
Moruzzi et al. [2.13]	0.59		
Anderson et al. [2.18]	0.63		
Fritsche et al. [2.19]	0.52	0.045	10.0
Strange et al. [2.21]	0.60	0.046	10.5 ± 7.0
Daalderop et al. [2.11]	0.60	0.051	− 0.5

2.1.2.1 Scalar-Relativistic Calculations

The first first-principles calculations of the electronic structure of ferromagnetic Fe, Co and Ni were those of *Connolly* [2.14]. These calculations are non-relativistic and employ the muffin-tin approximation for the potential. The exchange-only exchange-correlation potentials were obtained by optimizing them for the constituent atoms.

The non-relativistic calculations of *Callaway* and co-workers for Ni [2.15] and Fe [2.16] were done with a linear-combination of-atomic-orbitals (LCAO) basis and *Kohn–Sham* exchange-only exchange-correlation potentials. For Fe the potential was adjusted slightly by a multiplicative constant. The calculations employ no shape approximation for the potentials.

The non-relativistic calculations of *Moruzzi* et al. [2.13] for Fe, Co and Ni employ the muffintin approximation for the potential and use the *von Barth–Hedin* [2.17] exchange-correlation potential. The fcc structure, rather than hcp, was used the Co calculation. We note that the 29 other elemental metals investigated by *Moruzzi* et al. are predicted to be paramagnetic by the LDA.

Anderson et al. [2.18] have done both non-relativisitc and scalar-relativistic calculations for Ni using several exchange-correlation potentials. These calculations point out the nominal scalar-relativistic effects in Ni. The moment we quote is scalar-relativistic using the *von Barth-Hedin* potential.

What emerges from these calculations (including the full-relativistic calculations to be discussed below) is that the LDA correctly finds the spin moment of the elemental ferromagnets: with accuracies, in the later calculations which use one of the modern exchange-correlation potentials, verging on the spectacular. It is clear that the LDA handles the spin-spin interaction arising from exchange adequately. The significance for magnetic thin films is that calculations which solve the LDA *Kohn–Sham* equations with sufficient accuracy can be relied on to predict the presence of a magnetic moment and its magnitude. However, we note in this connection that approximations which work in the bulk, like the muffin-tin approximation, may be inadequate in a thin film because of reduced symmetry.

2.1.2.2 Full-Relativistic Calculations

Full-relativistic calculations offer the possibility of a more accurate calculation of $\langle S \rangle$ and the calculation of $\langle L_\parallel \rangle$. More important, however, is the possibility of determining the magnetocrystalline anisotropy: a property of considerable technical importance in magnetic thin films.

The magnetocrystalline anisotropy involves calculating the total energy of the system as a function of the direction of spin quantization. In such calculations the total energy is universally approximated by the sum of the eigenvalues of occupied one-electron states. In both bulk and thin film calculations this sum is an integral of an energy density over the k-space volume enclosed by the Fermi surface. The spin-orbit operator, because it is traceless, introduces a fundamental computational difficulty into this integration. To see the nature of the difficulty consider the simplest approach to the full-relativistic problem in which one starts with a solution of the scalar-relativistic problem and adds the spin–orbit interaction as a perturbation. This full problem is not solved by perturbation theory in practice but variationally in the basis of the eigenstates of the scalar problem. However perturbation theory makes manifest the nature of the difficulty with the spin–orbit operator. Since the angular momentum is quenched in the scalar problem, the matrix of the spin–orbit operator will have vanishing diagonal elements in the scalar basis. This means that, aside from points of degeneracy of the bands, contributions will only come from second and higher orders of perturbation theory which are strongly influenced by energy denominators. Large contributions of the spin-orbit interaction to the energy density in the Brillouin zone will come from points of degeneracy and points in the vicinity of the Fermi level where occupied and unoccupied bands are close together. These contributions which consist of sharp peaks are typically much larger than the final anisotropy. In the subtraction to get the energy difference between two quantization directions, the peaks from the two directions nearly cancel. The occurrence of these peaks in the energy density and their near cancellation in an anisotropy calculation means that an exceedingly fine mesh is required to insure adequate convergence of the integral representing the sum of one-electron energies. This makes calculation of the magnetocrystalline anisotropy a very laborious process even when only the one-electron part of the total energy is calculated.

The calculations of *Fritsche* et al. [2.19] for Fe and Ni were done as just described by adding the spin-orbit interaction as a perturbation to scalar relativistic calculations obtained by the linear rigorous cellular method. The muffin-tin approximation and a non-standard exchange-correlation potential which yields an exchange splitting in Ni closer to experiment [2.20] was used.

The calculations of *Strange* et al. for Ni [2.21] and Fe [2.22] use a relativisitic KKR method [2.23] which solves the Dirac equation (2.6) within a muffin-tin approximation. This approach is in principle superior because it does not give the exchange splitting precedence, but deals with all of the terms of (2.6) simultaneously. This may be important for heavy atoms. In the quoted calculations, however, the non-relativistic self-consistent potentials for Fe and Ni of *Moruzzi* et al. [2.13] were used in the Dirac equation. This means that exchange still has precedence because the other effects are not in those potentials.

The calculations of *Daalderop* et al. [2.11] for Fe, Co and Ni were done by adding the spin-orbit interaction as a perturbation to scalar relativistic calculations obtained by the atomic spheres approximation [2.24] which uses the muffin tin approximation. The *von Barth–Hedin* exchange-correlation potential [2.17] was used.

It is clear from the tables that the calculated magnetocrystalline anisotropies do not agree with experiment or with each other. This unhappy state of affairs has been studied exhaustively by *Daalderop* et al. [2.11]. They analyze with great care the sources of error in their work (and to the extent possible in the other calculations). With regard to their work, they conclude that numerical errors in evaluating the Brillouin zone integral and other errors, such as those due to potential shape approximations or the choice of exchange-correlation potential cannot account for the differences between their results and experiment. Stated another way, their conclusion is that calculations which use the exact self-consistent solutions to the scalar-relativistic LDA as a basis for the variational calculation would give magnetocrystalline anisotropies closer to theirs than to the experimental results. They surmise that the other calculations do not agree with their results because of incomplete convergence of the Brillouin zone integrals in those calculations.

Daalderop et al. conclude that their incorrect theoretical results are due either to a failure of the 'force theorem' [2.25], the theorem that says the eigenvalue sum is an adequate approximation to the total energy, or a failure of the LDA. They point out that the latter possibility cannot be completely dismissed since there are examples where the LDA makes total enery errors of the order of 10 meV/atom. However, they prefer the first possibility since they estimate that eigenvalue sum approximations can have errors large compared to the magnetocrystalline anisotropy. Such errors will have to cancel systematically if the magnetocrystalline anisotropy is to emerge from the subtraction of eigenvalue sums for two quantization directions. To go beyond the force theorem will require much greater computational effort.

We remark that the calculation of the LDA one-electron sums is a well-posed problem in itself. Thus two different calculations of these sums for the same system, both of which have no uncontrolled approximations so they can be systematically improved, must converge to the same answer. It is clear that the theory is not at that stage.

There is another possibility for the failure to agree with experiment which *Daalderop* et al. consider only in passing. That possibility is that the errors in the theoretical magnetocrystalline anisotropy are due to lack of self-consistency. Recall that all the full-relativistic calculations employ an effective Hamiltonian derived from a scalar-relativistic charge density that is currentless. However, if the full problem is iterated, subsequent charge densities would possess induced orbital moments and currents which provide another source of anisotropy through the first term of the third row of (2.7). This anisotropy is a manifestation of the spin–other–orbit interaction which is not negligible compared to the spin–orbit interaction in 3d atoms [see the discussion following (2.4)]. Since we expect that N iterations will require only something like N times the computational effort of a non-self-consistent calculation, this may be an easier possibility to check than a failure of the force theorem. This speculation does not contest the overall conclusion of *Daalderop* et al. that the magnetocrystalline anisotropy cannot be calculated reliably with present technqiues.

Jansen [2.26] has included orbital moment contributions in model calculations of the magnetocrystalline anisotropy in Fe. He concludes that such contributions are necessary based on the fact that he obtains values of the magnetocrystalline anisotropy two orders of magnitude too small with spin-orbit alone. Since the first-principles calculations we have cited, all obtain the correct order of magnitude of the magnetocrystalline anisotropy with spin-orbit alone, we assume *Jansen's* model does not reproduce the Fe band structure with sufficient accuracy.

The situation is less grim for the calculation of $\langle L_\parallel \rangle$. Calculated values agrees with one another and with experiment for Ni. Theoretical values are roughly half the experimental values for Fe and Co. The calculation of $\langle L_\parallel \rangle$ probably is less difficult than the calculation of the magnetocrystalline anisotropy for two reasons. The first is that L is a one-electron operator so that the calculation does not require calculating a total energy. The second is that $\langle L_\parallel \rangle$ contains a dominant part that is independent of the spin quantization direction. This is true for monolayers of the 3d metals [2.27] and should be more so for bulk materials. This isotropic part may be easier to calculate.

To summarize this section, calculated spin moments of Fe, Co and Ni are in good agreement with experiment, calculated orbital moments are in fair agreement and calculated magnetocrystalline anisotropies are in poor agreement. To the extent that firm experimental results are available, in the next section we shall find in the survey of calculations for thin films similar situation.

2.1.3 Calculation of the Magnetic Properties of Thin Films

The first workers to study ferromagnetism in systems with only two-dimensional periodicity were *Liebermann* et al. [2.28, 29] who measured the magnetic moment of Fe and Ni films as a function of thickness during deposition. It now appears on both experimental [2.30, 31] and theoretical grounds [2.32] that the 'dead' layers they observed were due to hydrogen contamination. Nevertheless, their work showed that the magnetic moment is changed by a surface and that the change can be measured, and it is generally regarded as the seminal research in interface magnetism.

A decade passed before theoretical techniques were sufficiently developed to permit first-principles LDA calculations of the surface electronic structure of ferromagnets. The techniques developed universally employ infinite slabs or films containing a few layers of atoms to represent a surface. Solving the *Kohn–Sham* equations self-consistently for such a metallic film is intrinsically more difficult than for the bulk. The loss of periodicity perpendicular to the film makes the simple muffin-tin approximation inappropriate. Further, because atoms in different planes are not equivalent, there is no symmetry requirement which prevents charge from sloshing between the center and surface of the film during successive iterations. Thus convergence to self-consistency is a delicate process.

2.1.3.1 Scalar-Relativistic Calculations

We will be mainly concerned with two film calculational schemes: the self-consistent local-orbital (SCLO) method [2.33] and the full-potential linearized-augmented-plane-wave (FLAPW) method [2.34]. The SCLO method is non-relativistic and uses a local basis consisting of the atomic orbitals of the constituent atoms augmented by 'polarization' orbitals. There are no potential shape approximations and the implementation of the method permits systematic improvement of the solution to the *Kohn–Sham* equations within the subspace of the local basis. The FLAPW method is scalar relativistic. Space is partitioned into muffin–tin spheres around the atoms, vacuum regions on either side of the film and the remaining interstitial region. Within each of the regions all quantities are expanded in basis functions which can be systematically made more complete. The *Kohn–Sham* equations can thus be solved to any accuracy in principle. This is the advantage of the method. The SCLO method is limited by its basis which cannot be systematically improved. Nevertheless it has proved itself accurate and reliable and has a considerable advantage in speed. An early indication of its accuracy came from a calculation for Cu {1 0 0} [2.35] which predicted a surface state band that was later confirmed experimentally [2.36].

The first spin-polarized calculations on surfaces were done by *Wang* and *Freeman* for Ni {1 0 0} [2.37] and Fe {1 0 0} [2.38]. These calculations, which used an atomic orbital basis, showed that the surface layers were not dead. However, because of limitations in the calculational approach [2.39], the calculations do not give layer magnetic moments that agree closely with later calculations.

Table 2.2 gives the work function and layer moments for Ni {1 0 0} found in three more recent calculations. *Jepsen* et al. [2.40] employ a non-relativistic spin-polarized linearized-augmented-plane-wave method similar to the FLAPW method, but with some shape approximations to the potential. The *von Barth-Hedin* exchange-correlation potential is used. *Zhu* et al. [2.41] use a spin-polarized version of the SCLO method and their own parameterization of the exchange-correlation potential. *Wimmer* et al. [2.39] use the spin-polarized FLAPW method and the *von Barth–Hedin* potential.

The three calculations are in close agreement with regard to the quantities in Table 2.2. This is a reflection of the excellent overall agreement of the energy bands in the three calculations. The agreement between the two five layer calculations is the more striking since the calculational approaches are so different. The calculations do not agree on the subtle question of whether the majority surface state at $\bar{M}$ is occupied; *Wimmer* et al. find it just occupied in agreement with experiment, whereas the others find it just empty. However, it is clear that the calculations have essentially produced the same electronic structure for Ni $\{100\}$.

The increased moment at the surface of Ni $\{100\}$ is presumably a consequence of the reduced coordination of the surface atoms[5], since further reduction in coordination to that of a monolayer of Ni increases the moment to $\sim 1\mu_B$ [2.40, 41]. This provokes the speculation that a ferromagnetic overlayer on a relatively inert substrate might also exhibit a greatly enhanced moment. This was confirmed by *Richter* et al. [2.43] who carried out a spin-polarized SCLO calculation for a monolayer of Fe on Ag $\{100\}$. The calculation, which modeled the system by a seven layer Ag film with a layer of Fe on each surface, obtained an Fe moment of $3.0\mu_B$. This is somewhat less than the $3.2\mu_B$ of an isolated monolayer at the Ag lattice constant [2.43], but much larger than the $2.2\mu_B$ of bulk Fe. The layer projected density of states of the Fe layers is shown in Fig. 2.1. The combination of large spin splitting and narrow bands combine to completely resolve the majority and minority bands.

Table 2.2. Calculated work function in eV and layer magnetic moments in μ_B/atom of Ni $\{100\}$ films. $S - n$ denotes the nth layer in from the surface of the film. The experimental value for the work function (ϕ) is 5.2 eV

	Layers	ϕ	S	$S - 1$	$S - 2$	$S - 3$
Jepsen et al. [2.40]	5	5.4	0.65	0.58	0.61	
Zhu et al. [2.41]	5	5.0	0.66	0.55	0.61	
Wimmer et al. [2.39]	7	5.4	0.68	0.60	0.59	0.56

---

[5] The reduced coordination will narrow the d-bands and in general will increase the density of states at the Fermi level. The importance of the density-of-states at the Fermi level for making a paramagnetic system unstable to spin polarization has been discussed by *Slater* [2.42]. An LDA calculation for a 3d atom will spin-polarize and have the maximum moment permitted by the exclusion principle (e.g. $2\mu_B$ for Ni). The intraatomic exchange which spin-polarizes the atom is the principal factor driving a system to ferromagnetism. The factor opposing ferromagnetism is that the creation of a moment in a paramagnetic system requires that a certain number of electrons be transported from one spin band across the Fermi level to the other spin band. The cost in energy to do this is inversely proportional to the density of states at the Fermi level. Thus reduced coordination will in general favor the onset of ferromagnetism and, because it makes the atoms of a system more atomic-like, will cause an increase in the moment per atom.

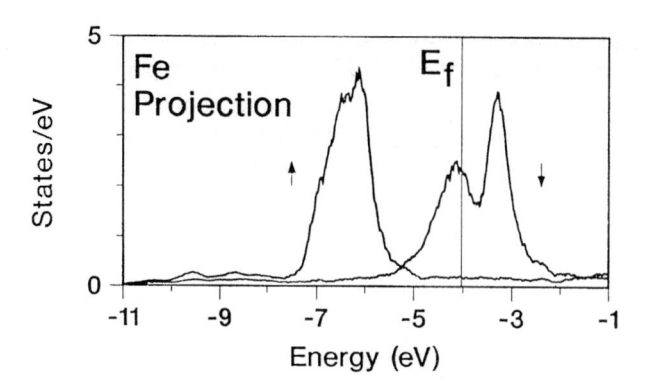

Fig. 2.1. Layer projected densities of states of the Fe layers in an Fe on Ag{1 0 0} film

Fu et al. [2.44] reported a number of FLAPW calculations on films containing layers of Cr, Fe and V adjacent to layers of Cu, Ag and Au. Among their predictions are that a monolayer of V would be ferromagnetic on any of the noble metal substrates and that a monolayer of Cr on Au {1 0 0} has a moment of $3.7\mu_B$ which is over six times the bulk antiferromagnetic moment.

One of their calculations modeled the Fe on Ag {1 0 0} system by a film consisting of a layer of Fe and a layer of Ag. They find a Fe moment of $2.96\mu_B$ in essentially exact agreement with *Richter* et al. Also the energy lowering upon spin polarization is almost the same. *Richter* et al. find -1.2 eV/atom while *Fu* et al. find -1.14 eV/atom (see Table 2.3). Although *Fu* et al. did not publish a band structure so that detailed comparisons could be made, the agreement of the moments indicates that the FLAPW and SCLO methods are, as for the Ni films, producing the same electronic structure for the Fe overlayer. Measurements of the magnetic moment per atom using polarized neutron reflection are discussed in chapter 6 by *Bland* for the Fe/Ag system.

The fact that films of fcc Fe can be grown on the Cu {1 0 0} surface has led to several theoretical studies in which independent calculations have been done on similar films. *Fu* and *Freeman* [2.45] have done spin-polarized FLAPW calculations on a five layer film of fcc Fe sandwiched between monolayers of Cu, and on a five layer film of Cu sandwiched between monlayers of Fe. *Fernando* and *Cooper* [2.46] have done a spin-polarized calculation for a five layer film of fcc Fe. Their scalar relativistic film linearized muffin-tin orbital method uses the same partitioning of space as the FLAPW method but different expansions in the various regions. *Richter* and *Gay* [2.47] have done a SCLO calculation on a seven layer film of Cu sandwiched between monolayers of Fe. Relevant comparisons of the layer moments obtained by these calculations are presented in Table 2.4. Note that both fcc Fe calculations predict antiferromagnetic coupling of the central plane. Agreement is good considering that the compared calculations are not on identical films.

2.1 Electronic Structure of Magnetic Thin Films

Table 2.3. Calculated magnetic moment in μ_B/atom of the Fe atom in the Fe/Ag{1 0 0} system and the energy decrease with spin-polarized in eV/atom

	Fe moment	E(ferro) $-$ E(para)
Richter et al. [2.43]	3.0	-1.2
Fu et al. [2.44]	2.96	-1.14

Table 2.4. Calculated layer magnetic moments in μ_B/atom for films representing the fcc Fe on Cu{1 0 0} system. $S - n$ denotes the nth layer in from the surface of the film

	Film	S	$S-1$	$S-2$
Fu and *Freeman* [2.45]	Cu on fcc Fe	2.68	2.31	-1.45
Fernando and *Cooper* [2.46]	fcc Fe	2.79	2.30	-1.68
Fu and *Freeman* [2.45]	Fe on Cu	2.85		
Richter and *Gay* [2.47]	Fe on Cu	2.75		

We believe the comparisons we have presented are persuasive that the various methods solve the scalar relativistic LDA *Kohn–Sham* equations with good precision. Coupling this with the fact that the LDA gives the experimental moments for the bulk ferromagnets suggests that theory can reliably predict the spin magnetic moments of thin transition metal films.

Despite a host of experimental studies on monolayer range films of Fe on Cu, Ag and Au, there has never been a determination of the magnetic moment per atom in an Fe monolayer on one of these noble metal surfaces. As has been discussed [2.48], it is easier to determine the easy direction of magnetization than the moment per atom. This is perhaps fortunate considering the presumed accuracy of the theoretical moment predictions and the difficulties encountered in anisotropy calculations.

However, recently there has a been a magnetometry study of monolayer range films of Fe(1 1 0) on W(1 1 0) [2.49] which determines a moment per atom for that system. The system is especially suited to the experiment because W(1 1 0) has a large surface energy which makes the Fe monolayer thermodynamically stable. Fe monolayers on the noble metal surfaces discussed above are unstable. To prevent contamination, the Fe films were overcoated with Ag after preparation. A moment per atom was found by measuring the magnetic moment of the film with a torsion magnetometer, dividing by the number of atoms in the film and extrapolating to zero temperature. The value obtained was $2.52 \pm 0.12 \mu_B$.

Hong et al. [2.50] have done a FLAPW calculation which models the W/Fe/Ag system by a film consisting of five layers of W(1 1 0) with a layer of Fe(1 1 0) and a layer of Ag(1 1 0) on each surface. They find, after minimizing the film energy with respect to the W–Fe and Fe–Ag layer spacing, a Fe moment per atom of $2.18\mu_B$. The 15% discrepancy between the experimental and theoretical values is larger than would be expected from our earlier discussion. We note that the theoretical result is the spin moment alone whereas the experimental value contains a spin–orbit induced orbital contribution. In a full-relativistic calculation [2.51] for an isolated Fe{1 1 0} monolayer the total moment is increased by 6% by the orbital contribution. Thus a significant part of the discrepancy may be neglect of the orbital contribution in the theoretical value.

2.1.3.2 Full-Relativistic Calculations

The first first-principles calculations of the magnetocrystalline anisotropy for magnetic thin films were carried out by the authors for isolated monolayers of Fe, Ni and V [2.27], and subsequently for bcc Fe films and Fe monolayer on Ag {1 0 0} [2.52]. These calculations were done as described above by solving for the eigenstates of a full-relativistic Hamiltonian variationally in a basis of non-relativistic eigenstates determined by the SCLO method. The Ceperley–Alder [2.53] exchange-correlation potential was used.

These calculations predated the unsuccessful attempts to calculate the magnetocrystalline anisotropy of the bulk ferromagnets so the ramifications of this failure were not appreciated. The spin-orbit matrix elements were calculated in an approximation, which in retrospect may not be sufficiently accurate, that considers only interactions among d-orbitals on the same atomic site. Nevertheless, the calculations do give the right order of magnitude for the magnetocrystalline anisotropy as we now demonstrate.

Figure 2.2 shows results for a Fe {1 0 0} monolayer with the Ag {1 0 0} lattice constant so as to represent as closely as possible a monolayer on Ag {1 0 0}. The plot shows the spin–orbit energy, E_{so}, for four quantization directions as a function of the number of k vectors used to approximate the integral over the Brillouin zone required to obtain E_{so}. $\hat{z}$ is perpendicular to the plane of the monolayer and $\hat{x}$ is along a nearest neighbor direction in the plane of the monolayer. The plot shows the large number of k vectors required for convergence and that, at convergence, the easy direction is perpendicular to the monolayer.

There is inconsequential variation in the plane of the monolayer and the energy of the monolayer relative to in-plane quantization can be accurately represented by the formula

$$E = \Delta E \cos^2 \theta,$$

where θ is the angle made by the quantization direction with the monolayer normal. From the converged results of Fig. 2.2, the anisotropy energy is $\Delta E = E_{so}(\hat{z}) - E_{so}(\hat{x}) = -0.38$ meV/atom.

The magnitude of this spin anisotropy is large: several hundred times the anisotropy of bulk Fe and six times the anisotropy of bulk Co. It is large enough

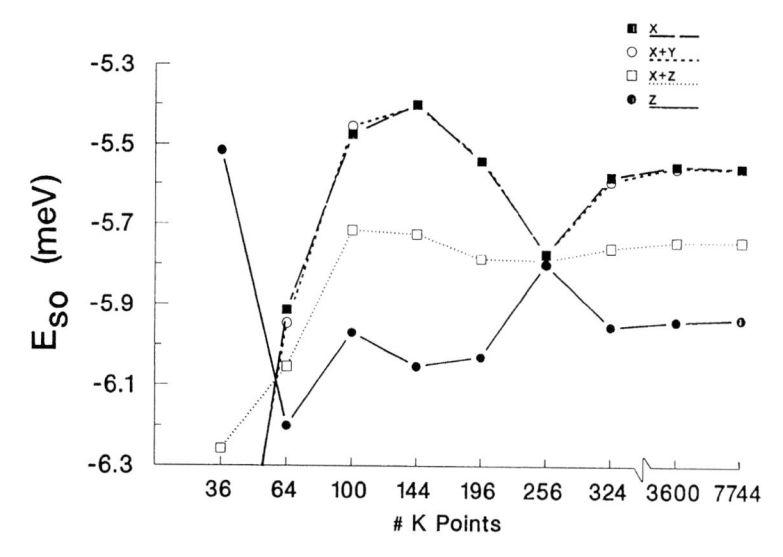

2.1 Electronic Structure of Magnetic Thin Films

Fig. 2.2. Spin–orbit energy E_{so} of a Fe monolayer for four directions of spin quantization as a function of the number of $k_\|$ vectors used to approximate the integral over the first Brillouin zone

to overcome the classical dipole depolarization energy which is 0.29 meV/atom for a monolayer on Ag {1 0 0} and 0.32 meV/atom for an isolated monolayer. The moment is therefore predicted to be perpendicular.

This result is consistent with the fact that, in the experiments of *Jonker* et al. [2.54], spin polarized bands but no spin polarization was observed for monolayer films of Fe on Ag {1 0 0}. This is because the experiment could not detect spin polarization when the moment is perpendicular. For thicker Fe films, spin polarization was observed indicating that the moment had rotated in-plane.

Similar behavior has been observed by others for Fe films on Ag {1 0 0} [2.55, 56], and on Cu {1 0 0} [2.57, 58] and Au {1 0 0} [2.59, 60] as well. The films on Ag and Au are bcc while those on Cu are fcc. Perpendicular moments are found below a critical film thickness which ranges from two to six layers with in-plane moments found beyond that thickness.

We can use these observations to estimate the experimental anisotropy ΔE existing in these Fe films. Since the major part of the spin anisotropy is a surface effect, it should be proportional, when measured in units of energy per film unit cell, to the unit cell cross sectional area rather than to its volume and be independent of thickness in first approximation [2.52]. On the other hand, the classical dipole depolarization energy, in the same units, is proportional to the volume of the film unit cell. This energy for bcc Fe films is shown in Fig. 2.3 as a plot of depolarization energy, D, the difference in energy when the dipoles are oriented perpendicular and in the plane of the film, versus film thickness. Since the transition to inplane moment occurs when $D + \Delta E = 0$, we conclude that the experimental spin anisotropy ranges between -0.4 meV/unit cell (two layer

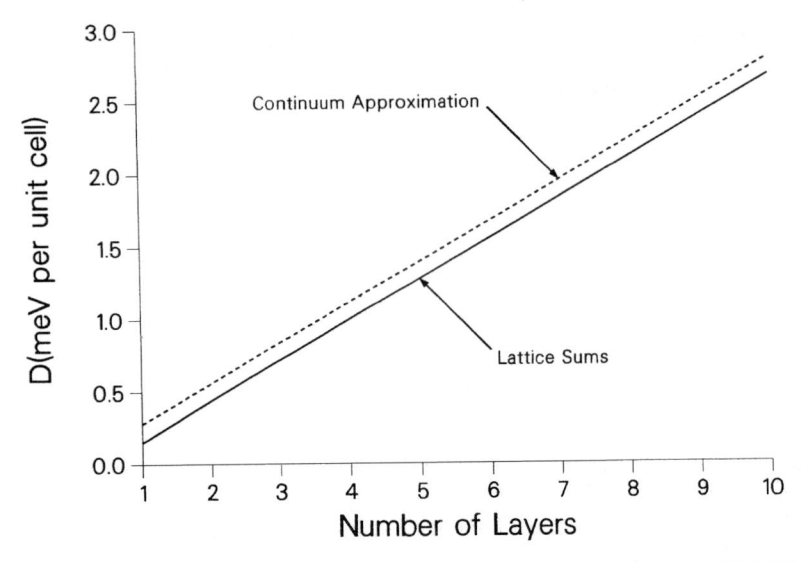

Fig. 2.3. Depolarization energy D of bcc Fe films as a function of the number of $\{100\}$ layers in the film. D was computed by carrying out lattice sums assuming the bulk Fe moment per atom of $2.2\mu_B$ at each site. The continuum approximation to D is shown for comparison. The almost constant difference is due to the fact that the continuum approximation assumes that every site in a film sees the bulk depolarization field, $[4\pi/3 - (8\pi/3)]M = 4\pi M$, whereas in the lattice sums sites near the surface of a film see less than this field

critical thickness) and -1.6 meV/unit cell (six layer critical thickness). This estimate is strictly applicable only to the bcc films on Ag$\{100\}$ and Au$\{100\}$.

These experimental anisotropies are comparable to, if somewhat larger in magnitude than, the anisotropies we calculated for bcc Fe films. The largest we found was -0.72 meV/(unit cell) for a nine layer film [2.52]. However, based on other film results and the unsuccessful bulk calculations, we now regard this agreement as fortuitous. When we did a more realistic calculation for Fe on Ag$\{100\}$ using the seven layer Ag film with Fe monolayers on each surface of [2.43], we found an anisotropy per monolayer of -0.07 meV/atom [2.52]: a value too small to overcome the monolayer shape anisotropy and at variance with experiment. A similar result, -0.10 meV/atom, was found for Fe on Cu$\{100\}$ [2.47].

There have been other full-relativistic calculations of the magnetocrystalline anisotropy of thin films. *Karas* et al. [2.61] have calculated the anisotropy of an Fe monolayer. *Li* et al. [2.62] have done a calculation on an Fe monolayer and calculations on bilayer films consisting of a layer of Fe and a layer of host metal designed to more accurately model Fe on Ag$\{100\}$, Fe on Au$\{100\}$ and Fe on Pd$\{100\}$. Both groups based their calculations on scalar-relativistic FLAPW calculations which used the *von Barth–Hedin* [2.17] exchange-correlation potential.

Table 2.5. Calculated anisotropy energy ΔE in meV/atom for monolayer films of Fe. The monolayer calculation of *Karas* et al. was done at the bcc Fe lattice constant. The other two monolayer calculations were done at the slightly larger Ag lattice constant

	Gay and Richter [2.27, 47, 52]	Karas et al. [2.61]	Li et al. [2.62]
Fe monolayer	− 0.38	3.4	0.043
Fe monolayer on Ag{1 0 0}	− 0.08		− 0.064
Fe monolayer on Cu{1 0 0}	− 0.10		
Fe monolayer on Au{1 0 0}			− 0.57
Fe monolayer on Pd{1 0 0}			− 0.35

Their anisotropy results, as well as ours, are collected in Table 2.5. As with the bulk calculations, the calculations generally do not agree with each other or with experiment. Note that only in our monolayer calculation and in the calculations of *Li* et al. for monolayers on Au and Pd is the spin anisotropy large enough to overcome the monolayer shape anisotropy. Ironically, in the one instance of reasonably close agreement (Fe on Ag{1 0 0}) the result does not agree with experiment.

There have recently appeared calculations of the magnetocrystalline anisotropy of several superlattices containing Co layers [2.63] using the calculational scheme of [2.11] as discussed in Sect. 2.2. All superlattices show perpendicular anisotropy. For $(Co)_n/(Pd)_m$ superlattices, $\Delta E = -0.85$ meV per unit cell for $n = 1$ and decrease with increasing n. These results are in agreement with experiment but, considering the failure of the identical calculational scheme to predict the easy axis of bulk Co [2.11], it is hard to see how this agreement can be anything but fortuitous.

2.1.4 Conclusions

We have presented a substantial body of evidence that first-principles LDA calculations can predict the magnetic moment per atom in magnetic thin films, although it may be necessary to calculate the orbital contribution to the moment induced by the spin-orbit interaction for best accuracy. On the other hand, LDA calculations are totally unable to predict the easy direction of magnetization in any ferromagnetic system at their current stage of refinement. This state of affairs, while unsatisfactory in principle, is in a practical sense complementary to the experimental situation for magnetic thin films where easy directions are relatively simple to determine but moments per atom are more difficult to ascertain [2.48].

We believe the appropriate arena in which to take up the matter of the failuer of the anisotropy calculations is in the bulk. It seems to us pointless to attempt to predict the magnetocrystalline anisotropy of magnetic thin films theoretically until a theory is developed that can reproduce the unassailable experimental bulk anisotropies.

2.2 Magnetic Anisotropy from First Principles

G.H.O. DAALDEROP, P.J. KELLY, and M.F.H. SCHUURMANS

Experimentally it has been shown for a variety of [1 1 1] oriented multilayers containing Co that their magnetization is oriented perpendicular to the multilayer planes if the Co layer is made as thin as one or a few monolayers [2.64]. An indication as to the origin of this phenomenon has been found through measurements of the anisotropy energy density, K, as a function of the thickness of the Co layers, t. The anisotropy energy density varies approximately with the inverse of the Co layer thickness and can be expressed by the relation $Kt \approx 2K_S + K_V t$. Because the volume anisotropy energy density K_V is negative and usually of the order of the anisotropy energy density of bulk hcp Co [2.64], perpendicular magnetic anisotropy (PMA) will only occur if the interface anisotropy energy density K_S is positive and sufficiently large.

Various explanations for the origin of the interface anisotropy have been put forward. *Néel* [2.65] has predicted that the anisotropy energy density will markedly change if the dimensions of magnetic particles are reduced to about 100 Å. The surface atoms of the particles will contribute differently to the anisotropy energy since the local symmetry of these atoms is different from that in the bulk. *Néel* developed a quantitative model based upon the assumption that the anisotropy energy can be expressed in pair interaction energies between localized moments on magnetic atoms, where the parameters can be obtained from *bulk* magnetoelastic and elastic constants. Using this model for Co/X multilayers, where X is a non-magnetic atom, the interface anisotropy is expected to depend on the X atom layers *only* by their influence on the structure of the Co layers.

Recently it was pointed out that magnetoelastic energy will manifest itself in the interface anisotropy if the lattice misfit strain of the Co layer is inversely proportional to its thickness [2.69]. However, experimentally the structures of the multilayers are not well known, so that the magnetoelastic contribution to the measured interface anisotropy is still unclear [2.64].

First-principles calculations of the magnetocrystalline anisotropy energy (MAE) have been reported for free standing monolayers [2.67, 68], overlayers on silver [2.69], $Co_n X_m$ multilayers where X = Cu, Ag, Pd [2.70], Ni [2.71] and thin slabs comprising a monolayer of Fe and a monolayer of Ag, Au or Pd [2.68]. Using model calculations [2.72] the surface anisotropy energy has been studied within the context of the electronic structure. From these calculations it can be concluded beyond doubt that the anisotropy energy depends on the type of X atom in the layer (substrate) adjacent to the magnetic layer. Apparently Néel's theory is too simple or incomplete. It is the purpose of this chapter to give an explanation within the itinerant electron picture for the perpendicular orientation of the magnetization of Co/Pd multilayers. After briefly discussing the calculational method, the band structure and anisotropy energy of a freestanding Co monolayer are analyzed. General conditions favorable for PMA are derived. Results obtained for multilayers are then reviewed, and finally, using the analysis of the monolayer, the origin of the anisotropy energy of a Co_1Pd_2 multilayer is discussed.

2.2.1 Method

The magnetic anisotropy energy equals the difference in the total energy when the magnetization is oriented along a direction $\hat{n} = \hat{n}(\Theta, \phi)$, and when it is oriented perpendicular to the Co plane. Θ and ϕ are polar coordinates with respect to a rectangular coordinate system which is defined with respect to the crystal structure. The z axis of the coordinate system is chosen normal to the plane of the Co layer, and the y axis is chosen along a nearest neighbor direction. The properties of the Co monolayer and multilayers are calculated within the framework of the local-spin-density approximation (LSDA) [2.73]. The one electron Kohn–Sham equations for the scalar-relativistic spin-polarized Hamiltonian are solved self-consistently using the linear muffin–tin orbital (LMTO) method in the atomic sphere approximation (ASA) [2.74]. The spin–orbit coupling is then included and new Kohn–Sham one electron eigenvalues are obtained by diagonalizing the full Hamiltonian. The magnetic anisotropy energy is well approximated by the difference in the sums of the Kohn–Sham eigenvalues [2.75, 76]

$$\Delta E(\hat{n}) = E(\hat{n}) - E(\hat{n} \parallel z) = \int^{\varepsilon_F(\hat{n})} \varepsilon D(\varepsilon, \hat{n}) d\varepsilon - \int^{\varepsilon_F(z)} \varepsilon D(\varepsilon, z) d\varepsilon, \qquad (2.8)$$

where $D(\varepsilon, \hat{n})$ is the density of states when the magnetization is directed along $\hat{n}$. We will not consider the dependence of the magnetic anisotropy energy on the orientation of the magnetization within the plane of the Co layer, i.e. neglect the dependence on the azimuthal angle ϕ, because it is expected to be very small (as will be shown later). We choose $\hat{n}$ in the x, z plane ($\phi = 0$). In general we are interested in the anisotropy energy when the magnetization is rotated from x to z, and this energy is denoted by $\Delta E \equiv \Delta E(x)$.

Using the band structure, the Fermi energy $\varepsilon_F(q, \hat{n})$, can be calculated as a function of the band-filling q_b of this band structure. An anisotropy energy curve $\Delta E(q)$ can then be obtained from 2.8 with $\hat{n} = x$. Alternatively we may consider the anisotropy energy curve as a function of the Fermi energy corresponding to a band-filling of q states.

2.2.2 Spin–Orbit Coupling Matrix

The dependence of the Hamiltonian on the direction of the magnetization can be implemented in several ways. The following choice is particularly convenient. The exchange splitting, $B(r)\hat{n}\cdot\boldsymbol{\sigma}$ with respect to the real space coordinate system, is diagonalized by minority spin ('up') and majority spin ('down') spinors which are the columns in the Wigner rotation matrices $D^{1/2}(\phi,\Theta,0)$. By this transformation, the z axis of the spin coordinate system is chosen along the direction of the magnetization; the exchange splitting is given by $B(r)\sigma_z$, and the spin-orbit interaction is given by $D^{1/2\dagger}(\phi,\Theta,0)\xi\boldsymbol{l}\cdot\boldsymbol{\sigma}D^{1/2}(\phi,\Theta,0)$.

For our later analysis it is helpful to note here that the latter matrix has the transparent form

$$\mathcal{H}^{SO}(\hat{n}) = \xi\begin{pmatrix} \hat{n}\cdot\boldsymbol{l} & \hat{n}_\perp\cdot\boldsymbol{l} + \tfrac{1}{2}(l_- - l_+) \\ \hat{n}_\perp\cdot\boldsymbol{l} - \tfrac{1}{2}(l_- - l_+) & -\hat{n}\cdot\boldsymbol{l} \end{pmatrix}, \qquad (2.9)$$

where $\hat{n}$ is chosen in the x,z plane ($\phi = 0$); $\hat{n} = (\sin\Theta, 0, \cos\Theta)$ and where $\hat{n}_\perp$ is normal to $\hat{n}$ in the x,z plane; $\hat{n}_\perp = (\cos\Theta, 0, -\sin\Theta)$. For clarity it was assumed here that the spin–orbit coupling parameters for the radial d wavefunctions are both energy and spin independent. Apart from the terms in the matrix elements that are independent of Θ, the matrix elements between states with opposite spins are identical to the matrix elements between corresponding states with equal spin, provided that the magnetization is rotated over 90°, i.e. from $\hat{n}$ to $\hat{n}_\perp$. Matrix elements of $\mathcal{H}^{SO}(\hat{n})$ on the basis of tesseral harmonics have been given in [2.77].

2.2.3 Co Monolayer

2.2.3.1 Details of the Calculation

The calculated properties of a monolayer will not be different from the calculated properties of a periodic sequence of monolayers, provided that the distance between the monolayers is large enough. This enables the use of a band structure method that is normally used for three dimensional crystal structures. We have described the monolayers as a sequence of [1 1 1] lattice planes of an fcc lattice. The Co layers are separated by five layers of vacuum which are filled with empty spheres serving as expansion centers for the muffin–tin orbitals. The stacking sequence of the spheres is chosen ABCABC, so that the lattice is close packed and has inversion symmetry. Due to the specific description of the potential and charge density in the vacuum using the atomic sphere approximation, and the chosen ABC stacking sequence the monolayer is *not* a mirror plane (the pointgroup symmetry is D_{3d}), However, this 'artificial' effect due to the ASA is very small.

We have performed calculations for three different lattice parameters. We denote these separate calculations as I, II and III. The corresponding atomic sphere radii are $S = 2.621$ a.u. (corresponding to the atomic sphere radius in hcp Co) for I, $S = 2.73$ a.u. for II, and $S = 2.845$ a.u. (corresponding to the in-plane lattice parameter used for calculations of the anisotropy energy of $Co_n Pd_m$ multilayers [2.70]) for III.

The basis consists of s, p, d, f partial waves on the Co site, s, p, d partial waves on the empty sphere sites adjacent to the Co sites, and s, p partial waves for the other empty sphere sites. For the latter spheres this small basis is sufficient since a negligible charge density is present on the inner spheres. Including higher order partial waves in the bases at the empty sphere sites do not influence the calculated properties. The large separation between the Co monolayers resulted in a dispersion along the z direction in reciprocal space which was less than 4 µeV for all energy bands within 8 eV of the Fermi energy. This very small dispersion can be safely disregarded in the discussion of all properties. As a result, only two dimensional Brillouin zone (BZ) integrals need to be, and have been performed.

2.2.3.2 Band Structure

The band structure along the high symmetry lines Γ–K–M in the two dimensional BZ is shown in Fig. 2.4 for the majority spin (dashed curves) and minority spin bands (solid curves). Only the energy range of the d-bands has been shown. The actual Fermi energy is denoted by the horizontal solid line. At the high symmetry points the angular dependence of the d partial waves contributing predominantly to the eigenstates has been listed. To a fairly good approximation the minority spin d-bands are simply displaced from the majority spin d-bands by an exchange splitting of about 1.6 eV, the d magnetic moment of $1.87\mu_B$ times the Stoner exchange integral of 0.86 eV [2.78]. As the radial d wavefunction contracts with increasing energy, the exchange integral is increased (decreased) going towards the top (bottom) of the d-band. Therefore, the width of the minority spin d-band is nearly 20% larger than that of the majority spin d-band [2.79].

The d-bands are only about 2 eV wide, which is much less than in bulk hcp Co. This is caused partly by the reduced number of nearest neighbors. Assuming the density of states to be rectangular, the bandwidth can be estimated from the second moment of the structure constants [2.80], and is proportional to the square root of the number of nearest neighbors. This results in a calculated bandwidth of 3.1 eV, compared to 4.4 eV for the bulk [2.81]. The actual band width is still considerably smaller, and this is related to the diminished, or even absent hybridization between d orbitals in the monolayer. As a result, the shape of the density of states strongly deviates from a rectangle.

The eigenstate of the singlet d-band at Γ which increases in energy going to K and M and which shows hardly dispersion between K and M, is mainly of

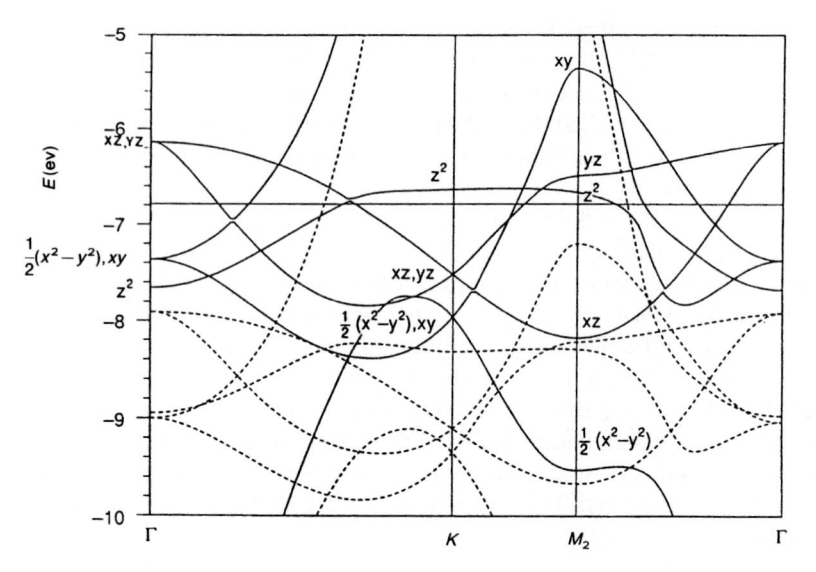

Fig. 2.4. Majority spin (dashed) and minority spin (solid) band structure of Co monolayer I along the high symmetry lines of the two dimensional Brillouin zone in the energy range of the d-bands. The actual Fermi energy is denoted by the horizontal line. The predominant character of the eigenstates has been listed at the high symmetry points

$(1/2\sqrt{3})(3z^2 - r^2)$ character ($l = 2$, $m = 0$). The small dispersion is caused by a small overlap of $(3z^2 - r^2)$ orbitals with those on neighboring Co atoms. In contrast, the bands with mainly $\frac{1}{2}(x^2 - y^2)$ and xy character ($l = 2$, $|m| = 2$) show the largest dispersion. If the plane of the monolayer were a mirror plane, the eigenstates of the bands would be either odd in z (with xz, yz character, $|m| = 1$) or even in z ($|m| = 0, 2$). These eigenstates would belong to different representations and would not hybridize; the band structure could almost be calculated by hand by diagonalizing a 2×2 and 3×3 Hamiltonian [2.79]. Although odd and even bands do hybridize in our representation of the monolayer (due to the ABC stacking sequence and the ASA), in practice they hardly do and the bands can still be considered as being of either odd or even character.

In Fig. 2.5 the band structure including spin–orbit coupling is shown along the high symmetry lines Γ–K–M_2–Γ–M_1, where $\hat{n} \parallel x$ (dashed) and $\hat{n} \parallel z$ (solid). Here $M_2 = \frac{1}{2}G_2$ and $M_1 = \frac{1}{2}G_1$, where $G_{1,2}$ are the reciprocal lattice vectors and $G_2 \parallel x$. These M points are equivalent in the absence of spin–orbit coupling. The main *visible* effect upon a rotation of $\hat{n}$ is the changing splitting between energy bands near accidental or true degeneracies (in the absence of spin–orbit interaction). However, *all* energy bands change on a scale of 1–10 meV and this will be important when we investigate the anisotropy energy.

2.2 Magnetic Anisotropy from First Principles

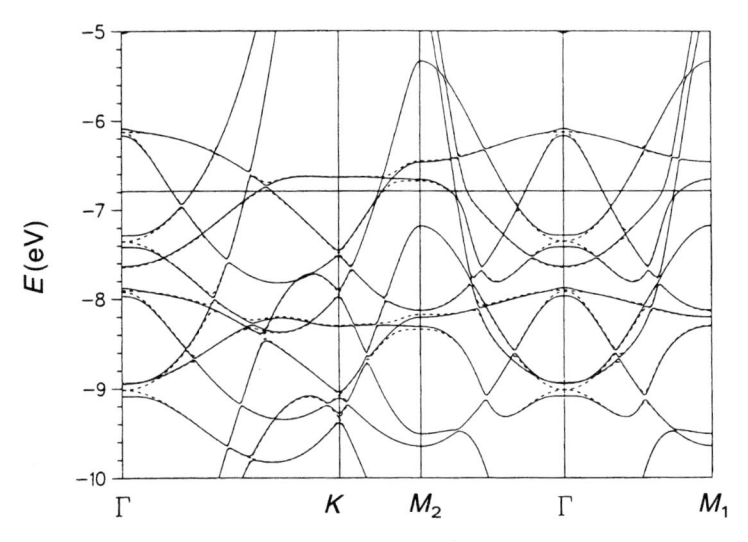

Fig. 2.5. Band structure of Co monolayer I along high symmetry lines of the two-dimensional Brillouin zone, where spin–orbit coupling has been included. Solid curve: magnetization parallel to z, dashed curve: parallel to x. M_1 and M_2 are the M points along the reciprocal lattice vectors G_1 and G_2 respectively, where $G_2 \parallel x$

2.2.3.3 Anisotropy Energy

The anisotropy energy as a function of the band filling is shown in Fig. 2.6 for the three monolayers, broadened using a Gaussian with a width of 0.2 eV. As opposed to the calculated perpendicular anisotropy of Co_1/Pd, Co_1/Ag and Co_1/Cu multilayers [2.70], an in-plane anisotropy is obtained for a Co monolayer. Although the anisotropy energy depends strongly on the lattice parameter and the position of the actual Fermi energy, *all* three anisotropy energy curves are very similar and oscillate with the bandfilling. To understand them, it is therefore sufficient to analyze one monolayer only. We will discuss monolayer I.

Assuming that the anisotropy energy is an analytical function of the direction cosines of the magnetization, it follows from the symmetry of the hexagonal lattice that the anisotropy energy can be expanded as $\Delta E(\hat{n}) = K_2 \sin^2 \Theta + K_4 \sin^4 \Theta = \Delta E(\Theta)$. K_2 and K_4 are second and fourth order anisotropy constants respectively. Sixth and higher order anisotropy constants (which depend on ϕ) have been ignored. The magnitude of K_4 has been investigated for the band fillings given above. In Fig. 2.7 $\Delta E(\Theta)/\sin^2 \Theta$ is shown as a function of $\sin^2 \Theta$. K_2 is given by the intercept with the vertical axis, K_4 is given by the slope of the line. For the band fillings of $8.2e$ and $9.4e$, K_4 is less than 3% of the value for K_2. This is of the same order of magnitude as the numerical accuracy of the calculation. At a band filling of $9.0e$ the expansion does not appear to be adequate. We have not investigated this further, but this may be related to the

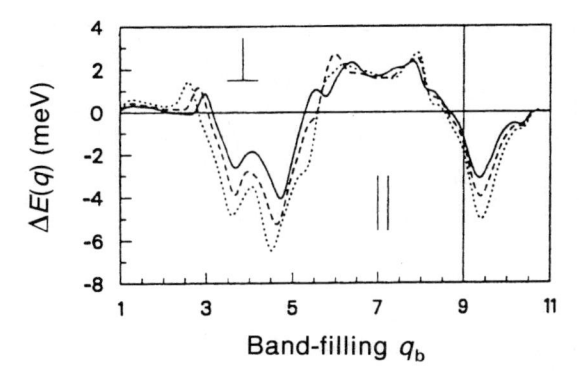

Fig. 2.6. Anisotropy energy versus band filling for Co monolayers at three different lattice parameters; monolayer I: solid curve, II: dashed curve and III: dotted curve. The subscript b is omitted from q_b in the labelling of the vertical axis

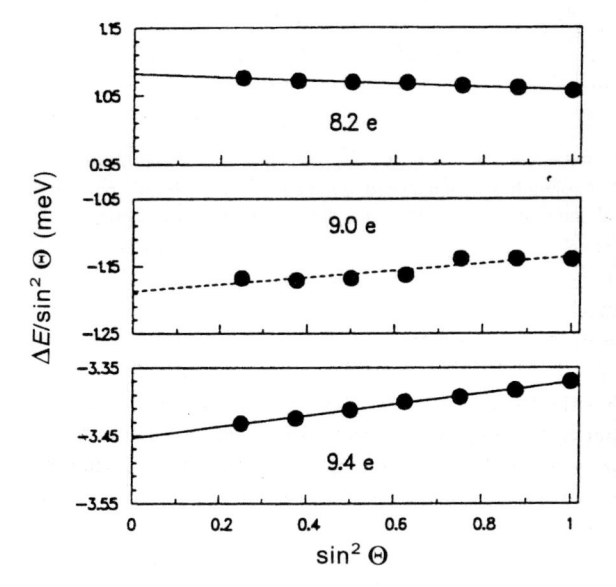

Fig. 2.7. $\Delta E(\Theta)/\sin^2\Theta$ as a function of $\sin^2\Theta$. The second order anisotropy constant K_2 is given by the intercept of the line, obtained by a least squares fit, with the vertical axis. The fourth order anisotropy constant K_4 is given by the slope of the line. At a band filling of $9.0e$ such an expansion does not appear to be adequate

existence of accidental degeneracies along the high-symmetry lines Γ–K and K–M [2.82]. Finally we note that using perturbation theory [2.83] the higher order anisotropy constants are expected to be smaller than K_4.

2.2.3.4 Analysis of the Anisotropy Energy

In Fig. 2.8 it is illustrated that the anisotropy energy curve calculated using a limited number of **k** points is similar to that obtained from a full calculation employing a large number of **k** points. The anisotropy energy is calculated using 144 (solid curve) and 6 (dashed curve) divisions of the reciprocal lattice vectors. In the latter calculation seven irreducible **k** points are used if $\hat{n} \parallel z$. The curve is

2.2 Magnetic Anisotropy from First Principles

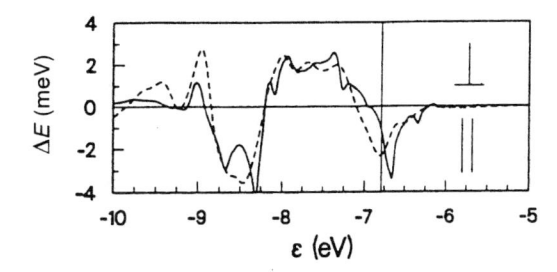

Fig. 2.8. Anisotropy energy curve as a function of the Fermi energy. Calculated using 144 divisions (solid curve) and 6 divisions (dashed curve) of the reciprocal lattice vectors. The latter curve is broadened using a Gaussian with a width of 0.2 eV. The actual Fermi energy is denoted by the vertical line

broadened using a Gaussian with a width of 0.2 eV to smear the discrete energies into bands. In this section it will be shown that the anisotropy energy curve of a monolayer can be understood by analyzing the eigenstates and energies at the high symmetry points Γ, K and M only. The anisotropy energy at each $\boldsymbol{k}$ point is broadened using a Gaussian with a width of 0.2 eV, and multiplied with a weightfactor w_k. The anisotropy energy curve of all $\boldsymbol{k}$ points together is obtained by adding the weighted contributions from each $\boldsymbol{k}$ point. The weightfactors for the $\boldsymbol{k}$ points Γ, K and M are determined by their number density in the BZ. If $\hat{\boldsymbol{n}} \parallel \boldsymbol{x}$ (parallel to reciprocal vector G_2), the two K points in the BZ, $\frac{1}{3}(G_1 + G_2)$ and $\frac{2}{3}(G_1 + G_2)$, are equivalent. One of the M points, $M_2 = \frac{1}{2}G_2$, is not equivalent with the other two M points, $\frac{1}{2}G_1$ and $\frac{1}{2}(G_1 + G_2)$, denoted by M_1. The weightfactors for the special points Γ, K, M_1 and M_2 are hence chosen as $1/6, 1/3, 1/3$ and $1/6$, respectively.

In Fig. 2.9 the contributions to the anisotropy energy from these $\boldsymbol{k}$ points are shown. The arrows in the figure indicate the positions of the energy levels at each $\boldsymbol{k}$ point, and a double arrow denotes degenerate energy levels (in the absence of spin–orbit interaction). Upward (downward) pointing arrows denote minority (majority) spin eigenstates. The eigenstates have mainly d character, and the predominant azimuthal quantum number $|m|$ of the d partial wave is indicated as well. The contributions from M_1 and M_2 are indicated by the solid and the dashed curve, respectively. In the bottom figure the dashed curve represents the added contributions from all $\boldsymbol{k}$ points. The solid curve is obtained by an additional broadening with a Gaussian with a width of 0.4 eV. Since the anisotropy energy curve is very similar to that obtained from a full calculation (see Fig. 2.8), it can be understood in essence by analyzing Γ, K and M only.

At a single $\boldsymbol{k}$ point, two types of contributions to the anisotropy energy can be distinguished. One type is caused by the existence of degenerate energy levels at high symmetry points. Degenerate energy levels at the Fermi energy are split by the spin–orbit interaction into energy levels lying above and below the Fermi energy. The total energy of the system is thereby reduced by an amount which depends on the direction of the magnetization since the spin-orbit splitting depends on the direction. Spin–orbit splitting of degenerate levels at the Fermi energy thus contributes to the anisotropy energy. Twofold degeneracies exist at Γ and K. We do not discuss accidental degeneracies here, although they are found to be particularly important along K–M at $E \sim -6.7$ eV. For the 'true'

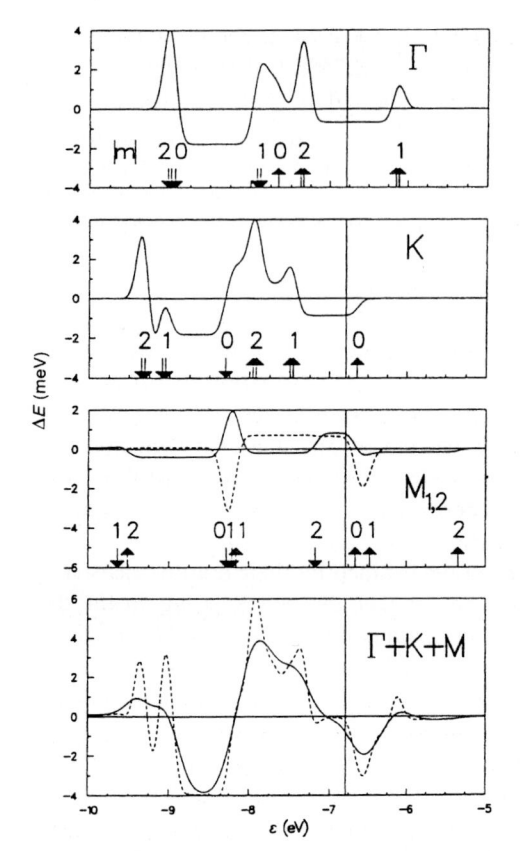

Fig. 2.9. Anisotropy energy contributed by Γ, K and M_1 (solid curve) and M_2 (dashed curve), as a function of the Fermi energy. Arrows indicate the position of the energy levels, a double arrow is used to denote degenerate eigenstates. Upward (downward) pointing arrows indicate minority (majority) spin eigenstates. The dashed curve in the bottom figure is the result of adding the contributions of these k points. The solid curve is obtained by broadening the dashed curve using a Gaussian with a width of 0.4 eV. The actual Fermi energy is indicated by the vertical lines

degeneracies, the corresponding eigenstates will involve partial waves with $l = 2$, m and $l = 2$, $-m$ character, where $m = 2$ or $m = 1$. The spin–orbit coupling splits the degenerate energy bands by $2m\xi|\cos\Theta|$. If perturbative coupling to other bands is neglected (to be discussed below), the degeneracy is not lifted when $\hat{n} \parallel x (\Theta = \pi/2)$. True degeneracies at the Fermi energy therefore give a contribution to the anisotropy energy which favors the perpendicular orientation ($\Theta = 0$).

The second type of contribution is due to the spin–orbit interaction, coupling eigenstates ψ_i and ψ_j with energies ε_i below the Fermi energy and ε_j above the Fermi energy. If the level splitting $\Delta_{ij} = \varepsilon_i - \varepsilon_j$ is much larger than the spin–orbit coupling parameter ξ one can use perturbation theory to deduce the contribution to the anisotropy energy. The contribution from each pair of states is given by $w_k \Delta E_{ij}$, where

$$\Delta E_{ij} = \frac{1}{\Delta_{ij}}(|H_{ij}^{SO}(x)|^2 - |H_{ij}^{SO}(z)|^2) \tag{2.10}$$

and $H_{ij}^{SO}(\hat{n}) = \langle \psi_i | \mathcal{H}^{SO}(\hat{n}) | \psi_j \rangle$. This contribution to the anisotropy energy can favor either a perpendicular orientation, or an in-plane orientation of the magnetization, depending on the spins and symmetries of the states i and j. In the bottom panels of Figs. 2.10(a)–(c) for Γ, K and M_2, respectively, the contributions ΔE_{ij} to the anisotropy energy from each pair of eigenstates i and j, calculated using perturbation theory, are denoted by bars connecting these eigenstates i and j. The thickness of each bar is taken proportional to ΔE_{ij}. Couplings between two eigenstates favoring an in-plane (perpendicular) orientation of the magnetization are indicated by horizontally (vertically) shaded bars. Note that the top panels of Figs. 2.10(a)–(c) have been calculated by full diagonalization of the Hamiltonian, and display the contribution of both the spin–orbit split degenerate levels at the Fermi energy, accounting for most of the structure in the panels, and the contributions of the spin–orbit coupled states lying energetically on opposite sides of the Fermi energy.

Let us illustrate by an example that perturbative coupling between states on either side of the Fermi energy can give a large contribution to the anisotropy energy. Consider the $m = 0\downarrow$ and $m = \pm 1\uparrow$ eigenstates at Γ. The relevant matrix elements for the anisotropy energy are $\langle m = 0 | \hat{n}_\perp \cdot l | m = \pm 1 \rangle = -\cos\Theta\sqrt{3/2}$. Therefore, $\Delta E_{0\downarrow,+1\uparrow} + \Delta E_{0\downarrow,-1\uparrow} = |p|^2(\xi^2/\Delta_{0\downarrow,\pm1\uparrow})\cos^2\Theta$, where $|p|^2 = 2 \times 3/2$, $\Delta_{0\downarrow,\pm1\uparrow} = 2.81$ eV and $\xi = 36$ meV. Despite the fact that the energy distance Δ_{ij} is larger than the d-band width, the contributed anisotropy energy is large, of the order of $w_\Gamma \times 1.4$ meV $= 0.23$ meV. In the following, each high symmetry point will be discussed briefly.

The Γ point. Degenerate anti-bonding states with xz and yz ($|m| = 1$) character are high in energy and degenerate bonding states with $\frac{1}{2}(x^2 - y^2)$ and xy ($|m| = 2$) character are low in energy. The easy axis is the z axis for Fermi energies close to the degeneracies ($\Theta = 0$). An in-plane easy axis is favored for Fermi energies located in between the energies of bonding (m even) and anti-bonding (m odd) states of the same spin, because these states are coupled by the spin–orbit interaction $\xi\hat{n}\cdot l$ if $\hat{n} \parallel x$, and are not coupled if $\hat{n} \parallel z$. The coupling between even and odd states of the same spin therefore favors the in-plane orientation of the magnetization. Even or odd states of opposite spin are coupled by the spin–orbit interaction $\xi\hat{n}_\perp \cdot l$ if $\hat{n}_\perp \parallel z$, i.e. if $\hat{n} \parallel x$. These interactions therefore favor an in-plane orientation of the magnetization as well.

Because the exchange splitting of 1.6 eV is about equal to the bandwidth of 2 eV, the top of the majority spin d-band, where the eigenstates have antibonding character (odd states), and the bottom of the minority spin d-band, where bonding states are situated (even states), coincide approximately in energy. They are coupled by the spin-orbit interaction $\xi\hat{n}_\perp \cdot l$ if $\hat{n}_\perp \parallel x$, i.e. if $\hat{n} \parallel z$. These interactions therefore favor a perpendicular orientation of the magnetization. Thus, at the band edges (bottom and top of each spin subband) a perpendicular orientation of the magnetization is expected, whereas in the middle of each spin subband an in-plane orientation is favored.

The K point. A similar analysis of the anisotropy energy curve of the K point, displayed in Fig. 2.10(b), can be given. Partial waves with $m = \pm 2$ and $m = \pm 1$ form degenerate eigenstates. The $|m| = 1$ eigenstates partially have a bonding character, due to the phase factor introduced in the Bloch wave. As a result,

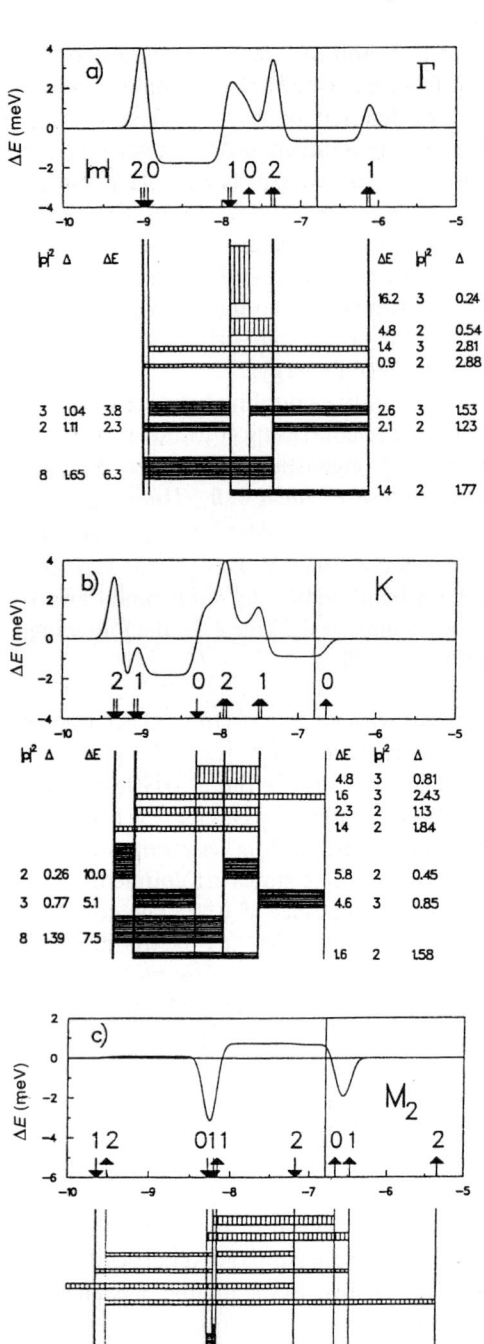

50   2. Magnetic Anisotrophy, Magnetization and Band Structure

Fig. 2.10. Calculated anisotropy energy as a function of the Fermi energy, analyzed using perturbation theory in the bottom panel. The anisotropy energy contributed by each pair of states i and j at each $\boldsymbol{k}$ point, ΔE_{ij}, is proportional to the thickness of bars connecting these states. Horizontally (vertically) shaded bars indicate in-plane (perpendicular) contributions to the anisotropy energy. Perpendicular contributions originating from degenerate eigenstates have not been indicated in the bottom panel of the figure. Only couplings with $\Delta E_{ij} > 1$ meV have been indicated. Both ΔE_{ij} (in meV), the separation in energy Δ_{ij} (in eV) and $|p_{ij}|^2$ have been indicated (the labels ij have been omitted in the figure). A spin–orbit coupling parameter ξ of 36 meV was assumed in the analysis. Note that the solid curve in the top of the figure was calculated by diagonalization of the Hamiltonian. The actual Fermi energy is denoted by the vertical lines at -6.8 eV

within each spin subband the lowest eigenstates consist of $|m| = 2$ and $|m| = 1$ orbitals, and the highest anti-bonding eigenstate consists of an $m = 0$ orbital. If the Fermi energy is situated in the lower part of each spin subband, the degenerate states will favor a perpendicular orientation of the magnetization which, however, is reduced by the coupling between $|m| = 2$ and $|m| = 1$ orbitals. In the middle of the spin subband, the in-plane orientation is favored, like at Γ. Finally, because bandwidth and exchange splitting are about equal, the couplings between the eigenstates at the top of the majority spin subband, and bottom of the minority spin subband strongly favor a perpendicular orientation of the magnetization.

The M point. The situation at the M points is more complex, as can be seen from Fig. 2.10(c). All representations at the M point are one dimensional, so that no (true) degeneracies exist. We simply note here that all couplings cancel to a large extent, except where eigenstates are close in energy.

Thus the only mechanisms by which perpendicular anisotropy occurs is through the lifting of degeneracies at Γ and K at the top and bottom of each spin band, and through the couplings by the spin–orbit interaction between eigenstates of opposite spins. The latter type of contribution is important when, as a result of a decrease in the band width and an increase in the exchange splitting, the top of the majority spin subband and the bottom of the minority spin subband come close in energy, and the actual Fermi energy is located in the bottom of the minority spin band. The magnetic anisotropy energy then has a large perpendicular contribution from the coupling by the spin–orbit interaction of even and odd states with opposite spin. These are conditions that can occur in layered structures, when the number of equal nearest neighbors is low, and when the center of the d bands of the nonmagnetic transition metal element is sufficiently low in energy. This special situation occurs in the compounds with the AuCu($L1_0$) structure, where the constituent elements are Fe, Co and Pd or Pt [2.84]. The perpendicular orientation is then favored if the Fermi energy is at or above the top of the majority spin subband and in the bottom of the minority spin subband. In the middle of each spin subband, an in-plane anisotropy is generally favored, because the couplings between bonding and anti-bonding eigenstates within one spin subband favor the in-plane orientation. The anisotropy energy is thus expected to oscillate with the bandfilling. On general grounds this has been proven for a tight-binding d band [2.85].

2.2.4 Co/Pd, Co/Ag and Co/Cu Multilayers

2.2.4.1 Structure

Co/Pd multilayers deposited on a polycrystalline Pd base with [1 1 1] texture form films comprising close packed planes of Pd and Co [2.86]. Their precise structure is not known experimentally. Epitaxial [0 0 1] Co/Pd multilayers have been obtained by vapor deposition in ultra high vacuum (UHV) onto cleaved [0 0 1] NaCl with a Pd buffer layer [2.87]. In this case the Co and Pd layers were found to be coherent with each other. Typically the Pd layer is 20 Å thick.

We will construct a model for the structure which is appropriate for multilayers where the Pd layer is sufficiently thick that it may be assumed to determine the in-plane lattice parameter. The parameters in this model are evaluated by performing first-principles calculations of the total energy using the full-potential-linear-augmented-plane-wave (FLAPW) method [2.88].

In our model for the structure of [1 1 1] $Co_n Pd_m$ multilayers, we assume an ABCABC stacking sequence of close packed planes. The resulting crystal structure has trigonal symmetry (pointgroup D_{3d}), a basis consisting of 3 or 6 atoms, and primitive vectors that can be chosen as hexagonal lattice vectors with lengths a and c. The latter is expressed as $c = 2c_{Co-Pd} + (n-1)c_{Co-Co} + (m-1)c_{Pd-Pd}$, where c_{A-B} denotes the perpendicular distance between neighboring A and B planes (Fig. 2.11). The in-plane nearest-neighbor (nn) distance, a, and the nn separation of Pd atoms in neighboring planes is assumed to be equal to that calculated for fcc Pd; $a = \sqrt{3/2}c_{Pd-Pd} = a_{fcc}^{Pd}/\sqrt{2}$. Thus the lattice mismatch between Co and Pd (which is $\sim 9\%$ between $[0001]_{hcp}$-Co and $[1 1 1]_{fcc}$-Pd) is assumed to be compensated for by elastic deformation of the Co lattice. The two remaining structural parameters, c_{Co-Co} and c_{Co-Pd}, were calculated by minimizing the total energy for trigonal Co_3 and $Co_1 Pd_2$ respectively, keeping the in-plane nn distance a constant ($= a_{fcc}^{Pd}/\sqrt{2}$). The c/a ratio which we then find for strained trigonal Co_3 is 2.03, compared to $(3/2)\sqrt{8/3} \simeq 2.45$ for fcc Co. The resulting inter planar nn distances are only slightly smaller than those obtained from close packing of hard spheres. This model for the nn distances was also applied to the other $Co_n X_m$ multilayers.

The structure of [0 0 1] $Co_n Pd_m$ multilayers is a stacking of [0 0 1] planes of an fcc lattice. The multilayer has tetragonal symmetry and a is taken to be equal to the calculated fcc Pd cubic lattice constant. The interplanar separations are found from calculations for body centered tetragonal (bct) cobalt and $Co_1 Pd_1$, yielding $c_{Co-Co} = 1.40$ Å and $c_{Co-Pd} = 1.71$ Å. The results to be discussed below are not sensitive to small changes in the structure parameters.

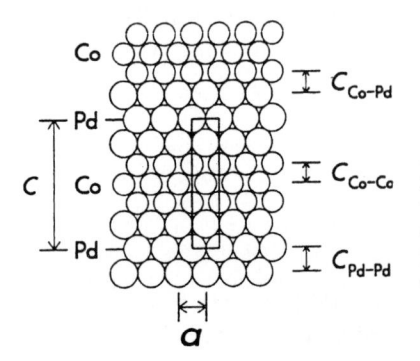

Fig. 2.11. Model for the structure of $Co_n Pd_m$ multilayers containing three structure parameters: c_{Co-Pd}, c_{Co-Co} and $a = \sqrt{\frac{3}{2}}c_{Pd-Pd} = a_{fcc}^{Pd}/\sqrt{2}$. The values of the parameters used are (in Å), $c_{Co-Co} = 1.84$, $c_{Co-Pd} = 1.99$ and $c_{Pd-Pd} = 2.22$.

2.2.4.2 Anisotropy Energy of [1 1 1] Multilayers

In previous studies of the MAE the convergence of the BZ integral of the Kohn–Sham eigenvalues has been an important source of uncertainty. In Fig. 2.12 the calculated MAE of Co_1X_2 and Co_1X_5 multilayers, with X = Pd, Cu and Ag, is shown as a function of the volume element, v, used to evaluate the three-dimensional integral. An infinitely dense integration mesh corresponds to $v \rightarrow 0$. For the Co_1X_2 multilayers satisfactory convergence is achieved and the estimated numerical accuracy of the MAE is ± 0.03 meV/unit-cell. The anisotropy energy of the Co_1X_5 multilayers (open symbols) is less well converged. Because the unit cell is now twice as large, the number of bands is doubled and the time required to diagonalize the Hamiltonian increases by a factor of ~ 8. Computing time limitations [2.89] restrict the maximum number of k points used to evaluate the BZ integral.

The anisotropy energy of multilayers containing Cu and Ag is seen to be essentially equal and does not depend on the thickness of the Cu and Ag layers. The magnetic moments of Cu and Ag are $\lesssim 0.01 \mu_B$ because their d-bands are filled. The anisotropy energy of Co_1Pd_5 is about 10% smaller than that of Co_1Pd_2. In Co_1Pd_5 the calculated spin moment on the Pd atoms decreases from $0.30\mu_B$ for the interface layer, to $0.13\mu_B$ for the next layer and is only $0.06\mu_B$ for the central Pd layer. For Pd layers thicker than five monolayers we expect the MAE to change by less than its difference between Co_1Pd_2 and Co_1Pd_5 because the Pd atoms in the additional layers are neutral and have a magnetic moment of $\lesssim 0.04\mu_B$.

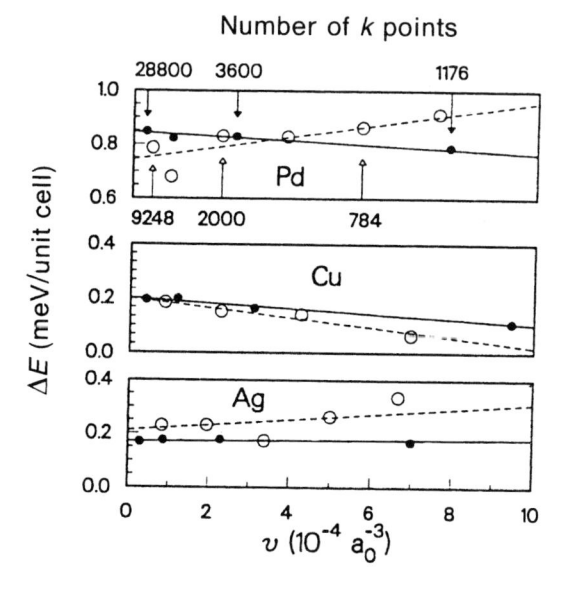

Fig. 2.12. The magnetocrystalline anisotropy energy per unit cell, ΔE, of Co_1X_2 (●) and Co_1X_5 (○) multilayers for X = Pd, Cu and Ag, plotted as a function of the tetrahedron volume, v, used to perform the integration over the Brillouin zone. The sign of ΔE is such that $\Delta E > 0$ favors a perpendicular orientation of the magnetization. The straight lines indicate the extrapolation of the anisotropy energy to an infinitely dense mesh $v \rightarrow 0$. The number of k points used in the BZ integration for Co_1Pd_2 (Co_1Pd_5) is indicated at the top (bottom) of the upper panel

The magnetocrystalline anisotropy (*without* contributions from dipolar interactions) is seen to favor an orientation of the magnetization perpendicular to the multilayer plane ($\Delta E > 0$) for all multilayers considered. We want to check how sensitive this prediction is to various approximations which have been made. Since only the Kohn–Sham eigenvalues enter into the calculation of the MAE, we need to know (i) how a given approximation affects the energy bands and, in particular, the energy bands around the Fermi energy and (ii) how such shifts in the energy bands around the Fermi energy affect the anisotropy energy. We study Co_1Pd_2 in particular detail.

We first examine the sensitivity of the MAE to the position of the Fermi energy itself. By using the (LMTO-ASA) band structure calculated self-consistently for Co_1Pd_2 (with $n = 29$ valence electrons) and filling the energy bands to an energy $E_F(q_b)$ (where q_b is a non-integral number of electrons) the anisotropy energy is calculated as a function of q_b. For $|q_b - n| < 0.25$, corresponding to $|E_F(q_b) - E_F(n)| < 0.08$ eV the change is less than 5%.

We next modify the band structure of Co_1Pd_2 by changing the parameters in the structure model. For each set of parameters the MAE is calculated from first principles. In Fig. 2.13(a) the fractional change of the MAE, $\delta(\Delta E)/\Delta E$, is shown as a function of the fractional change in the volume of the unit cell, $\delta\Omega/\Omega$. Here ΔE and Ω refer to the multilayer Co_1Pd_2 for which we have shown results in Fig. 2.13

A uniform compressive strain was applied, i.e., *all* structure parameters were changed with the same fraction $\sim \delta\Omega/3\Omega$. Decreasing the volume causes a reduction of the MAE. For example, on changing the volume by -8%, the change in the MAE is only about -10%. Assuming $E_{Me} = -\frac{3}{2}\lambda\sigma$ for stress induced magnetoelastic energy, experiments on sputtered Co/Pd multilayers and Co/Pd alloys indicate a value for the magnetostriction constant $\lambda_M = -1.5 \times 10^{-4}$ [2.90]. The calculated results yield a magnetostriction constant of about -6×10^{-5} assuming an elastic constant of 2×10^{11} N/m^2.

We might expect the anisotropy energy to depend differently on c_{Co-Pd} than on a. In Fig. 2.13(b) the change of the MAE is shown as a function of the fractional change in c_{Co-Pd}. Here a was kept fixed. The change of the MAE is about a factor of three larger than in Fig. 2.13(a) for corresponding changes in the volume, indicating that the MAE depends less on a than on c_{Co-Pd}. From Fig. 2.13 it is concluded that for a wide range of values of the structure parameters, a perpendicular orientation of the magnetization is predicted.

We also examine the effect on the MAE of treating the spin–orbit interaction self-consistently. Because the spin–orbit coupling parameter for Pd d states is so large ($\xi_d^{Pd} = 0.23$ eV), it is not a priori clear that this is unnecessary. We first solve the Kohn–Sham equations, including the spin–orbit coupling and with the spin-quantization axis chosen perpendicular to the multilayer plane, self-consistently. The charge and magnetization densities thus obtained were then used as input to a calculation of the MAE using the force theorem. The MAE was unchanged. We have thus verified that our prediction of a perpendicular orientation of the magnetization is not sensitive to a number of approximations

2.2 Magnetic Anisotropy from First Principles

Strain dependence ΔE

Fig. 2.13a, b. The fractional change in the calculated magnetocrystalline anisotropy energy of [1 1 1] Co_1Pd_2 as a function of the fractional change in the volume, (**a**), and as a function of the fractional change in c_{Co-Pd}, the separation between interface Co and Pd layers, (**b**) The solid lines are a guide to the eye. The prediction of a perpendicular orientation of the magnetization is not sensitive to the particular values of the structure parameters

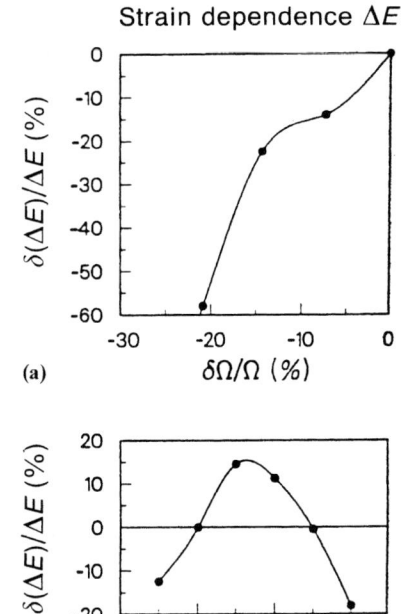

made in the calculations. The MAE does not depend strongly on shifts of order 0.1 eV in the bands close to the Fermi energy.

Finally, we compare the energy bands calculated with the LMTO-ASA method with those obtained using the FLAPW-method in which no shape approximation to the potential is made. Differences between the energy bands at **k** points that contribute significantly to the anisotropy energy are found to be $\lesssim 0.05$ eV relative to ε_F. We conclude that use of a full-potential method will not lead to qualitatively different predictions. In a calculation of the MAE with such a method, it would be very difficult to demonstrate the convergence of the Brillouin zone integral for a unit cell containing six atoms because of the computational time required.

Experimental [2.87] and calculated results for Kt of Co_nPd_m multilayers are compared in Fig. 2.14. The error bars indicate the degree of convergence of the calculations; for six atoms per unit-cell it is ~ 0.1 meV. The demagnetization energy of ~ -0.09 meV/Co atom which favors an in-plane magnetization is included in the calculated values shown. A perpendicular orientation of the magnetization and decreasing anisotropy energy with increasing Co thickness is predicted, in agreement with experiment. (Increasing the number of Pd layers in Co_4Pd_2 may lead to a reduced MAE, as found on going from Co_1Pd_2 to Co_1Pd_5). However, the decrease of Kt with increasing t is about a factor three larger than that caused by the demagnetization energy only. For thick Co layers

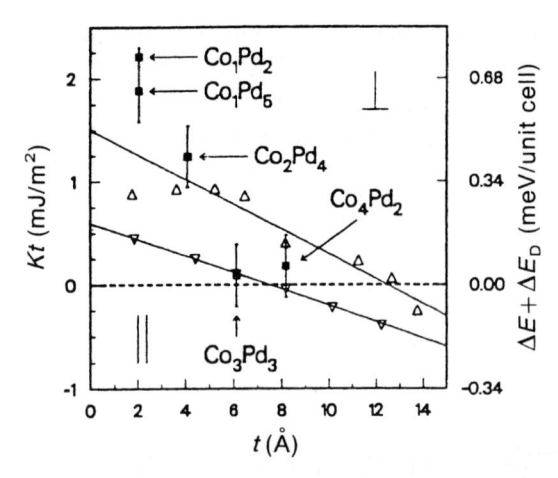

Fig. 2.14. The anisotropy energy density, K, times Co-thickness, t, as a function of t for polycrystalline $[1\,1\,1]_{fcc}$ multilayers deposited at $T_s = 50\,°C$ ($\triangledown$) and $T_s = 200\,°C$ ($\triangle$), compared with the ab initio calculated values for Co_1Pd_2, Co_1Pd_5, Co_2Pd_4, Co_3Pd_3 and Co_4Pd_2. The (room temperature) experimental data are taken from *den Broeder* et al. [2.87]. $K > 0$ corresponds to a perpendicularly oriented magnetization. The magnetocrystalline anisotropy energy plus demagnetization energy per unit-cell, $\Delta E + \Delta E_D$, is equal to KtA where A, the Co cross sectional area, is taken so as to be consistent with [2.87]. The straight lines are fits to the experimental data points. Error bars indicate the estimated numerical accuracy of the calculation

this slope will be determined by the sum of $\Delta E + \Delta E_D$ for strained trigonal Co. We have calculated the MAE, ΔE, for Co_3 for various c/a ratios in a range from 2.45 (fcc Co) to 2 keeping a constant. We find it to be less than 0.03 meV/Co atom so that asymptotically the slope will be determined by ΔE_D. We conclude that the preference for a perpendicular orientation of the magnetization must be attributed to the presence of the Co/Pd interface and that the influence of the Co/Pd interface extends beyond the Co interface layer.

Of all the multilayers considered, the anisotropy energy is largest for Co_1Pd_2. The large MAE may not be attributed to the strained Co layer because the mismatch between Co and Pd is intermediate between the Co–Ag and Co–Cu mismatches. The large spin–orbit coupling on the Pd site does play an important role by introducing large splittings in the strongly hybridized Co and Pd d bands within 1 eV of the Fermi energy. Setting ξ_d^{Pd} to zero leads to reduced splittings and the MAE is approximately halved.

2.2.4.3 Anisotropy Energy of [001] Multilayers

In Fig. 2.15 the calculated results for Kt of [0 0 1] Co_nPd_m multilayers are compared with experiment [2.87]. Compared to the [1 1 1] oriented multilayers there are two essential differences in the experimental data. Firstly, the interface anisotropy energy is strongly reduced. Secondly, an additional easy-plane an-

2.2 Magnetic Anisotropy from First Principles

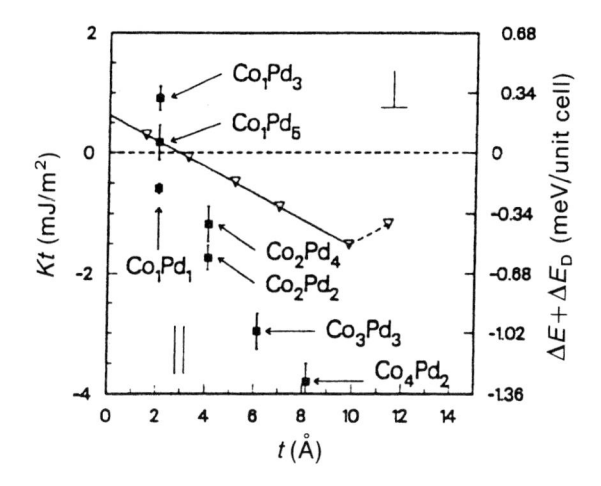

Fig. 2.15. Kt as a function of t for coherent epitaxial [0 0 1] multilayers deposited at $T_s = 50\,°C$ ($\triangledown$) [2.87] compared with the ab initio calculated values

isotropy is present in the volume anisotropy, which is nearly as large as the demagnetization energy. Hence [0 0 1] multilayers are only perpendicularly magnetized if they contain monolayers of cobalt.

Our calculated anisotropy energies reflect these two differences. Only Co_1Pd_3 and Co_1Pd_5 are predicted to be perpendicularly magnetized, and a strong decrease of the anisotropy energy with increasing t is also found. However, as in the calculations for [1 1 1] oriented multilayers, the slope is about a factor of three larger than is found experimentally.

2.2.5 Co/Ni Multilayers

In this section we describe the prediction, from *ab initio* calculations, of a perpendicularly oriented magnetization in a novel multilayer (ML) comprising *two* magnetic elements, Co and Ni, together with its subsequent experimental varification. From an analysis made possible by the calculation we find that there are two factors which contribute to this result: (i) the presence of an interface between ultrathin close-packed layers of the magnetic elements Co and Ni, which differ by only one valence electron, is sufficient to give rise to a large magnetic anisotropy energy and (ii) the total number of valence electrons in Co_1/Ni_2 positions the Fermi energy close to bands with $x^2 - y^2$ and xy character (the z axis is chosen to be normal to the interface) whose spin–orbit interaction favors a perpendicular orientation of the magnetization.

The ML structure used in the calculations was chosen to consist of close-packed Co and Ni planes stacked in an ABCABC... sequence, i.e. [1 1 1] planes of an fcc lattice. The unit cell then consists of multiples of three atoms. Because of the small lattice mismatch (0.8%) between bulk Co (in its fcc form) and Ni the same lattice constants were used for the Co and Ni layers. We have verified that our predictions are not sensitive to changes in the lattice parameters of a few percent.

2.2.5.1 Anisotropy Energy

The results of our calculations of the anisotropy energy (including ΔE_D) for Co_1/Ni_2, Co_1/Ni_5 and Co_2/Ni_1 ML's (where the subscripts denote the number of atomic layers per unit-cell) are shown in Fig. 2.16. The anisotropy energy is plotted as a function of the volume element, v, used to evaluate the three-dimensional BZ integral. An infinitely dense integration mesh corresponds to $v \to 0$. The calculations are seen to be well converged. A perpendicular orientation of the magnetization is predicted for Co_1/Ni_2, whereas an in-plane orientation is found for Co_2/Ni_1. For Co_1/Ni_5 we find an anisotropy energy of $+0.01 \pm 0.03$ meV where the error bar is derived from the convergence of the integral, shown in Fig. 2.16. Thus a definite prediction cannot be made for Co_1/Ni_5. The differences in the calculated anisotropy energies between either Co_1/Ni_2 and Co_1/Ni_5 or between Co_1/Ni_2 and Co_2/Ni_1 cannot be attributed to the demagnetization energy which is about -0.08 meV per Co atom and -0.01 meV per Ni atom. Thus *both* a Co *and* a Ni thickness dependence of the magnetocrystalline anisotropy energy is predicted. The calculated magnetization of Co_1/Ni_2 is 1.1 T.

These predictions have been investigated experimentally. Polycrystalline Co/Ni ML's were prepared by electron beam (e-beam) evaporation onto oxidized Si and glass substrates at room temperature. Deposition rates were controlled at 0.1–1.0 Å/s for Co and 0.2–2.0 Å/s for Ni using oscillating quartz sensors. Chemical analysis proved the nominal thicknesses to be correct within about 15%. Prior to multilayer fabrication a polycrystalline Au base layer with

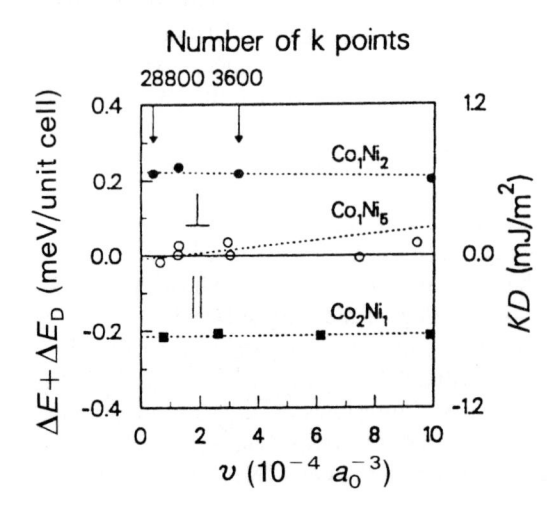

Fig. 2.16. The anisotropy energy per unit cell, $\Delta E + \Delta E_D$, of Co_1/Ni_2 (●), Co_1/Ni_5 (○) and Co_2/Ni_1 (□) multilayers as a function of the volume element, v, used to perform the Brillouin zone integration. The right hand vertical axis denotes KD, the anisotropy energy constant K times the layer period D. The sign of KD is such that $KD > 0$ favors an out-of-plane orientation of the magnetization. Extrapolation of the anisotropy energy to an infinitely dense mesh $v \to 0$ is indicated by straight lines

2.2 Magnetic Anisotropy from First Principles

[1 1 1] texture was deposited on the substrates to induce the same texture for the multilayer, as was verified by X-ray diffraction (XRD). The periodic structure was confirmed by small-angle XRD. Transmission electron microscopy showed an ABC stacking sequence of the atoms as in an fcc lattice [2.91]. The magnetic properties of the samples were investigated by means of vibrating sample magnetometry (VSM) and by torque magnetometry in fields up to $H = 1300$ kA/m at room temperature.

For Co/Ni ML's where the Co thickness, t_{Co}, is 6 Å or more, the preferred orientation of the magnetization is in the plane of the film for all Ni thicknesses t_{Ni}. For $t_{Co} = 4$ Å the easy axis is out-of-plane for $t_{Ni} = 6$–8 Å, and for $t_{Co} = 2$ Å the easy axis is out-of-plane for t_{Ni} between 2 and ~ 12 Å. The in-plane and out-of-plane hysteresis loops of $20 \times (2$ Å Co $+ 4$ Å Ni) or $20 \times Co_1/Ni_2$ are shown in Fig. 2.17. The perpendicular loop shows a remanence of 100% of the saturation magnetization value of ~ 1.0 T. The anisotropy energy constant K, calculated from the area between the two magnetization curves is in this case 0.57 MJ/m^3. The maximum value of KD (where D is the bilayer period) which we found was about 0.40 mJ/m^2, in fair agreement with the theoretical prediction of 0.65 ± 0.1 mJ/m^2 (Fig. 2.16). That the perpendicular anisotropy is undoubtedly due to the layered structure is demonstrated by the magnetic torque curves shown in Fig. 2.18 for a $100 \times (2$ Å Co $+ 4$ Å Ni) ML and for a 600 Å thick $Co_{33}Ni_{67}$ alloy, both deposited on glass with a 1000 Å Au underlayer. In contrast to the ML, the alloy film has a (normal) in-plane magnetization.

Our calculations indicate that Co_1/Pd_2 and Co_1/Ni_2 ML's are particularly favorable for obtaining perpendicular anisotropy. To understand why, we use the band structures calculated self-consistently with n valence electrons to calculate the anisotropy energy as a function of the band-filling q_b. The resulting function, $\Delta E''(q_b)$, is shown in Fig. 2.19 for $Co_1/Ni_2(n = 29)$, $Co_2/Ni_1(n = 28)$ and $Co_1/Pd_2(n = 29)$. We also calculated $\Delta E''(q_b)$ within the virtual crystal approximation for a number of different values of n and the same atomic positions. The maximum value of the MAE found attainable with this model of

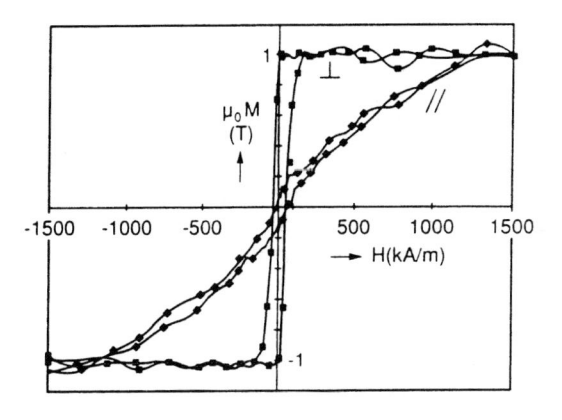

Fig. 2.17. Hysteresis loops of a $20 \times (Co_1/Ni_2)$ multilayer with applied field H oriented parallel and perpendicular to the film plane. The remanent magnetization in the out-of-plane direction is 100%, with $\mu_0 M_s \sim 1.0\ T$

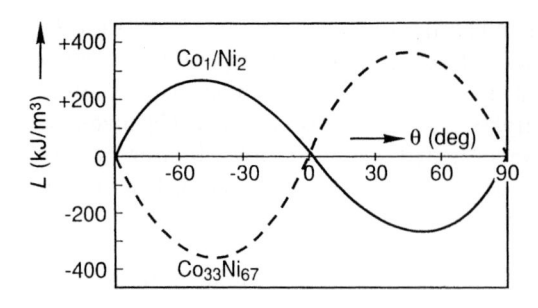

Fig. 2.18. Magnetic torque curves of a $100 \times (\mathrm{Co_1/Ni_2})$ ML yielding $K = +0.27\ \mathrm{MJ/m^3}$, and of a 600 Å thick $\mathrm{Co_{33}Ni_{67}}$ alloy thin film with $K = -0.36\ \mathrm{MJ/m^3}$

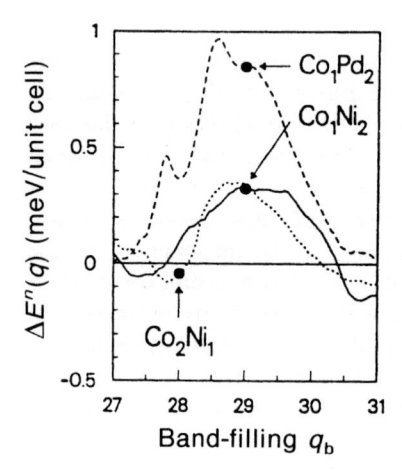

Fig. 2.19. $\Delta E^n(q_b)$ as a function of the band filling, calculated using the self-consistent band structure of $\mathrm{Co_1/Ni_2}$ (solid, $n = 29$), $\mathrm{Co_1/Pd_2}$ (dash, $n = 29$) and $\mathrm{Co_2/Ni_1}$ (dot, $n = 28$). The magnetocrystalline anisotropy energy of each of these multilayers, $\Delta E \equiv \Delta E^n(n)$, is denoted by a solid circle. The demagnetization energy is not included. The subscript b is omitted from q_b in the labelling of the vertical axis

a random alloy is much smaller than when there is an interface, even when this is as "weak" as the interface between Co and Ni.

The similarity of $\Delta E^n(q_b)$ for $\mathrm{Co_1/Ni_2}$ and $\mathrm{Co_2/Ni_1}$ indicates that the in-plane magnetization of $\mathrm{Co_2/Ni_1}$ is essentially due to the different position of the Fermi level in the d band structure. This similarity is a consequence [2.76] of the fact that Ni and Co are strong ferromagnets differing by only one valence electron so that the electronic structure may in a first appoximation be described by a rigid band model. A number of valence electrons of less than ~ 28 or more than ~ 30 is unfavorable for obtaining large positive anisotropy energies. Therefore, the MAE of ML's is not solely due to the reduced symmetry of atoms at the interfaces, as in *Néel's* model of surface anisotropy [2.65], but also depends on the interface electronic structure. The importance of the band filling aspect is further emphasized by the fact that the magnetization of a free-standing monolayer of Co has been shown to be oriented in-plane. The q_b (band filling) dependence of ΔE^n is qualitatively very similar for $\mathrm{Co_1/Ni_2}$ and $\mathrm{Co_1/Pd_2}$. However, for $\mathrm{Co_1/Pd_2}$ the amplitude of the variation is larger. This is mainly a consequence of the strong hybridization of the Co and Pd d-bands in combination with the large spin–orbit coupling constant, ζ_d^{Pd}, of Pd. If we set ζ_d^{Pd} to zero, then the MAE is approximately halved.

In the following section it will be shown that the out-of-plane magnetization can be attributed to the presence of states with $x^2 - y^2$ and xy character close to the Fermi level. The spin–orbit interaction of these states, which are extended in the plane of the multilayer favors a perpendicular orientation of the magnetization.

2.2.6 Analysis of the Anisotropy Energy of Co_1Pd_2 Multilayers

In this section the origin of the PMA of Co_1Pd_2 multilayers will be analyzed. In Fig. 2.20 the anisotropy energy is calculated as a function of the Fermi energy, using a grid employing 9408 k points in the BZ, corresponding to 28 divisions along the in-plane reciprocal lattice vectors and 12 divisions along the perpendicular reciprocal lattice vector, which we denote a 28-28-12 grid (solid curve). An oscillating dependence on the band filling is found, and a maximum occurs at the actual Fermi energy, denoted by the vertical line. The dotted curve indicates the results of a calculation using a 6-6-1 grid, which was broadened using a Gaussian with a width of 0.4 eV. Although an oscillatory dependence on the band filling is indeed found, in-plane anisotropy occurs over too large of a region. A calculation using a 6-6-2 grid, in which two divisions along the perpendicular reciprocal lattice vector were chosen gives much better results. Hence, the dispersion of the energy bands in the k_z direction can not be neglected.

We have examined the contributions to the anisotropy energy curve from the six k points Γ, K, M, A, H and L. The latter three high symmetry points correspond to the Γ, K and M points with $k_z = \pi/c$. The anisotropy energy contributed by each k point is shown in Fig. 2.21. Adding the contributions of all k points, the dashed curve in Fig. 2.21(a) is obtained. Broadening using a Gaussian with a width of 0.4 eV yields the dashed curve in Fig. 2.21(b), which has a similar functional dependence on the band filling as the full calculation,

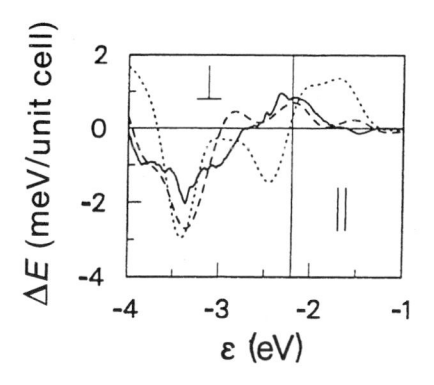

Fig. 2.20. Anisotropy energy curve of a Co_1Pd_2 multilayer as a function of the Fermi energy, calculated using 28 and 12 divisions along the in-plane and perpendicular reciprocal lattice vectors respectively (solid curve), 6 and 1 divisions (dotted curve), and 6 and 2 divisions (dashed curve). The latter two curves are broadened with a Gaussian with a width of 0.4 eV. The actual Fermi energy is denoted by the vertical line

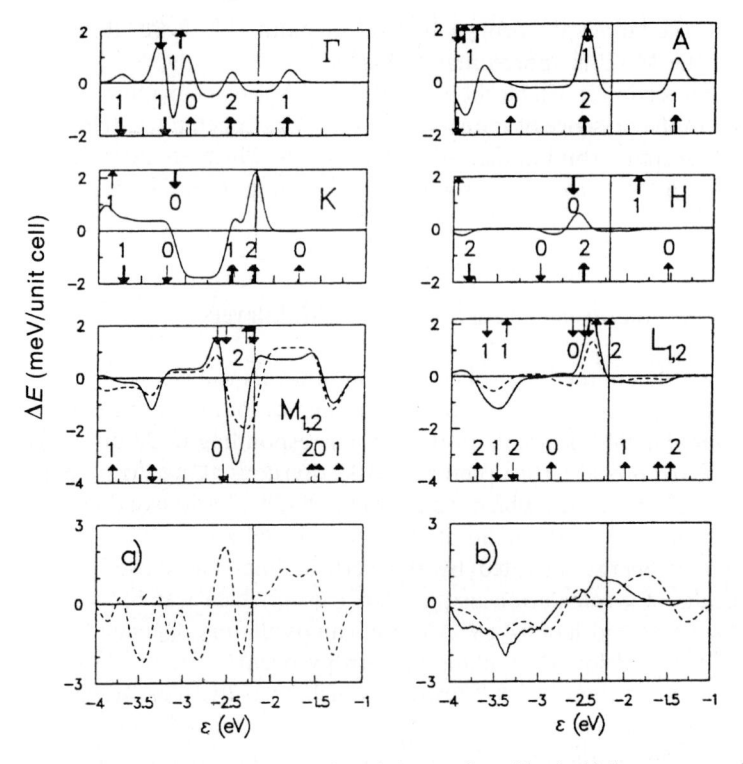

Fig. 2.21a, b. Anisotropy energy of $Co_1 Pd_2$ contributed by the high symmetry points Γ, K, M, A, H and L as a function of the Fermi energy. The actual Fermi energy is denoted by the vertical line at -2.2 eV. Arrows in the top (bottom) of the figures indicate the position of the energy levels with predominant Pd (Co) character, a double arrow is used to denote degenerate eigenstates. Upward (downward) pointing arrows indicate minority (majority) spin eigenstates. The predominant m partial wave character is denoted by the labels. The dashed curve in the bottom figure (**a**) is the result of adding the contributions of these k points. The dashed curve in figure (**b**) is obtained by an additional broadening using a Gaussian with a width of 0.4 eV. The solid curve indicates the result from a calculation employing 9408 k points in the BZ

represented by the solid curve. Only at the actual Fermi energy the correspondence is not so good, and this will be explained later.

At each k point the position of the majority and minority spin d-bands is indicated by arrows pointing downwards or upwards respectively. Arrows in the top (bottom) of the figure indicate eigenstates which consist of more than 50% of Pd (Co) character. Because of the low symmetry of the Bravais lattice (there is no mirror-plane parallel to the multilayer planes), the eigenstates contain, in general, a mixture of odd and even (in z) states. If a m partial wave is predominantly present in the eigenstate, the absolute value of m is indicated in the figure as well.

At Γ the anisotropy energy curve is similar to that of the monolayer. The amplitudes caused by the degenerate $|m| = 1$ and $|m| = 2$ eigenstates have decreased however, because of the (by a factor of two) reduced weight of Γ, and because the eigenstates have partly Pd character. At A the amplitudes of the anisotropy energy curve have increased strongly. As will be shown later in more detail, this is partly caused by the contribution from degenerate Pd eigenstates. As opposed to Γ, the hybridized Co and Pd character of the eigenstates is found to enlarge the amplitudes here. At K the anisotropy energy curve is reminiscent of that of the monolayer. However, important differences are the reversed ordering of the degenerate eigenstates of predominantly Co $|m| = 2$ and $|m| = 1$ character, as well as the location of the actual Fermi energy: in the multilayer at the degenerate $|m| = 2$ energy bands. This also occurs [2.70] in the band structure calculated using the full potential LAPW method [2.88, 74]. At H a small contribution from $|m| = 2$ degeneracies just below the Fermi energy is found to contribute. At M and L the analysis is complicated because of the large number of non-degenerate eigenstates that are present. The eigenstates around the actual Fermi energy have mainly Pd character. The anisotropy energy curve from about 0.5 eV below the actual Fermi energy to the top of the d-bands is due to the Pd character in the eigenstates together with the large Pd d spin–orbit coupling parameter of 0.1 eV. This is clearly shown by the results of a calculation in which the Pd spin–orbit coupling parameter was set equal to zero (Fig. 2.22). By adding the contributions of all k points the dashed curve in Fig. 2.22(a) is obtained. In Fig. 2.22(b) a broadening using a Gaussian with a width of 0.4 eV has been applied, and good agreement with the full calculation using a 28-28-12 grid, shown by the solid curve, is achieved. At the actual Fermi energy the anisotropy energy is halved. At the M points, the positive value of the anisotropy energy above the actual Fermi energy in Fig. 2.21 has been changed to a small negative anisotropy energy in Fig. 2.22. In addition, all amplitudes of about 2 meV in the anisotropy energy curves at all k points are reduced.

We return to a discussion of Fig. 2.22. The plateau in the anisotropy energy at the M_2 point has a similar origin as in the monolayer. The first Pd majority spin eigenstate that is shown in the top of the M_2 figure has 13% and 19% yz ($m = 1$) and xy ($m = 2$) character per Pd atom, respectively. The middle state of the highest three minority spin d eigenstates that are shown in the bottom of the M_2 figure has 13% $1/2\sqrt{3}(3z^2 - r^2)$ ($m = 0$) character per Pd atom. The energy separation between these two states is 1.2 eV. With $\xi_{Pd} = 0.1$ eV a perpendicular anisotropy of about $\frac{1}{12} \times 10$ meV is contributed by the coupling between these two eigenstates. In the monolayer, part of the plateau was caused by the interaction of an $yz \uparrow$ with a $1/2\sqrt{3}(3z^2 - r^2) \downarrow$ Co eigenstate.

Finally, we mention that the negative and positive amplitudes near the Fermi energy at M and L respectively cause a shifted maximum in the anisotropy energy curve obtained from the six k points. Because of the large dispersion of the energy bands responsible for these amplitudes, only a small number of electrons actually contribute to the anisotropy energy in the full calculation so that these eigenstates have a much smaller influence on the MAE. This explains the different location of the maximum in the anisotropy energy curve of Fig. 2.21(b).

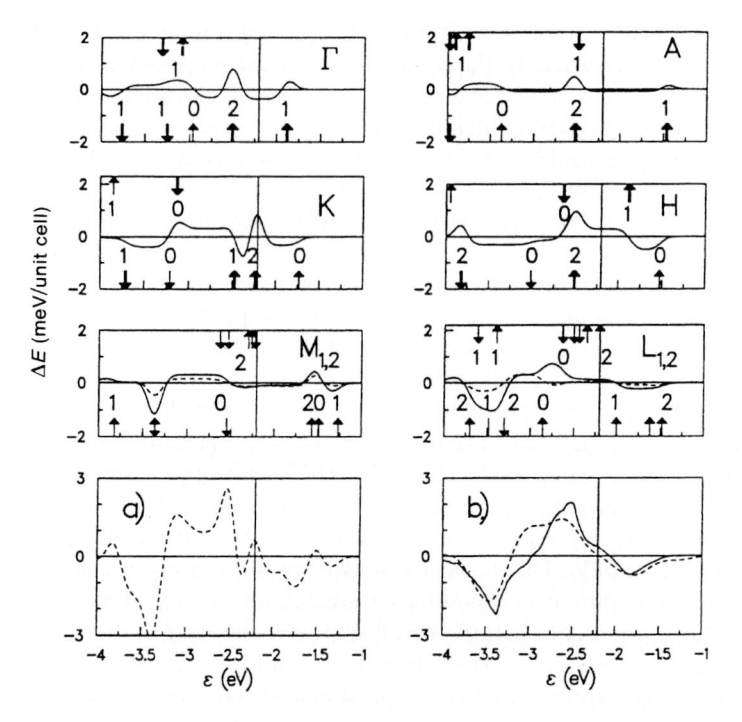

Fig. 2.22. Results of the calculation of the anisotropy energy curve at the high symmetry points, in which the Pd spin–orbit coupling parameter has been put equal to zero. The figure caption is the same as in Fig. 2.21

Summarizing, the perpendicular magnetic anisotropy is enhanced by the Pd character in the eigenstates because the Pd spin–orbit coupling parameter is so large. Eigenstates close to the actual Fermi energy with mainly $\frac{1}{2}(x^2 - y^2)$ and xy Co character are the cause of the PMA. This is partly a band filling effect because the actual Fermi energy is at a lower energy with respect to the Co d-band structure than in a Co monolayer, but a reordering in energy between these eigenstates and xz and yz eigenstates at the high symmetry points K and H is involved as well.

2.2.7 Conclusions

We have shown that calculations of the magnetocrystalline anisotropy energy of Co/X multilayers (where X is Ni, Cu, Pd, Ag) within the local-spin-density-approximation are in good agreement with experimental trends. By analyzing the anisotropy energy of a Co monolayer as well as of a $Co_1 Pd_2$ multilayer we could identify the origin of the preferred perpendicular orientation of the magnetization of the latter multilayer. It can be attributed to the existence of the eigenstates with a predominant $\frac{1}{2}(x^2 - y^2)$ and xy Co d character close to the Fermi level. Such eigenstates are extended in the plane of the Co layer. The large value of the anisotropy energy of Co_1Pd_2 multilayers is found to be due to hybridization of Co and Pd d states, in combination with the large value of the Pd d spin–orbit coupling parameter.

Acknowledgements. The authors have benefitted from numerous discussions with F.J.A. den Broeder. Also, discussions with R. Coehoorn, S.T. Purcell and W.B. Zeper were very stimulating.

2.3 Experimental Investigations of Magnetic Anisotropy

W.J.M. de JONGE, P.J.H. BLOEMEN, and F.J.A. den BROEDER

Magnetic anisotropy (MA) measures the dependence of the ground state energy on the direction of the magnetization. The anisotropy defines preferential (easy), intermediate and hard directions for the magnetization and is of technological importance for information storage and retrieval in, for instance, magneto-optical recording.

As was demonstrated already in the preceding chapter of this volume, the magnetic anisotropy in thin films or multilayered systems can be markedly different from bulk materials. The particular layered shape of the systems as well as the prominent presence of symmetry breaking elements such as surfaces and interfaces are the basic ingredients for this behavior. By varying the thicknesses of the individual layers and choosing appropriate elements, it appeared possible to manipulate the magnetic anisotropy. The most dramatic manifestation in this respect is the change of the preferential direction of the magnetization from the commonly observed in-plane orientation to the direction perpendicular to the plane. This phenomenon is usually referred to as Perpendicular Magnetic Anisotropy (PMA).

In this contribution we will, in contrast to the foregoing chapters of this volume, focus on the *experimental* research on the magnetic anisotropy in layered materials containing magnetic 3d transition metals. Rare earth transition metal multilayers are not included. We will discuss the analysis of the experimental data in terms of surface and volume contributions, discuss their physical origin and introduce some measuring techniques. Furthermore we will briefly comment on the experimental data published so far.

2.3.1 Origin of the Magnetic Anisotropy in Thin Films

Basically the two main sources of the magnetic anisotropy are the magnetic dipolar interaction and the spin–orbit interaction. Due to the long range, the dipolar interaction generally results in a shape-dependent contribution to the anisotropy, which is of particular importance in thin films and is largely responsible for the in-plane magnetization usually observed. In the absence of spin–orbit interaction the total energy of the electron–spin system does not depend on the direction of the magnetization. With spin–orbit interaction, a small orbital momentum is induced which couples the total (spin plus orbital) magnetic moment to the crystal axes. This results in a total energy which depends on the orientation of the magnetization relative to the crystalline axes and which reflects the same symmetry as the crystal. This is known as the magnetocrystalline contribution to the anisotropy. The lowered symmetry at a surface strongly modifies this contribution as compared to the bulk, yielding a so-called surface anisotropy as pointed out by *Néel* [2.92]. In conjunction with the overlap in wave functions between neighboring atoms, the spin–orbit interaction is also responsible for the magnetoelastic or magnetostrictive anisotropy induced in a system when strained, a situation which is frequently encountered in multilayers due to the lattice mismatch between the constituent materials.

Before we discuss each of these anisotropy terms in somewhat more detail in the next paragraphs, we will, for convenience, first introduce the phenomenological description of the anisotropy in terms of volume and surface contributions usually employed in multilayered thin films.

2.3.1.1 Surface and Volume Contributions

Since the magnetic anisotropy is strongly connected to the crystalline symmetry and the shape of the samples, a general expression of the anisotropy energy, for a given orientation of the magnetization and orientation of the surface relative to the crystal axes, will be a complex function reflecting the overall symmetry of the system. In the analysis and discussion of thin film anisotropy, however, it appears that a uniaxial description

$$E = -K \cos^2 \theta \tag{2.11}$$

is often sufficient. In this equation E is the orientational dependent energy of the magnetization, θ denotes the angle between the magnetization and the film normal (Fig. 2.23) and K is an anisotropy constant determining the strength of the anisotropy. A second-order uniaxial $K_2 \cos^4 \theta$ term, which is in some cases needed to account for experimental observations, is usually very small. By definition, a positive K describes the case of a preferred direction of the magnetization perpendicular to the layer plane. The anisotropy energy K (defined per unit volume) includes all contributions from various sources and is therefore sometimes indicated as the effective anisotropy K^{eff}.

In the spirit of *Néel's* prediction, it has been found useful in the analysis of the anisotropy data on thin films and multilayers, to distinguish between contributions from the surface or interface (K_S, per unit area) and contributions from the volume or bulk (K_V, per unit volume). This yields for the average

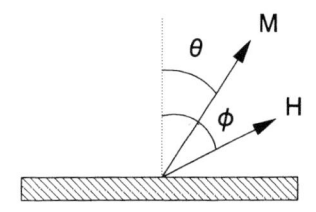

Fig. 2.23. Definition of the angles θ and ϕ subtended by the magnetization and the field, respectively, with respect to the film normal

magnetic anisotropy K of a magnetic layer of thickness t,

$$K = K^{\text{eff}} = K_V + 2K_S/t. \tag{2.12}$$

The prefactor 2 accounts for the fact that we assumed that the magnetic layer is bounded by two identical interfaces. In the case of more complicated layered structures (an example of which we will discuss later on) (2.12) should be modified accordingly. Note that K_S/d (with d the thickness of one monolayer) does not represent the magnetic anisotropy of the surface or interface atoms but represents the *difference* between the anisotropy of the surface or interface atoms and the inner (bulk) atoms. Equation (2.12) is commonly used in experimental studies, and the determination of K_V and K_S can be obtained by a plot of the product Kt versus t. Figure 2.24 shows a typical example of such a plot for Co/Pd multilayers [2.93]. The negative slope indicates a negative volume anisotropy K_V, favoring in-plane magnetization, while the intercept at $t = 0$ indicates a positive interface anisotropy K_S, favoring perpendicular magnetization. Below a certain thickness $t_\perp$ ($= -2K_S/K_V$, in this case 13 Å) the surface anisotropy contribution outweighs the volume contribution, resulting in a perpendicularly magnetized system.

It may be interesting to note that the surface anisotropy can be large. In order to get an impression of the magnitude, we have compared in Table 2.6 the magnetic anisotropy of the 2 Å Co/11 Å Pd multilayer of Fig. 2.24, with values for several other compounds. The anisotropy of the Co/Pd multilayer is significantly larger than for bulk Co. Moreover, it is not much less than the anisotropies reported for permanent magnet materials such as $Nd_2Fe_{14}B$.

Although the method of plotting Kt versus t as in Fig. 2.24 is frequently used and has contributed appreciably to the understanding of magnetic anisotropy in thin layers, (2.12), should be used with care, since the results might otherwise be rather misleading. Therefore, a few comments are in order:

- The description is based on the assumption that the "local" anisotropy at the surface or interface is felt, as it were, by the system as a whole, and that the system behaves as one magnetic entity with the individual magnetic moments aligned. This is only true when the anisotropy is much smaller than the intralayer exchange or when the layer is thinner than the so-called exchange length $\sqrt{2A/\mu_0 M_S^2}$ (see the introduction to this volume). For Co for instance, this exchange length is ≈ 30 Å.

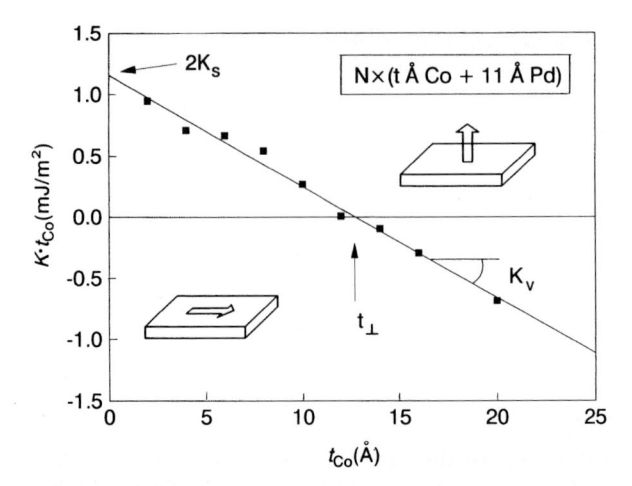

Fig. 2.24. Magnetic anisotropy per unit area per Co layer versus the Co layer thickness of Co/Pd multilayers. The y axis intercept equals twice the surface anisotropy, whereas the slope gives the volume contribution. Data are taken from [2.93]

Table 2.6. Magnetic anisotropy energies (MAE), calculated per unit volume of the magnetic material or magnetic transition metal (TM) atom, of multilayers compared with several other compounds

System	MAE [MJ/m^3]	MAE [μeV/TM atom]
Fe	0.017	1.4 μeV/Fe
Ni	0.042	2.7 μeV/Ni
Co	0.85	65 μeV/Co
Multilayers		
Co/Ni	2	
Co/Pd, Co/Pt	5	300 μeV/Co
Permanent magnets		
YCo$_5$	7	760 μeV/Co
Nd$_2$Fe$_{14}$B	12	
SmCo$_5$	30	

- The separation of the total anisotropy in terms of a volume contribution K_V and an interface contribution K_S for thin layers of a few monolayers thick is somewhat questionable.
- Volume contributions are not necessarily independent of t. As we will see, stress, for instance, can induce a $1/t$ dependence in K_V which, according to common analysis may be erroneously associated with a surface contribution.

2.3.1.2 Magnetic Dipolar Anisotropy (Shape Anisotropy)

Among the most important sources of the magnetic anisotropy in thin films is the long range magnetic dipolar interaction, which senses the outer boundaries of the sample.

Neglecting the discrete nature of matter, the shape effect of the dipolar interaction in ellipsoidal ferromagnets can be described, via an anisotropic demagnetizing field H_d given by $H_d = -\mathcal{N}M$. Here $\mathcal{N}$ is the shape-dependent demagnetizing tensor. For a thin film, all tensor elements are zero except for the direction perpendicular to the layer: $\mathcal{N}^\perp = 1$. Since the magnetostatic energy can be expressed as

$$E_d = -\frac{\mu_0}{2V} \int M \cdot H_d \, dv \tag{2.13}$$

it results in an anisotropy energy contribution per unit volume V of a film of

$$E_d = \tfrac{1}{2} \mu_0 M_s^2 \cos^2 \theta. \tag{2.14}$$

Here we have assumed the magnetization to make an angle θ with the film normal. According to this expression, the contribution favors as in-plane orientation for the magnetization rather than a perpendicular one. The thickness of the film does not enter into the continuum approach employed above, and thus it contributes only to K_V. It is this contribution which is primarily responsible for the negative slope of the Kt vs. t plot in Fig. 2.24. This continuum approach is common in the analysis of the experimental data.

However, when the thickness of the ferromagnetic layer is reduced to only a few monolayers (ML), the film should not, in principle, be considered as a magnetic continuum, but has to be treated as a collection of discrete magnetic dipoles on a regular lattice. Calculations made on the basis of discretely summing the dipolar interactions for films in the range of 1–10 monolayers lead to the following results [2.94]. Depending on the symmetry of the surface, the outer layers experience a dipolar anisotropy which can be appreciably lower than the inner layers. For the inner layers, the dipolar anisotropy is rather close to the value based on the continuum approach. Consequently, the average dipolar anisotropy can be phenomenologically expressed by a volume contribution K_V and a surface contribution K_S. However, the magnitude of the surface contribution is of minor importance, and other sources of anisotropy, such as spin–orbit coupling, are apparently dominant.

2.3.1.3 Magnetocrystalline Anisotropy

As stated before, the microscopic origin of the magnetocrystalline anisotropy is the spin–orbit interaction. In principle also the exchange interaction and the dipolar interaction could contribute to the magnetocrystalline anisotropy. However, the exchange interaction cannot give rise to anisotropy since it is proportional to the scalar product of the spin vectors and therefore independent of the angle between the spins and the crystal axes. The dipolar interaction on the other hand, does depend on the orientation of the magnetization relative to the crystal axes. In principle it results, apart from the shape contribution already discussed in Sect. 2.3.2(b), in a magnetocrystalline contribution. However, for cubic crystals it can be shown from symmetry arguments that the sum of the dipole–dipole energies cancels. For structures with lower symmetry, such as hexagonal crystals, this is generally not the case. However, for bulk hcp cobalt this contribution is negligible. This is essentially a consequence of the fact that the deviation of the c/a ratio from the ideal value $\sqrt{8/3}$ is relatively small (-0.67%) [2.95]. Thus basically only the spin–orbit interaction will be responsible for the magnetocrystalline anisotropy in Fe, Ni (both cubic) and Co.

Before a good understanding of itinerant electron behavior was achieved, *van Vleck* discussed the magnetocrystalline anisotropy (in the case of bulk) in a pair interaction model assuming localized magnetic moments [2.96]. *Néel* [2.92] extended this model for surfaces and showed that the reduced symmetry at the surface should result in magnetic anisotropies at the surface differing strongly from the bulk atoms. For this *surface anisotropy* energy he derives for fcc(1 1 1) and fcc(1 0 0) surfaces, for instance, a relation of the same form as (2.11): $E = -K_S \cos^2 \theta$, with K_S differing for [1 1 1] and [1 0 0] surfaces. Although the pair interaction model also played a role, as we will see later in the discussion about roughness and interdiffusion, and contributed significantly to the understanding, it is fundamentally incorrect. In some cases it predicts the wrong sign. Moreover, it does not give a dependence of K_S on the adjacent (non-magnetic) metal. A thorough understanding of the magnetocrystalline anisotropy can now be obtained from *ab initio* band structure calculations. As shown in the preceding chapters, the symmetry and location with respect to the Fermi level of spin–orbit split or shifted states are of major importance. The symmetry of the state, for instance, determines whether or not the state is split if the direction of the magnetization is, e.g., perpendicular or parallel to the film plane. In other words, it determines the sign of the contribution of the state to the magnetocrystalline anisotropy. For further and more detailed discussions we refer the reader to Chap. 2.2.

2.3.1.4 Magnetoelastic Anisotropy

Strain (ε) in a ferromagnet changes the magnetocrystalline anisotropy, and may thereby alter the direction of the magnetization. This effect is the reverse of magnetostriction which is the phenomenon that the sample dimensions may change if the direction of the magnetization is changed. The energy per unit volume associated with this effect can, for an elastically isotropic medium with isotropic magnetostriction, be written as

$$E_{me} = \tfrac{3}{2} \lambda_m \sigma \cos^2 \theta. \tag{2.15}$$

Here σ is the stress which is related to the strain via the elastic modulus E by $\sigma = \varepsilon E$. Alternatively (2.15) thus reads in terms of strain as

$$E_{\text{me}} = \tfrac{3}{2} \lambda_m E\varepsilon \cos^2 \theta. \tag{2.16}$$

The magnetostriction constant λ_m can be either positive or negative. The angle θ measures the direction of the magnetization relative to the direction of uniform stress. If the strain ε in the film is non-zero, the magnetoelastic coupling contributes in principle to the effective anisotropy. When the parameters are constant (not depending on t) this contribution can be identified with a volume contribution K_V (compare (2.11)).

Strain in films can be induced by various sources. Among them is thermal strain associated with differences in thermal expansion coefficients, and intrinsic strain brought about by the nature of the deposition process. Of particular interest in the present context, is strain due to the lattice mismatch between adjacent layers. Currently this problem is described in terms of the *van der Merwe* model in which elastic as well as dislocation energies are considered [2.97]. Two regimes should be distinguished. If the lattice mismatch $\eta = (a_A - a_B)/a_A$ between materials A and B is not too large, minimizing the total energy leads to a situation whereby, below a critical thickness t_c, the misfit can be accomodated by introducing a tensile strain in one layer and a compressive strain in the other such that ultimately the two materials A and B adopt the same in-plane lattice parameter. This regime is called the *coherent* regime (the lateral planes are in full lattice-registry).

The strain, as well as t_c, depends strongly on the specific geometry (bilayer, sandwich, film on substrate, multilayer, etc.). For a general multilayer A/B in the coherent regime, minimization of the elastic energy yields, for instance,

$$\varepsilon_A = -\eta/(1 + q_r t_A/t_B) \tag{2.17}$$

with q_r being the ratio between the elastic moduli E_A and E_B of layer A and B, $q_r = E_A/E_B$. The in-plane strain in layer B is given by $\varepsilon_B = \eta + \varepsilon_A$. For other geometries, analogous relations can be derived. Assuming layer A to be the magnetic layer, substitution of ε_A in (2.16) gives the magnetoelastic contribution to the anisotropy $K_\sigma^{\text{coh}} = -\tfrac{3}{2}\lambda_m E_A \varepsilon_A$. In principle, this contribution contains the thickness of the magnetic layer, and therefore may obscure the simple analysis in terms of volume and surface contributions (2.12). This problem does not exist when $t_A \ll t_B$ or $t_A/t_B = $ constant. In these cases K_σ^{coh} contributes only to K_V.

The elastic energy associated with the coherent situation is proportional to the strained volume. Increasing the thickness of one of the layers will therefore increase the elastic energy. At a certain critical thickness t_c, already mentioned above, it becomes energetically more favorable to introduce misfit dislocations which partially accomodate the lattice misfit, allowing the uniform strain to be reduced. The lattice-registry is then lost and the layers become partially coherent, or briefly *"incoherent"*.

In general, it is not an easy task to calculate the strain in the partially coherent regime. In the special case of a single layer A on a rigid substrate it has been shown [2.98, 99], by minimization of the sum of the elastic and dislocation energies, that the residual strain ε_A, which is assumed to be uniform within the layer, can be written as:

$$\varepsilon_A = -\eta t_c/t_A. \tag{2.18}$$

Consequently, the contribution to the magnetoelastic energy (2.16) also contains the $1/t$ dependence. Following the common analysis of anisotropy data as introduced by (2.12), this contribution, which is essentially generated in the volume, will emerge as a surface or interface contribution $K_{S,\lambda}^{inc} = \frac{3}{2}\lambda_m E_A \eta t_c$. It should be noted that $K_{S,\lambda}^{inc}$ does not depend on η because t_c is in the first approximation given by $t_c = G_s b/2|\eta| E_A$ [2.92]. Here G_s is the shear modulus and b is the Burgers vector. An alternative expression for the magnetoelastic surface contribution can thus be given in terms of G_s and b [2.93]:

$$K_{S,\lambda}^{inc} = -\frac{3}{4}\lambda_m G_s b. \tag{2.19}$$

Figure 2.25 illustrates the transition between the coherent and incoherent regime and the resulting effect observed in the magnetic anisotropy.

2.3.1.5 Effect of Roughness and Interdiffusion

So far we have assumed the layers to have ideal flat interfaces. In practice, films cannot be grown perfectly. Roughness and/or interdiffusion will be present and will modify the magnetic properties. Here we will briefly comment on their effects on the magnetic anisotropy.

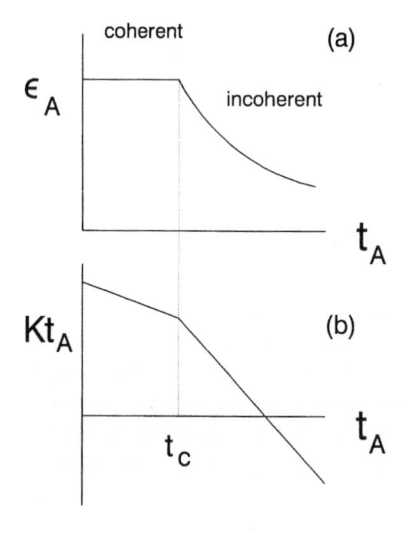

Fig. 2.25a, b. Thickness-dependence of (a) the strain and (b) the magnetic anisotropy in the coherent and incoherent regime, for the case of a magnetic layer on a rigid substrate

The effect of roughness on the dipolar anisotropy has been studied theoretically by *Bruno* [2.100]. The roughness can be characterized by an average fluctuation amplitude σ_t and a mean lateral size ξ_t of terraces and craters. It creates, for in-plane magnetization, local demagnetizing fields at the surface thereby reducing the shape anisotropy. The anisotropy contribution resulting from the roughness will therefore always be positive (favoring PMA) and scale as $1/t$. The magnitude of the corresponding dipolar surface anisotropy is a complicated function of σ and ξ, for which we refer to [2.100]. However, under "normal" conditions ($\sigma_t = 1$–2 ML, $\xi_t = 10$ ML) the contribution appears to be small (≈ 0.1 mJ/m^2 for Co(0 0 0 1)).

Roughness also introduces step atoms at the surface. It has been derived [2.101], using the pair interaction model, that these step atoms should reduce the surface anisotropy contribution of a magnetocrystalline origin. The extent of this reduction will be determined by the change of the anisotropy of the step atoms relative to terrace or crater atoms and by the number of step atoms relative to the number of terrace and crater atoms. The former depends on the geometry of the surface; the latter is determined by the height of the steps σ_t, their length ξ_t and the number of steps per unit length ($1/\xi_t$). *Bruno* [2.101] has derived that the relative reduction in K_S for a simple cubic (sc) (1 0 0) surface is given by $\Delta K_S/K_S = -2\sigma_t/\xi_t$, which can be substantial (20% for $\sigma_t = 1$ ML and $\xi_t = 10$ ML).

As mentioned, interdiffusion might occur during the deposition of the layers. It is clear that diffuse interfaces introduce randomness in the magnetic pair bonds according to *Néel*'s model, which obviously reduces the surface anisotropy. *Draaisma* et al. [2.102] have paid attention to this problem and have demonstrated via calculations based on the pair interaction model a strong dependence of K_S on the degree of mixing.

Although the application of the pair interaction model in the latter two cases demonstrated the importance of the topology of the interface, it would be interesting to compare the results with the outcome of more advanced calculations such as the one discussed in the preceding chapters. However, such calculations have not been performed yet.

2.3.2 Experimental Methods

The magnetic anisotropy can be deduced from the dynamic response of the magnetic system or from the static response. The *dynamic* response of the magnetic layers can be measured with ferromagnetic resonance (FMR) and Brillouin light scattering (BLS). Since these techniques will be discussed in the Volume 2, we refer the reader to these contributions. The static response can be measured by torque magnetometry, torsion oscillating magnetometry (TOM), the magneto-optic Kerr effect (MOKE) and various techniques which measure the magnetic moment, such as vibrating sample magnetometry (VSM), superconducting quantum interference device (SQUID) magnetometry, fluxgate magnetometry, alternating gradient magnetometry (AGM), pendulum magnetometry, Faraday balance, etc.

Most of the experimental data thus far are obtained from static measurements. In particular, magnetization and torque measurements are fairly common. Due to limited space we will restrict ourselves to an introduction of the magnetization experiments. For discussions on the analysis of torque measurements we refer to [2.103]. Furthermore we would like to introduce the use of the Kerr effect for local anisotropy measurements.

2.3.2.1 Magnetization Methods

Most commonly the magnetic anisotropy is determined from the information provided by *field dependent* measurements along two orthogonal directions of the magnetic field relative to the sample. An example of such a measurement, with the field parallel and perpendicular to the film plane, is shown in Fig. 2.26, where the preferential direction is clearly perpendicular. The *strength* of the MA can be determined from the area enclosed between the parallel and perpendicular loops. This is based on elementary electromagnetic considerations which show that the energy needed to change the sample magnetization in an applied field H by an amount dm is given by $\mu_0 H \, dm$. In some cases, as we will see later on, the strength of the MA can also be obtained from the fields at which saturation occurs.

The angle-dependent part of the energy E of the magnetization of the thin film can be written as

$$E = (-K_i + \tfrac{1}{2}\mu_0 M_s^2)\cos^2\theta - \mu_0 M_s H \cos(\phi - \theta) \tag{2.20}$$

In this expression, K_i contains all first order (intrinsic) anisotropy energy contributions except the shape anisotropy or magnetostatic energy contribution, which, as we saw before, equals $\tfrac{1}{2}\mu_0 M_s^2$ for a saturated film. The last term describes the interaction between the applied field and the resulting magnetization; θ and ϕ denote the angles subtended by the magnetization and field,

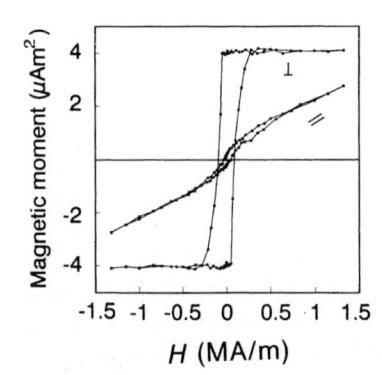

Fig. 2.26. Hysteresis loops measured with the applied field parallel and perpendicular to the layer plane of a 30 × (2 Å Co + 4 Å Ni + 2 Å Co + 10 Å Pt) multilayer [2.104]

2.3 Experimental Investigations of Magnetic Anisotropy

respectively, with respect to the film normal. Energy minimization as a function of the applied field H yields the field dependence of the equilibrium angle $\theta_{eq}(H)$ and the field component of the magnetization $M = M_S \cos(\theta_{eq} - \phi)$.

For easy-plane samples ($K^{eff} = K_i - \frac{1}{2}\mu_0 M_S^2 < 0$) this procedure gives the magnetization curves shown in Fig. 2.27a. The area enclosed between the two curves clearly equals the effective magnetic anisotropy energy (MAE), K^{eff}. As the figure shows, the MAE can also be obtained from the hard axis saturation field, the so called anisotropy field $H_A = -2K^{eff}/\mu_0 M_S$.

For the case of perpendicular easy samples ($K^{eff} > 0$) one should be careful in applying this analysis. We take as a starting point ($H = 0$) a situation in which the film consists of up and down domains of equal size, as shown schematically near the origin of Fig. 2.27b. The magnetostatic energy contribution used in (2.20) is then not valid in general because it depends on the size of the domains.

For very small domains, the magnetostatic interaction between the domains is such that the magnetostatic term can be described by the continuum result: $\frac{1}{2}\mu_0 M_\perp^2$, with $M_\perp$ the average perpendicular component of the magnetization. The perpendicular magnetization curve can then be obtained by minimizing $E = \frac{1}{2}\mu_0 M_\perp^2 - \mu_0 M_\perp H$ with respect to the magnetization. Here the $-K_i \cos^2\theta$ term is left out because it is constant since the magnetization either points up

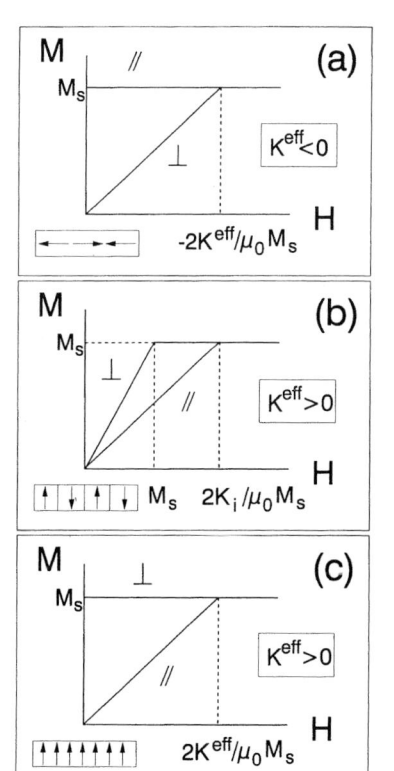

Fig. 2.27a–c. Magnetization curves for magnetic layers having an in-plane (**a**) or perpendicular (**b, c**) preferential orientation; (**b**) situation with small domains; (**c**) situation with large domains

($\theta = 0$) or down ($\theta = \pi$) in perpendicular applied fields, with $K^{\text{eff}} > 0$. Setting $\partial E/\partial M_\perp = 0$ yields $M_\perp = H$. For in-plane applied fields, $\frac{1}{2}\mu_0 M_\perp^2$ is always zero and the magnetization curve is obtained by minimization of $-K_i \cos^2\theta - \mu_0 M_S H \cos(\pi/2 - \theta)$ with respect to θ. The resulting curves are shown in Fig. 2.27b. Again the area between the curves gives the MAE, K^{eff}, including the shape anisotropy. In this case the saturation fields have no direct relation to K^{eff}.

In the case of very large domains, the effect of the interaction between the domains is negligible and, although at $H = 0$ the average perpendicular component of the magnetization is zero, the magnetostatic energy term is, for both perpendicular and in-plane applied fields, correctly described as in (2.20). In magnetizing the layer perpendicularly, the magnetostatic energy does not have to be overcome. Minimization of (2.20) therefore gives a perpendicular curve saturating immediately and an in-plane curve saturating at the field $2K_i/\mu_0 M_S - M_S$ or $2K^{\text{eff}}/\mu_0 M_S$ (Fig. 2.27c). As in the previous cases, the area between the in-plane and perpendicular magnetization curve gives the effective MAE, K^{eff}, including the shape anisotropy energy.

The general case of arbitrary domain size has been treated by *Kooy* and *Enz* [2.105]. Their analysis shows that the dimensionless parameter $\tau = \frac{1}{2}\mu_0 M_S^2 \cdot 4t/\pi\gamma$, with γ the domain wall energy per unit area, largely determines the domain size. The continuum approximation appears valid only for $\tau > 10^3$, whereas $\frac{1}{2}\mu_0 M_S^2$ may be used for $\tau < 1$. For intermediate situations, $10^3 < \tau < 1$, no simple expression for the magnetostatic energy can be given. It depends in a complicated way on the applied field via the (field dependent) domain size [2.105]. The saturation fields vary continuously with τ, but the area between the in-plane and perpendicular curves still gives K^{eff}.

Summarizing, in principle the MAE can be determined from the field dependence of the magnetization by measuring the saturation fields or the area between the in-plane and perpendicular magnetization curves. However, in general, no one to one relation exists between the hard axis saturation field and the total effective magnetic anisotropy K^{eff}. The area between the loops, on the contrary, yields K^{eff} in all cases.

In practice, problems are encountered using the "area method". For instance, as Fig. 2.26 illustrates, the experiments can show a considerable hysteresis and the available field is in some cases not large enough to saturate the sample in the hard direction. In order to determine the MAE in such cases, the hysteresis is removed by averaging the hysteresis loop branches, and the hard-axis loop is extrapolated, which is often rather arbitrary for non-linear curves. These problems can be circumvented by measuring the angular dependence of the magnetization. In this method the applied field is set at a constant value but its angle with respect to the film normal is varied by rotating the sample. During this rotation, the component of the magnetization along the field is measured. The magnetic anisotropy is then determined by fitting the obtained angular dependence by minimization of (2.20). For an example of such a study we refer to [2.106].

2.3.2.2 Anisotropy Determined from Magneto-optical Kerr Effect (MOKE) Measurements

The magneto-optical Kerr effect will be the subject of a separate contribution in Vol. 2. We will only introduce the effect here in relation to the measurement of the magnetic anisotropy [2.107]. The magneto-optical Kerr effect is the change in the polarization state of light by reflection at the surface of magnetic materials. This change originates from a difference in the complex Fresnel reflection coefficients for left and right circularly polarized light. The difference or optical anisotropy is only present for a non-zero magnetization, and the magnitude is determined by the off-diagonal terms in the permittivity tensor, which are odd linear functions of the magnetization. It is the latter relation which makes MOKE suitable to study changes in the magnitude or direction of the magnetization. No *direct* information can be obtained regarding the magnitude of the magnetization. However, since the magnetization is only monitored in that region of the sample which is illuminated by a (sharply focused) light beam, (typically 100 µm in diameter, in present studies [2.108]), a highly localized analysis of magnetic behavior is possible. For example, this allows a positional scan along a wedge-shaped (magnetic) layer, which enables one to investigate magnetic properties as a function of the layer thickness, in one single sample.

In a polar Kerr effect measurement, where the light beam is at normal incidence to the sample, one usually applies the field perpendicular to the sample. The Kerr rotation and the ellipticity in this geometry are proportional to the perpendicular component of the magnetization [2.109]. For easy-plane samples one thus obtains, as we saw in the previous section, a linear increase of the magnetization (and thus of the Kerr signal) up to saturation. From the saturation field or anisotropy field one can determine the MAE. However, for perpendicular easy samples one may obtain a rectangular loop from which the MAE cannot be obtained. Applying the field at a non-zero angle with respect to the film normal, creates a torque between M and H allowing determination of the MAE. The perpendicular component is given by $M_\perp = M_S \cos\theta_{eq}(H)$, and can be calculated relatively easy by solving the equilibrium angle $\theta_{eq}(H)$ from minimization of (2.20) for given anisotropy. The thus obtained $M_\perp/M_S$ versus H curve can be compared directly with the normalized experimental Kerr signal $I_S(H)/I_{S0}$, where $I_S(H)$ and I_{S0} are the measured Kerr rotations or ellipticities at field H and field zero, respectively [2.110]. A least squares fitting process routinely gives the magnetic anisotropy. Results of such a procedure are the measurements and the corresponding fits shown in Fig. 2.28 [2.108].

A disadvantage of the method is the fact that the Kerr effect does not measure the magnetization directly: the fits only provide the anisotropy *fields*. To obtain the anisotropy *constants*, a value for the saturation magnetization has to be assumed. Usually, experimental accuracies of 10% are obtained. However, this accuracy cannot be obtained for large anisotropies because, in those cases, the drop in the Kerr signal, using maximum available field, is too low to reliably

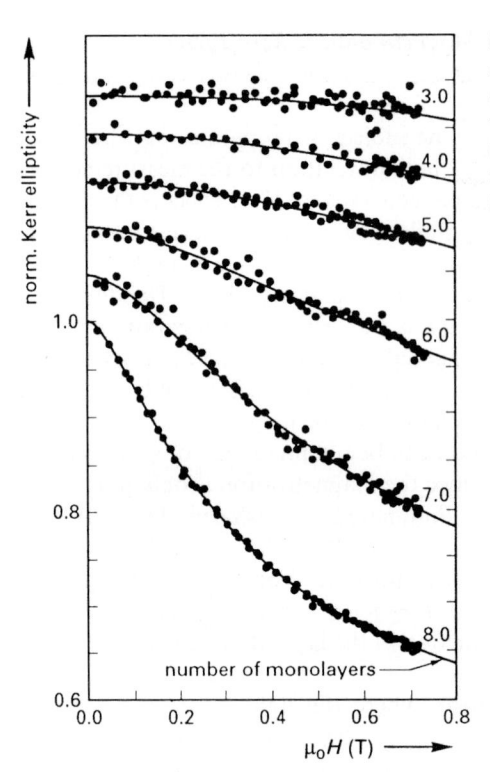

Fig. 2.28. Measurements of the Polar Kerr ellipticity at various Co thicknesses along a Co wedge, sandwiched between Pd, with the applied field at an angle of 60° with respect to the film normal. The solid lines represent fits using first and second order anisotropy constants [2.108]

determine the anisotropy. The latter is best demonstrated in Fig. 2.28 by the upper curves. Nevertheless, due to its local character and its monolayer sensitivity, MOKE provides a powerful tool for investigating anisotropies of (single, wedge-shaped) ultrathin ferromagnetic layers.

2.3.3 Experimental Results

Tables 2.7, 8 contain an up-to-date selection of experimentally observed anisotropy data in Fe, Co and Ni based multilayers. In the discussion of the experimental results, it should be realized that the structure of the magnetic layers is extremely important for the actual observed magnetic anisotropies. We already mentioned the importance of the topology of the interfaces and the crystallographic structure of the magnetic layers. The structural properties are strongly determined by the complex interplay between the employed growth technique (e.g., sputtering, evaporation, laser ablation), the preparation conditions (temperature, growth rate, pressure), the elements which are grown, their thicknesses and lattice mismatch, the symmetry, lattice spacing and quality of the substrates which are used and the resulting growth mode (layer by layer, Volmer–Weber, Stranski–Krastanov). Moreover, actual data on the structural

2.3 Experimental Investigations of Magnetic Anisotropy

Table 2.7. Out-of-plane surface anisotropies of sandwiches and multilayers containing Fe or Ni layers. K_S is defined positive for favoring a perpendicular easy axis. X denotes that a single crystal was used

Sample	K_s [mJ/m²]	Substrate	Reference
Fe(0 0 1)/UHV	0.96	X	[2.111]
Fe(0 0 1)/Ag	0.81, 0.69, 0.79	X	[2.111]
	0.64	X	[2.112]
	0.8		[2.113]
Fe(1 1 0)/Ag(1 1 1)	0.3		[2.114]
	1.45		[2.115]
Fe(0 0 1)/Au	0.47, 0.40, 0.54	X	[2.111]
Fe(1 1 0)/Au(1 1 1)	0.51		[2.116]
Fe(0 0 1)/Cu	0.62	X	[2.111]
Fe/Cu(1 1 1)	0.29		[2.117]
Fe/Mo(1 1 0)	0.55		[2.118]
Fe(0 0 1)/Pd	0.17	X	[2.111]
Fe(1 1 0)/Pd(1 1 1)	0.14		[2.119]
	0.15	X	[2.120]
	0.30	X	[2.118]
W(1 1 0)/Fe(1 1 0)/UHV	1.0	X	[2.121]
W(1 1 0)/Fe(1 1 0)/Ag	1.1	X	[2.121]
W(1 1 0)/Fe(1 1 0)/Au	1.2	X	[2.121]
W(1 1 0)/Fe(1 1 0)/Cu	1.4	X	[2.121]
Ni(1 1 1)/UHV	−0.48	X	[2.123]
Ni/Au(1 1 1)	−0.15	X	[2.122]
Ni(1 1 1)/Cu	−0.22	X	[2.123]
	−0.3, −0.12		[2.124, 125]
Ni/Cu(1 0 0)	−0.23		[2.125]
Ni/Mo	−0.54		[2.126]
Ni(1 1 1)/Pd	−0.22	X	[2.123]
Ni/Pd(1 1 1)	0		[2.93]
Ni(1 1 1)/Re	−0.19	X	[2.123]

parameters of the layers which have been realized are scarce and generally hard to obtain, or require at the least a series of dedicated experiments. The fact that, in Tables 2.7, 8, various (sometimes quite different) numbers for K_S and K_V are cited for one particular system indicates the sensitivity of the anisotropy parameters to the conditions mentioned above. At the same time it makes comparison of the results and the search for trends and systematics quite hazardous. Therefore, conclusions drawn from Tables 2.7, 8 should be considered with care. In this spirit, we will comment in what follows on some of the features apparent from the data tabulated in Tables 2.7, 8.

2.3.3.1 Fe versus Ni versus Co

Comparing the observed interface anisotropies, one notes that for Co and Fe these are often positive, i.e., favoring a perpendicular easy direction, whereas for

Table 2.8. Out-of-plane surface and volume anisotropies of Co based sandwiches and multilayers. K_S and K_V are defined as positive for favoring the perpendicular easy axis. X denotes that a single crystal was used

Sample	K_S [mJ/m^2]	K_V [MJ/m^3]	Substrate	Reference
Co/Au(1 1 1)	0.42	− 0.43		[2.127]
	0.53			[2.99]
	0.45			[2.128]
	1.28			[2.129]
	0.34	− 0.73		[2.130]
Co/Ag	0.2, 0.3	− 0.97		[2.93]
	0.16	− 0.93		[2.114]
Co/Al	0.25	− 0.76		[2.131]
Co/Cu(1 1 1)	− 0.02	− 1.19		[2.132]
	0.53			[2.129]
	0.53			[2.99]
	0.10, 0.12	− 0.8		[2.93]
Co/Ir(1 1 1)	0.8	− 1.20		[2.93]
Co(1 1 1)/Mo(1 1 0)	0.3, 0.2	− 0.84, − 0.87		[2.93]
Co/Ni(1 1 1)	0.31			[2.133]
	0.20			[2.134]
	0.22			[2.135]
	0.1			[2.136]
Co/Os(1 1 1)	0.7	− 0.90		[2.137]
Co/Pd(1 1 1)	0.26	− 0.72		[2.137]
	0.40, 0.56	− 0.86		[2.93]
	0.58, 0.74	− 0.91		[2.93]
	0.31	− 0.80		[2.138]
	0.25	− 0.64		[2.138]
	0.16			[2.139]
	0.5	− 1.2		[2.130]
	0.63	− 0.5	X	[2.140]
	0.92	− 1.00	X	[2.108]
Co/Pd(1 0 0)	0.32	− 2.19		[2.141]
	0.63	− 4.5	X	[2.140]
Co/Pd(1 1 0)	0.63	− 1.82	X	[2.140]
Co/Pt(1 1 1)	0.42	− 0.63		[2.142]
	0.50, 0.58	− 0.10, − 0.7		[2.93]
	0.76	− 0.92		[2.144]
	0.27	− 0.7		[2.145]
	1.15	− 0.77	X	[2.143]
Co/Pt(1 0 0)	0.31	− 1.0		[2.144]
	0.20	− 0.73		[2.145]
Co/Pt(1 1 0)	0.37	− 0.91		[2.144]
Co/Ru(0 0 1)	0.4	− 0.45		[2.130]
	0.5	− 0.68		[2.146]
Co/Ti	0.23	− 0.92		[2.147]
Co/V	1.05	− 1.1		[2.137]

the Ni based multilayers they are usually negative. There is no simple argument from which one can understand this difference in sign or the fact that a negative K_S should always be the case for Ni. As discussed in the previous chapters of this volume, the MAE is determined by large cancellations of various contributions which depend on the actual location of the Fermi level and the detailed shape and position of the energy bands. This makes a prediction of the sign of the MAE, without detailed calculation, extremely difficult.

The tables do not show the thicknesses $t_\perp$ at which the multilayers change their preferential orientation from perpendicular to along the film plane. Typical values for Co based multilayers are in the range of 0–18 Å, depending on the non-magnetic metal and the structural quality of the layers. For Fe based systems, the observed values for $t_\perp$ are generally smaller, notwithstanding the often larger positive surface anisotropies. This is mainly due to the shape contribution, which is larger for the Fe layers because of the higher saturation magnetization, and to the absence of significant bulk magnetocrystalline contributions. As we will see below for Co layers containing the hexagonal phase, a bulk magnetocrystalline contribution can increase the volume term K_V substantially [2.149].

2.3.3.2 Magnetoelastic Effects

The data as tabulated in Tables 2.7, 8 give, as such, no direct clue about the physical background. K_V and K_S may, as we saw before, include contributions of magnetic dipolar, magnetocrystalline and magnetoelastic origin. The relevance of the latter contribution, in particular that induced by incoherent behavior, is currently not well established. Calculations of the strain, based on expressions given in Sect. 2.3.2(d) show, in some cases, that magnetoelastic contributions to K_S could be expected to be in the same range (sign and magnitude) as the experimentally observed data (Co/Au [2.99], Co/Pd [2.93]). On the other hand, dedicated experiments in Ni/Pd and NiFe/Pd multilayers demonstrated a marked disagreement between the expected magnetoelastic contribution and the experimental data [2.93].

To analyze the data, one of the crucial tasks is thus determining whether or not the layers are incoherent. The observation of a coherent-incoherent transition at t_c (see Fig. 2.27) might of course be indicative but, as we will see below, other mechanisms may also induce such non-linear behavior. Another possibility is to rely on the theoretical prediction for t_c. Although the specific numbers depend on the geometry of the multilayer system, it appears that for Co systems with a large mismatch such as Co/Au and Co/Pd ($\eta = 0.15$, $\eta = 0.09$) respectively, a t_c of at most 1 or 2 ML is predicted for symmetric multilayers [2.148]. Consequently, incoherent behavior in the complete thickness range is expected. For Co/Cu or Co/Ni multilayers ($\eta = 0.02$, $\eta = -0.01$) respectively, t_c is much larger, and coherent behavior might be anticipated.

Direct observations of ε are not consistent with expectations in all cases. Indeed for Co/Pd an almost immediate relaxation towards the bulk lattice spacing was concluded from Reflection High-Energy Diffraction (RHEED) [2.150] and Transmission Electron Microscopy (TEM) [2.151] experiments. For Ni on Cu (1 0 0) the expected behavior was observed also [2.152]. On the other hand, nuclear magnetic resonance (NMR) experiments showing a $1/t$ dependence of the Co strain in Co/Ni and Co/Cu multilayers, do not seem to agree with the expected coherent behavior [2.153].

In a few studies the actual observed strain is used to calculate the magnetoelastic anisotropy. The results are conflicting. *Awano* et al. [2.154] report that the calculated stress-induced anisotropy is 10 to 100 times smaller than the measured values of the magnetic anisotropy, whereas *Lee* et al. [2.155] show that the measured Kt vs. t behavior can be explained largely by a magnetoelastic contribution.

A final comment concerns K_V. Inspection of Table 2.8 shows that a rather broad range of values is reported. In layered systems, both hcp Co and fcc Co can be stabilized at room temperature. The shape anisotropy for both equals -1.27 MJ/m^3. Magnetocrystalline anisotropy for bulk fcc Co is negligible and for hcp Co 0.41 MJ/m^3. Without magnetoelastic contributions, one therefore expects $K_V = -1.27$ MJ/m^3 (fcc Co) or $K_V = -0.8$ MJ/m^3 (hcp Co). Comparison with Table 2.8 shows important deviations which generally are attributed to strain or mixtures of hcp and fcc phases [2.149]. In particular a strain contribution is invoked in order to explain K_V values outside the range spanned by the hcp and fcc K_V values.

2.3.3.3 Interface Quality

It is worth noting that the interface anisotropies observed for samples grown on single crystals are generally larger than those observed for polycrystalline samples. This indicates the importance of the interface quality. Examples expressing this are a study on Co/Au multilayers [2.128] and work on a wedge-shaped Co layer [2.108]. In the experiment with the Co/Au multilayers, the samples were prepared by ion beam sputtering, resulting in rather diffuse interfaces due to the Ar-ion bombardment during growth. The interface quality was manipulated by an annealing treatment. The anisotropy was measured before and after the annealing. The results rendered in Fig. 2.29 show a drastic increase of K_S after heat treatment. This increase in K_S, accompanied by a strong increase in the intensity of the multilayer reflections observed in X-ray diffraction experiments, was interpreted as due to a sharpening of the Co/Au interfaces.

In the experiments on a Co wedge [2.108], an attempt was made to study the magnetic anisotropy in a Co/Pd bilayer grown as perfectly as possible on a Pd(1 1 1) single crystal. The use of a Co wedge ensured identical preparation conditions for each Co thickness. Positional scans along the wedge using the

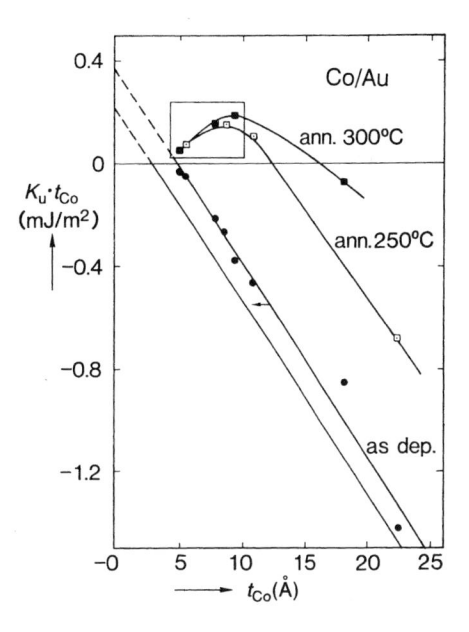

Fig. 2.29. Magnetic anisotropy density versus the Co thickness for ion-beam sputtered Co/Au multilayers before and after annealing at 250 °C and 300 °C [2.128]

Kerr effect yielded the magnetic anisotropies. The results are depicted in Fig. 2.30. Note that the surface anisotropy (0.92 mJ/m²) and the thickness range (18 Å) for which perpendicular magnetization occurs are significantly larger than observed for polycrystalline Co/Pd(1 1 1) multilayers (compare Table 2.8 and Fig. 2.24). Unfortunately, few studies of this type have been performed. Apart from their relevance from fundamental point of view, they also yield an upper limit for the best achievable properties, which is of importance for application-oriented research.

2.3.3.4 Non-Linear Kt vs. t Behavior

Figure 2.29 shows a deviation from the linear behavior at small Co thicknesses. This feature is often encountered in the anisotropy studies of transition metal multilayers. Apart from a possible coherent–incoherent transition, with the accompanying changes in the magnetoelastic anisotropy contributions as discussed in Sect. 2.3.2(d), several explanations can be given. Assuming for instance that, at small Co thickness, the film is no longer a continuous flat layer but is broken up into islands, necessarily yields a lower effective magnetic/non-magnetic interface, a lower interface contribution and a correspondingly lower total anisotropy than expected from the relation $K^{\text{eff}} = K_V + 2K_S/t$. Apart from interdiffusion, which can also account for non-linear behavior if the magnetic layer thickness becomes comparable to the thickness of the diffusion zone, a lowering of the Curie temperature with the magnetic layer thickness, which is a well known finite size effect, can also play a role in the case of room

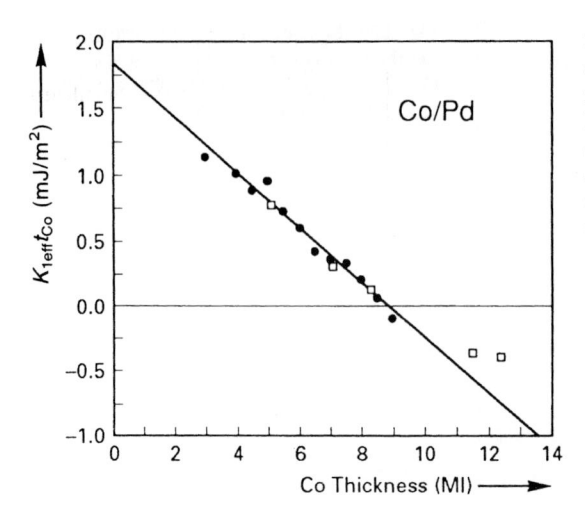

Fig. 2.30. Magnetic anisotropy density versus the Co thickness measured on a wedge-shaped Co layer on Pd(1 1 1) using the Kerr effect (solid circles). The open squares represent FMR measurements on separate samples with homogeneous Co thicknesses [2.108]

temperature measurements. Moreover, one should realize that for ultrathin magnetic layers it is not at all apparent that one is allowed to separate the MAE into surface and volume contributions, a remark already made in Sect. 2.3.2(a).

2.3.3.5 Co/Ni Multilayers

To date, the magnetic anisotropy studies have been mainly focused on magnetic films coated with a non-magnetic metal and multilayers with nonmagnetic spacer layers. Multilayers in which both types of layers are ferromagnetic have received very little attention, as may be inferred from Tables 2.7, 8. In order to achieve perpendicular anisotropy, these systems do not seem, *a priori*, to be the most advantageous, since the interface anisotropy has to overcome an additional shape anisotropy, namely that of the second ferromagnetic layer. On the other hand, using several magnetic layers increases the degrees of freedom to obtain certain desired properties. For example, the use of Ni in combination with Co layers allows manipulation of the effective shape anisotropy and thus of the thickness range with PMA which is, as mentioned before, important for magneto-optic recording applications [2.104]. Co/Ni multilayers are in this respect interesting systems to discuss here.

As reported in the previous Chap. 2.2, first-principles band structure calculations of the magnetic anisotropy have led to the prediction of PMA a [1 1 1] Co_1/Ni_2 superlattice and subsequent experimental confirmation [2.133, 134]. These calculations give the total MAE, characteristic for the specific superlattice. In principle, no information is given on the partitioning of the MAE over the several contributions. It is therefore interesting to investigate whether the perpendicular contribution, which is apparently present in 2 Å Co/4 Å Ni multilayers, is also already intrinsically contained as a surface contribution in

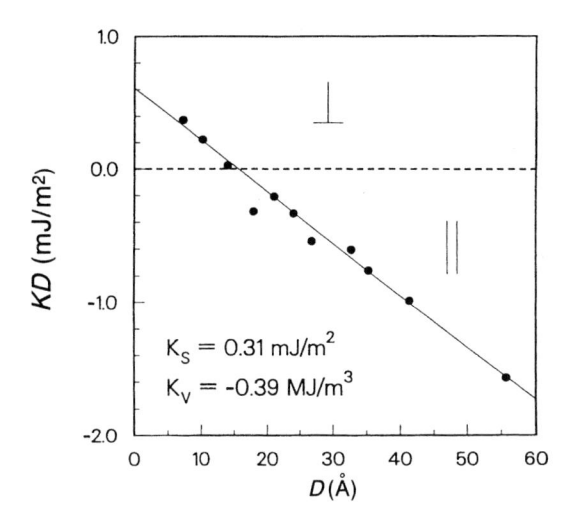

Fig. 2.31. Magnetic anisotropy density versus the multilayer period measured for Co/Ni multilayers with fixed ratio $\alpha = t_{Ni}/t_{Co} = 2.2$ [2.133]

Co/Ni multilayers with much thicker Co and Ni layers. To answer this question, a systematic study of the MAE as a function of the Co and Ni thicknesses is required. To describe the MAE, (2.12) has to be modified. In this case, *two* magnetic layers contribute to the MA. Per unit area per Co/Ni bilayer, the MAE can be written as

$$KD = K_V^{Co} t_{Co} + K_V^{Ni} t_{Ni} + 2K_S^{Co/Ni}, \qquad (2.21)$$

with $D = t_{Co} + t_{Ni}$. Two experiments, one at fixed Co thickness as a function of the Ni thickness and one at fixed Ni thickness with varying Co thickness, allow the determination of the interface and volume contributions [2.135].

Alternatively, the interface anisotropy can be obtained from experiments in which the *ratio* of the Ni and Co thicknesses is fixed at a certain value α. Equation (2.21) can then be rewritten as

$$KD = \frac{1}{1+\alpha_r}(K_V^{Co} + \alpha_r K_V^{Ni})D + 2K_S^{Co/Ni}, \qquad (2.22)$$

with $\alpha_r = t_{Ni}/t_{Co}$. Examples of such experiments, with $\alpha_r \approx 2$, are shown in Fig. 2.31 [2.133]. In this case, a positive interface contribution $K_S = 0.31$ mJ/m^2 was obtained, showing that the perpendicular contribution as calculated for the Co$_1$/Ni$_2$ superlattice is not restricted to samples in the monolayer range but is already contained as a surface contribution in thick layers.

2.3.4 Concluding Remarks

A discussion was given of the important contributions to the anisotropy of magnetic thin films, such as the magnetic dipolar, the magnetocrystalline and the magnetostrictive contributions. It was shown that the interfaces played a key role and were predominantly responsible for observed perpendicular preferential directions. Experimental data available at present were commented on. An important observation is the significant scatter in the data making comparison with theory and the search for trends rather difficult. Most of the scatter is believed to originate from differences in the structure of the layers which is often not well known. For future research, it is therefore important to perform magnetic as well as structural investigations on the same samples. A recent promising experimental development in this respect is the use of wedge-shaped magnetic layers in combination with (*in situ*) local probes such as the Kerr effect, Low Energy Electron Diffraction (LEED) and Reflection High Energy Electron Diffraction (RHEED) allowing a thickness dependent study of magnetic and structural properties on one single relative simple and ideal system (only magnetic layer on a well-prepared single crystal substrate). Apart from the saving of an enormous amount of time and effort accompanied with the preparation of single crystals and the structural investigations, several uncertainties which are commonly present in most studies, are removed, such as varying preparation conditions and unwanted gradients in properties across the multilayer stack.

Although much research has been performed to date, a number of interesting issues such as the dependence of the interface anisotropy on the growth orientation and the temperature dependence of the magnetic anisotropy, have received very little attention and still need thorough experimental investigations.

Acknowledgement. The authors gratefully acknowledge the valuable discussions with R. Coehoorn, G.H.O. Daalderop, H.A.M. de Gronckel and K. Kopinga.

References

Section 2.1

2.1 L.I. Schiff: *Quantum Mechanics* (McGraw-Hill Book Company, New York, 1955), Chapter 12
2.2 H.A. Bethe, E.E. Salpeter: *Quantum Mechanics of One- and Two-Electron Atoms* (Springer, Berlin, Heidelberg 1957)
2.3 J.C. Slater: *Quantum Theory of Atomic Structure*, Vol. II (McGraw-Hill Book Company, New York 1960), Chapter 24
2.4 M. Blume, R.E. Watson: Proc. Roy. Soc. **A270**, 127 (1962); **A271**, 565 (1963)
2.5 P. Hohenberg, W. Kohn: Phys. Rev. **136**, B864 (1964)
2.6 W. Kohn, L.J. Sham: Phys. Rev. **136**, A1113 (1965)
2.7 G.D. Mahan, K.R. Subbaswamy: *Local Density Theory of Polarization* (Plenum Press, New York, 1990)
2.8 A.K. Rajagopal: J. Phys. C **11**, L943 (1978)
2.9 A.H. MacDonald, S.H. Vosco: J. Phys. C **12**, 2977 (1978)
2.10 M.V. Ramana, A.K. Rajagopal: J. Phys. C **14**, 4291 (1981)
2.11 G.H.O. Daalderop, P.J. Kelly, M.F.H. Schuurmans: Phys. Rev. B41, 11919 (1990)
2.12 L.D. Landau, E.M. Lifshitz: *Quantum Mechanics* (Pergamon Press Ltd., London, 1958) p. 79

References

2.13 V.L. Moruzzi, J.F. Janak, A.R. Williams: *Calculated Electronic Properties of Metals* (Pergamon Press Inc., New York, 1978) Chapter V
2.14 J.W.D. Connolly: Int. J. Quantum Chem. **2**, 257 (1968)
2.15 J. Callaway, C.S. Wang: Phys. Rev. B **7**, 1096 (1973)
2.16 R.A. Tawil, J. Callaway: Phys. Rev. B **7**, 4242 (1973)
2.17 U. von Barth, L. Hedin: J. Phys. C **5**, 1629 (1972)
2.18 J.R. Anderson, D.A. Papaconstantanopoulos, L.L. Boyer, J.E. Schirber: Phys. Rev. B **20**, 3172 (1979)
2.19 L. Fritsche, J. Noffke, H. Eckardt: J. Phys. F **17**, 943 (1987)
2.20 H. Eckardt, L. Fritsche: J. Phys. F **18**, 925 (1987)
2.21 P. Strange, H. Ebert, J.B. Staunton, B.L. Gyorffy: J. Phys: Condens. Matter **1**, 3947 (1989)
2.22 P. Strange, J.B. Staunton, B.L. Gyorffy, H. Ebert: Proceedings of the Conference *Electronic Structure in the 1990's*, in press
2.23 P. Strange, J.B. Staunton, B.L. Gyorffy: J. Phys. C **17**, 3355 (1984)
2.24 O.K. Andersen: Phys. Rev. B **12**, 3060 (1975)
2.25 A.R. Mackintosh, O.K. Andersen: *Electrons at the Fermi Surface*, ed. by M. Springford (Cambridge University Press, Cambridge, 1980)
2.26 H.J.F. Jansen: J. Appl. Phys. **64**, 5604 (1988)
2.27 J.G. Gay, R. Richter, Phys. Rev. Lett. **56**, 2728 (1986)
2.28 L. Liebermann, D.R. Fredkin, H.B. Shore, Phys. Rev. Lett. **22**, 539 (1969)
2.29 L. Liebermann, J. Clinton, D.M. Edwards, J. Mathon: Phys. Rev. Lett. **25**, 232 (1970)
2.30 U. Gradmann: J. Magn. Magn. Mater. **6**, 177 (1977)
2.31 W. Göpel: Surf. Sci. **85**, 400 (1979)
2.32 M. Weinert, J.W. Davenport: Phys. Rev. Lett **54**, 1547 (1985)
2.33 J.R. Smith, J.G. Gay, F.J. Arlinghaus: Phys. Rev. B **21**, 2201 (1980)
2.34 A.J. Freeman, C.L. Fu, S. Ohnishi, M. Weinert: *Polarized Electrons in Surface Physics*, ed. by R. Feder (World Scientific, Singapore, 1985)
2.35 J.G. Gay, J.R. Smith, F.J. Arlinghaus: Phys. Rev. Lett. **42**, 332 (1979)
2.36 P. Heimann, J. Hermanson, H. Miosga, H. Nedermeyer: Phys. Rev. Lett. **42**, 1782 (1979)
2.37 C.S. Wang, A.J. Freeman: Phys. Rev. B **21**, 4585 (1980)
2.38 C.S. Wang, A.J. Freeman: Phys. Rev. B **24**, 4364 (1980)
2.39 E. Wimmer, H. Krakauer, M. Weinert, A.J. Freeman: Phys. Rev. B **26**, 2790 (1982)
2.40 O. Jepsen, J. Madsen, O.K. Andersen: Phys. Rev. B **26**, 2790 (1982)
2.41 Xue-yuan Zhu, J. Hermanson, F.J. Arlinghaus, J.G. Gay, R. Richter, J.R. Smith: Phys. Rev. B **29**, 4426 (1984)
2.42 J.C. Slater: *The Self-Consistent Field for Molecules and Solids: Quantum Theory of Molecules and Solids* Vol. IV (McGraw-Hill Book Company, New York, 1974) Chapter 8
2.43 R. Richter, J.G. Gay, J.R. Smith: J. Vac. Sci. Technol. A3, 1498 (1985); Phys. Rev. Lett. **54**, 2704 (1985)
2.44 C.L. Fu, A.J. Freeman, T. Oguchi: Phys. Rev. Lett. **54**, 2700 (1985)
2.45 C.L. Fu, A.J. Freeman: Phys. Rev. B **35**, 925 (1987)
2.46 G.W. Fernando, B.R. Cooper: Phys. Rev. B **38**, 3016 (1988)
2.47 R. Ritcher, J.G. Gay in *Growth, Characterization and Properties of Ultrathin Magnetic Films and Multilayers*, MRS Symposium Proceedings, **151**, ed. by B.T. Jonker, J.P. Heremans, E.L. Marinero, (Materials Research Society, Pittsburgh, 1989) p. 3
2.48 S.D. Bader, C. Liu: J. Vac. Sci. Technol. A9, 1924 (1991)
2.49 H.J. Elmers, G. Liu, U. Gradmann: Phys. Rev. Lett. **63**, 566 (1989)
2.50 S.C. Hong, A.J. Freeman, C.L. Fu: Phys. Rev. B **38**, 12156 (1988)
2.51 J.G. Gay, R. Richter: J. Magn. Magn. Mat. **93**, 315 (1991)
2.52 J.G. Gay, R. Richter, J. Appl. Phys. **61**, 3362 (1987)
2.53 H. Vosko, L. Wilks, M. Nusair: Can. J. Phys. **58**, 1700 (1980)
2.54 B.T. Jonker, K.H. Walker, E. Kisker, G.A. Prinz, C. Carbone: Phys. Rev. Lett. **57**, 142 (1986)
2.55 B. Heinrich, A.S. Arrott, J.F. Cochran, K.B. Urquhart, K. Myrtle, Z. Celinski, Q.M. Zhong: Phys. Rev. Lett. **59**, 1756 (1986)

2.56 N.C. Koon, B.T. Jonker, F.A. Volkening, J.J. Krebs, G.A. Prinz: Phys. Rev. Lett. **59**, 2463 (1987)
2.57 C. Liu, E.R. Moog, S.D. Bader: Phys. Rev. Lett. **60**, 2422 (1988)
2.58 J.F. Cochran, W.B. Muir, J.M. Rudd, B. Heinrich, Z. Celenski, T.-T. Le Tran, W. Schwarzacher, W. Bennett, W.F. Egelhoff, Jr., J. Appl. Phys. **69**, 5206 (1991)
2.59 J. Araya-Pochet, C.A. Ballentine, J.L. Erskin: Phys. Rev. B **38**, 7846 (1988)
2.60 C. Liu, S.D. Bader: J. Vac. Sci. Technol. **A8**, 2727 (1990)
2.61 W. Karas, J. Noffke, L. Fritsche: J. Chim. Phys. **86**, 861 (1989)
2.62 C. Li, A.J. Freeman, H.J.F. Jansen, C.L. Fu: Phys. Rev. B **42**, 5433 (1990)
2.63 G.H.O. Daalderop, P.J. Kelly, M.F.H. Schurmans: Phys. Rev. B **42**, 7270 (1990)

Section 2.2

2.64 F.J.A. den Broeder, W. Hoving, P.J.H. Bloemen: J. Magn. Magn. Mat. **93**, 562 (1991) and references therein
2.65 L. Néel: J. Phys. Rad. **15**, 225 (1954)
2.66 C. Chappert, P. Bruno: J. Appl. Phys. **64**, 5736 (1988)
2.67 J.G. Gay, R. Richter, Phys. Rev. Lett. **56**, 2728 (1986); W. Karas, J. Noffke, L. Fritsche: J. Chim. Phys. Phys.-Chim. Biol. **86**, 861 (1989)
2.68 C. Li, A.J. Freeman, C.L. Fu: Phys. Rev. B **42**, 5433 (1990)
2.69 J.G. Gay, R. Richter, J. Appl. Phys. **61**, 3362 (1987)
2.70 G.H.O. Daalderop, P.J. Kelly, M.F.H. Schuurmans: Phys. Rev. B **42**, 7270 (1990)
2.71 G.H.O. Daalderop, P.J. Kelly, F.J.A. den Broeder: Phys. Rev. Lett. **68**, 682 (1992)
2.72 A.J. Bennett, B.R. Cooper: Phys. Rev. B **3**, 1642 (1971); H. Takayama, K.P. Bohnen, P. Fulde: Phys. Rev. B **14**, 2287 (1976); M. Kolar, Phys. Stat. Sol. (b) **96**, 683 (1979)
2.73 O. Gunnarsson, R.O. Jones: Rev. Mod. Phys. **61**, 689 (1989) and references therein
2.74 O.K. Andersen, Phys. Rev. B **12**, 3060 (1975)
2.75 A.R. Mackintosh, O.K. Andersen: in *Elecrons at the Fermi Surface*, ed. by M. Springford (Cambridge University Press, Cambridge, 1980); M. Weinert, R.E. Watson, J.W. Davenport: Phys. Rev. B **32**, 1215 (1985)
2.76 G.H.O. Daalderop, P.J. Kelly, M.F.H. Schuurmans: Phys. Rev. B **41**, 11919 (1990)
2.77 E. Abate, M. Asdente: Phys. Rev. **140**, A1303 (1965)
2.78 J.F. Janak: Phys. Rev. B **16**, 255 (1977); O. Gunnarsson: Physica B + C (Amsterdam) **91B**, 329 (1977)
2.79 O. Jepsen, J. Madsen, O.K. Andersen: Phys. Rev. B **26**, 2790 (1982)
2.80 O.K. Andersen, H.L. Skriver, H. Nohl, B. Johansson: Pure and Applied Chemistry **52**, 93 (1979)
2.81 See the previous reference and the potential parameters given by O.K. Andersen, O. Jepsen, D. Glötzel: in *Highlights of Condensed Matter Theory*, ed. by F. Bassani, F. Fumi, M.P. Tosi (North-Holland, Amsterdam, 1985), p. 59
2.82 This is discussed for Ni by R. Gersdorf: Phys. Rev. Lett. **40**, 344 (1978)
2.83 J. Friedel: in *The Physics of Metals*, ed. by J.M. Ziman (Cambridge University Press, Cambridge, 1969)
2.84 G.H.O. Daalderop, P.J. Kelly, M.F.H. Schuurmans: Phys. Rev. B **44**, 12054 (1991)
2.85 V. Heine, J.H. Samson: J. Phys. F: Met. Phys. **10**, 2609 (1980); V. Heine, J.H. Samson, J. Phys. F: Met. Phys. **13**, 2155 (1983); V. Heine, W.C. Kok, C.M.M. Nex: J. Magn. Magn. Mat. **43**, 61 (1984)
2.86 P.F. Carcia, A.D. Meinholdt, A. Suna: Appl. Phys. Lett. **47**, 178 (1985); H.J.G. Draaisma, W.J.M. de Jong, F.J.A. de Broeder: J. Magn. Magn. Mat. **66**, 351 (1987)
2.87 F.J.A. den Broeder, D. Kuiper, H.C. Donkersloot, W. Hoving: Appl. Phys. A **49**, 507 (1989)
2.88 H.J.F. Jansen, A.J. Freeman: Phys. Rev. B **29**, 5965 (1984)
2.89 CPU time on an IBM3090 computer with vector-processor for the Co_1Pd_5 multilayer is ~ 20 hours for the most dense mesh used

2.90 S. Hashimoto, Y. Ochiai, K. Aso: J. Appl. Phys. **66**, 4909 (1989)
2.91 W. Coene, F. Hakkens: private communication

Section 2.3

2.92   L. Néel: J. de Phys. et le Rad. **15**, 225 (1954)
2.93   F.J.A. den Broeder, W. Hoving, P.J.H. Bloemen: J. Magn. Magn. Mater. **93**, 562 (1991)
2.94   H.J.G. Draaisma, W.J.M. de Jonge: J. Appl. Phys. **64**, 3610 (1988)
2.95   G.H.O. Daalderop, P.J. Kelly, M.F.H. Schuurmans: Phys. Rev. B **41**, 11919 (1990)
2.96   J.H. van Vleck: Phys. Rev. **52**, 1178 (1937)
2.97   J.H. van der Merwe: J. Appl. Phys. **34**, 123 (1963)
2.98   J.W. Matthews: in *Dislocations in Solids*, ed. by F.R.N. Nabarro (North-Holland Publishing Company, city, year) p. 463 ff
2.99   C. Chappert, P. Bruno: J. Appl. Phys. **64**, 5736 (1988)
2.100 P. Bruno: J. Appl. Phys. **64**, 3153 (1988)
2.101 P. Bruno: J. Phys. F **18**, 1291 (1988)
2.102 H.J.G. Draaisma, F.J.A. den Broeder, W.J.M. de Jonge: J. Appl. Phys. **63**, 3479 (1988)
2.103 E.g., J. Burd, M. HuQ, E.W. Lee: J. Magn. Magn. Mater. **5**, 135 (1977); S. Swaving, G.J. Gerritsma, J.C. Lodder, Th.J.A. Popma: J. Magn. Magn. Mater. **67**, 155 (1987); J.O. Artman: IEEE Trans. Magn. **21**, 1271 (1985)
2.104 P.J.H. Bloemen, W.J.M. de Jonge: J. Magn. Magn. Mater., **116**, L1 (1992)
2.105 C. Kooy, U. Enz: Philips Res. Rep. **15**, 7 (1960)
2.106 P.J.H. Bloemen, E.A.M. van Alphen, W.J.M. de Jonge, F.J.A. den Broeder: Mater. Res. Soc. Symp. Proc. **231**, 479 (1991)
2.107 P. Wolniansky, S. Chase, R. Rosenvold, M. Ruane, M. Mansuripur: J. Appl. Phys. **60**, 346 (1986)
2.108 S.T. Purcell, M.T. Johnson, N.W.E. McGee, W.B. Zeper, W. Hoving: J. Magn. Magn. Mater. (1992), to be published
2.109 E.g., M. Mansuripur: J. Appl. Phys. **67**, 6466 (1990)
2.110 Preceding the measurement, the sample is saturated perpendicularly so that one starts with a situation of maximum perpendicular remanence
2.111 B. Heinrich, Z. Celinski, J.F. Cochran, A.S. Arrott, K. Myrtle: J. Appl. Phys. **70**, 5769 (1991)
2.112 J.F. Cochran, B. Heinrich, A.S. Arrott, K.B. Urquhart, J.R. Dutcher, S.T. Purcell: J. Phys. C **8**, 1671 (1988)
2.113 A. Bartélemy, A. Fert, P. Etienne, R. Chabanel, S. Lequien: J. Magn. Magn. Mater. **104–107**, 1816 (1992)
2.114 R. Krishnan, M. Porte, M. Tessier: J. Magn. Magn. Mater. **103**, 47 (1992)
2.115 H. Hurdequint: J. Magn. Magn. Mater. **93**, 336 (1991)
2.116 S. Araki, T. Takahata, H. Dohnomae, T. Okuyama, T. Shinjo: Mater. Res. Soc. Proc. **151**, 123 (1989)
2.117 L. Smardz, B. Szymański, J. Barnaś, J. Baszyński: J. Magn. Magn. Mater. **104–107**, 1885 (1992)
2.118 Y. Obi, Y. Kawano, Y. Tange, H. Fujimori: J. Magn. Magn. Mater. **93**, 587 (1991)
2.119 H.J.G. Draaisma, F.J.A. den Broeder, W.J.M. de Jonge: J. Magn. Magn. Mater. **66**, 351 (1987)
2.120 B. Hillebrands, A. Boufelfel, C.M. Falco, P. Baumgart, G. Güntherodt, E. Zirngiebl, J.D. Thompson: J. Appl. Phys. **63**, 3880 (1988)
2.121 H.J. Elmers, T. Furubayashi, M. Albrecht, U. Gradmann: J. Appl. Phys. **70**, 5764 (1991)
2.122 J.R. Childress, C.L. Chien, A.F. Jankowski: Phys. Rev. B **45**, 2855 (1992)
2.123 U. Gradmann: J. Magn. Magn. Mater. **54–57**, 733 (1986); H.J. Elmers, U. Gradmann: J. Appl. Phys. **63**, 3664 (1988)
2.124 E.M. Greorgy, J.F. Dillon, Jr., D.B. McWhan, L.W. Rupp, Jr., L.R. Testardi: Phys. Rev. Lett. **45**, 57 (1980)
2.125 G. Xiao, C.L. Chien: J. Appl. Phys. **61**, 4061 (1987)

2.126 M. Pechan, I.K. Schuller: Phys. Rev. Lett. **59**, 132 (1987); M. Pechan: J. Appl. Phys. **64**, 5754 (1988)
2.127 T. Takahata, S. Araki, T. Shinjo: J. Magn. Magn. Mater. **82**, 287 (1989)
2.128 F.J.A. den Broeder, D. Kuiper, A.P. van de Mosselaar, W. Hoving: Phys. Rev. Lett. **60**, 2769 (1988)
2.129 F.J. Lamelas, C.H. Lee, Hui Le, W. Vavra, R. Clarke: Mater. Res. Soc. Symp. Proc. **151**, 283 (1989)
2.130 M. Sakurai, T. Takahata, I. Moritani: J. Magn. Soc. Jpn. **15**, 411 (1991)
2.131 T. Mitsuzuka, A. Kamijo, H. Igarashi: J. Appl. Phys. **68**, 1787 (1990)
2.132 C.D. England, W.R. Bennet, C.M. Falco: J. Appl. Phys. **64**, 5757 (1988)
2.133 G.H.O. Daalderop, P.J. Kelly, F.J.A. den Broeder: Phys. Rev. Lett. **68**, 682 (1992)
2.134 F.J.A. den Broeder, E. Janssen, W. Hoving, W.B. Zeper: IEEE Trans. Magn. **28**, 2760 (1992)
2.135 P.J.H. Bloemen, W.J.M. de Jonge, F.J.A. den Broeder: J. Appl. Phys. **72**, 4840 (1992)
2.136 M.P.M. Luykx, C.H.W. Swüste, H.J.G. Draaisma, W.J.M. de Jonge: J. Phys. (Paris) Colloq. **49**, C8-1769 (1988)
2.137 P.J.H. Bloemen, H.A.M. de Gronckel, W.J.M. de Jonge: to be published
2.138 S. Hashimoto, Y. Ochiai, K. Aso: J. Appl. Phys. **67**, 4909 (1989)
2.139 P.F. Carcia, A.D. Meinhaldt, A. Suna: Appl. Phys. Lett. **47**, 178 (1985)
2.140 B.N. Engel, C.D. England, R.A. van Leeuwen, M.H. Wiedmann, C.M. Falco: Phys. Rev. Lett. **67**, 1910 (1991)
2.141 F.J.A. den Broeder, D. Kuiper, H.C. Donkersloot, W. Hoving: Appl. Phys. A **49**, 507 (1989)
2.142 W.B. Zeper, P.F. Carcia: IEEE Trans. Magn. **25**, 3764 (1989)
2.143 N.W.E. McGee, M.T. Johnson, J.J. de Vries, J. aan de Stegge: J. Appl. Phys. **73**, 3418 (1993)
2.144 C.-J. Lin, G.L. Gorman, C.H. Lee, R.F.C. Farrow, E.E. Marinero, H.V. Do, H. Notarys, C.J. Chien: J. Magn. Magn. Mater. **93**, 194 (1991)
2.145 J.V. Harzer, B. Hillebrands, R.L. Stamps, G. Güntherodt, D. Weller, Ch. Lee, R.F.C. Farrow, E.E. Marinero: J. Magn. Magn. Mater. **104–107**, 1863 (1992)
2.146 A. Dinia, K. Ounadjela, A. Arbaoui, G. Suran, D. Muller, P.Panisod: J. Magn. Magn. Mater. **104–107**, 1871 (1992)
2.147 R.A. van Leeuwen, C.D. England, J.R. Dutcher, C.M. Falco, W.R. Bennet, B. Hillebrands: J. Appl. Phys. **67**, 4910 (1990)
2.148 J.H. van der Merwe, W.A. Jesser: J. Appl. Phys. **63**, 1509 (1988)
2.149 H.A.M. de Gronckel, P.J.H. Bloemen, A.S. van Steenbergen, K. Kopinga, E.A.M. van Alphen; W.J.M. de Jonge: Phys. Rev. B, to be submitted
2.150 S.T. Purcell, H.W. van Kesteren, E.C. Cosman, W. Hoving: J. Magn. Magn. Mater. **93**, 25 (1991)
2.151 F. Hakkens, W. Coene, F.J.A. den Broeder: Mater. Res. Soc. Symp. Proc. **231**, 397 (1991)
2.152 P. Bruno, J.-P. Renard: Appl. Phys. A **49**, 499 (1989)
2.153 H.A.M. de Gronckel, B.M. Mertens, P.J.H. Bloemen, K. Kopinga, W.J.M. de Jonge: J. Magn. Magn. Mater. **104–107**, 1809 (1992)
2.154 H. Awano, O. Taniguchi, T. Katayama, F. Inoue, A. Itoh, K.Kawanishi: J. Appl. Phys. **64**, 6107 (1988)
2.155 C.H. Lee, Hui He, F.J. Lamelas, W. Vavra, C. Uher, R. Clarke: Phys. Rev. B **42**, 1066 (1990)

3. Thermodynamic Properties of Ultrathin Ferromagnetic Films

D.L. MILLS

We explore the finite temperature properties of ultrathin ferromagnetic films, with attention to influence of various interactions between spins in the film on the ground state spin configuration, the spin wave spectrum, and the ordering temperature. We comment also on the nature of the temperature renormalization of the physical parameters that characterize the film at non-zero temperatures.

3.1 Introduction

The field of condensed matter physics continues to be lively and rapidly evolving as new materials are discovered whose properties challenge our basic understanding of dense matter. The past two decades have led to the ability to synthesize new artificial materials in the laboratory not realized in nature. Prominent among these are multilayer and superlattice structures, synthesized from films whose thickness is in the range of a few atomic layers. These structures can have interfaces that are sharp on virtually the atomic scale, thanks to the appearance of growth techniques such as molecular beam epitaxy. We first saw superlattice structures synthesized from semiconductors, and in the last decade, a remarkably wide variety of metallic superlattices of very high quality have been fabricated. Among these are materials which contain films of magnetically ordered metals, such as the 3d ferromagnetic Fe, Co, and Ni, and rare earth metals of complex spin structure as well.

In order to understand the magnetic properties of the multilayers, we first must have in hand a full description of those of the individual films, considered as isolated entities, possibly deposited on a non-magnetic substrate. This chapter is devoted to the simplest possible example of an ultrathin film: a few atomic layer film of ferromagnetically coupled spins. The prototype we have in mind is the ultrathin film of ferromagnetic Fe. Such films have been grown on the Cu(1 0 0) surface, the Ag(1 0 0) surface, and also on W(1 1 0). In addition, they have been incorporated in numerous multilayer and superlattice structures. The growth and characterization of such films is covered elsewhere in this volume. We shall have the luxury of adopting the theorist's view of such a film, which will

J.A.C. Bland and B. Heinrich (Eds.)
Ultrathin Magnetic Structures I
© Springer-Verlag Berlin Heidelberg 1994

consist of perfect layers and surfaces that are perfectly smooth and flat on the atomic scale.

Such a film at first glance sounds rather boring in nature. Clearly, the very strong exchange interactions between the spins will strongly couple them together to form a ferromagnet rather similar to bulk Fe; we just have a very thin slice of a Fe crystal, and our previous knowledge of bulk Fe, along with extensive knowledge gained earlier from the study of films of ferromagnetic metals with thickness in the micron range will allow us to understand the new ultrathin films quite straightforwardly. In fact, the contrary is the case. Very interesting new physics has been revealed in the ultrathin films prepared in the past few years.

The anisotropies encountered in these new materials are larger than those in the bulk materials by one to two orders of magnitude in the cubic ferromagnets Fe and Ni. We shall see that dipolar interactions play a role very different in quasi- two-dimensional ferromagnets than in three dimensions, and finally that the onset of long-range magnetic order itself is a topic that requires considerable discussion. Thus there is much to be learned from these materials, and as a consequence a considerable amount of theoretical activity has been in evidence in recent years. This chapter will review the basic concepts that have been introduced, and their consequences.

Whether or not the theoretical models and conclusions discussed here apply to actual materials remains an open question in the mind of the present author. The issue is the extent to which the real samples look like the "theorist's film", with its perfect structure and perfect surfaces. Terraces are surely present in all real films, and there must be interdiffusion across any interface, so the interface between the films and its substrate is in fact not truly perfect on the atomic scale. Unfortunately, the experimental information we have in hand on the microstructure of ultrathin films is limited at present. There is currently a very considerable experimental effort devoted to these questions, and meanwhile it is important to set down the basic theoretical description of the "theorist's film".

Throughout this chapter, we shall have ultrathin films of the 3d transition metals in mind as we proceed with our discussion. Various numerical estimates will use parameters that we believe are characteristic of these systems. While it is the case that much of the current experimental and theoretical research has focused on these systems, we must keep in mind that there have been earlier studies of well characterized two-dimensional magnetic systems. There is an extensive literature, not reviewed here, on crystals within which one finds two-dimensional planes of spins, seperated by appreciable distances by virtue of the crystal structure. Such materials allow one to study magnetism in two dimensions, until the temperature becomes sufficiently low that weak interplanar couplings necessarily present allow the actual three-dimensional character of the material to assert itself. Of course, such interplanar couplings surely influence the magnetic properties at temperatures well above the onset of three-dimensional long range order.

Of particular interest are studies of true two-dimensional magnetic films that may be synthesized by attaching magnetic ions to long molecules that form Langmuir–Blodgett films. Such films have been prepared, characterized and studied in detail by *Pomerantz* [3.1]. The Langmuir-Blodgett films were formed from the fatty acid $C_{17}H_{35}COOH$, and the Mn ions were attached to the ends of these molecules, to form a truely two-dimensional layer of high structural quality. The films appear to act as antiferromagnets, though a weak ferromagnetic moment is observed in the low temperature ordered state.

3.2 Interactions Between Spins: A Basic Spin Hamiltonian

If we wish to obtain a description of any physical system, we must begin by writing down a Hamiltonian which will form the basis of the analysis. If we begin the task of describing an ultrathin film of one of the 3d ferromagnetic metals, we have a serious problem. These materials in the bulk are itinerant ferromagnets; the magnetic moment bearing electrons are band-like in nature [3.2]. Our theoretical understanding of the magnetic properties of itinerant magnets remains primitive at present, in the sense that fully quantitative descriptions at finite temperature based on the actual electronic structures remain to be developed. So it is difficult to envision erecting a full and complete description of ultrathin films, within the framework of a proper itinerant electron theory. (These concerns apply only to the problem of describing the finite temperature properties of such films. Their ground state electronic properties can be and have been calculated within the framework of density-functional theory. These impressive studies are summarized in Sect. 2.1 by *J.G. Gay* and *R. Richter*.)

Virtually all theoretical discussions of ultrathin films at finite temperature proceed by adopting the Heisenberg model, within which the film is envisioned to consist of layers of spins, with a spin of length S localized on each lattice site. These spins experience anisotropy, ad described below, interact by means of dipolar interactions, and are coupled by phenomenological exchange interactions usually assumed to be nearest neighbor in character. General considerations given many years ago [3.3] insure that such a procedure, with the exchange strength adjusted to account for the exchange stiffness appropriate to the material at long wavelengths, gives a proper account of the long wavelength spin waves that control the low temperature thermodynamics, and which are probed in many experiments that excite spin waves of long wavelength. Examples are ferromagnetic resonance and Brillouin light scattering. Also, if we wish to discuss the onset of long-range order, and the film properties near the critical temperature at which long-range order sets in, we have a Hamiltonian associated with the appropriate universality class, and the critical properties are uninfluenced by its microscopic nature. Thus, such a Heisenberg picture proves an acceptable phenomenology, though we must keep its limitations in mind. For example, it does not describe the magnetic excitations of short wavelength properly; in the itinerant materials, Stoner excitations (electron–hole excitations, in which the electron flips its spin upon being excited) play an important role at short wavelengths.

With these remarks in mind, we turn to a description of our basic Hamiltonian consisting of four terms: an anisotropy term H_A, a dipolar term H_D, the exchange term H_X. Finally, we shall suppose a spatially uniform magnetic field is applied. The interaction of the spins with this field is described by the Zeeman term H_Z. The last three terms are straightforwardly written down; their form is familiar from the theory of magnetism in bulk materials. We simply apply them to the new geometrical structure of the ultrathin film, and examine the consequences of the reduced dimensionality. We keep in mind that at present we understand little about the strength of the effective exchange interactions in the films of interest here. In few atomic layer films, particularly in their surfaces and interfaces, these may assume values quite different than found in bulk materials. We ignore this complication here, and suppose all exchange couplings assume identical values in the film; for numerical estimates, we shall assume the exchange stiffness constant A encountered below has the value appropriate for bulk Fe, which is roughly $2.5 \times 10^{-9}\,\mathrm{G\,cm^2}$.

The anisotropy term H_A requires discussion. The physical origin of intrinsic anisotropy (there is a shape anisotropy we will encounter later, with origin in the sample shape-dependent internal macroscopic field set up by the ordered spin array) lies in the spin orbit interaction. This allows the spins to sample the microscopic potential of the crystal lattice and sense its symmetry. Let us consider an ion with spin S, in a localized description where the electrons are localized to a lattice site. The spin orbit term in the Hamiltonian has the form

$$V_{SO} = \xi \mathbf{L} \cdot \mathbf{S}. \tag{3.1}$$

Here L is the orbital angular momentum, and ξ is the strength of the spin–orbit coupling. We can use perturbation theory to treat the influence of this term on the energy levels of the spin for a 3d ion. We can write out the results of the various orders of perturbation theory, integrating over the spatial coordinates of the electron, then expressing the result as an operator in spin space. By these means, we generate an operator referred to as the spin Hamiltonian [3.4]. Time reversal allows only even powers of the spin S to appear.

Consider a spin at a bulk site in a cubic crystal, such as bulk Fe or Ni. The second order terms in the energy, expressed in terms of the Cartesian coordinates of S, can only have the form

$$\Delta E^{(2)} = \gamma_a \left(\frac{\xi^2}{\Delta E} \right)(S_x^2 + S_y^2 + S_z^2). \tag{3.2}$$

Here, ΔE is a typical energy denominator in the perturbation theory, and γ_a is a dimensionless constant of the order unity in value. We need to perform a proper microscopic calculation to find its magnitude. The second order term is isotropic; spin orbit coupling cannot produce magnetic anisotropy in second order. We need to turn to the fourth order. There are only two cubic invariants in fourth order, so one has

$$\Delta E^{(4)} = \left(\frac{\xi^4}{(\Delta E)^3}\right)\{\gamma_b(S_x^4 + S_y^4 + S_z^4) + \gamma_b(S_x^2 S_y^2 + S_x^2 S_z^2 + S_y^2 S_z^2)\} \quad (3.3)$$

which may be rewritten to read

$$\Delta E^{(4)} = \left(\frac{\xi^4}{(\Delta E)^3}\right)\{(\gamma_c - 2\gamma_b)(S_x^2 S_y^2 + S_x^2 S_z^2 + S_y^2 S_z^2) + \gamma_a(S_x^2 + S_y^2 + S_z^2)^2\}. \quad (3.4)$$

The first term in (3.4) describes true magnetic anisotropy, while the second term is again isotropic. It follows that in the cubic material, anisotropies in the magnetic energy appear first in the fourth order of perturbation theory; and the bulk anisotropy is rather weak.

Now consider a thin film, and a spin in the surface layer or in the first layer just above the substrate. Assume the site has fourfold symmetry. Now the z direction, taken normal to the surface, is inequivalent to the x and y directions. The second order term now has the form

$$\Delta E^2 = \left(\frac{\xi^2}{\Delta E}\right)(\gamma_d S_z^2 + \gamma_e(S_x^2 + S_y^2))$$

$$= \frac{\xi^2}{\Delta E}([\gamma_d - \gamma_e]S_z^2 + \gamma_e(S_x^2 + S_y^2 + S_z^2)). \quad (3.5)$$

We now have anisotropy from the second order term. The low site symmetry in the surface or at an interface "turns on" the second order terms. In ultrathin films, this is a source of the very strong anisotropy. If the coefficient of the S_z^2 term in (3.5) is negative, the axis normal to the surface is an easy axis, and the spin orbit anisotropy wants the spins oriented perpendicular to the surface. If the coefficient is positive, then this term renders the axis normal to the surface a hard axis, and the magnetization lies in-plane. Our simple argument cannot decide which of these alternatives are realized in any particular case, and in practice one encounters both. If only the second order terms are present, then the xy plane is magnetically isotropic, and we have an "xy model". In practice, the fourth order terms are present, and these produce weaker in-plane anisotropy that singles out easy directions within the plane.

The above argument allows one to make rough order of magnitude estimates of the strength of the anisotropy from the sources just considered. For Fe, one may extract values for the spin orbit coupling constant ξ from atomic spectra, and this should be roughly correct for the solid state, since the 3d electrons are tightly bound. In spectroscopic units, ξ is about 200 cm^{-1}, or 0.025 eV [3.5]. For ΔE, we recognize that the 3d electrons actually do reside in energy bands whose width is 4 eV. We take ΔE to be about half of this, or 2 eV. If $S = 2$, then one estimates

$$\frac{\xi^2 S^2}{\Delta E} \cong 1.2 \text{ meV}. \tag{3.6}$$

This is approximately correct. If there are 10^{15} atoms cm^{-2} in the surface, this corresponds to an anisotropy energy of 2 erg cm^{-2} which is larger than that typically found by about a factor of two. The energy required to rotate the spin through 90° can also be expressed as an effective magnetic field; the anisotropy field defined in this way is roughly 120 kG.

As we have seen, this picture shows the bulk anisotropy to be much weaker than that experienced by spins in the surface or interface, by the factor

$$r = \left(\frac{\xi S}{\Delta E}\right)^2 \cong 6 \times 10^{-4}. \tag{3.7}$$

This gives a bulk anisotropy energy of about 0.7×10^5 erg cm^{-3}, which is somewhat smaller than that found in bulk Fe, though the estimate is in the correct region.

In what follows, we shall see that the magnetic anisotropy enters the discussion of the thermodynamic properties of the ultrathin film in a central manner. In bulk magnets, the anisotropy per spin is always small compared to $k_B T$ for all temperatures above a few Kelvin. As a consequence, it has a negligible influence on the thermodynamics, except at rather low temperatures. The same statement on the order of magnitude of the anisotropy remains true in the ultrathin films, as one sees easily from the numbers given above. However, this weak anisotropy (weak in the sense that the energy per spin is small compared to $k_B T$) influences the thermodynamics of the quasi-two-dimensional ultrathin films very importantly, as we shall come to appreciate. For this reason, we have spent some time on the discussion of its order of magnitude in these materials.

In what follows, we shall introduce into our spin Hamiltonian the term

$$H_A = -K \sum_l S_z^2(l), \tag{3.8}$$

where the sum ranges over the sites in the outer surface of the film, and over sites in the interface between the film and its substrate. In general, the strength of the anisotropy at the two kinds of sites will differ, but we have little experimental information on this question at present.

There are additional sources of spin orbit anisotropy in ultrathin films. One must keep these in mind when addressing data. The spacing between adjacent layers may not be the same as realized in the bulk. The consequence is that even in the film interior, the local site symmetry may not be cubic, in a film of material whose bulk structure is cubic. This interlayer relaxation will "switch on" the quadratic terms for sites in the film interior. An important class of ultrathin films studied by the *Gradmann* group [3.6], Fe on W(1 1 0), are grown on a substrate whose unit cell is rectangular, not square. The Fe magnetization lies in the plane in these films. There is a second order contribution to the in-plane anisotropy for such a system; there is thus an easy axis within the plane, and the in-plane anisotropy is strong in the sense used here.

We now turn to the consequences of the above spin Hamiltonian, applied to the ultrathin film.

3.3 Properties of Ultrathin Ferromagnetic Films at Low Temperatures: The Ground State and the Spin Wave Regime

We now consider the properties of an ultrathin ferromagnetic film which is described by the Hamiltonian, as summarized in the previous section. The discussion will be divided into two parts: the first concerns the nature of the ground state, and then we argue that at low temperatures, the thermodynamic properties of the films may be described by spin wave theory. We summarize the predictions of this picture.

3.3.1 The Ground State

Suppose we first consider a film in the absence of an applied external magnetic field, in the presence of very strong exchange interactions which lock all spins into ferromagnetic alignment. The limit of very strong exchange is in fact appropriate to the 3d transition metal ferromagnets. A measure of the strength of the exchange is the Curie temperature of the bulk material, which ranges from a few hundred (Ni) to nearly 1500 K (Co). The dipolar fields are easily estimated to be in the range of 1 K, when their energy is expressed in temperature units, and the anisotropy fields only a bit more at best.

With the spins locked together in ferromagnetic alignment, the next question is the issue of the direction the net magnetization points. Since the Heisenberg exchange is invariant under rigid rotations of the spins, the magnetization direction is controlled by the combination of the anisotropy and dipolar couplings, in the absence of an external field. Consider a very thick film, suppose it is cubic, and ignore the very weak bulk anisotropy for the moment. If the magnetization is aligned perpendicular to the surface, the direction of the z axis, then if the magnetization points along the $+z$ direction, elementary magnetostatics requires the existence of an internal H field of strength $4\pi M_0$ in the $-z$ direction. This arises because each spin carries a magnetic moment; the microscopic origin of this field and the resulting interaction energy $+2\pi M_0^2$ lies in the magnetic dipole–dipole interaction. Here we have a situation in which each spin resides in a field anti-parallel to its magnetic moment. Thus perpendicular orientation is unstable. If the magnetization lies parallel to the film surfaces, there is no macroscopic field generated by the dipoles and the state is stable. This conclusion is not affected, for a thick film, by surface or interface anisotropy, since this influences only the outer layers and contributes little to the energy of the magnetic configuration. Thus, for such a simple film, one finds the magnetization in the film plane.

When the film is one, or perhaps a small number of atomic layers in thickness, then the surface or interface anisotropy enters importantly in controlling the orientation of the magnetization. A measure of the strength of the dipole–dipole interaction is, as we have just seen, $4\pi M_0$. In Fe, $4\pi M_0$ is roughly 20 kG. The estimates of the previous section show that the surface or interface anisotropy, expressed as an effective field experienced by a spin there, can be several times 20 kG. Thus if the sign of the surface anisotropy is such that the axis normal to the surface is an easy axis, a circumstance often realized, the anisotropy energy will overwhelm the dipolar energy and the film magnetization will lie normal to the surface, rather than parallel to it. If the surface normal is a hard axis, then both interactions favor the in-plane geometry. We therefore may realize both possibilities.

The energetics read as follows. Suppose the axis normal to the surface is an easy axis. Let the anisotropy in one outer layer of spins have the magnitude H_A, in magnetic field units; this defines a model studied in detail by *Erickson* and *Mills* [3.7]. Orient the magnetization normal to the surface. The energy gain, relative to the parallel state, is $H_A - 4\pi M_0 c_1 N$, where the first term represents the energy gain from anisotropy, and the second the unfavorable dipolar energy. Here N is the number of layers in the film, and c_1 a constant which follows from microscopic theory [3.7]. For a monolayer, $c_1 = 0.762$ if the unit cell is square, while c_1 is unity for a thick film. It follows that if $H_A > 4\pi M_0 c_1$, the magnetization will be directed normal to the surface. When this condition is realized, one sees that as the film thickness is increased, when $N > 4\pi M_0 c_1 / H_A$, suddenly the parallel magnetization case becomes stable. Thus, as film thickness is increased, the film switches from perpendicular to parallel in orientation. The dipolar energy overwhelms that provided at the surface. The estimates of the previous section indicate that the critical thickness is of the order of four or five layers, which is what is realized in practice.

In what follows, we shall be interested in the influence of an external magnetic field. Suppose such a field of strength H is applied to a monolayer, and the direction of H is parallel to the surface. Suppose further that the surface anisotropy favors the perpendicular orientation. Then the field will cant the magnetization away from the normal. If θ is the angle between the magnetization and the film normal, then a simple calculation shows that [3.7] if $H < \tilde{H}_A$, where $\tilde{H}_A = H_A - 4\pi M_0 c_1$, one has $\sin\theta = H/\tilde{H}_A$ while if $H > \tilde{H}_A$, $\theta = \pi/2$. Thus there is a critical field above which the magnetization lies strictly parallel to the surface. We shall be very interested in the behavior of the film near the critical field in the discussion below. Such a critical field will exist not only for the monolayer, of course, but for any multilayer film whose surface or interface anisotropy is sufficiently strong and of such a sign that the perpendicular state is the low field ground state.

The above discussion assumes that the exchange is so strong that all spins in the film are locked rigidly into ferromagnetic alignment. In a multilayer film where surface anisotropy favors the perpendicular orientation, one might suppose spin canting is present, with spins in the interior canted toward the parallel state, while those in the surface wish to twist toward the normal to take advantage of the favorable anisotropy localized there. This possibility has been studied in detail by *Erickson* and *Mills* [3.7]. Numerical studies show that in thin films with parameters characteristic of the 3d ferromagnetic metals, the magnetization lies *strictly* parallel to the surface until the surface anisotropy field exceeds a certain critical value. As long as the film is thinner than the exchange length $t_{ex} = (A/\pi M_0 a_0^2)^{1/2}$, with a_0 the lattice constant and A the exchange stiffness, then the critical surface anisotropy field required to create the perpendicular orientation is accurately given by $H_A - 4\pi c_1 M_0 N$. For parameters characteristic of Fe, the thickness t_{ex} is estimated to be about 26 layers. For films whose thickness is in the range of ten layers or less, in the perpendicular state, the amount of canting in the film is very small, with the spins in the center of the film canted away from the normal by less than one degree.

If the film is treated as a continuum whose thickness is N layers, then it is possible to derive an analytical expression for the critical value of the surface anisotropy field required to induce spin canting. One finds [3.7] this critical value is $4\pi M_0(t_{ex} \tanh(N/2t_{ex}) + 1)$; this formula fits the full calculations quite nicely down to the two or three layer level. The derivation of the expression proceeds by generalizing an earlier treatment given by *Mills* [3.8] for the critical surface anisotropy field required to induce spin canting at the surface of a semi-infinite ferromagnet. As the film thickness N becomes large compared to t_{ex}, notice the critical value of the surface anisotropy field approaches $4\pi M_0(t_{ex} + 1)$, a value very large compared to those found in practice, and given by the crude estimates provided above. Thus, for films is the thickness range where N is comparable to or larger than t_{ex}, the parallel state is realized (at least for materials that are the focus of this chapter) and once again the spins in each atomic layer are strictly parallel to the surface in this configuration.

From the above remarks, for ultrathin films of 3d transition metals which are a small number of atomic layers in thickness, the magnetization in each layer may be regarded to an excellent approximation as either strictly perpendicular or strictly parallel to the surfaces; exchange is strong enough to lock them together very tightly, so in the perpendicular state there is a negligible amount of spin canting. It is only in films whose thickness becomes to comparable to t_{ex} that canting is appreciable. These remarks apply to the case where no external magnetic field is present. Application of an external field parallel to the surface will cant the net magnetization, very much as described above for the monolayer.

3.3.2 The Nature of Spin Waves in Ultrathin Films; Low Temperature Thermodynamic Properties

We now turn to the dynamic response of the ultrathin film, and its thermodynamic properties at low temperatures. Given the ground state spin arrangement as discussed above, the low temperature properties of the film may be described through use of spin wave theory. A general formulation of spin wave theory appropriate to ultrathin ferromagnetic films, possibly with a spatially non-uniform ground state, and with dipolar coupling and anisotropy included, has been given by *Erickson* and *Mills* [3.7]. A second paper applies this formalism to the study of the thermodynamics of ultrathin films, in the spin wave domain [3.9].

In this chapter, we shall confine our attention to a very simple case, which contains the essential physics. We consider a single layer of ferromagnetically coupled spins, which also interact via dipolar interactions. These spins also feel uniaxial anisotropy of the form given in (3.8), where the constant K can be either positive (easy axis normal to the surface), or negative (easy plane case). In addition, an external magnetic field of strength H is applied parallel to the surface, canting the film magnetization with respect to the film normal as described earlier, when $K > 0$. Of course, when $K < 0$, the magnetization lies in the plane of the film. This field configuration was used in the elegant light scattering study of an easy axis, ultrathin Fe film on Cu(100), reported by *Dutcher* et al. [3.10]. We comment on these experiments below.

For this case, we quote two expressions that have been derived elsewhere [3.11]. The first is the dispersion for a spin wave of wave vector k, which is long compared to the lattice constant. We have

$$\Omega_0(k) = [A_1(k)A_2(k)]^{1/2}, \tag{3.9a}$$

where

$$A_1(k) = H\cos\theta + \tilde{H}_A\sin^2\theta + \frac{\pi M_0 k a_0}{\sqrt{2}}(1 - \cos 2\phi) + Ak^2 \tag{3.9b}$$

and

$$A_2(k) = H\cos\theta + \tilde{H}_A\cos 2\theta - \sqrt{2}\pi M_0 k a_0 \cos^2\theta$$
$$+ \frac{\pi M_0 k a_0}{\sqrt{2}}(1 + \cos 2\phi)\sin^2\theta + Ak^2 \tag{3.9c}$$

In these expressions, H is the strength of the external magnetic field, again applied parallel to the surface of the film. The quantity $\tilde{H}_A$ is given by $H_A - 4\pi c_1 M_0$, where $H_A = 2KS$ with K the strength of the uniaxial anisotropy as defined in (3.8). The angle θ is that made by the magnetization with the normal to the plane of the film; the prescription for calculating this angle has been given above. If $\tilde{H}_A > 0$, then the magnetization is perpendicular to the film's surface when $H = 0$ ($\theta = 0$), and $\tilde{H}_A < 0$, one has $\theta = \pi/2$ always. If the normal to the film is in the z direction, and the external magnetic field is applied along the x direction, then the angle ϕ is the angle between the wave vector and the x axis. Thus, $\phi = 0$ is propagation parallel to the magnetization. The exchange stiffness constant, mentioned earlier, is A, and a_0 is the lattice constant. The lattice has its basic structure a face centered square, with a_0 the side of the square.

The terms proportional to M_0 have their origin in the long ranged dipole–dipole interaction. Of very considerable interest in what follows, as noted first by Yafet, et al. [3.12], are the terms linear in the wave vector k. Notice that these terms depend not only on the magnitude of the wave vector, but also on its direction. These contributions are a reflection of the long-ranged nature of the dipole–dipole interaction, and are unique to two dimensions. It is interesting to comment on their mathematical origin. The theory of spin waves, for a lattice of any dimensionality, leads one to the dipole lattice sums of the form

$$d_{\alpha\beta}(\mathbf{k}) = \sum_{l}{}' \left[\frac{3 l_\alpha l_\beta - \delta_{\alpha\beta} l^2}{l^5} \right] e^{i\mathbf{k} \cdot \mathbf{r}} \qquad (3.10)$$

where the sum ranges over all sites in the lattice, save for that at the origin, which is omitted. In an infinite three dimensional lattice, the sum diverges if the wave vector k is set to zero. For finite k the sum converges because of the oscillatory exponential factor, and can be evaluated in the limit of small wave vector by replacing the sum over lattice sites by an integration, excluding a small sphere centered at the origin. The result is that in the limit of small wave vector, the dipole sum depends on the direction but not the magnitude of the wave vector. One finds

$$d_{\alpha\beta}(\mathbf{k}) = \frac{4\pi}{V_c} \left[\delta_{\alpha\beta} - \frac{k_\alpha k_\beta}{k^2} \right], \qquad (3.11)$$

where V_c is the volume of the unit cell of the crystal. The dipole sums are non-analytic functions of the wave vector, in that as the wave vector approaches zero, their value for the infinite lattice depends on the direction of approach of the wave vector to the origin.

For the two-dimensional lattice, as the wave vector approaches zero, the dipole lattice sums converge and are quite finite. The "renormalization" of the surface anisotropy from the "bare" value H_A to the "renormalized" value $\tilde{H}_A$ has its mathematical origin in the two dimensional dipole sums evaluated at zero wave vector. Yafet and co-workers [3.12] pointed out that although the two-dimensional dipole sums are finite in the limit of zero wave vector, the leading corrections are in fact linear, with magnitude dependent on the direction as well as the magnitude of the two dimensional wave vector. This is the origin of the linear terms in (3.9a, b, c).

Given the dispersion relation of the spin waves, one may generate an expression for the temperature dependence of the magnetization $M(T)$ of the film. If M_0 is the value of the magnetization at $T = 0$, one has [3.11]

$$\frac{M(T)}{M_0} = 1 - \frac{1}{N_S S} \sum_k \frac{A_3(k)}{\Omega_0(k)} n(k), \qquad (3.12a)$$

where N_S is the number of unit cells in the quantization area of the monolayer, and $n(k)$ is the Bose–Einstein function $n(k) = [\exp(\hbar\Omega_0(k)/k_B T) - 1]^{-1}$,

$$A_3(k) = H\cos\theta + \tilde{H}_A(1 - \tfrac{3}{2}\sin^2\theta)$$
$$+ \sqrt{2\pi} M_0 k a_0 (1 - \tfrac{3}{2}\cos^2\theta - \tfrac{1}{2}\cos 2\phi \cos^2\theta) + Ak^2. \qquad (3.12b)$$

One can learn a great deal by considering various special limits of the expressions in [3.9a, b, c] and [3.12]. We look at various special cases next.

3.3.2.1 The Isotropic Two-Dimensional Heisenberg Model

Here we have a two dimensional lattice, with only exchange interactions between the spins. The external magnetic field H is set to zero, along with M_0 in the spin wave dispersion relation so the magnetic dipole interaction is "turned off", and finally we ignore the uniaxial anisotropy. The dispersion relation then assumes the well known long wavelength form

$$\Omega_0(k) = Ak^2 \qquad (3.13)$$

and for the magnetization we have, after converting the sum on the wave vector to an integral, with A_C the area of the unit cell,

$$\frac{M(T)}{M_0} = 1 - \frac{A_C}{2\pi S} \int_0^\infty dk\, k n(k). \qquad (3.14)$$

The integral in (3.14) diverges as a consequence of the singularity at the lower limit. The spin wave picture thus breaks down completely in two dimensions for the isotropic Heisenberg model. Physically, the long wavelength spin fluctuations are strongly excited at finite temperatures, and break up the long range order present in the fully aligned ground state. One may rigorously prove that in this model, there is no long range order at any finite temperature. This result is known as the Mermin–Wagner Theorem and the difficulty we have encountered is a reflection of this general result. We can proceed no farther within the framework of a simple approach such as spin wave theory.

3.3.2.2 The Role of Surface or Interface Anisotropy; the Case of an Easy Axis Normal to the Surface

One supposes, with the external field H still equal to zero, we "turn on" the surface anisotropy by allowing $\tilde{H}_A$ to be non-zero. We suppose that $\tilde{H}_A > 0$, so the axis normal to the surface is an easy axis, and $\theta = 0$. We still have no dipolar coupling, so that $M_0 = 0$. Now the dispersion relation reads

$$\Omega_0(k) = \tilde{H}_A + Ak^2, \tag{3.15}$$

so we have a gap in the dispersion relation at zero wave vector. The magnetization is still given by the expression in (3.14), the integral is evaluated straightforwardly, and the gap in the spin wave dispersion relation renders the spin wave expression for the magnetization finite at non-zero temperatures. We have

$$\begin{aligned}\frac{M(T)}{M_0} &= 1 - \frac{a_0^2 k_B T}{8\pi AS} \ln\left[\frac{1}{1 - \exp(-\tilde{H}_A/k_B T)}\right] \\ &\cong 1 - \frac{a_0^2 k_B T}{8\pi AS} \ln\left[\frac{k_B T}{\tilde{H}_A}\right]. \end{aligned} \tag{3.16}$$

where the second form applies when $k_B T$ is large compared to the gap, a limit applicable at all but the lowest temperatures for the materials of interest here.

We may inquire how effective the anisotropy is in suppressing the large amplitude thermal fluctuations that break up the long range order in the isotropic Heisenberg model. This may be assessed by evaluating the expression in (3.16) for parameters characteristic of the ultrathin films of 3d transition metals. If for A we take the exchange constant of bulk Fe ($A = 2.5 \times 10^{-9}$ G cm^2), $S = 2$ and $a_0 = 3 \times 10^{-8}$ cm, while the temperature T is expressed in K, we find

$$\frac{M(T)}{M_0} = 1 - (5.5 \times 10^{-5}) T \ln\left[\frac{k_B T}{\tilde{H}_A}\right]. \tag{3.17}$$

The rough estimates of the strength of the surface anisotropy fields given in Sect. 3.2 suggest the gap is in the range of 1 K in magnitude. Thus, at room temperature one has the rough estimate $M(T)/M_0 = 0.9$. If we suppose the gap to be an order of magnitude smaller, say 0.1 K, then we still have $M(T)/M_0$ in the range of 0.85 at room temperature.

It is the view of the author that there is a message in the above numbers. While it is indeed true that the perfectly isotropic Heisenberg model has its long-range order broken up at finite temperature by long wavelength low frequency thermal excitations, as we saw in the previous subsection, in fact small perturbations are extremely effective in suppressing these fluctuations. The system wants to order, and any small deviation from isotropy locks in long-range order very efficiently. We see that even the extraordinarily small gap of 0.1 K gives a magnetization at room temperature smaller than the saturation value at $T = 0$ by only 15%, if the above numbers are trusted. (The reader must keep in mind that the estimates just given assume that the film is described by an exchange stiffness equal to that in bulk Fe. If the actual exchange stiffness in the ultrathin film is substantially smaller, an issue on which we have little information at present, then the conclusion may be altered and the domain of validity of spin wave theory may be more limited than we suggest.) Thus spin wave theory, possibly corrected by the effects of interactions between spin waves, can provide an adequate description of the film thermodynamics over a rather wide range of temperature. We shall comment on the nature of the lowest order corrections below. Spin wave theory, elaborated by interaction effects, cannot describe the phase transition to the paramagnetic state. We also shall discuss how this is done below. We will argue below that some schemes found in the literature based on the use of renormalization group methods overestimate the influence of spin fluctuations in films with the properties assumed here.

3.3.2.3 The Exchange and Dipolar Coupled Two-Dimensional Heisenberg Ferromagnet

In the previous subsection, we saw that a rather modest amount of anisotropy stabilizes long-range order in the two-dimensional Heisenberg model, and in fact suppresses the large fluctuations expected for the purely isotropic case. It is interesting to inquire about the role of the long-ranged dipolar interactions, in the absence of any single site anisotropy.

This may be done within the framework of the formulae given above. If we have isotropic exchange supplemented by only dipolar interactions in the model Hamiltonian, then for reasons given earlier, the magnetization lies in the plane, and $\theta = \pi/2$. Also, $\tilde{H}_A = -4\pi M_0 c_1$. This gives for the two factors that enter the spin wave dispersion relation

$$A_1(\mathbf{k}) = \frac{\pi M_0 k a_0}{\sqrt{2}}(1 - \cos 2\phi) + Ak^2 \tag{3.18a}$$

and

$$A_2(\mathbf{k}) = 4\pi M_0 c_1 - \sqrt{2}\pi M_0 k a_0 + Ak^2. \tag{3.18b}$$

We see from (3.9a) that there is no gap in the spin wave spectrum. However, the density of low lying spin wave excitations is very much lower than in the "pure" isotropic Heisenberg model; for a general direction of propagation the spin wave frequency vanishes as $k^{1/2}$, rather than k^2, as displayed in (3.13). It follows that the magnetization as given in (3.12a) remains finite at non-zero temperatures.

We must keep in mind that symmetry requires single site anisotropy of the form given in (3.8) to exist for all spins which reside in the surface of a film, or in the interface layer with a substrate. However, we see that even in its absence, the dipolar interactions always present stabilize long-range order in two dimensions. In the presence of dipolar couplings and anisotropy, we may use spin wave theory to discuss the low temperature properties of the system. In any given case, one must assess the limits of validity of spin wave theory, and the task of extending its description to higher temperatures remains. We comment on this question in Sect. 3.5. The notion that dipolar interactions stabilize long-range order in the two-dimensional Heisenberg ferromagnet was put forth some years ago by *Maleev* [3.13]. In the context of ultrathin films of the sort considered here, the question was examined also by *Yafet* and collaborators. [3.12]

3.3.2.4 Some General Remarks

Erickson and *Mills* [3.9] have carried out detailed studies of the thermodynamic properties of ultrathin films, based of the model outlined above. Spins in the outer surface of these few layer films experience anisotropy described by (3.8), with easy axis normal to the surface. We comment on several trends in the results.

First of all, one may inquire how many layers are required before the three-dimensional character of the film asserts itself. Clearly, the monolayer is an example of a two-dimensional spin system, but as one adds layers, there must be a transition to three-dimensional behavior. One may answer this question as follows. Consider a three layer film. For each value of the wave vector k parallel to the surface, one has three spin wave branches. The lowest lying branch has the character of an "acoustical" spin wave, wherein spins in all layers precess in phase, in the limit of wavelength long compared to the lattice constant. The two higher branches are of "optical" character, and in the long wave limit spins in different layers precess out of phase. The two optical branches have finite frequency as the wave vector approaches zero. The calculations reported in [3.9] show that for parameters characteristic of Fe, the two higher branches made negligible contributions to the thermodynamic properties of the film, even for temperatures as high as 400 K. Thus, the model film behaves as a quasi-two-dimensional magnetic structure, at least in theory.

One may derive a rough estimate for the number of layers required for the film to make a transition from two- to three-dimensional behavior, as follows. If A is the exchange stiffness, then one expects the separation in frequency between different branches to be roughly $\Delta\Omega = A(\pi/Na_0)^2$. If this frequency splitting is large compared to $k_B T$, then the thermodynamics will be dominated by the long wavelength portion of the acoustical spin wave branch, at least at low temperatures, and we are in the quasi-two-dimensional limit. When $\Delta\Omega$ becomes comparable to or smaller than $k_B T$, then we move to the quasi-three-dimensional regime. Numbers characteristic of transition metal films suggest that we begin to make the transition for rather small values of N, using this criterion. One finds near room temperature that $\Delta\Omega$ and $k_B T$ are comparable for N as small as two, though as mentioned above the explicit calculations reported in [3.9] show the lowest spin wave branch still dominates the thermodynamics of the film at room temperature, for a three layer example. This issue is discussed in more detail from the theoretical point of view in [3.9].

The very interesting experiments reported by *Gradmann* and his collaborators [3.6] bear directly on this question. As noted earlier, these authors studied Fe films grown on the W(1 0 0) surface. Through use of Mossbauer spectroscopy, these authors deduce values for the ratio $\langle S_z \rangle / S$, the expectation value of the spin for a particular Fe spin in the sample, over fairly wide range of temperature. If we write

$$\langle S_z \rangle / S = 1 - \Delta(T), \tag{3.19}$$

then for films in the thickness range of one to four layers, the data shows that $\Delta(T)$ scales inversely with film thickness.

This result can be understood by invoking two assumptions. The first is that the acoustical spin wave branch dominates the thermodynamics of the film, and the second requires more discussion. Note first that the hyperfine field measurements in [3.6] probe the behavior of $\langle S_z \rangle / S$ at the site of *one spin* in the sample. The contribution of one particular spin wave mode to $\langle S_z \rangle / S$ is proportional to $|e(\mathbf{k}; l)|^2 n(\mathbf{k})$, with $n(\mathbf{k})$ the number of spin waves of wave vector $\mathbf{k}$ excited thermally (this factor also enters (3.12a)), and $|e(\mathbf{k}; l)|^2$ is the eigenvector of the acoustic spin wave of wave vector $\mathbf{k}$, evaluated at lattice site l. Normalization requires $\sum_l |e(\mathbf{k}; l)|^2 = 1$, with the sum ranging over all sites in the film. Now for the long wavelength, low lying acoustic spin wave, $e(\mathbf{k}; l) \sim \exp(i\mathbf{k} \cdot \mathbf{l}_{\parallel})$ independent of l_z, so the normalization condition requires $|e(\mathbf{k}; l)|^2 = 1/N_S N$, with N_S the number of atoms in a particular layer, and N the number of layers in the film. For the monolayer, $N = 1$, and we recover (3.12a). For thicker films, where the acoustic spin wave dominates the thermodynamics, one finds $\Delta(T)$ scales inversely with N. Physically, as N increases, the energy density associated with a given spin wave mode is spread over a larger number of spins. The energy stored in the one mode is fixed at the value $\hbar\Omega(\mathbf{k})$, so its contribution to the amplitude of the thermal motions of a particular spin decreases, with increasing film thickness.

If this were the only dependence of $\Delta(T)$ on the number of layers, we could conclude the data provides clear proof that the films are quasi-two-dimensional in the temperature region explored, for films as thick as four layers. However $\Delta(T)$ also depends on the spin wave dispersion relation, as we see from (3.12a), and this must be independent of film thickness for the conclusion to hold. If the anisotropy felt by the spins is dominated by that experienced by spins in the film surface or those at the interface with the substrate, then the analog of $\tilde{H}_A$ for these films will depend on film thickness. Thus, $\Delta(T)$ will have a dependence on film thickness more complex than that given by the normalization of the eigenvector. If the anisotropy is the same in all layers, then $\tilde{H}_A$ will be independent of film thickness, and we expect $\Delta(T)$ to scale inversely with film thickness. The site symmetry in these films grown on a (1 1 0) surface requires there to be anisotropy at each site, with origin in the fact that the Fe film is distorted from the square symmetry to rectangular, to fit on the lower symmetry W(1 1 0) substrate. If we accept this as the origin of the uniaxial, in-plane anisotropy, then from the data in [3.6] we may conclude that the films are quasi-two-dimensional, up to four layers in thickness.

Gradmann et al. [3.6] also find that their data on $\Delta(T)$ can be fitted by a power law, with $\Delta(T) = aT^n$. These authors find the exponent n to lie in the range 1.3–1.5. The numerical calculations in zero external magnetic field reported in [3.9], carried out for a three layer Fe film, are consistent with n in the range of 1.3, and thus display behavior rather similar to that found in the experiments, though no attempt was made to fit the actual data. Certainly the data, set alongside the calculations, suggest that spin wave theory indeed can describe these films, at least at room temperature and below, as we have argued above.

As we mentioned earlier, *Dutcher* et al. [3.10] have studied long wavelength spin waves in an ultrathin film of Fe on Cu(1 0 0) by means of Brillouin light scattering. This film, three monolayers thick, has an easy axis normal to the surface. These authors applied a magnetic field parallel to the surface, thus canting the film magnetization with respect to the film normal, and studied the frequency of spin waves with wave vectors k very near zero as a function of canting angle. They find, at the critical value of magnetic field required for the magnetization to just become parallel to the surface (just as θ becomes $\pi/2$ in our earlier notation), the spin wave frequency vanishes, to "stiffen" with further increase in field. Precisely this behavior is expected from the dispersion relation in (3.9a, b, c). There is thus a "soft mode" in this system, as a magnetic field applied parallel to the surface is increased to the point where the magnetization just touches the surface. It would be of great interest to see studies of the thermodynamic properties of such a film near the critical value of the field. *Erickson* and *Mills* [3.9] have calculated the field dependence of $\Delta(T)$ in this case, within spin wave theory, to find a divergence at the critical field. Simple considerations suggest that when the gap vanishes, in fact long ranged order disappears. We shall see in Sect. 3.3.2e that in fact the properties of the film in the near vicinity of the critical field are more complex and more interesting than this result suggests.

When the authors of [3.10] analyze the dependence of the spin wave frequency in their film as a function of applied magnetic field, they find a model with anisotropy given by (3.8) proves inadequate. They argue the data also requires a term proportional to the fourth power of the perpendicular components of spin. The anisotropy realized in actual ultrathin films is thus more complex in nature than assumed in much of our discussion here, which we intend to be pedagogical in orientation. The reader must keep in mind that we have also ignored in-plane anisotropy, which for a film with a fourfold axis normal to the surface generates terms in $S_x^4 + S_y^4$. The presence of such terms is required by symmetry, though the arguments of Sect. 3.2 suggest they are weaker than the principal terms displayed in (3.8). Such in-plane quartic terms have been measured through analysis of light scattering data and ferromagnetic resonance [3.14]. If we consider a film magnetized in-plane, in zero external magnetic field, by virtue of the fact that (3.8) supplemented by the dipolar fields render the axis normal to the film surface a hard axis, then if the quartic in-plane anisotropy is ignored, we see from (3.9a, b, c) that there is no gap in the spin wave dispersion relation. Inclusion of the quartic terms will introduce a finite gap, and this may influence the thermodynamic properties importantly in the low temperature regime. Thus in zero external magnetic field the quartic terms should be included in a complete theory.

3.3.2.5 Instabilities in the State of Uniform Magnetization in Ultrathin Ferromagnetic Films

We have proceeded so far by assuming that for our model of the ultrathin film, the ferromagnetic ground state is always the lowest energy configuration. If we have anisotropy as described by (3.8), combined with ferromagnetic exchange couplings between spins and a spatially uniform applied magnetic field, then very clearly this assumption is obviously correct. However, *Yafet* and *Gyorgy* [3.15] argued that dipolar couplings between spins can render the state of uniform ferromagnetic alignment unstable with respect to a configuration within which a linear domain structure is realized. This state can occur when the effective anisotropy field $\tilde{H}_A$ is a small fraction of $4\pi M_0$. In the non-uniform state proposed by *Yafet* and *Gyorgy*, the magnetization cants out of the plane within the domains.

One thus may expect to observe the non-uniform state in films whose thickness is very close to the critical thickness for transition from the state of perpendicular magnetization to that where the magnetization lies parallel to the surface. We saw in the discussion above that in films where anisotropy at the surface or in interfaces favors the perpendicular orientation, as the film thickness increases eventually the dipolar anisotropy (shape anisotropy) favors the parallel state. In recent experiment [3.16]. *Allenspach* et al. observed a domain structure in such an ultrathin film, controlled by a length scale quite similar to that which emerges from the analysis in [3.15].

Subsequently, *Erickson* and *Mills* noted a second circumstance wherein the state of uniform magnetization is unstable [3.11]. We have already discussed the ultrathin film with an easy axis normal to its surface, and in addition with a magnetic field applied parallel to the surface. The field cants the magnetization with respect to the film normal, and there is a certain critical field at which the magnetization just becomes parallel to the surface. From the spin wave dispersion relation given in (3.9a, b, c), one sees that the $k = 0$ spin wave frequency just vanishes at this critical field. This field induced "soft mode" has been studied in light scattering experiments [3.10].

Suppose the magnetic field is just a bit above the critical field which drives the angle θ to $\pi/2$. Then notice the term linear in wave vector, proportional to $\sin^2\theta$ in (9c), leads to negative dispersion in the spin wave frequency in the long wavelength region. If the external magnetic field H is just a small amount above the effective anisotropy field $\tilde{H}_A$, and $\theta = \pi/2$, this linear term can drive $A_2(k)$ negative, and a consequence is that the spin wave frequency becomes complex. This means the state of uniform magnetization is unstable once again, in a field regime just above the critical field. Further examination of the dispersion relation shows that the uniform state is unstable also in a field "window" just below the critical field.

The width in magnetic field within which the instability resides is easily seen to be $\Delta H = \pi^2 M_0^2 a_0^2 / A$. The numbers above show that ΔH is quite small, the order of 20 G or so. It would be of very great interest to examine the dynamic response of the films explored in [3.10], for magnetic fields in the very near vicinity of the field where the spin wave mode "goes soft". One expects a domain structure in this field regime with a length scale in the range of a micron [3.11]. There should be substantial quasi-elastic scattering of light when the domain pattern is present. The length scale is expected to be comparable to the wavelength of light, so such quasi-elastic scattering should be observable.

The considerations above should bear on experiments reported by *Pappas* and co-workers [3.17]. These authors monitor the magnitude and direction of the magnetization of few layer films of Fe on Cu(1 0 0) as a function of temperature. At low temperatures, the film magnetization is normal to the surface; the effective anisotropy field $\tilde{H}_A$ is thus positive. The films are just below the critical thickness for a transition to the parallel state, so in fact $\tilde{H}_A$ is quite small. Incidentally, in the experiment, the direction and magnitude of the magnetization is monitored by measuring the spin polarization of secondary electrons emitted from the Fe d-bands, and detecting their polarization, both magnitude and direction, through use of a Mott detector. As the temperature is raised, the perpendicular component of the magnetization decreases and disappears at a certain temperature. One then encounters a temperature regime where no long-range magnetic order is detected, but above this a magnetization develops parallel to the surface and grows with increasing temperature. One may understand this behavior by assuming the effective anisotropy field $\tilde{H}_A$ first decreases and then in fact passes through zero, so that at higher temperatures the parallel state is favored. We shall discuss theoretical modeling of the temperature dependence of this parameter in the next section. It is significant, in view of the earlier remarks of this section, that there is a gap in temperature within which no long-range order is observed. *Erickson* and *Mills* suggest this has its origin in the dipolar induced instability of the uniform state [3.11].

The gap in temperature found by *Pappas* et al. [3.17] is significantly larger than that estimated (crudely) from the spin wave dispersion relation; it is possible that non-uniformities in the film thickness can broaden the temperature range within which the uniform state is unstable [3.11].

It would be of very great interest to see further experimental study of ultrathin films in regions of magnetic field, temperature and thickness where dipolar interactions are expected to render the uniform state unstable. So far, the only direct observation of the non-uniform state is that reported by *Allenspach* and co-workers [3.16]. While experiments have been carried out in the appropriate regimes, as we have just seen, in fact the studies do not probe the nature of the order within the regions where the uniform state is expected to be unstable.

3.4 Beyond Spin Wave Theory: The Intermediate Temperature Regime

The discussion of the thermodynamics of ultrathin films was based on the application of spin wave theory. We argued that modest amounts of anisotropy efficiently suppresses fluctuations in two-dimensional spin systems, with the consequence that spin wave theory provides a starting point for the discussion of the finite temperature properties of the films of interest to this chapter. If our numbers for 3d transition metal films are accepted, then spin wave theory may work for temperatures as high as room temperature even though the anisotropy which induces long-range order is very much smaller than $k_B T$ at such temperatures.

It is of great interest to inquire if one may construct an improved theoretical description, valid at temperatures higher than the domain of simple spin wave theory. This is a classical problem in the theory of magnetism in three dimensions as well; spin wave theory works impressively well at low temperatures for a wide range of magnetically ordered materials, and modern theoretical methods such as the renormalization group approach allows one to discuss in fully quantitative terms the near vicinity of the critical temperature. There is, as far as we know, no generally accepted means of constructing a quantitative and rigorous theory in the intermediate temperature regime, in general.

One may extend spin wave theory by generating a description of interactions between spin waves in the model Hamiltonian, then use these as a basis for generating corrections to the low temperature limit forms provided by spin wave theory. When this is done for a three-dimensional array of exchange coupled spins interacting also by exchange interactions, in the end the leading corrections allow one to describe the system through use of the spin wave dispersion relation applicable at low temperatures. The temperature dependent parameters (magnetization, exchange stiffness) replace those applicable at low temperatures [3.18].

Erickson has carried out such an analysis for quasi-two-dimensional films described by the model Hamiltonian considered here; the results and their implications are described in [3.11]. The results are summarized most easily if we ignore the terms linear in wave vector in the spin wave dispersion relation introduced by the dipolar interactions. These generally are quite unimportant, unless one is in or near the parameter regime where the state of uniform magnetization is unstable. The conclusion of *Erickson*'s study is that at finite temperature, the spin waves remain described by (39a,b,c), provided the effective anisotropy field $\tilde{H}_A$ is replaced by a renormalized effective anisotropy field $\tilde{H}_A(T) = \tilde{H}_A + \Delta\tilde{H}_A(T)$, where the temperature correction $\Delta\tilde{H}_A(T)$ has the form

$$\frac{\Delta\tilde{H}_A(T)}{\tilde{H}_A} = -2\Delta(T), \qquad (3.20)$$

3.4 Beyond Spin Wave Theory: The Intermediate Temperature Regime

where $\Delta(T)$ measures the deviation of the reduced film magnetization from unity, at low temperature, as in (3.12a), and also (3.19). It would be of great interest to see measurements of both the magnetization of ultrathin films, and the temperature variation of the effective anisotropy field to see if the factor of two expected from theory, and displayed in (3.20), is evident in the data. In the dispersion relation for spin waves applicable at finite temperatures, the exchange constant A is also renormalized, so we have $A(T) = A + \Delta A(T)$, where

$$\frac{\Delta A(T)}{A} = -\frac{a_0^2}{4NS}\sum_k k^2 \frac{A_3(k)}{\Omega_0(k)} n(k). \tag{3.21}$$

The quantity $A(k)$ is defined in (3.12b); again we ignore the influence of the terms linear in wave vector on the spin wave dispersion relation. If we consider temperatures high enough that the Zeeman and anisotropy contributions to the dispersion relation may be ignored (we use (3.13) to describe the thermally excited spin waves), then one finds that

$$\frac{\Delta A(T)}{A} = -\frac{a_0^4}{32\pi S}\left(\frac{k_B T}{\hbar A}\right)^2 \zeta(2), \tag{3.22}$$

where $\zeta(2)$ is the Riemann zeta function of argument 2.

These results allow us to extend our discussion of the nature of spin waves, and the material parameters on which the spin wave dispersion relation is based, to temperatures beyond the low temperature regime. The results have implications for the very interesting observations of *Pappas* et al. [3.17], who found the magnetization in an ultrathin film to rotate from perpendicular to parallel to the surface, with increasing temperature. These questions are discussed in [3.11].

We conclude this section with remarks on an alternate scheme for the discussion of the temperature variation of material parameters in ultrathin ferromagnetic films. This is a renormalization group approach described by *Pescia* and *Pokrovsky* [3.19]. It is the view of this author that this scheme recently argued [3.20] to apply quantitatively to ultrathin films of Co on Cu(1 0 0), is based on rather unphysical assumptions. In the end, this scheme overestimates the finite temperature corrections to the $T=0$ material parameters.

A difficulty with the renormalization group scheme used in [3.19, 20] is that it is based on the application of classical statistical mechanics to the spin system. In fact, the ultrathin films discussed in this chapter are quantum dominated systems. We may see this by estimating the excitation energy of spin fluctuations in the system, and comparing this to $k_B T$. The excitation energy must be small compared to $k_B T$ before one may apply classical statistical mechanics to the system. If we consider a spin fluctuation of wave vector k, then a rough estimate of its energy is provided by Ak^2, where A is the spin wave exchange stiffness. We have in mind wavelengths sufficiently short that contributions from dipolar interactions and anisotropy may be ignored. Suppose we take $k = \pi/a_0$, where a_0 is the lattice constant. Then if A in our ultrathin film roughly equals that of bulk Fe, we have in temperature units, $Ak^2 = 2 \times 10^3$ K, a value very large compared to any temperature at which we want to apply our thermodynamic description of the film. Clearly, the influence of the short wavelength spin fluctuations must be described within the framework of a quantum theory. One may inquire what fraction of the total spectrum of spin fluctuations can be described by classical statistics; we call this fraction f. We may estimate f by defining the wave vector k_T of a spin wave whose energy is equal to $k_B T$. Then we have $f = (k_T a_0)^2$ as a rough estimate. At room temperature, one estimates $f = 10^{-2}$, so in fact only a small fraction of modes very near the zone center can be described classically. We thus indeed have a quantum dominated system.

Now if the renormalization of the various physical parameters is controlled only by the long wavelength, classically describable modes, then a quantum description would not be required, despite the above comments. However, it is the case that the short wavelength, high energy fluctuations enter importantly. As an example, consider renormalization of the effective anisotropy field, as stated in (3.20). A description based on classical mechanics may be obtained by replacing the Bose–Einstein function by its classical limit, so we have

$$\frac{\Delta \tilde{H}_A}{\tilde{H}_A} = -\frac{2k_B T}{NS} \sum_k \frac{A_3(k)}{\Omega_0(k)^2} . \tag{3.23a}$$

We apply this to a film with easy axis normal to the surface, in zero external magnetic field. The spin wave dispersion relation is given by (3.15), and (3.23) then becomes

$$\frac{\Delta \tilde{H}_A}{\tilde{H}_A} = -\frac{A_c k_B T}{\pi S} \int_0^{\pi/a_0} \frac{dk\, k}{\tilde{H}_A + Ak^2} , \tag{3.23b}$$

where A_c is the area of the unit cell. The integral on wave vector in (3.23b) diverges at large wave vectors; it is thus necessary to introduce a short wavelength cutoff for the renormalization effect to remain finite. The short wavelength fluctuations thus enter very importantly; in our earlier discussion, the Bose–Einstein function provides a natural cutoff. In the classical theory, we must proceed by introducing a short wavelength cutoff, which may be taken to be π/a_0. Then we find

$$\frac{\Delta \tilde{H}_A}{\tilde{H}_A} = -\frac{A_c k_B T}{\pi A S} \ln\left[\frac{\pi \xi}{a_0}\right], \tag{3.24}$$

where $\xi = (A/\tilde{H}_A)^{1/2}$ is the correlation length that controls the scale of spin correlations in the system. This length has entered an earlier discussion on the nature of spin correlations in ultrathin films [3.9].

The result in (3.24) is virtually identical to that obtained by *Pescia* and *Pokrovksy* [3.19]. From the reasoning that leads to it, we see that the influence of short wavelength spin fluctuations is overestimated, and in fact the renormalization of material parameters is more modest than these authors

3.4 Beyond Spin Wave Theory: The Intermediate Temperature Regime

suggest. Correct results in the limit of low temperatures is provided by spin wave theory supplemented by interaction effects, though extension of such results to higher temperatures is problematical, as mentioned earlier. When this is done, the right hand side of (3.24) is replaced by $-(A_c k_B T/2\pi AS)\ln(k_B T/\tilde{H}_A)$ [3.11].

In [3.19], the temperature renormalization of the exchange constant is also discussed. Here the classical theory produces a rather substantial overestimate of the effect. *Pescia* and *Pokrovsky* find that $\Delta A(T)/A$ follows a temperature dependence identical to the reduced magnetization, given in classical statistical mechanics by half of the right hand side of (3.24). This is very different behavior indeed from that provided by spin wave theory, as displayed in our (3.21, 22). In spin wave theory, when the temperature renormalization of the exchange is generated, there are two contributions which provide the first low temperature correction. These may be described in language appropriate to a gas of interacting bosons: one has the direct or Hartree contribution, and an exchange contribution. These partially cancel against each other, to produce in the end a correction such as that displayed in (3.21), which scales as the energy stored in the gas of thermally excited spin waves, rather than the magnetization. The factor of $(a_0 k)^2$ in (3.21) is the "residue" that remains after the partial cancellation. If one includes only the Hartree term, and ignores the exchange, then the renormalized exchange constant is (incorrectly) found to scale with the magnetization, not the energy density of the thermal spin waves. We argue that the classically based theory of (3.19) fails to include the exchange, and thus overestimates renormalization of the exchange.

Nearly three decades ago, the issue of how to correctly renormalize the effective exchange constants, or spin wave excitation energies in the 3D Heisenberg ferromagnet was a topic of active discussion. A Green's function decoupling scheme associated with Tyablikov [3.21], used widely by various authors to extend spin wave theory to intermediate temperatures, yielded effective exchange constants which scale as the magnetization, while in fact the earlier exact treatment of *Dyson* [3.22] showed that the effective exchange scales as the energy density of thermal spin waves, as in (3.21). We review this issue in the Appendix, and trace the issue to the cancellation between Hartree and direct contributions, as discussed in the previos paragraphs.

It would be of great interest to see simple spin wave theory extended beyond the low temperature limiting form summarized in the previous section, within a quantum theoretic scheme. As long as the ordering temperature of the ultrathin quasi-two-dimensional film is small compared to the mean field ordering temperature, a quantum theory will be necessary. It should be noted that arguments virtually identical to those advanced here have been put forward earlier by *Chakravarty* et al. [3.23] in their analysis of the dynamical properties of two dimensional Heisenberg ferromagnets.

Note Added in Proof: Recently *Wang* and *Mills* have examined spin wave interaction effects more completely, to find higher order terms modify Eq. (3.22) importantly at low temperatures. See R.W. Wang and D.L. Mills, Phys. Rev. B **48**, 3792 (1993)

3.5 The Transition Temperature of Ultrathin Films

Quite clearly, from the discussion above, we see that the transition temperature of ultrathin ferromagnetic films is controlled by the combination of anisotropy and dipolar interactions. If the only interaction between the spins is the isotropic exchange interaction, then the Mermin–Wagner theorem assures us that there is no long-range order present at finite temperatures. Symmetry breaking interactions such as single ion anisotropy, or dipolar interactions thus control the transition temperature.

Bander and *Mills* [3.24] by means of a renormalization group method, have studied the nature of the phase transition in a monolayer film in which the spins are coupled by isotropic exchange interactions, and single ion anisotropy of the form given in (3.8) above. From this analysis, an analytic expression for the transition temperature is obtained. For spins of length $S = 1$, the transition temperature T_2 of the film is given by (correcting an error of a factor of two in the original publication [3.25])

$$T_2 = 2T_3/\ln(\pi^2 J/K), \tag{3.25}$$

where T_3 is the transition of the three-dimensional, isotropic Heisenberg ferromagnet. Furthermore, the analysis in (3.24) shows the character of the phase transition should be Ising-like, i.e. the critical exponents are predicted to be those characteristic of the two dimensional Ising model.

There is a message in (3.25) which echoes that which emerged from our analysis of the ultrathin film in the spin wave regime. First, as just noted, the Mermin–Wagner theorem tells us that there is no long-ranged order in the isotropic two dimensional Heisenberg model in two dimensions. At finite temperatures, very large amplitude spin fluctuations of long wavelength break up the long-range order. However, we saw that a very modest amount of anisotropy suppresses these fluctuations efficiently, to the point where, in the view of the author, spin wave theory can be applied over a wide temperature range. The same message is contained in (3.25). The transition temperature indeed vanishes in the limit of vanishing anisotropy, but since the dependence on K is logarithmic, even when the anisotropy strength K is very small, T_2 can be a very substantial fraction of T_3. Suppose, for example, that $K/J = 10^{-3}$. Then (3.25) gives $T_2 = 0.22T_3$! A very tiny amount of anisotropy drives the transition temperature up to almost a quarter of that realized in three dimensions.

As remarked above, the treatment of (3.24) predicts that the critical exponents in the two-dimensional films assume Ising values. From magneto-optical studies of ultrathin films, *Bader* and his collaborators [3.26] have inferred the critical exponent which describes the temperature dependence of the order parameter as the transition temperature is approached from below, to find values quite close to the Ising limit.

One may inquire how many layers of material are required before the transition temperature of the film rises to its full three-dimensional value. This issue was addressed by *Erickson* and *Mills* within the framework of a classical Monte Carlo study of Heisenberg films, with spins in one surface subject to uniaxial anisotropy such as that given in (3.8). Only very few layers are required to pass from two- to three-dimensional behavior according to this study. For example, when $K/J = 0.1$, (3.25) gives $T_2 = 0.6 T_3$. By the time six layers were present in the simulation, it was found that $T_2 = 0.94 T_3$, which is close to T_3 itself. In our exploration of the spin wave regime, we argued above that here the transition to quasi-three-dimensional behavior will occur by the time only a few layers are present, for parameters characteristic of the ferromagnetic transition metals. The Monte Carlo studies of the thickness dependence of the transition temperature thus reach a conclusion compatible with that from our exploration of the low temperature regime.

The above discussion applies to an ultrathin film with the easy axis perpendicular to the surface. It is interesting to inquire about the nature of the transition for the in-plane case. Here, in our discussion of the spin wave regime, we saw that even in the absence of single ion anisotropy, the dipolar interactions necessarily present stabilize the ferromagnetically aligned in-plane state, within the spin wave picture. While we know of no study of the phase transition in a two-dimensional array of dipole and exchange coupled spins, very interesting data and comments are found in a paper by *Bozler* and his collaborators [3.27]. These authors have studied a very different system than those we have discussed in this chapter. They examine the magnetic properties of liquid He^3 in contact with a graphite substrate. For many years, it has been known that a very thin layer of He^3 spins very near the substrate exhibit a magnetic susceptibility very strongly enhanced over that found in the liquid. The authors in [3.26] have strong experimental evidence that at ultralow temperatures, long-range order is established in the He^3 surface layers. This is perhaps surprising because on physical grounds one expects the exchange couplings between He^3 atoms within the layer to be quite isotropic. These authors propose that the (very weak) nuclear dipole–dipole interactions are responsible for inducing long-range order. Through use of a simple "hand waving" physical argument, these authors arrive at an expression very similar in structure to (3.25), with the anisotropy constant K replaced by $4\pi M_0$, which serves as a measure of the strength of the dipole-dipole coupling. They argue, as we do here, that even though in the He^3 case, $4\pi M_0$ is orders of magnitude smaller than kT, the weak symmetry breaking interactions strongly suppresses the large amplitude fluctuations expected for the perfectly isotropic two-dimensional Heisenberg model, and lead to the onset of long-range order at fairly high temperatures.

3.6 Concluding Remarks

One message that emerges from the discussion in this chapter is the central role played by both anisotropy and dipolar couplings between spins, in the thermodynamic properties of two-dimensional magnetic films. This is so, as we have seen, even though these couplings, when expressed as a contribution to the energy per spin of the system, have energies very small compared to $k_B T$ for temperatures above a few Kelvin. Such anisotropic terms in the spin Hamiltonian suppress the large amplitude, long wavelength spin fluctuations which the Mermin–Wagner theorem tells us destroy long-range order in the two-dimensional Heisenberg ferromagnet with isotropic exchange coupling only. We have argued here that the fluctuations are suppressed also at finite temperatures, with the consequence that spin wave theory supplemented with corrections from spin wave interactions suffices to describe the ordered state over a rather wide range of temperatures, very much as in three-dimensional magnetism.

This situation is still very different than that which is found in three dimensions, however. Save for materials with ordering temperatures in the range of a few Kelvin or less, the spin excitations that contribute to the thermodynamic properties of the film have energies dominated by exchange, above the few Kelvin region. In three dimensions, the long wavelength modes whose excitation energy is influenced importantly by anisotropy or dipolar couplings occupy such a small fraction of the relevant phase space that their contribution is quite negligible in this temperature region. Theories of the thermodynamics of three-dimensional magnets thus place primary focus on only the exchange portion of the Hamiltonian. Experiments which probe spin waves in the ordered state such as ferromagnetic resonance spectroscopy and light scattering excite modes with wavelengths very long compared to a lattice constant. These waves have energies influenced only very modestly by exchange, and more importantly by the combination of anisotropy and dipolar coupling. A consequence is that such experiments offer us our primary means of obtaining quantitative information about such terms in the Hamiltonian. However, once again such modes have little influence on the thermodynamic properties of the material, save at very low temperatures. In two dimensions, as we have seen the combination of dipolar couplings and anisotropy enter in a central manner at all temperatures.

Our discussion was based entirely on the notion that the basic spin Hamiltonian consists entirely of isotropic exchange couplings, supplemented by single site anisotropy and dipolar terms. We know very little about the nature of the effective exchange in ultrathin transition metal films at present. It is entirely possible there is an anisotropic, layer dependent contributions to the exchange. Of course, as noted in our early remarks, ultimately the materials of central interest to the present chapter are itinerant electron systems. Our use of a localized spin Hamiltonian is thus a phenomenological procedure, and improved descriptions of the basic physcis may provide important modifications to the above conclusions. These considerations ensure the topic of this chapter will remain lively and interesting in the near future.

Appendix: Temperature Renormalization of the Exchange in Heisenberg Ferromagnets; The Role of Quantum Mechanical Exchange

We consider an array of ferromagnetically aligned spins on a lattice with sites designated be l. We then examine the Green's function defined as

$$G(l - l'; t) = i\theta(t)\langle [S_-(l, t), S_+(l', 0)]\rangle, \quad (3.26)$$

where $S_+ = S_x + iS_y$ and $S_- = S_x - iS_y$ are the spin raising and lowering operators. In (3.26), these are placed in the Heisenberg representation. The spin wave frequencies are found from the appropriate Fourier transform (3.26) with respect to time and with respect to $l - l'$. The Hamiltonian is the simple Heisenberg exchange Hamiltonian, which we write as

$$H = -\frac{1}{2}\sum_{l,l'} J(l - l') S(l) \cdot S(l'). \quad (3.27)$$

Note that $J(0) \equiv 0$.

We shall study the temperature variation of the spin wave frequencies of the system, and use these to define effective temperature dependent exchange couplings. In units where $\hbar = 1$, a short calculation gives

$$i\frac{\partial}{\partial t}G(l - l'; t) = 2\langle S_z\rangle \delta(t)\delta_{ll'} + i\theta(t)\sum_{l''} J(l - l'')\langle [S_z(l, t)S_-(l'', t), S_+(l', 0)]\rangle$$
$$- i\theta(t)\sum_{l''} J(l - l'')\langle [S_z(l'', t)S_-(l, t), S_+(l'; 0)]\rangle. \quad (3.28)$$

As is usual, the equation of motion for the desired Green's function leads us to a more complex object. We proceed by writing

$$S_z(l, t) = S - \Delta(l, t), \quad (3.29)$$

which introduces the operator $\Delta(l, t)$. Then if N is the number of sites in the sample, we let

$$G(l - l'; t) = \frac{1}{N}\sum_q G(q; t)e^{iq \cdot (l - l')}. \quad (3.30)$$

One then has

$$\left[i\frac{\partial}{\partial t} + \omega(q)\right]G(q; t) = 2\langle S_z\rangle\delta(t) + i\theta(t)\sum_{ll''} J(l - l'')e^{-iq\cdot(l - l')}$$
$$\times \{\langle[\Delta(l'', t)S_-(l, t), S_+(l'; 0)]\rangle$$
$$- \langle[\Delta(l, t)S_-(l'', t), S_+(l', 0)]\rangle\}, \quad (3.31)$$

where

$$\omega(q) = S[\mathscr{J}(0) - \mathscr{J}(q)] \quad (3.32a)$$

and

$$\mathscr{J}(\mathbf{q}) = \sum_\delta J(\boldsymbol{\delta})e^{i\mathbf{q}\cdot\boldsymbol{\delta}}. \tag{3.32b}$$

At very low temperatures, the spins are very nearly aligned along the z direction, and $\langle S_z \rangle$ is very close to S. The influence of the operator $\Delta(l,t)$, which describes that of fluctuations in the z components of the spins, may be overlooked, and the terms involving $\Delta(l,t)$ on the right-hand side of (3.31) may be set to zero. The poles of the time Fourier transform of $G(\mathbf{q},t)$ then occur at the frequency $\omega(\mathbf{q})$, which we recognize as the well known formula for the spin wave frequency of the Heisenberg ferromagnet.

As the temperature is increased, the influence of the terms involving $\Delta(l,t)$ is felt, and the spin waves are modified. In general, one must introduce an approximation scheme to proceed. A commonly used scheme, due initially to Tyabliakov [3.21], if referred to often as the random phase approximation (RPA). One makes the replacements

$$\Delta(l'',t)S_-(l,t) \rightarrow \langle \Delta(l'',t) \rangle S_-(l,t) \equiv (S - \langle S_z \rangle) S_-(l,t) \tag{3.33a}$$

and

$$\Delta(l,t)S_-(l'',t) \rightarrow \langle \Delta(l,t) \rangle S_-(l'',t) \equiv (S - \langle S_z \rangle) S_-(l'',t). \tag{3.33b}$$

A short calculation then gives

$$\left[i\frac{\partial}{\partial t} + \omega_{\mathrm{RPA}}(\mathbf{q})\right] G(\mathbf{q},t) = 2\langle S_z \rangle \delta(t), \tag{3.34}$$

where

$$\omega_{\mathrm{RPA}}(\mathbf{q}) = \langle S_z \rangle [\mathscr{J}(0) - \mathscr{J}(\mathbf{q})]. \tag{3.35}$$

We now have temperature dependent spin wave frequencies. In this picture, at finite temperatures, each exchange interaction $J(\boldsymbol{\delta})$ in the system is renormalized by the reduced magnetization. That is, one calculates the spin wave frequency by renormalizing each exchange coupling by the rule

$$J(\boldsymbol{\delta}) \rightarrow \frac{\langle S_z \rangle}{S} J(\boldsymbol{\delta}). \tag{3.36}$$

A more careful discussion, presented next, shows that renormalizing the exchange constants as in (3.36) provides the incorrect result for the leading low temperature correction to the spin wave frequency. When applied to the three-dimensional Heisenberg ferromagnet, the RPA then shows the exchange has the same $T^{3/2}$ dependence on temperature as the magnetization itself. The proper answer is a $T^{5/2}$ variation. Furthermore, as the temperature is raised toward the Curie temperature, the RPA predicts that all spin wave frequencies vanish; this result is in fact quite unphysical. Appreciable short range order is present near the Curie temperature, and there is a finite "restoring force" for short wavelength spin fluctuations.

Appendix

While no satisfactory simple scheme gives a proper description of spin fluctuations at all temperatures, the correct behavior at low temperatures follows by noting that we may use the Holstein–Primakoff transformation here. One has the boson annihilation and creation operators $a(l,t)$ and $a^+(l,t)$; the operator $\Delta(l,t) = a^+(l,t)a(l,t)$, and the first correction to the Green's function is found by using this form for $\Delta(l,t)$, then replacing $S_-(l,t)$ by $(2S)^{1/2}a^+(l,t)$. Noting that the only non-zero terms involving Δ on the right hand side of (3.31) have $l'' \neq l$, we proceed as follows to generate the leading low temperature corrections:

$$\Delta(l'',t)S_-(l,t) \cong \sqrt{2S}\,a^+(l,t)a^+(l'',t)a(l'',t)$$

$$\cong \sqrt{2S}\,\{\langle a^+(l'')a(l'')\rangle a^+(l,t) + \langle a^+(l)a(l'')\rangle a^+(l'',t)\}$$

$$\equiv \langle a^+(l'')a(l'')\rangle S_-(l,t) + \langle a^+(l)a(l'')\rangle S_-(l'',t). \tag{3.37}$$

The expectation values produced by the decoupling scheme are independent of time; $\langle a^+(l'')a(l'')\rangle$ is independent of l'' as well, but for convenience we retain reference to the lattice site. We apply a similar approximation to the combination $\Delta(l,t)S_-(l'',t)$.

At low temperatures, the spin waves may be viewed as a weakly interacting gas of bosons. Equation (3.37) is the Hartree–Fock approximation applied to this boson gas. The first term is the direct term, and the second term is the exchange term. Both are the same order of magnitude at low temperatures, as one sees by letting l'' approach l in (3.37). If we ignore the second term (the exchange term) in (3.37), and proceed in the same manner with the term $\Delta(l,t)S_-(l'',t)$, then one easily sees that we reproduce the (incorrect) results of the RPA.

When (3.37) and its companion are introduced into (3.31), one finds

$$\left[i\frac{\partial}{\partial t} + \omega(q)\right]G(q,t) = 2\langle S_z\rangle\delta(t) - \sum_{l,l'}S\Delta J(l-l')e^{-iq\cdot(l-l')}$$

$$\times \{G(l-l';t) - G(l''-l';t)\}, \tag{3.38}$$

where $\Delta J(l-l')$ is a temperature dependent contribution to the effective exchange coupling between a spin at site l, and that at l'. Setting $\boldsymbol{\delta} = l - l'$, one obtains

$$\Delta J(\boldsymbol{\delta}) = -\frac{J(\boldsymbol{\delta})}{NS}\sum_k[1 - \cos(k\cdot\boldsymbol{\delta})]\bar{n}(k) \tag{3.39}$$

where $\bar{n}(k) = [\exp(\beta\omega(k)) - 1]^{-1}$ is the Bose–Einstein function, which gives the number of thermally excited spin waves of wave vector k. The term in $\cos(k\cdot\boldsymbol{\delta})$ in (3.39) is the exchange term.

It is a short exercise to show from (3.13) that we again have spin waves with renormalized frequencies, calculated from (3.32a, b). However, now $J(\boldsymbol{\delta})$ is replaced by

$$J(\boldsymbol{\delta}) + \Delta J(\boldsymbol{\delta}) = (1-r)J(\boldsymbol{\delta}), \tag{3.40}$$

with

$$r = \frac{1}{NS}\sum_{k}(1 - \cos[\boldsymbol{k}\cdot\boldsymbol{\delta}])\bar{n}(\boldsymbol{k}). \tag{3.41}$$

At low temperatures, the wavelength of spin waves is long compared to the range of the exchange couplings, so we may write $1 - \cos(\boldsymbol{k}\cdot\boldsymbol{\delta}) \cong (\boldsymbol{k}\cdot\boldsymbol{\delta})^2/2$. If we consider the three-dimensional Heisenberg ferromagnet with an isotropic dispersion relation, then the leading low temperature behavior is calculated by choosing $\omega(\boldsymbol{k}) = Ak^2$. One finds

$$r = \frac{V_c \delta^2}{24\pi^2}\zeta\left(\frac{5}{2}\right)\Gamma\left(\frac{5}{2}\right)\left(\frac{k_B T}{A}\right)^{5/2}, \tag{3.42}$$

with V_c the volume of the unit cell, and $\zeta(5/2)$, $\Gamma(5/2)$ the Riemann zeta function and gamma function of argument cited.

We could choose to treat our gas of weakly interacting bosons by diagrammatic perturbation theory, after generating the terms quartic in boson annihilation and creation operators, in the spin Hamiltonian. Then the lowest order corrections to the spin wave frequencies are generated by the two self-energy diagrams given in Fig. (3.1). The first, which is the Hartree diagram, generates a low temperature result which coincides with that provided by the RPA. The exchange diagram, when included, generates the correct results summarized in (3.40, 41).

From this discussion, we can appreciate that exchange diagrams, whose origin lies in quantum mechanical nature of spin fluctuations in the Heisenberg

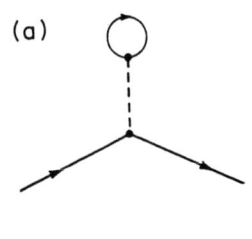

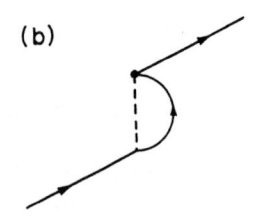

Fig. 3.1a, b. The two self-energy diagrams which, when combined, yield renormalization of spin wave energies consistent with the results displayed in (3.40) and (3.41). We have (a) the direct, or Hartree diagram, and (b) the exchange diagram

ferromagnet, must be included in any theory of the temperature dependence of material parameters below the ordering temperature. Such features are missing from theories based on the application of classical statistical mechanics, such as that described in [3.19].

References

3.1 M. Pomerantz: Surf. Sci. **142**, 556 (1984)
3.2 For a general discussion of the magnetism of itinerant electron system: C. Herring: *Magnetism*, ed. by G. Rado, H. Suhl, Vol. 4 (Academic, New York 1966)
3.3 C. Herring, C. Kittel: Phys. Rev. **81**, 869 (1951)
3.4 See the article by K.W.H. Stevens: *Magnetism*, ed. by G. Rado and H. Suhl, Vol. 1 (Academic, New York 1963) p. 1
3.5 See Table 6-2, page 185 of M. Tinkham: *Group Theory and Quantum Mechanics* (McGraw-Hill, New York 1964) p. 185
3.6 M. Przybylski, I. Kaufmann, U. Gradmann: Phys. Rev. B **40**, 8631 (1989)
3.7 R.P. Erickson, D.L. Mills: Phys. Rev. B **43**, 10715 (1991)
3.8 D.L. Mills: Phys. Rev. B **39**, 12306 (1989)
3.9 R.P. Erickson, D.L. Mills: Phys. Rev. B **44**, 11825 (1991)
3.10 J.R. Dutcher, J.F. Cochran, I. Jacob, W.G. Egelhoff, Jr.: Phys. Rev. B **39**, 10430 (1989); J.R. Dutcher, B. Heinrich, J.F. Cochran: J. Appl. Phys. **63**, 3464 (1988)
3.11 R.P. Erickson, D.L. Mills: preprint entilted "Magnetic Instabilities in Ultra Thin Ferromagnets, to be published; R.P. Erickson: Title, PLD. Press University of California, Irvine (1991) (unpublished)
3.12 Y. Yafet, J. Kwo, E.M. Gyorgy: Phys. Rev. B **33**, 6519 (1986)
3.13 S.V. Maleev: Sov. Phys. **43**, 1240 (1976)
3.14 A summary of experimental determinations of the various anisotropy constants of ultrathin films, by means of light scattering and ferromagnetic resonance studies of ultrathin films may be found in Chap. 3, Vol. 2
3.15 Y. Yafet, E.M. Gyorgy: Phys. Rev. B **38**, 9145 (1988)
3.16 R.A. Allenspach, M. Stampanoni, A. Bischof: Phys. Rev. Lett. **65**, 3344 (1990)
3.17 D.P. Pappas, K. Kämper, H. Hopster: Phys. Rev. Lett. **64**, 3179 (1990)
3.18 Talat S. Raman, D.L. Mills: Phys. Rev. B **20**, 1173 (1979)
3.19 D. Pescia, V.L. Pokrovsky: Phys. Rev. Lett. **65**, 2599 (1990)
3.20 M.G. Pini, A. Rettori, D. Pescia, N. Majlis, S. Selzer: Phys. Rev. B (to be published)
3.21 S.V. Tyablikov: Ukr. Mat. Zh. **11**, 287 (1959)
3.22 F.J. Dyson: Phys. Rev. **102**, 1271 (1956); Phys. Rev. **102**, 1230 (1956)
3.23 S. Chakravarty, B. Halperin, D.R. Nelson: Phys. Rev. Lett. **60**, 1057 (1988)
3.24 M. Bander, D.L. Mills: Phys. Rev. B **38**, 12015 (1988)
3.25 R P. Erickson, D.L. Mills: Phys. Rev. B **43**, 11527 (1991)
3.26 Z.Q. Qiu, J. Pearson, S.D. Badel: Phys. Rev. Lett. **67**, 1646 (1991)
3.27 L.J. Friedman, A.L. Thomson, C.M. Gould, H.M. Bozler, P.B. Werchman, M.C. Cross: Phys. Rev. Lett. **62**, 1635 (1989)

4. Spin-Polarized Spectroscopies

The use of spin-polarized electron scattering techniques is discussed in this chapter. Polarized electron spectroscopies are of particular importance in the subject of surface and ultrathin film magnetism because of the surface sensitivity of these techniques. In the first section *Hopster* describes elastic scattering, spin-resolved secondary electron spectroscopy, inelastic scattering and photoemission techniques. The section surveys studies of magnetic structure, critical magnetic properties and magnetic excitations such as electron–hole pair excitations. In the second section *Siegmann* and *Kay* discuss the use of spin-polarised electron beam techniques in extracting information specifically on the magnetization as a function of field and temperature, and discuss recent experiments which test key ideas in two-dimensional magnetism. These studies relate to the theoretical discussion of thermodynamic properties in Chap. 3.

4.1 Spin-Polarized Electron Spectroscopies

H. HOPSTER

Spin-polarized electron spectroscopies offer the opportunity to simultaneously study the magnetic properties and the underlying electronic structure of surfaces and thin films. In this chapter various spin-polarized spectroscopic techniques and their application to ultrathin films are described. Polarized electron spectroscopies rely on the use of a polarized electron beam or on polarization analysis of electrons emitted from the sample. Recent developments in spin polarized electron sources and spin polarimeters will be discussed.

4.1.1 Introduction

Electron spectroscopies have played a major role in modern surface and thin film research. The principal reason for this is the short mean free path (MFP) of energetic electrons in solids which leads to an escape depth, or penetration depth, of only a few atomic layers for electron energies between 10 eV and 1 keV, which is the typical energy range used. Hence, in electron spectroscopy the properties of the topmost atomic layers are probed. The short MFP is due to strong scattering, elastic and inelastic, of energetic electrons in solids. The price for this high surface sensitivity in electron spectroscopy, compared to neutron or light scattering, for example, is that a direct quantitative interpretation of experimental results is often difficult since multiple scattering has to be taken into account.

Experimental progress over the last decade permits all of the commonly applied electron spectroscopies [4.1] to be performed in a spin–polarized version by adding an electron spin polarization analyzer or a polarized electron beam to the experiment (and adding an "SP" prefix to the corresponding acronym). The spin polarization of an ensemble of electrons (e.g., in a beam) is defined as

$$P = (N_+ - N_-)/(N_+ + N_-), \tag{4.1}$$

where N_+, N_- are the number of electrons with spin up and spin down, respectively, measured with respect to a given quantization axis [4.2]. The spin polarization P and the total intensity $N = N_+ + N_-$ are related to the spin resolved intensities N_+, N_- by

$$N_+ = N \cdot (1 + P)/2, \qquad N_- = N \cdot (1 - P)/2. \tag{4.2}$$

The natural quantization axis in a ferromagnet is given by the magnetization direction. Spin-up is, by definition, the direction of majority-spin, which is opposite to the magnetization direction since electron spin and magnetic moment are antiparallel.

Most of the development and application in polarized electron spectroscopy techniques was directed toward surfaces of single crystals. With many of the techniques reaching a certain stage of maturity the emphasis has shifted towards ultrathin films. It is obvious that all techniques are equally applicable and useful when applied to ultrathin magnetic structures. Articles on spin polarized techniques applied to surfaces can be found in [4.3, 4].

4.1.1.1 Overview

The most common excitation sources in electron spectroscopies are electrons and photons. Figure 4.1 schematically shows the electron spectrum upon bombardment of the sample with a monochromatic electron beam at primary energy E_p. In the different parts of the energy spectrum the spin polarization carries quite different information related to the magnetism. On a single crystal, the elastically scattered electrons are concentrated in diffraction spots. In low energy electron diffraction (LEED) studies the intensity versus energy (I/V) curves are used for detailed structural analyses, including surface relaxation and reconstruction. On a ferromagnetic surface the reflected intensities depend on the relative orientation of the spin of the incoming electron and the magnetization because of the spin dependence (exchange potential) of the scattering

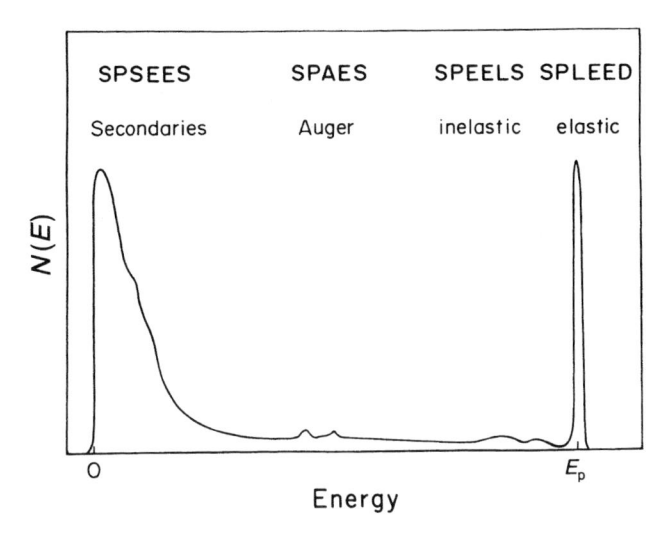

Fig. 4.1. Schematic electron energy spectrum

potential. The scattering asymmetries can be used as a sensitive probe of long-range ferromagnetic order. SPLEED asymmetry (A/V) curves can be used for magnetic structure determination.

The collective and single-particle excitations at the surface are probed in electron energy loss spectroscopy (EELS). Cross sections for inelastic scattering due to electron–hole pair creation in a ferromagnet exhibit large spin dependences due to exchange scattering. These lead to scattering asymmetries, which signal the presence of long-range magnetic order. In addition, the spin dependent energy loss spectrum reveals the exchange splitting of the electronic structure in magnetic materials. In its "complete" form, i.e. using a polarized beam combined with polarization analysis, the SPEELS experiment does not rely on the presence of magnetic long-range order. Spin exchange scattering can be detected on paramagnetic (e.g. ferromagnets above the critical temperature T_c) or antiferromagnetic surfaces.

In the next part of the spectrum, lines resulting from Auger transitions are superimposed on a smooth background due to electrons that have undergone multiple energy losses. The spin polarization of the Auger lines yields element specific magnetic information, a feature of great value in studies of alloy surfaces, adsorbates, or in the case of ferrimagnetic or antiferromagnetic coupling.

At very low energies one gets into the realm of the "true" secondary electrons. These are created mainly by excitations from the valence band. The spin polarization carries information on the average magnetization at the surface and can be used for magnetometry (Sect. 4.2). In addition, structures in intensity and polarization are related to the spin-split electronic structure above the vacuum level. When a finely focused beam of primary electrons is used for excitation and scanned across the surface one obtains a direct map of the surface magnetization. This is realized in Secondary Electron Microscopy with Polarization Analysis (SEMPA). Spatial resolutions of a few tens of nm have been achieved.

In photoemission spectroscopy the surface is irradiated with monochromatic light and the energy spectrum of the emitted electrons is recorded. One distinguishes between *UPS* (UV) and *XPS* (X-ray) photoemission spectroscopy, although with the increasing use of continuously tunable synchrotron radiation this distinction has become blurred. In the XPS regime, deeper core levels are accessible whose spin polarization yields information on the *local* magnetic environment, due to intra-atomic exchange effects. Angle resolved (UV) photoemission spectroscopy (ARPES) allows for the determination of the electronic band structure. Combined with spin analysis, the exchange-split band states, including surface and interface induced states, can be observed.

Inverse photoemission spectroscopy (IPES) is the time-reversed process of photoemission. An incoming electron drops into an empty state above the Fermi energy E_F by emitting a photon. Inverse photoemission probes the band structure above E_F. Use of a polarized electron beam therefore probes the spin dependent band structure (e.g., exchange splittings) above E_F.

In all polarized spectroscopies one has to distinguish between effects due to long-range and short-range magnetic order. The measured spin polarization in an experiment, e.g., in photoemission, is an average over the area probed (usually macroscopic). On the other hand, in the energy spectra, features such as spin-split peaks in photoemission are due to short-range order, independent of the macroscopic sample magnetization.

4.1.1.2 Information Depth

Energetic electrons in a solid are strongly scattered. This leads to an intensity attenuation usually described by an exponential decay $\exp(-z/\lambda)$ (z is path length and λ the attenuation length) [4.5]. Experimental electron spectra are therefore often described phenomenologically by an exponentially weighted superposition of contributions from layers normal to the surface. The attenuation length is often equated with the *inelastic* mean free path (MFP) taken from the so-called "universal curve" [4.6], although elastic scattering may play a role. In ferromagnetic materials the scattering processes are expected, at least in principle, to be spin dependent. Hence, spin dependent MFPs and attenuation lengths are expected. The extent to which measured spin polarization values are influenced by spin dependent scattering processes has been of concern since the early days of spin polarized electron spectroscopies [4.7, 8].

In recent experiments by *Pappas* et al. [4.9] the question of spin dependent electron attenuation is addressed using a new experimental approach. These authors performed spin–polarized photoemission spectroscopy on a Cu (1 0 0) sample with ultrathin ferromagnetic epitaxial Fe overlayers. The electrons emitted from Cu 3d states (originally unpolarized) can be clearly distinguished

4.1 Spin-Polarized Electron Spectroscopies

from the Fe 3d states by their spectral behavior. It was found that the electrons emitted from the Cu become polarized as they traverse the ferromagnetic film. From the measured polarization and film thickness and the intensity attenuation compared to the bare Cu, the spin dependent attenuation length can be directly determined. These experiments were performed using synchrotron radiation so that electron energy can be continuously varied. Appreciable spin polarizations (up to 20%) are found at low energies while at higher energies ($>40\,\text{eV}$) the attenuation becomes spin independent. At low energies the attenuation for spin-down electrons is stronger than for spin-up electrons. This leads to preferential emission of spin-up electrons. The measured attenuation lengths obtained for the fcc Fe layers used are shown in Fig. 4.2. The inset shows the spin averaged attenuation length and for comparison also the universal curve. The values from the universal curve are too large at low energies while at higher energies the agreement becomes quite good.

The spin dependent attenuation was attributed to preferential inelastic scattering of spin-down electrons. However, it should be noted that an alternative interpretation in terms of spin dependent *elastic* scattering has been given by *Gokhale* and *Mills* [4.10]. The experimental data available so far cannot rule out either model and both processes, elastic and inelastic, may play a role. However, the polarization enhancement of secondary electrons (Sect. 4.1.3.1) suggests that a very general mechanism independent of the structure is at work. If SPLEED effects dominate the spin dependence of the electron attenuation then large variations with energy and also direction should be expected in single crystal samples. More experimental and theoretical investigations are needed to resolve this question.

A short probing depth at very low energies (zero kinetic energy) was first reported by *Abraham* and *Hopster* [4.11] in measurements of the spin polarization of secondary electrons from Ni (1 1 0). It was found that the measured

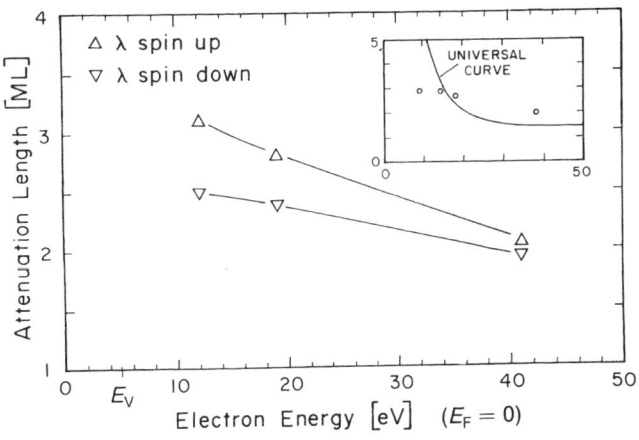

Fig. 4.2. Spin dependent and spin averaged (inset) attenuation length in ferromagnetic fcc Fe layers [4.9]

polarization as a function of temperature deviated significantly from the bulk magnetization and followed a curve more characteristic of a surface magnetization. A value of 3–4 layers for the effective magnetic information depth was deduced. Secondary electron experiments measuring the polarization as a function of film thickness [4.12, 13] agree with this value. *Zhang* et al. [4.14] addressed the probing depth at higher electron energies (30 eV) by SPEELS and found a value of only one layer, consistent with the universal curve since in SPEELS the electrons have to go into and out of the surface, unlike secondary electrons. Also the effective path length inside the solid is increased by the non-normal incidence and exit angles of the electrons in SPEELS. In general, one should expect a higher surface sensitivity for scattering methods, such as SPLEED and SPEELS, compared to emission or injection methods, like photoemission and inverse photoemission, because of this effective path difference. However, no systematic comparison has been made so far.

The probing depth in magnetic materials other than transition metals has barely been touched upon. For Gd films, *Paul* [4.15] arrives at values similar to the ones for transition metals, although the mechanisms leading to the spin polarization spectrum seem more complicated due to the presence of the localized 4f levels. From a practical standpoint, the short probing depth even at very low kinetic energies makes spin polarized electron spectroscopies, e.g., secondary electrons, ideally suited for the study of ultrathin films.

4.1.2 Instrumentation

Spin–polarized electron spectroscopies require a polarized electron beam or a spin polarization analyzer, or both. A polarization detector is usually coupled to an electrostatic energy analyzer. Spin polarization analysis is based on a scattering experiment. This makes electron polarimeters intrinsically inefficient whereas polarized electron sources deliver currents comparable to unpolarized sources. Recent progress in polarimetry and source development is summarized in [4.16, 17]. Since low energy electrons probe the topmost atomic layers, ultra-high vacuum (UHV) is required as in conventional electron spectroscopy. Standard surface analysis techniques (LEED and AES) are usually incorporated into the UHV system.

4.1.2.1 Spin Polarization Detectors

Most of the electron spin polarization analyzers in use are based on the spin–orbit interaction in electron scattering. The typical detector geometry is shown in Fig. 4.3. The electron beam is scattered off of a target and two intensities are measured under equivalent scattering angles. The detector is sensitive to the spin polarization component normal to the scattering plane. For a polarized beam the intensities to the "left" and "right", I_L and I_R, are different

4.1 Spin-Polarized Electron Spectroscopies

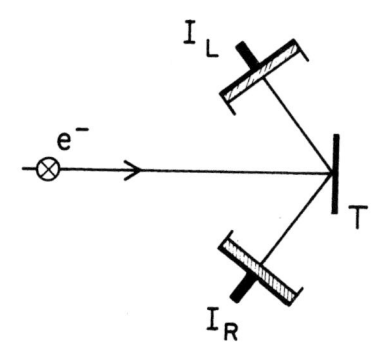

Fig. 4.3. Scattering geometry in an electron spin polarimeter

due to the spin–orbit interaction. The spin polarization P is given in terms of the intensities as

$$P = 1/S_{\text{eff}}(I_L - I_R)/(I_L + I_R). \tag{4.3}$$

Here, S_{eff} (effective Sherman function) determines how well the electrons are spin separated by the scattering. This calibration constant of the detector depends on the scattering parameters, target material, energy, scattering angle, angle acceptance of detectors, suppression of unwanted electrons (e.g., inelastic) [4.2] and has to be determined experimentally for every detector. Ideally, S_{eff} would be unity. In practice, S_{eff} typically ranges between 0.1 and 0.3. By adding a second pair of detectors, 90° rotated around the incoming beam axis, one can measure the other transverse polarization component. In order to measure the longitudinal component one can deflect the beam 90° electrostatically to transform the longitudinal polarization into transverse and then measure this component in a separate detector [4.18].

In terms of counting statistics, the quality of a spin detector is given by the figure of merit FM = $S_{\text{eff}}^2 * (I/I_0)$ where I/I_0 is the ratio of detected intensity, $I_L + I_R$, to the total incident intensity I_0. Typical figures of merit are on the order of 10^{-4}. For a given figure of merit the Sherman function should not be too small, because then the detector becomes too sensitive to beam deflection effects, which can lead to spurious polarization measurements [4.2].

The classical spin polarimeter is the high energy Mott detector. The electron beam is accelerated to 100 keV and scattered from a thin Au foil. This detector is quite bulky due to the high voltage. Much of the effort in detector improvements over the last decade has gone into developing smaller detectors operating at lower voltages. A driving force behind this was the desire to couple spin detectors to conventional electron energy analyzers and make the whole arrangement movable in a UHV chamber for angle dependent studies. In the medium energy Mott detector the energy has been lowered to 20–40 keV and only the scattering target is at high potential while the electron detectors are at ground potential [4.19–21].

Two low energy (100 eV) detectors based on the spin–orbit interaction have been developed. The LEED detector developed by *Kirschner* and coworkers uses the spin asymmetry of equivalent beams (the (20) beams) in elastic diffraction from a W (1 0 0) crystal [4.22]. The other detector is based on diffuse scattering from a polycrystalline Au film [4.18, 23]. Since the spin–orbit interaction increases with increasing atomic number of the target material, some improvements in S_{eff} have been achieved by using Th or U in medium energy detectors [4.24–26].

Recently, *Kisker* and collaborators [4.27, 28] have introduced a new type of detector based on spin dependent scattering from a magnetic surface. The scattering geometry is shown in Fig. 4.4. The low energy electron beam impinges on a remanently magnetized Fe surface. The specularly reflected intensity is detected by a channeltron. Inelastic electrons and secondaries are suppressed by retarding grids. To determine the polarization, one measures the intensity change upon the reversal of the detector magnetization. A figure of merit of 3.5×10^{-3} is reported, i.e. more than an order of magnitude improvement over previous detectors.

Spin detectors are usually calibrated within about a 10% relative accuracy. The Mott detector can be calibrated by using different scattering foil thicknesses and extrapolating to zero thickness [4.29], for which the Sherman function is well established (single atom scattering). With some experimental effort, calibrations within 5% can be obtained this way [4.30]. This method is not available for the low energy detectors. The LEED detector was self-calibrated by performing a double-scattering experiment [4.31]. This tedious experimental procedure can be simplified by use of a polarized beam from a GaAs source, as was shown by *Abraham* and *Hopster*, who achieved calibrations of a high energy Mott detector on the percent level [4.32]. Measuring the circular polarization of light emitted from He atoms upon electron impact reaches the same level of accuracy [4.30]. *Oro* et al. [4.26] used electrons of known polarization from polarized metastable He atoms to calibrate their medium energy Mott detector. Thus, with some experimental effort absolute spin polarization values can be determined with fairly good accuracy. However, spin detectors, because of their low efficiency, remain the weakest part in any polarized electron experiment.

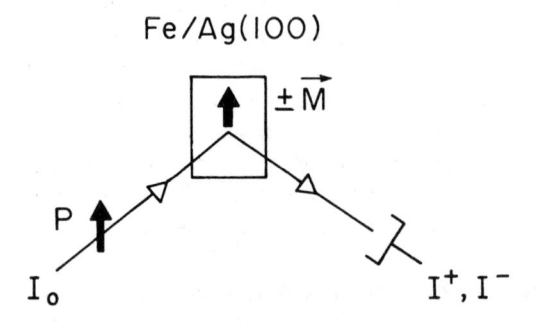

Fig. 4.4. Principle of the low-energy SPLEED detector

4.1.2.2 Spin-Polarized Electron Sources

All of the polarized sources in use today in electron spectroscopies are based on the "GaAs source", in which electrons are polarized by optical pumping with circularly polarized light of photon energy just above the optical band gap energy. The first successful operation of this source was demonstrated by *Pierce* et al. [4.33, 34]. A detailed account of the operation of the source was given by *Pierce* et al. [4.25]. An AlGaAs laser diode with a wavelength around 800 nm is used for the GaAs source while HeNe lasers can be used in other materials (GaAsP or AlGaAs) with a larger band gap matched to the HeNe wavelength [4.36–38]. A typical setup of the source is shown in Fig. 4.5. The light is circularly polarized by an electro-optic modulator (Pockels cell) operated as a quarter-wavelength plate and usually computer controlled. The light is then focused through a viewport onto the source crystal in the UHV system. The emitted electrons (longitudinal polarization) are accelerated by an extraction voltage and in most cases electrostatically deflected by 90° in order to obtain a transversely polarized beam. However, the requirement for transverse polarization is largely of historical origin when the sample magnetizations were all in–plane (on bulk crystal surfaces). For many interesting ultrathin film systems with perpendicular magnetizations, longitudinal polarizations are actually desirable.

The source material has to be cleaned in UHV by heating. The work function is then lowered by Cs and oxygen adsorption. Negative electron affinity, i.e. when the vacuum level is below the conduction band edge, can be achieved in this way. This requires heavily p-doped material (approx. 10^{19} cm^{-3}). The maximum theoretical polarization for the GaAs source is 50%

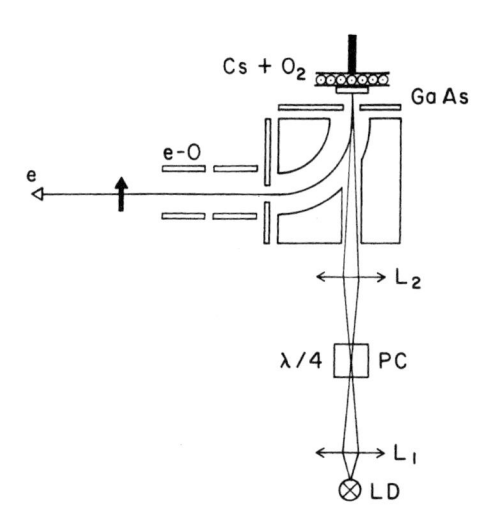

Fig. 4.5. GaAs polarized electron source; LD = laser diode, L_1, L_2 = lenses, PC = Pockels cells, e-O = electron optics

due to the cubic symmetry [4.39]. However, scattering of the excited electrons in the conduction band leads to depolarization and values obtained in the range of 25–30%, somewhat dependent on the surface conditions [4.40]. Quantum yields, i.e. number of electrons emitted per incoming photon, range up to a few percent. Thus, with semiconductor laser diodes of 100 mW continuous wave (cw) output power several hundred microamp dc currents can easily be extracted. Space charge effects may limit the maximum usable current. The polarized source is therefore comparable in intensity to unpolarized sources. The energy spread, typically on the order of 150 meV, actually offers an advantage over thermal electron guns. The lifetime of the source critically depends on the vacuum conditions in the source chamber. Usually, 1/e lifetimes of many hours are readily achieved so that a source has to be prepared at most only once a day. Repeated activation is easily achieved by cleaning and new Cs/O_2 treatment.

The efficiency (figure of merit) of polarized electron sources in terms of counting statistics is given by $P^2 * I$. Obviously it would be highly desirable to increase the polarization beyond the present values. An increase of P to close to 100% would correspond to more than an order of magnitude increase in the efficiency and therefore cut down correspondingly on the data acquisition time, providing of course that the current can be maintained at the same level. Attempts along these lines have been of rather limited success until recently. The effects of depolarization due to scattering in the conduction band can be reduced by using thin films instead of bulk samples. In this way polarizations close to the theoretical limit of 50% have been achieved [4.38, 41]. The increased polarization is achieved, however, at the expense of intensity since the films have to be thinner than the depolarization length.

With a reduction of the symmetry of the photoemitter material, polarizations of 100% would become possible due to the lifting of the degeneracy of the valence band. Attempts to use materials other than the III–V compounds have failed so far to produce usable sources due to difficulties with surface preparation [4.42]. However recently the breakthrough towards high polarization has been achieved from MBE grown strained InGaAs [4.43] and GaAs [4.44] layers and also AlGaAs–GaAs superlattices [4.45]. These sources are presently under development for use in high energy physics experiments at particle accelerators.

4.1.3 Secondary Electrons (SPSEES)

Secondary electrons are excited mainly from the valence band upon impact of energetic particles. The secondaries form a broad intensity distribution with a maximum at low energies. The secondary electron yield, i.e. the number of emitted secondaries per incident electron, shows a maximum at a few hundred eV primary energy. The maximum is due to the competing effects of excitation and escape depth of secondaries. Even though more electrons are excited with increasing primary energy fewer secondaries can escape because they are created increasingly deeper in the sample.

4.1.3.1 Polarization in Secondary Electron Spectra

Chrobok and *Hofmann* [4.46] showed that the total secondary electron yield from EuO films is indeed spin-polarized. Energy resolved measurements were then performed on 3d transition metal surfaces [4.47–49]. It was expected that the spin polarization would equal the average valence band polarization. However, the polarization was found to depend on the secondary kinetic energy with a maximum at the lowest energies. Figure 4.6 shows a spin polarization spectrum from Ni (1 1 0) excited by 400 eV primary electrons [4.49]. The polarization is enhanced at zero kinetic energy over the average valence band polarization by a factor of three (17% vs. 5.5.%) and rapidly drops with increasing kinetic energy, approaching the net valence band polarization around 10 eV. At higher energies the polarization decreases slowly, which can be explained by the increasing admixture of re-emerging un-polarized primary electrons that have undergone multiple energy losses. The enhancement is attributed to preferential inelastic scattering of spin-down electrons leading to a higher escape probability for spin-up electrons. This model was originally proposed by *Bringer* et. al. [4.50] to explain the polarization enhancement in the total UV photoyield from Ni.

Beside the strong enhancement there are additional structures in the polarization spectrum, e.g., the prominent peak around 16 eV. These structures are due to spin dependent elastic scattering (SPLEED), as shown by *Tamura* and *Feder* [4.51]. This explanation as a diffraction effect is corroborated by the face

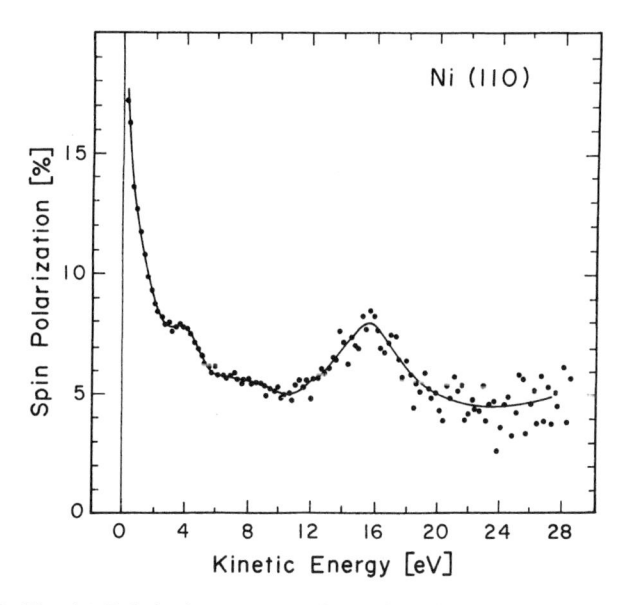

Fig. 4.6. Polarization spectrum of secondary electrons from Ni (1 1 0) [4.49]

dependence, e.g., the prominent peak at 16 eV is missing for emission from the Ni (1 0 0) surface [4.52] and by the dependence on emission angle [4.53, 54]. Similar structures are also absent in spectra from amorphous samples [4.55]. However, the enhancement is always present and must therefore be due to a general effect. The most probable explanation indeed seems to be a spin dependent mean free path [4.50, 56]. In addition, Stoner excitations increase the number of spin-up electrons at the expense of spin-down electrons [4.57]. The inelastic scattering acts as a spin filter, preferentially attenuating the down electrons. The polarization enhancement in this model is approximately given by $P = P_0 + A$ where P_0 is the valence band polarization and A is the spin asymmetry of the MFP: $A = (\lambda_+ - \lambda_-)/(\lambda_+ + \lambda_-)$. For Fe P_0 is 25%. In the overlayer experiments (Sect. 4.1.1.2) A was found to be on the order of 20% at low energies. Thus, the measured polarization (45%) from Fe surfaces agrees with the expectations based on the inelastic spin filter effect.

The secondary polarization from all 3d ferromagnets shows the same general features and seems to be reasonably well understood, at least on a qualitative basis. In other materials, e.g., rare earths, very little has been done to date. It is not even clear what the expected secondary polarization should be because it depends on the relative contribution of the 4f and conduction band electrons to the secondary electron yield. Also the electronic structure around the Fermi level is very different from the 3d metals. This leads to different energy loss mechanism in the 4f materials. Polarized secondary measurements on Gd surfaces [4.58] indeed indicate the absence of a large polarization enhancement at low energies.

4.1.3.2 Applications to Ultrathin Films

The sample magnetization direction determines the direction of the spin polarization vector [4.53]. Therefore, with a multicomponent polarization analyzer one can directly measure the magnetization direction at the surface. Due to the high surface sensitivity, a few monolayers of ferromagnetic material give a strong polarization signal. *Pappas* et al. [4.59, 60] applied *SPSEES* to epitaxial Fe films on Cu (1 0 0) and Ag(1 0 0), systems in which the magnetization direction switches between perpendicular and in-plane as a function of film thickness and temperature. Figure 4.7 shows the measured polarizations at 125 K as a function of film thickness. The onset of ferromagnetic perpendicular order and the switching of the magnetization direction to in-plane at higher thickness can be directly seen. In the switching region the remanent polarization vanishes over a narrow thickness range. This behavior is similar to the switching as a function of temperature in films of intermediate thickness, where a loss of magnetization over a range of 20 K is observed [4.59]. The nature of this transition is not clear at present (see Chap. 3). It could be due to the break up of the magnetization into domains (the measurements are done in zero external field) or due to a loss of ferromagnetic order in the transition region. Further applications of spin

4.1 Spin-Polarized Electron Spectroscopies

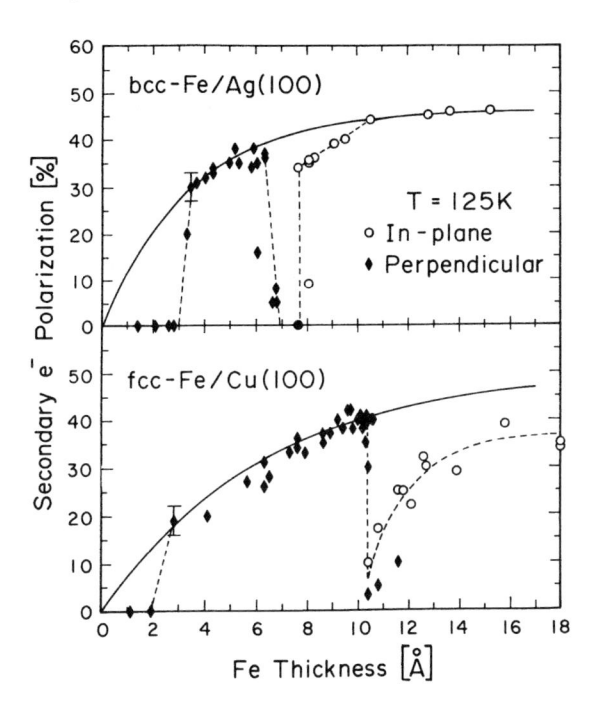

Fig. 4.7. Spin polarization of low energy secondary electrons from Fe films on Ag (100) and Cu (100) [4.60]

polarized secondary electrons are discussed in Sect. 4.2 for surface magnetometry and Vol. 2 Sect. 2.3 for magnetic domain microscopy.

4.1.4 Elastic Scattering (Spin–Polarized Low-Energy Electron Diffraction: SPLEED)

The spin-orbit interaction, which is always present, leads to spin dependences in elastic scattering (the basis of most spin detectors). On magnetic surfaces the exchange potential adds an additional spin dependence. Before the GaAs source was available the polarization of scattered electrons was measured using an unpolarized primary beam. These studies were restricted to spin-orbit effects on high-Z materials (W, Au, Pt) where resulting spin polarizations can be very large [4.61]. With the GaAs source the experiments can be performed by measuring intensity changes upon reversal of the beam polarization [4.62]. The equivalence of the two approaches has been demonstrated by combining a polarized source and polarization analysis [4.63]. The use of a GaAs source permits much higher sensitivities since no polarization analysis is required. Spin dependences below the percent level can readily be resolved, which opened up the field of surface magnetism to SPLEED [4.64]. Experimentally, two scattering intensitities, I_+, I_-, are measured for incoming spin-up and spin-down electrons, respectively. From these the asymmetry A is defined as

$$A = 1/P_0 (I_+ - I_-)/(I_+ + I_-). \tag{4.4}$$

The measured asymmetry has contributions from spin–orbit and exchange effects. These can be distinguished experimentally by reversing the sample magnetization [4.65] because the spin–orbit contribution is independent of magnetization whereas the exchange part changes sign. Thus, at least to a good approximation, one can measure the exchange asymmetry individually. While the spin–orbit induced asymmetries are of little interest in magnetic studies they are valuable for geometric structure analysis in LEED as additional information to the I/V curves. Since the asymmetries are normalized intensity differences, large values are usually associated with intensity minima, obviously a very undesirable situation. Exchange asymmetries from 3d ferromagnets are on the average only a few percent [4.66], however in the Bragg minima values of 20–30% can be obtained. Multiple scattering somewhat relaxes the close relationship between asymmetry maxima and intensity minima and can lead to additional asymmetry structures [4.67].

The SPLEED exchange asymmetries depend on the magnetic structure at the surface, i.e. the magnetic moments and the magnetization profile. Therefore, a comparison between a large experimental data set of asymmetry data, e.g., as a function of angle and energy, and theoretical asymmetry calculations should lead to a magnetic structure determination, equivalent to LEED studies for the geometric structure. An attractive feature of SPLEED is that it contains magnetic and geometric information at the same time.

While the quantitative interpretation of SPLEED exchange asymmetries is difficult [4.68] and requires a full dynamical scattering calculation, non-zero values prove the presence of long-range magnetic order. Thus, it is very useful, e.g., for determining the critical temperature T_C. For example, *Dürr* et al. [4.69] used SPLEED to determine the Curie temperature of ultrathin films of Fe on Au(1 0 0) as a function of film thickness. The antiferromagnetic coupling in Fe/Cr/Fe [4.70, 71] and Co/Cu/Co [4.72] sandwich structures were demonstrated by SPLEED. Oscillations in the exchange asymmetry as a function of film thickness were reported by *Kerkmann* et al. [4.73] on the Co/Cu (1 0 0). These oscillations result from interference between reflections at the surface and at the film substrate interface and are proposed for thickness calibration, similar to RHEED oscillations.

Spatially resolved SPLEED asymmetries can be used for imaging magnetic domains by using a polarized beam in the low energy electron microscope (LEEM) [4.74]. The feasibility of spin polarized LEEM (SPLEEM) has recently been demonstrated by imaging magnetic domains on Co [4.75]. This technique is still in its infancy.

4.1.4.1 Magnetic Structure Determination

Electronic structure calculations have predicted enhanced surface magnetic moments (Sect. 2.1). A quantitative analysis of SPLEED asymmetry data on Ni(1 1 0) favors indeed a small enhancement of the surface moment by about 5%

[4.76]. *Porter* and *Matthew* [4.77] analyzed a different data set [4.63] and favor a similar conclusion but also point out that more data are needed. The most extensive SPLEED data set available was taken on thick Fe (1 1 0) films grown on W (1 1 0) by *Waller* and *Gradmann* [4.78]. An analysis by *Tamura* et al. [4.79] on parts (at one energy) of this data set favors a large enhancement of the surface moment M_1 by 35% as well as a possible reduction of the second layer moment M_2 by 15%. *Ormeci* et al. [4.80, 81] analyzed this data set in more detail using data at different energies. An example of their calculation is shown in Fig. 4.8 for the specular beam at 52 eV. The overall agreement between experiment and calculated asymmetries is quite impressive, especially considering that the calculations are based on a scattering potential from first principles electronic structure calculations, thus containing no free parameters. However, it was found that the model of *Tamura* et al. does not give the best fit at all energies. Also, it was pointed out that the asymmetries are sensitive to the surface moments only for certain scattering parameters. The conclusion is that although the data seem to favor enhanced surface moments the magnetization profile at the surface is not well established at present.

No attempt has yet been made at quantitative SPLEED analyses on ultrathin film systems. Future experiments should be guided by theoretical predictions on the scattering parameters in order to optimize sensitivity to the magnetic moments. Also, theoretical progress will have to be made, especially at low energies where the surface barrier potential becomes important. The question of a possible influence of a spin dependent mean-free path on SPLEED, especially at low energies, will have to be resolved.

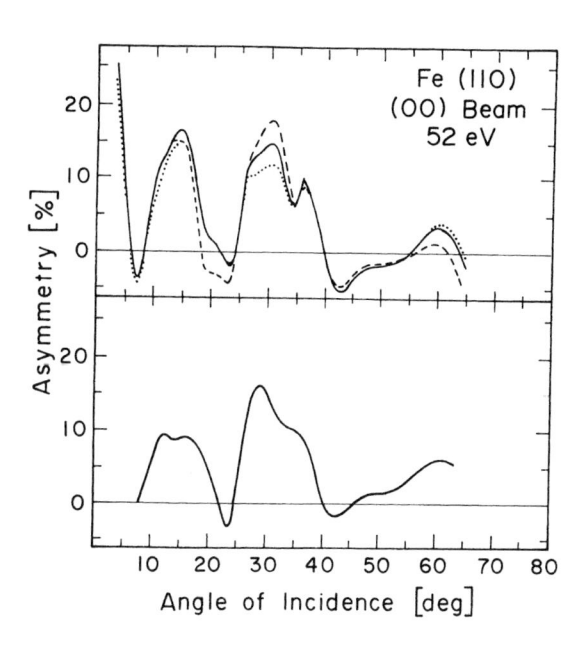

Fig. 4.8. Angle dependence of exchange asymmetry from Fe (1 1 0); lower panel: experiment; upper panel: calculations for different surface magnetizations; solid line: 19% surface enhancement (M_1), 6.8% in second layer (M_2); dashed line; model of *Tamura* and *Feder* [4.51]: M_1: + 35%, M_2: − 15%; dotted line: bulk moments throughout [4.81]

4.1.4.2 Critical Behavior

The exchange asymmetry has to go to zero at the critical temperature T_C. Although, in general, exchange asymmetries are not proportional to the magnetization, close to T_C (for small asymmetries) the proportionality becomes correct so that critical exponents can be determined by SPLEED [4.82]. The first temperature dependent SPLEED measurements to include the critical region were performed on Ni surfaces [4.83, 84]. The critical exponents on Ni(1 0 0) and Ni(1 1 0) were found to be in the range 0.79–0.8, in good agreement with theoretical values for a semi-infinite Heisenberg system. As an example, Fig. 4.9 shows temperature dependent SPLEED asymmetries from thick EuS(1 1 1) films [4.85]. The surface critical temperature is found to be equal to the bulk Curie temperature T_{Cb}, which was independently determined by magneto-optic Kerr effect (MOKE) measurements. From a log–log plot of the data in the critical region (14–16.7 K) a critical exponent of 0.72 ± 0.03 was determined, somewhat smaller than the value found for Ni and the theoretical values for Heisenberg or Ising model surfaces. The reason for this slight discrepancy is not clear.

A system which shows a dramatically different surface critical behavior is the Gd(0 0 0 1) surface. *Rau* and coworkers [4.86, 87] found evidence for surface magnetizations above T_{Cb} using electron capture spectroscopy. *Weller* et al. [4.88, 89] performed SPLEED on thick (500 Å) epitaxial Gd(0 0 0 1) films grown on W(1 1 0). Representative temperature dependent SPLEED data are shown in Fig. 4.10. The bulk magnetization was again monitored by MOKE. Exchange asymmetries were found up to 22 K above T_{Cb}. The structure around T_{Cb} is surface sensitive and disappears upon contamination. The behavior was interpreted as surface enhanced magnetic order (SEMO) due to enhanced coupling in the Gd(0 0 0 1) surface layer. The dip in the SPLEED asymmetries below T_{Cb} is atttributed to a compensation feature, due to antiferromagnetic coupling of the

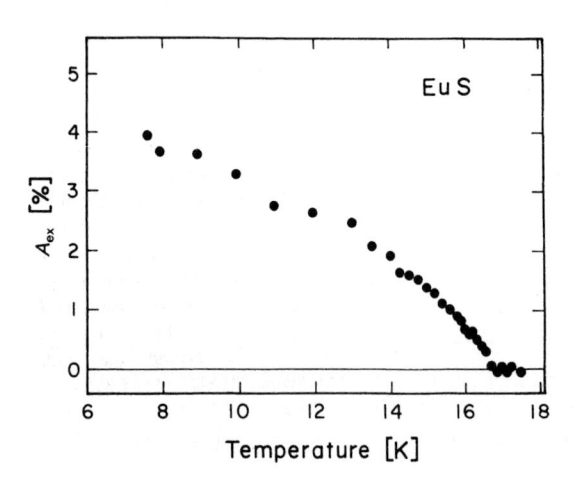

Fig. 4.9. Temperature dependence of exchange asymmetry from EuS films [4.85]

4.1 Spin-Polarized Electron Spectroscopies

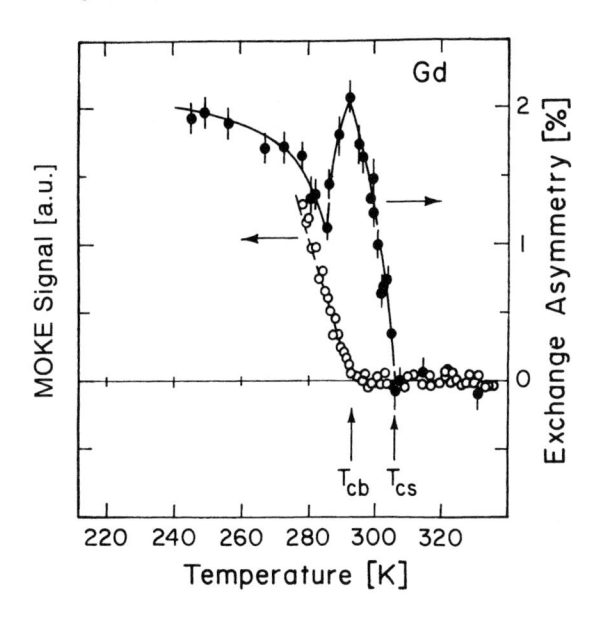

Fig. 4.10. Temperature dependence of exchange asymmetry from Gd (0 0 0 1) films on W (1 1 0) [4.88]

surface layer to the bulk. It is very interesting that structures very similar to these SPLEED structures have been found by electron capture spectroscopy on Tb by *Rau* et al. [4.90], suggesting that similar behavior is more general in rare earth systems. The possibility of a first order surface transition on Gd (0 0 0 1) was suggested by *Weller* and *Alvarado* [4.91] based on the fact that the presence of the surface SPLEED feature depends on the presence of an applied magnetic field when the sample is cooled through the surface transition temperature.

The Curie temperature of monolayer Fe films on W (1 1 0) was monitored by *Weber* et al. [4.92] using SPLEED. The temperature dependence of the SPLEED asymmetries for a clean Fe film and an Fe film to which 0.1 ML of Pd has been added are shown in Fig. 4.11. The increase in T_C is most striking. It was found that the increase is about the same as if Fe had been added instead of Pd. A similar increase is found also for submonolayer amounts of Ag and even for oxygen. No attempt at determining the critical exponent was made.

4.1.5 Inelastic Scattering (Spin-Polarized Electron Energy-Loss Spectroscopy: SPEELS)

In electron energy loss spectroscopy (EELS) a mono-energetic beam is scattered off a surface and the energy spectrum of the scattered electrons is recorded. EELS is widely used to study the vibrational excitations at a surface [4.93]. Depending on the scattering conditions, two different excitation mechanisms can be distinguished. In specular geometry (or in a diffracted beam) the inelastic scattering is a two-step process: large-angle elastic diffraction preceded or

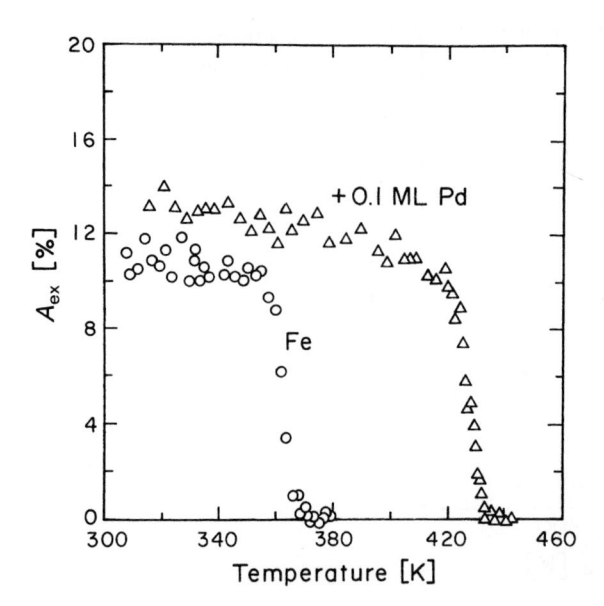

Fig. 4.11. Temperature dependence of exchange asymmetry from 1.3 ML Fe on W (1 1 0) and with 0.1 ML Pd added [4.92]

followed by the energy loss event with small ($k \approx 0$) momentum change (dipole scattering). The scattered electron interacts with the long-range electric fields above the surface and the interaction is therefore spin independent. The second excitation mechanism is of short range ("impact scattering") involving large momentum transfers ($k \neq 0$). This regime is of primary interest in SPEELS since one is dealing with e–e scattering and its spin dependence.

On ferromagnetic surfaces magnon and electron–hole pair excitations are expected to show spin dependences. So far, the energy resolution in SPEELS experiments has allowed the study of electron–hole pair excitation spectra only. In general, in a SPEELS experiment magnetic excitations are distinguished by a change of the spin direction of the electron (spin-flip scattering). It should be noted that the term spin flip is somewhat misleading. The electron–hole pair excitations studied in SPEELS are exchange processes in which the incoming electron transfers energy to an electron of opposite spin, which is then observed as the outgoing electron. Therefore, exchange scattering can be directly measured when a polarized primary beam is used and the polarization in the energy loss spectrum is analyzed.

4.1.5.1 Exchange Scattering

As an example of exchange scattering involving discrete energy losses the energy loss spectrum from a Cr_2O_3 surface [4.94] is shown in Fig. 4.12. A thick, disordered oxide layer in this experiment was grown by heating a Cr(1 0 0) surface in 10^{-6} Torr of oxygen for a few minutes. The incoming electron beam

4.1 Spin-Polarized Electron Spectroscopies

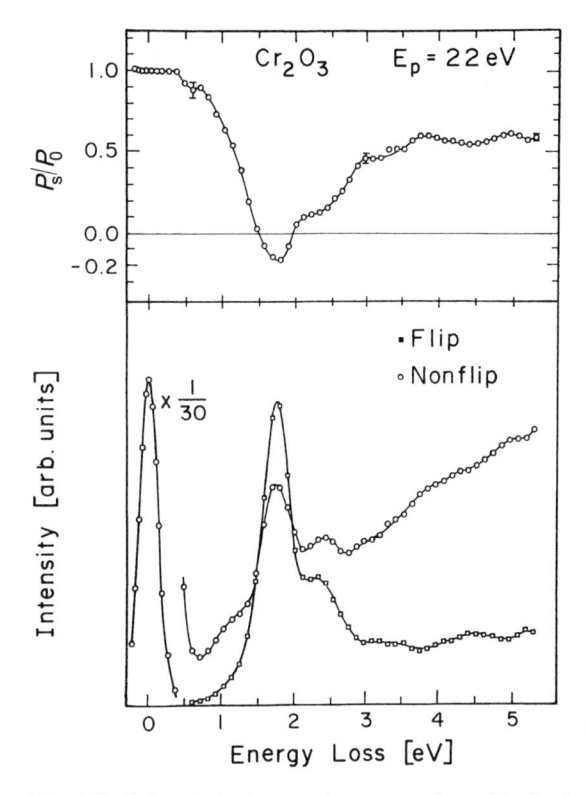

Fig. 4.12. Spin-polarized energy loss spectra from thin Cr_2O_3 films on Cr (1 0 0) [4.94]

has polarization P_0 (26%) and the intensity and polarization P_s are measured as a function of energy loss. The upper panel shows the polarization normalized to the primary polarization. Large changes in the polarization are observed in the inelastic region. Combined with the scattering intensity I, these data can be decomposed into scattering processes with spin-flip (F) and non-flip processes (N):

$$N = (1 + P_s/P_0)*I/2, \qquad F = (1 - P_s/P_0)*I/2. \tag{4.5}$$

These scattering rates are shown in the lower panel of Fig. 4.12. The strong peak around 1.75 eV energy loss and a weaker shoulder around 2.4 eV are mainly spin-flip processes. These energy losses correspond to the well known excitations of the Cr^{3+} ion. The ground state of Cr^{3+} (three 3d electrons) is a quartet state ($S = 3/2$). Transitions from the quartet state to excited doublet states ($S = 1/2$) proceed via exchange scattering. In addition to these exchange processes there are losses within the same spin multiplet. These are dipole allowed and spin conserving giving rise to the broad non-flip energy loss spectrum. The spin-flip losses are replicated with lower intensity in the non-flip spectrum because the Cr^{3+} ions on this surface do not have a preferential spin alignment

(Cr_2O_3 is antiferromagnetic). Exchange scattering is possible between the different Zeeman levels of the ground state ($S_z = \pm 3/2$ or $\pm 1/2$) to the excited states ($S_z = \pm 1/2$) without apparent spin flip, i.e. S_z from 1/2 to 1/2 and $-1/2$ to $-1/2$. These exchange events show up in the non-flip loss spectrum. The SPEELS spectra in this example are independent of the primary beam polarization direction. If the Cr moments were ferromagnetically aligned flip processes would be possible in one direction only.

4.1.5.2 SPEELS on Ferromagnetic Transition Metal Surfaces

On a ferromagnetic surface the SPEELS spectra depend on the alignment between incoming spin and majority-spin direction of the sample. Thus, one defines non-flip (N_+, N_-) and flip (F_+, F_-) rates, where $\pm$ now stand for the incoming spin. For example, F_- (flip down) is a process in which a spin-down electron comes in and the outgoing electron is spin-up. The scattering intensities (I_+, I_-) show spin asymmetries. As in SPLEED, one defines an asymmetry A, which is given in terms of the individual scattering rates as $A = (N_+ + F_+ - N_- - F_-)/(N_+ + N_- + F_+ + F_-)$. Conversely, when an unpolarized beam is used the scattered electrons will become polarized with $P_s = (N_+ + F_- - N_- - F_+)/(N_+ + N_- + F_+ + F_-)$.

The first experimental demonstrations of spin effects in inelastic scattering from ferromagnets were done by measuring asymmetries in inelastic scattering from Ni [4.95] and polarizations [4.96] from an Fe-based metallic glass. The asymmetries on Ni were found to be negative while the polarization on the Fe-based sample was positive. The combination of polarized beam and polarization analysis, by *Kirschner* and collaborators [4.97–99] on Fe (1 0 0), revealed that spin-flip processes are a significant part of the energy loss processes, with flip down dominating, thus explaining the negative asymmetries and the positive polarization. SPEELS data on Ni (1 1 0) [4.100] are shown in Fig. 4.13 over an energy loss range of about 600 meV. While the elastic scattering is purely non-flip, spin-flip forms a large contribution in the inelastic regime. The flip down rate shows a pronounced structure with a maximum around 300 meV loss energy. Figure 4.14 schematically shows the flip down process (exchange) in a ferromagnetic electronic structure. The incoming electron falls into an empty state above E_F and a spin-up electron from below E_F is emitted. If the states involved are at the same k point one expects a well defined energy loss equal to the exchange splitting. In essence, one has flipped a spin by creating an up-hole and a down-electron (Stoner excitation). The flip down intensity maximum corresponds to the exchange splitting. However, the shape of the observed spectrum is not well understood theoretically. The width can possibly be attributed to variations of the exchange splitting over the Brillouin zone and to transitions involving electron states at different k values ($k \neq 0$). The flip up rate (F_+) is found to be very small (Fig. 4.13). Within the 3d states there should be no flip up scattering in Ni since the spin-up bands are completely filled. The small

4.1 Spin-Polarized Electron Spectroscopies

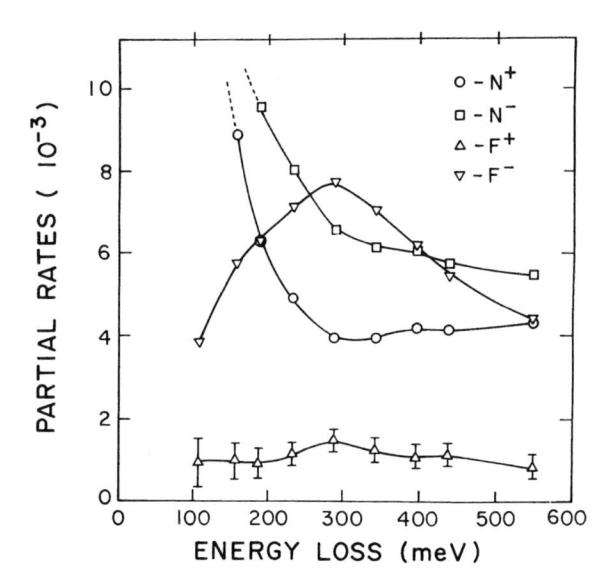

Fig. 4.13. Flip and non-flip scattering rates on Ni (1 1 0) [4.100]

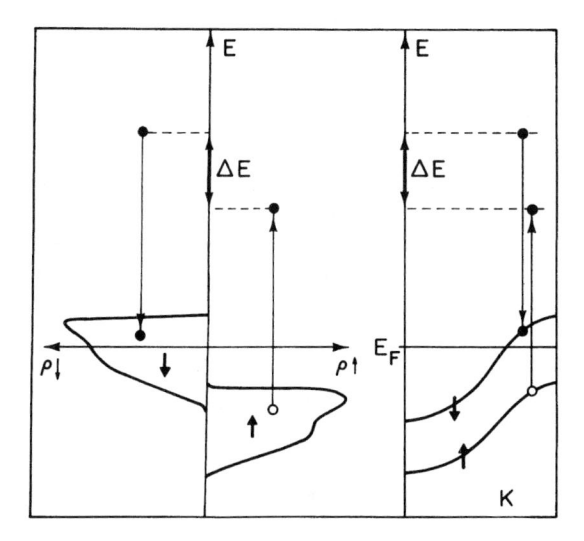

Fig. 4.14. Schematic of exchange scattering process (flip-down)

measured flip up rate can be attributed to thermal fluctuations (the measurements were done at room temperature) or to the involvement of free-electron 4sp states [4.101]. In addition to the large difference of the flip rates the spin down non-flip rate is higher than the up-spin rate leading to very large (negative) asymmetry values of up to -55% in Ni.

The large spin asymmetry found in Ni and other 3d ferromagnets is a possible reason for the enhancement of the spin polarization of secondary electrons

(Sect. 4.1.3) leading to a shorter inelastic mean free path for spin-down electrons. In addition, the large imbalance between flip down and flip up scattering increases the number of up electrons at the expense of down electrons. While this model can qualitatively explain the experimental observations, quantitative estimates from the SPEELS data are difficult since one has to integrate over all energy losses and over scattering angles.

Studies of the angle dependence in SPEELS on Ni (1 1 0) [4.102] have shown that in specular geometry the dipole losses (non-flip) dominate the exchange processes. Experimentally, it is sufficient to go 10–20° off specular to suppress most of the dipolar losses and the exchange processes can be seen most clearly and have been found to be quite insensitive to angle variations. With increasing primary energy the exchange scattering contributions are found to decrease, although no systematic studies over a large energy range have been performed.

While the existence of inelastic scattering asymmetries proves the presence of long-range ferromagnetic order (just as in SPLEED) the absolute magnitudes are difficult to interpret, e.g., in terms of surface magnetizations, since they depend on details of the electronic structure close to E_F. However, information on relative magnetizations can be obtained. For example, a decrease upon oxygen and CO adsorption was found on Ni [4.103]. However, the shape of the asymmetry spectrum remains largely unaffected. This was taken as evidence of an exchange unchanged splitting in the subsurface layers. The same behavior is found when the temperature is increased [4.102, 104], consistent with a largely temperature independent spin-split electronic structure (local band theory). The temperature dependence of the SPEELS asymmetries on Ni(1 1 0) was shown to follow a surface magnetization curve [4.102], in agreement with expectations based on the short probing depth.

4.1.5.3 SPEELS on Ultrathin Films

The only SPEELS study on ultrathin films to date was done by *Kämper* et al. [4.105, 106] on epitaxial fcc Co films on Cu (1 0 0). The most interesting question is whether the SPEELS spectra change in going from thick films to monolayer films due to changes in the electronic structure, e.g., two-dimensional vs. three-dimensional behavior or an enhancement of magnetic moments. The SPEELS spectra from thick (20 ML) Co films are shown in Fig. 4.15. The spectra were taken at 25 eV primary energy and 20° off specular. The qualitative features are very similar to the Ni spectra. The flip-down rate is much larger than the flip-up rate and shows a structure around 0.8 eV energy loss superimposed on a smoothly increasing "background". The maximum at 0.8 eV is taken as the average exchange splitting. With decreasing Co film thickness the spectra do not change down to about 4 ML thickness. On thinner films the asymmetries decrease mainly due to scattering contributions by the Cu substrate. However, the shape of the asymmetry curve and the shape of the flip and non-flip rates do not change significantly even for 2 ML films. Specifically, no energetic shift in

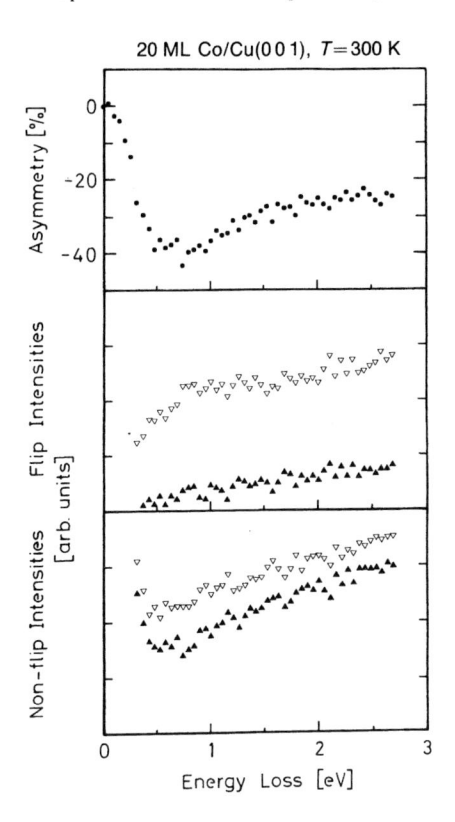

4.1 Spin-Polarized Electron Spectroscopies 145

Fig. 4.15. Flip and non-flip spectra from epitaxial 20 ML fcc Co films on Cu (1 0 0). Top panel shows scattering asymmetry [4.106]

the hump of the flip-down rate at 0.8 eV could be detected. Thus, there is no evidence for a thickness dependent exchange splitting due to surface enhanced moments in the monolayer Co films. The SPEELS experiments were extended into the submonolayer range where the films do not show ferromagnetic long-range order. Nevertheless, the flip and non-flip rates can still be measured individually and the shape of the flip spectrum was found unchanged from the thicker films. This shows the persistence of local magnetic moments in these films in the absence of ferromagnetic long-range order, i.e. in their paramagnetic state.

4.1.6 Photoemission Techniques

In photoemission spectroscopy an electron is optically excited from a state below E_F to a final state above the vacuum level E_V by absorption of a photon of energy hv as shown schematically in Fig. 4.16. Measuring the energy spectrum of the emitted electrons allows the determination of the band structure $E(k)$ ([4.107] for overview articles). With additional spin polarization analysis the spin character of the bands can be determined. Spin-polarized angle resolved

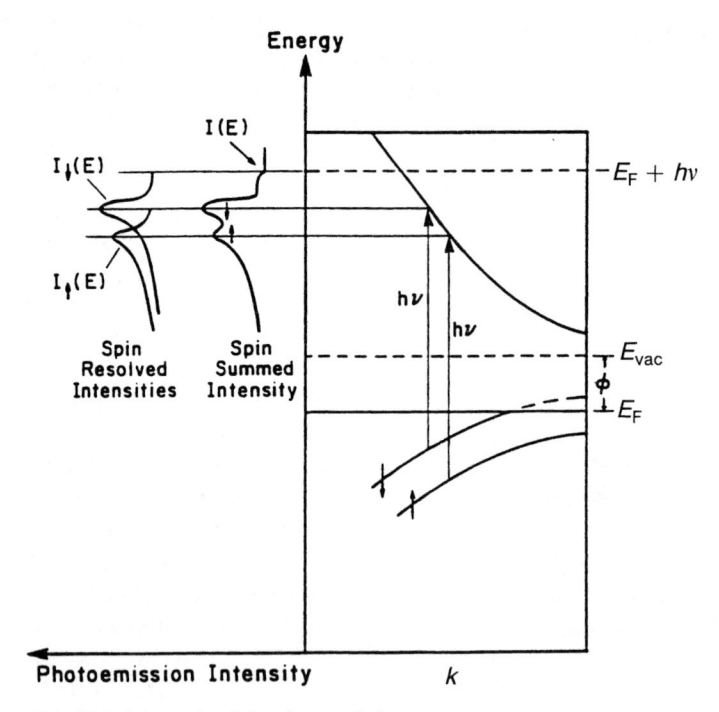

Fig. 4.16. Schematic of the photoemission process

photoemission spectroscopy is the most direct way of studying the spin-split electronic structure of ferromagnetic surfaces and thin films.

The first polarization of photoemitted electrons from a ferromagnetic sample (polycrystalline Gd films) was observed by *Busch* et al. [4.108]. In these experiments the polarization of the total photoyield excited by the full spectrum of a Xe–Hg lamp was measured. Thus, one averages over a range of states below E_F given by the difference between the maximum photon energy and the work function of the sample. The sample was in a strong magnetic filed along the surface normal (longitudinal geometry). In this set up the photoemitted electrons are extracted from the magnetic filed region and then electrostatically deflected by 90° (to convert to transverse polarization) and fed into a Mott detector for spin analysis. These measurements were extended to 3d metals and other magnetic materials [4.109, 110]. The technique was refined by using monochromatic light and tuning the photon energy from the photothreshold to higher energies thereby allowing for a crude spectroscopy close to E_F. For instance, it was shown by using single crystal samples that in Ni the spin polarization at E_F is highly negative [4.111, 112]. The external magnetic field precludes angle resolved measurements, and energy resolution is also difficult to implement.

A major achievement was the successful use of remanently in-plane magnetized samples (transverse geometry) [4.113, 114]. In order to minimize the effects of stray fields on the electron trajectories, "picture frame" single crystal samples were used. Thus, the full potential of angle resolved photoemission spectroscopy could be exploited, i.e. angle-dependent measurements can be performed combined with high energy resolution [4.115]. The use of polarized light allows the selective probing of electron states of different symmetry. However, it should be noted that the longitudinal geometry is the only polarized spectroscopy technique which allows measurements to be performed in strong external magnetic fields.

4.1.6.1 Valence Band Studies of Ultrathin Films

A study on bcc Fe films on Ag (1 0 0) by *Jonker* et al. [4.116] in the transverse geometry (including angle and energy resolution) showed spin resolved energy distribution curves on films of 5 ML thickness, identical to spin resolved photoemission spectra from bulk Fe (1 0 0) [4.117]. Interestingly, in this study no remanent in-plane polarization was detected on thinner films (< 5 ML), although the energy distribution curves showed the same features indicative of a spin-split ferromagnetic electronic structure as in the thicker films. It was suggested that the thin films might be magnetized perpendicular. This polarization component could not be measured in the experiment. Polarization measurements by *Pescia* et al. [4.118] on the total photoyield in the longitudinal geometry on fcc Fe on Cu (1 0 0) and by *Stampanoni* [4.119] on bcc Fe on Ag(1 0 0) directly showed the perpendicular magnetization in the very thin (< 5 ML) films.

The evolution of the spin-split electronic structure in ultrathin films as a function of film thickness has been addressed in a number of studies [4.120–124]. There is general agreement that the three-dimensional 3d band structure is already established in films of three to five layers thick. Thickness dependent exchange splittings were found in Ni (1 1 1) films on W (1 1 0), with the thickness dependence being different for different k vectors in the Brillouin zone [4.120]. This behavior is similar to the temperature dependence found on bulk Ni (1 1 1) [4.125] and could be explained by finite temperature effects due to reduced Curie temperatures in the thin films.

One obvious question to be addressed by spin-polarized photoemission is the existence of enhanced moments and exchange splittings in ultrathin films. However, experimentally, this is a very difficult task since for monolayer films the intensity is quite small. Also, the substrate contributions pose the problem of background subtraction. In addition, the Curie temperature in the very thin films might be too low, leading to the loss of long-range order (and spin polarization). *Heinen* et al. [4.122] report an enhanced exchange splitting of 2.4 ± 0.2 eV in a monolayer Fe on Au (1 0 0) versus 2.2 eV in bulk Fe. *Clemens* et al. [4.124] find a thickness dependent exchange splitting in fcc Co/Cu (1 0 0) changing from about 1.9 eV in the monolayer to the bulk value (1.5 eV), which is attained at 4–5 ML. This is in contrast to the findings of a thickness independent exchange splitting in SPEELS (Sect. 4.1.5.3).

Although electron spectroscopies can only probe the few top-most layer it is possible to address the interesting problem of interfaces between two materials, as long as sufficiently thin overlayers are used. *Weber* et al. [4.126] found the evolution of a two-dimensional interface state when Pd was deposited on Fe(1 1 0), with the strongest intensity occurring for Pd thicknesses between 1–2 ML. The exchange splitting of this state was determined as 130 meV, but inverted with respect to the Fe substrate, i.e. the minority state is at a higher binding energy. No indication of spin polarization was measured in Pd films above 5 ML. *Brookes* et al. [4.127] found the evolution of two-dimensional states when Ag was grown on Fe (1 0 0). The intensity spectra are shown in Fig. 4.17. A sequence of states was found with the binding energy shifting (1.7 eV for the first layer, 1 eV for the second layer, 0.3 eV for the third layer) for each completed layer. Spin polarization analysis shows that these states are of

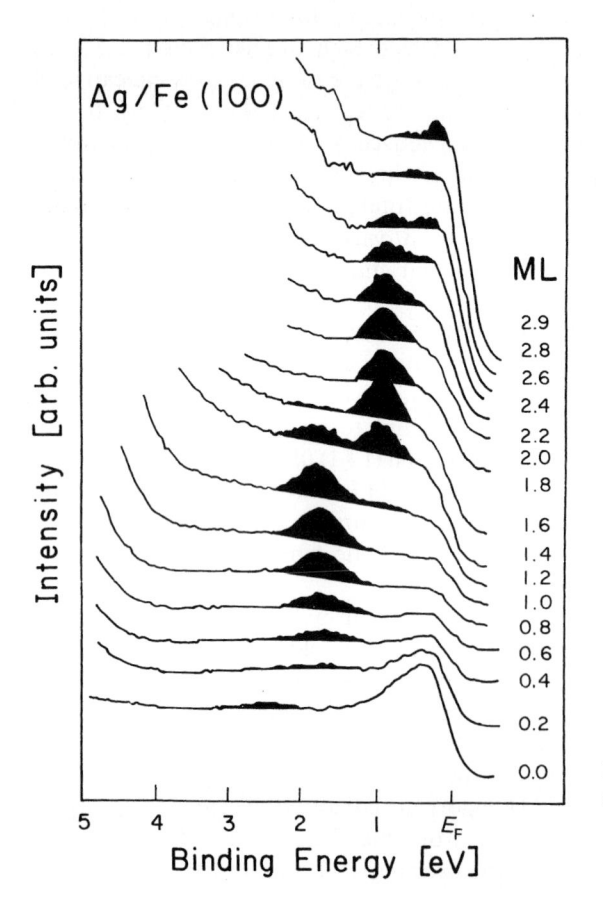

Fig. 4.17. Photoemission spectra for Ag layers on Fe (1 0 0) [4.127]

predominantly minority spin character. These states are attributed to interface states localized at the Fe/Ag interface.

4.1.6.2 Core Level Studies

The spin polarization from core levels carries information on the *local* magnetic order. The first polarization measurements on Fe 3p and 3s levels [4.28, 128, 129] have been performed only recently. The splitting of the 3s level in 3d transition metals and compounds has been known for quite some time. It was assumed to be of magnetic origin due to exchange interaction of the core hole state with the 3d electrons, i.e. it is a final state effect. Figure 4.18 shows spin resolved energy spectra from the Fe 3s states [4.28]. A high spin polarization is indeed found, proving the magnetic origin of the splitting. The majority and minority spin intensity curves have very different shapes. Unfortunately, the present theoretical understanding does not permit a quantitative determination of the magnetic moment, e.g., from the net polarization of the core levels.

Sincovic et al. [4.128] studied the initial stage of oxidation of the Fe (1 0 0) surface by measuring polarized 3p core levels. Chemical shifts were observed upon oxide formation. The initial oxide layer was identified as Fe_2O_3 and found to be coupled antiferromagnetically to the Fe substrate. Upon annealing the oxide layer a transformation was observed in the oxidation state to Fe_xO and the spin polarization was lost, which was interpreted as the formation of an antiferromagnetic compound. Thus, this study combines chemical specificity with magnetic information.

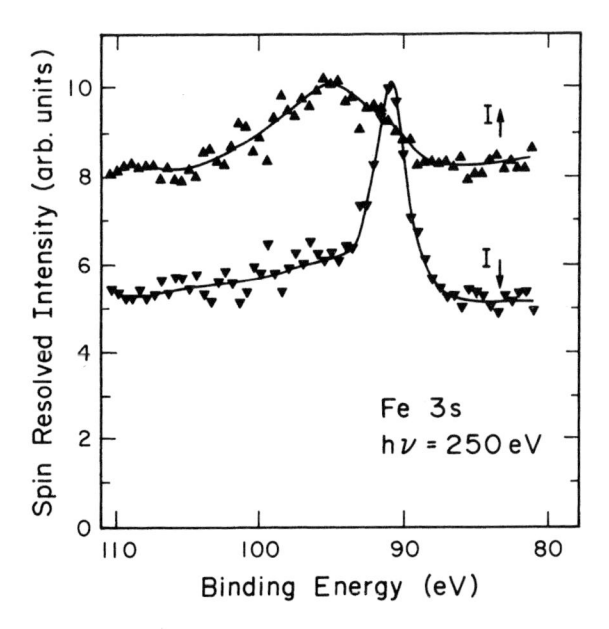

Fig. 4.18. Spin resolved Fe 3s photoemission spectrum [4.28]

The method was applied to Cr overlayers on Fe (1 0 0), a system of special interest due to the oscillatory exchange coupling in Fe/Cr/Fe sandwich structures and the giant magnetoresistance effect. *Jungblut* et al. [4.130] measured the spin polarization of the Fe 3p and Cr 3p levels with increasing Cr thickness. For one ML the Cr 3p level shows polarizations opposite to the Fe 3p states, directly showing the antiferromagnetic coupling of the first Cr layer to the Fe. With increasing Cr thickness a gradual decrease of the spin polarization was found. Based on the net 3p polarization a rough estimate of the monolayer Cr moment was given. It was concluded that the moment is much smaller than the theoretically predicted "giant moment" of 3.6 μ_B.

The 4f levels in the rare earths are highly localized levels. However, they carry most of the magnetic moment. Spin-polarized photoemission from these levels should therefore be an ideal tool to study the magnetization. An especially clear-cut case should be Gd with its fully polarized half-filled 4f shell. For perfectly aligned moments one expects the 4f intensity to show 100% polarization. However, *Weller* et al. [4.88] found only small 4f polarizations on thick epitaxial Gd (0 0 0 1) films. In combination with their SPLEED data shown earlier, this was interpreted as being due to an antiferromagnetically coupled surface layer. *Carbone* et al. [4.131] studied Gd films grown on Fe (1 0 0). A spin resolved spectrum from 1 ML Gd is shown in Fig. 4.19. The strong Gd 4f peak at 8 eV binding energy is seen to be highly polarized with the minority spin (with respect to Fe) intensity dominating. This shows directly that the Gd couples antiferromagnetically to the Fe. After background subtraction a net polarization of -60% is determined for the 4f levels. Although a high polarization is indeed observed it is still much smaller than the expected 100%. The reason for this reduction is not clear. It could be related to intermixing, which is known to

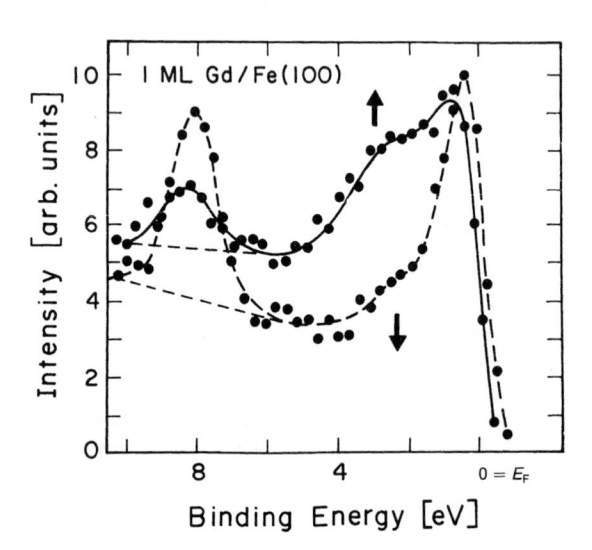

Fig. 4.19. Spin resolved photoemission spectrum from 1 ML Gd on Fe (1 0 0) [4.131]

be a problem. In this respect it is also interesting that the polarization in the Fe 3d part of the spectrum (between E_F and the 4 eV binding energy) decreases upon Gd deposition whereas the spectral features do not change from the bare Fe. This could be interpreted as magnetic disorder introduced by the Gd overlayer, very similar to finite temperature effects. For thicker Gd films the polarization was found to decrease dramatically (10% at 35 Å). The reason for this is not clear at present. One can speculate that the free Gd surface might be intrinsically unstable towards magnetic surface reconstruction, e.g., an antiferromagnetic surface coupling. The antiferromagnetic coupling of rare earth monolayers to Fe has also been seen with Tb, Dy, and Nd [4.132]. The antiferromagnetic coupling of Gd layers on Fe surfaces was first seen in spin resolved Auger spectroscopy [4.133–135]. This technique was developed by *Landolt* and collaborators [4.136]. In many respects it yields information similar to core level spectroscopy, i.e. element specific magnetization. One obvious advantage is that it can be performed in the laboratory using a electron gun as an excitation source without the need of synchrotron radiation as in core level photoemission. However, quantitative interpretation, e.g., determining magnetic moments, is even more complicated theoretically and beyond present capabilities.

4.1.6.3 Inverse Photoemission

In inverse photoemission the empty levels above E_F are probed by injecting electrons into these states and detecting the photons emitted (the time-reversed process of Fig. 4.16). *Pendry* [4.137] proposed using polarized electrons to probe the spin splitting of the states above E_F. Although no polarization analysis is needed, the experimental difficulties with inverse photoemission stem from the fact that the cross sections are about 10^{-4} times smaller than typical photoemission cross sections due to phase space restrictions [4.137]. Nevertheless, shortly after the proposal, the first spin resolved inverse photoemission spectra were obtained on Ni [4.138] using a GaAs source. Experimentally, it is most convenient to keep the photon energy fixed, using Geiger–Müller counters sensitive in the UV (9.4–9.6 eV) and sweeping the electron energy. The high energy cutoff in these detectors is given by the window material (CaF_2 or SrF_2) while the low energy threshold is determined by the ionization energy of the active gas in the counter (iodine in most cases). The onset of photon emission is given by $E_K = h\nu - E_V$, where E_K is the electron kinetic energy and E_V the vacuum energy.

The angle dependence and temperature dependence of spin-polarized inverse photoemission has been studied on Fe (1 0 0) [4.139, 140]. Adsorption studies on Ni(1 1 0) showed a reduction of the surface magnetization [4.141, 142]. On the other hand, *Donath* [4.143] resolved exchange splittings of adsorbate levels (O 2p) for oxygen on Ni (1 1 0) showing the existence of adsorbate magnetic moments. Also, spin splittings of empty surface states have been investigated [4.143]. While it is clear that spin-polarized inverse photoemission is a very valuable tool, it has not yet been applied to ultrathin films. *Himpsel* [4.144] studied epitaxial Fe films on Cu (1 0 0) and Ag (1 0 0) by inverse photoemission using unpolarized electrons and reports differences between the two systems and changes of the spectra upon annealing. These changes are attributed to changes of the magnetic properties of the films. However, the magnetic state of the films was not characterized directly in this study. It will be very interesting to study these systems with the added spin information in the future.

4.1.7 Concluding Remarks

The large number of epitaxial ultrathin magnetic systems with interesting magnetic properties offers a wide area of research for the application of polarized electron spectroscopies. These techniques will contribute to our understanding of the magnetic properties in terms of the spin-polarized electronic structure. Some of the polarized electron techniques are in a very advanced state, experimentally as well as where theoretical description is concerned (e.g., valence band photoemission, SPLEED) while others (e.g., core level polarizations, energy loss spectroscopy, Auger electrons, secondary electrons) are understood on a semiquantitative level, at best. For example, determining surface or interface magnetic moments remains a challenge for all techniques.

In this chapter, we have discussed the most widely applied polarized spectroscopies using electrons or photons as excitation sources. Other excitation sources offer even higher surface sensitivities. Electron capture spectroscopy (ECS), pioneered by *Rau* [4.145], uses a high energy grazing-incidence ion beam [4.146]. Metastable atom de-excitation [4.147] offers another way to induce electron emission. Finally, spin-polarized electron tunneling between a magnetic tip and magnetic surfaces has been reported recently in scanning tunneling microscopy (STM) [4.148, 149]. The spin-polarization of the tunneling current from a Ni tip has been demonstrated [4.150]. Spin-polarized field emission [4.151–153], which was abandoned when spin-polarized photoemission proved feasible and yielded much more detailed information, should become of interest again in conjunction with polarized STM studies in order to gain a better understanding of the vacuum tunneling process and as a possible magnetic characterization technique for magnetic STM tips.

4.2 Probing Magnetic Properties with Spin-Polarized Electrons

H.C. Siegmann and E. Kay

Ultrathin Magnetic Structures (UMS) are built from monoatomic layers of atoms. Ideally, the layers should be perfect over extended areas. In practice, one attempts to minimize faults such as voids, steps or chemical and crystalline defects. To avoid serious obscuring of primary magnetic properties by the phenomenon of superparamagnetism, to be discussed below, one typically has to have the ideal layers in patches exceeding 100 atomic distances. Under such conditions, some striking new magnetic phenomena can be observed which provide a new basis for the understanding as well as the applications of magnetism. Preparation and measurement of UMS pose great challenges to the experimentalist and require the most advanced material characterization and magnetization detection techniques. Electron beams play a crucial rôle in accomplishing this task. In the present chapter, the possibilities offered by spin polarized electron beams in obtaining information on magnetic properties of UMS is explored.

The quantity of primary interest in magnetism is the magnetization M, the magnitude of which is defined as the magnetic moment/volume of magnitude:

$$M = n_A \cdot n_B \cdot \mu_B, \qquad (4.6)$$

where n_A is the density of atoms and n_B the number of Bohr magnetons μ_B carried by each atom. Experimentally, one has to determine $M(H, T)$ for each atomic layer of UMS, as it depends on the external magnetic field H and on the temperature T. From $M(H,T)$ one obtains the spontaneous magnetization $M_0(T)$ which is generated in the absence of an external field by the quantum mechanical exchange interaction within one single Weiss domain. The temperature at which M_0 vanishes is the Curie point T_C. In the following chapter it is shown that very general theoretical arguments can predict how M_0 should vary at low temperatures $T/T_C \lesssim 0.3$, that is in the spin wave regime, and also with which "critical exponent" M_0 should vanish if $T/T_C \to 1$. Therefore, one important experimental task is to determine the layer dependent spontaneous magnetization $M_0(T)$. Further basic information, yet theoretically not as well founded, is contained in the magnitude of T_C. With UMS, T_C depends strongly on structure and spurious contamination below 1% of one atomic layer.

4.2.1 Magnetic Information from Measurement of Spin Polarization or Spin Asymmetry

First, one needs to discuss the question of what information on $M(H,T)$ may be obtained with spin–polarized electron beam techniques. In these techniques, electrons are extracted from the UMS and emitted into a vacuum. These electrons are then focused into a beam and the spin polarization P of the beam measured usually in a scattering experiment [4.154]. It will be shown below that P can in most cases be directly related to the magentization M within UMS. It is however important to be aware of the complexities that might arise.

The vector of the spin polarization is defined as $P = \{\langle \sigma_x \rangle, \langle \sigma_y \rangle, \langle \sigma_z \rangle\}$ where $\langle \sigma_v \rangle$ are expectation values of the spin direction along the three coordinates. The definition is meaningful in the non-relativistic limit which applies to the experiments discussed here. We have $|P| = (N^+ - N^-)/(N^+ + N^-) \leqslant 1$ where $N^+(N^-)$ is the density of spin-up (spin-down) electrons in the beam. It should be noted that the experiment can also be inverted: first, a beam of spin-polarized electrons is formed, for instance with a spin–polarized GaAs electron gun, and the emission of light or scattered electrons is measured when the spin-polarized electron beam strikes the solid. The dependence of the light or electron emission on the direction of the spin polarization of the incident beam is called the spin asymmetry A. It provides the relevant information in the inverted experiment [4.155].

P should directly reflect that part of the magnetization M which is generated by the spontaneous alignment of the electron spins. This arises because conservation of spin is the dominant rule in most processes of electron emission. The orbital part of M will not contribute to P as it disappears in the process of electron emission and beam formation. Generally, orbital magnetization is small, of the order of a few percent of the total magnetization in 3d transition metals, because the orbital moment is quenched in a cubic crystal field. In the bulk the orbital moment can be well accounted for by measurement of the g-factor. In UMS, the crystal field is generally distorted at the interfaces of the layers which may lead to relatively large changes of the orbital moment compared to conventional bulk material. Therefore, it is important to keep in mind that P obtained in electron beam techniques accounts only for the spin part M_σ of the magnetization. This yields

$$\boldsymbol{P} = f \cdot \boldsymbol{M}_\sigma / (n_e \cdot n_A \cdot \mu_B), \tag{4.7}$$

where f is a dimensionless function discussed below, and n_e the total number of electrons in the open shells of the atoms. In the following, we will omit the index σ for simplicity. Induction, torque and force magnetometers determine of course the total magnetization including the orbital part.

The factor f in (4.7) cannot be generally predicted. The amount and parentage of the electrons emitted from any surface depends on how the energy is supplied to induce electron emission. The most common source of energy are photons or primary electrons, that is one has photoemission or secondary electron emission. In photoemission for example, the energy, polarization and angle of incidence of the photons determines the relative probability of electron emission from the various electron states in the solid. Hence, the contribution of each type of electron state to the total electron beam is variable. Therefore, the polarization of the emitted electron beam does not represent the same average as the polarization of electrons in equilibrium within the solid.

One basic assumption is that P and M are collinear as postulated in (4.7). This is founded on the quantum mechanical result that once the quantization axis is fixed by the direction of the total magnetization M, the spin polarization of the electrons must be collinear, either parallel or antiparallel to M. However, this is strictly applicable only to the case of highly symmetric objects such as atoms. In solids, the scattering on selected crytal planes or the crystal structure itself can define a new quantization axis which might destroy the collinearity of P and M. However, for the sake of simplicity and in the context of the present applications, we shall assume that such complications are absent and that P is either antiparallel or parallel to M. The fact that P can have either sign with respect to M is best explained with the example of ferromagnetic Ni, where the electrons close to the Fermi level E_F are polarized opposite to the average spin polarization. Hence if the electron emission conditions favor states from E_F, the observed polarization changes sign. Similarly, if one has a ferromagnet with two opposite sublattice magnetizations M_A and M_B, electron emission can occur predominantly from sublattice A or from B, and the sign of P will change accordingly while the total magnetization $M = M_A + M_B$ remains fixed. However, if the conditions of electron emission are kept constant, P must change sign when M changes sign. This leads to

$$P = C_1 M_\sigma + C_2 M_\sigma^3 + C_3 M_\sigma^5 + \ldots, \tag{4.8}$$

where $C_1, C_2, \ldots$ are constants. Equation 4.8 shows that P is proportional to M_σ if the changes of M_σ are not large. Nonlinear phenomena affecting the proportionality include multiple spin dependent scattering of electrons in the emission process and, in an energy resolved electron emission experiment, energy shifts of spectral features that can occur when the temperature T changes. In summary, the following general statements apply:

1) Absolute magnetometry is not yet possible in spin–polarized electron beam experiments. However, with very high energy photons, one might be able to suppress the scattering phenomena discussed in Chap. 4.1. Therefore, absolute magnetometry could become possible with future synchrotron radiation sources. Møller scattering of high energy spin–polarized electrons is one existing example of absolute magnetometry frequently used in high energy physics [4.156].
2) P and M are collinear in most applications. Therefore, magnetic structures, e.g., in domain walls, or magnetic hysteresis loops including magnetic remanence and coercivity may be determined by measuring P.
3) P and M are proportional in many cases, particularly if M is smaller or if small changes of M are considered only. This means that one can determine the T-dependence of M_0 by measuring $P(T)$ in the spin wave regime where the changes of M_0 are small. Similarly, close to T_C, M is small and the critical behavior of M_0 as well as T_C can be determined from $P(T)$.

4.2.2 Unique Features of Magnetometry with Spin Polarized Electrons

Magnetometry based on emission or scattering of electrons has several decisive advantages over other conventional and nonconventional methods. Particularly, laser induced photoelectrons as well as low energy secondary electrons come in large quantities which makes it possible to obtain a large quantity of highly precise data in a short time. This reduces artifacts due to adsorption of residual gas molecules on the sample surface. Furthermore, there are the technical advantages that scattering or emission of photo- or secondary electrons is compatible with other electron spectroscopies used to structurally and chemically define UMS, and can be done "on line", that is, while the sample grows. Each UMS can also be spectroscopically defined before and after each measurement. Additional chemicals can be deposited readily at the surface and their effect on the magnetic properties of UMS can be ascertained on the spot. The central feature common to electron spectroscopies is the fact that the probing depth is very small. UMS play an important role in the assessment of the probing depth as pointed out by a number of authors, since they provide well defined magnetization profiles at the surface. With threshold photoelectrons as well as low energy cascade electrons, the spin polarization P of the emitted electron beam represents the average magnetization M over a probing depth of only ~ 0.5 nm in the case of the transition metals [4.157, 158, 159]. The small probing depth has the disadvantage that thicker magnetic films cannot be studied. However, one key advantage of electron beam techniques is that the magnetism of ~ 3 surface layers can be measured irrespective of whether the substrate is also magnetic or not. This is a unique feature of magnetometry with spin–polarized electrons which is of particular interest when studying UMS. Other such basic features include:

Time Resolution. Pulsed lasers or synchrotron light sources provide short photon pulses; pulse durations of $\sim 10^{-11}$ s are common. With such a pulse, enough photoelectrons can be emitted from a surface to perform an accurate measurement of P. The pulse of photoemitted electrons is space charge limited, but this does not affect P as the spin of the total bunch of emitted electrons must be conserved.

Spatial Resolution. A primary electron beam may be focussed into an extremely small spot at the surface; typically, the diameter of a suitable focus is ~ 10 nm at present. Secondary electrons are emitted from the close neighborhood of the focus, and their polarization can be measured. With UMS, the lateral resolution is often sufficient to directly measure the spontaneous polarization within one domain without applying an external field. Unique images of magnetic domains and the internal structure of domain walls are also obtained with this technique when used in a scanning mode as demonstrated in Vol. II [4.157].

Element Specificity. Electrons excited from preselected atomic shells yield a measure of the local magnetization in the atom from which they are emitted. This is critical in ascertaining the contribution of each element in a magnetic alloy or from nearest neighbor atoms in the form of overlayers or substrate layers. With UMS, one obtains for instance the magnetization induced in the spacer layer between the magnetic layers.

Magnetization of Specific Electron States. Spin–polarized electron spectroscopies permit the selection of electrons emitted from specific states and determine their contribution to the total magnetization. This provides the most rigorous test of the theory of metallic magnetism. Furthermore, it shows which electron states are modified by the interface or by chemical reactions.

Magnetism in Unoccupied or Fully Occupied Electron States. Unoccupied or fully occupied states do not contribute to the magnetization, yet the states of spin–up electrons may have a different energy compared to the states of spin–down electrons. This exchange splitting can be determined by analyzing the spin of photoemitted electrons [4.160] or by observing the Bremsstrahlung emitted when a spin–polarized electron beam strikes the surface [4.161].

4.2.3 Field Dependence of the Magnetization

Magnetization curves, that is the dependence of the magnetization M on the external field strength H, provide important information on various magnetic phenomena. In studies utilizing spectroscopies with spin–polarized electrons, external magnetic fields cause a number of special problems. The reason is that stray magnetic fields created outside of the sample deflect the electrons due to the Lorentz force. This may have unfavorable effects on the location of the incident as well as emerging electron beam and its focusing. In this section we show that it is nevertheless possible to observe $M(H)$ and that important information on UMS is obtained from such measurements.

4.2.3.1 Perpendicular Magnetic Fields

If H is applied perpendicular to the layers of UMS, electrons photoemitted from the layers can be extracted parallel to the field lines [4.154]. The sample is located for instance in the bore of a superconducting coil. The light beam strikes the surface of the sample thereby causing the emission of electrons. The electrons emerging near the axis of the coil can be extracted from the external magnetic field by suitable electrical acceleration parallel to the axis of the coil, and an electron beam can be formed for the measurement of P. Those electrons that emerge off of the center spiral away along the field lines and cannot be extracted. This configuration has the disadvantage that an angular resolved measurement of the electrons cannot be made and that energy analysis is difficult. However, it has the advantage that the sample may be exposed to very high magnetic fields, in fact spin polarization measurements have been performed up to a field strength of $H = 5 \times 10^6$ A/m [4.154]. As an example, Fig. 20 shows the degree of the spin polarization vs. the external magnetic field for 2.6 and 1 monolayer (ML) of Fe grown epitaxially on Cu(0 0 1) measured at $T = 30$ K [4.159]. Fe on Cu(0 0 1) is a complex system, and controversial data have appeared in the literature. With the preparation conditions applied in [4.159], 1 ML exhibits no magnetic saturation, no remanence and no coercivity, while 2.6 ML clearly

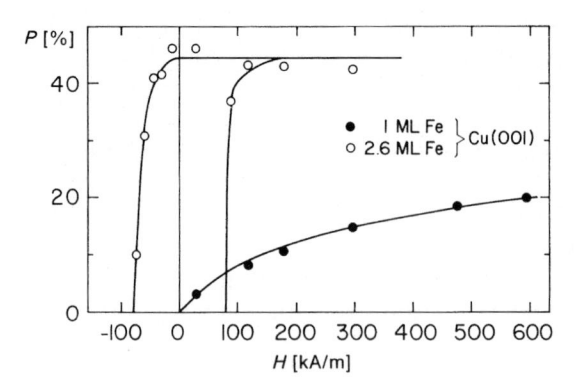

Fig. 4.20. Spin polarization P along the magnetic field H applied perpendicular to the surface of epitaxial Fe on Cu(0 0 1). The photoelectrons are excited with Photons of 5 eV from a xenon high pressure arc, the temperature of the measurement is $T = 30$ K, from [4.159]

shows the celebrated phenomenon of complete perpendicular magnetic remanence with a coercivity of $H_C \cong 80$ kA/m.

The question is: why does 1 ML not show perpendicular remanence? The uniaxial anisotropy K_s should appear at any surface whenever the cubic symmetry is broken, and the thinner the film, the larger the influence of K_s on the total anisotropy should be. The shape of $M(H)$ in the case of the 1 ML film and the absence of coercivity and remanence are the clear signature of superparamagnetism, that is the 1 ML film consists most likely of islands, not or only weakly coupled to one another, for example at terrace ledges (steps) on the Cu substrate. $M(H,T)$ in fact allows one to estimate the anisotropy and the average diameter D of the superparamagnetic islands. The physical idea behind this is the following: The magnetic anisotropy energy E of one island which is assumed not to interact with the others, is given by

$$E = C t_0 D^2 \sin^2 \theta, \tag{4.9}$$

where $C = (K_s - M^2/2\mu_0)$. t_0 the thickness of 1 ML and θ the angle between surface normal and magnetization M. If $C > 0$, E has minima at $\theta = 0$ and $\theta = \pi$ separated by an energy barrier $C t_0 D^2$. Thermal agitation may cause the magnetization to fluctuate between 0 and π. The time τ elapsed between two fluctuations has been calculated [4.162] to be

$$1/\tau = 10^9 \exp(-C t_0 D^2/2k_B T), \tag{4.10}$$

where k_B is the Boltzmann constant. As one measurement of P typically takes $\tau = 100$ s, one obtains from (4.10) the "blocking" temperature $T_B = C t_0 D^2/(25.3 k_B)$. For $T > T_B$, the observed remanence will be zero since M fluctuates between $\theta = 0$ and $\theta = \pi$. With increasing H, the fluctuations are suppressed as $M t_0 D^2 \cdot H$ grows. Assuming that M is as in the bulk, and that C for 1 ML is the same as the one observed on 2.6 ML, namely $C = M \cdot H_C$, one obtains $T_B = 28$ K with an island size of $D = 10$ nm. $M(H)$ can be calculated from the Brillouin function in good agreement with the experiment. All of the assumptions are reasonable and consistent with the experimental conditions [4.159]. However, as will be discussed below, large spin blocks are also a feature of the ideal 2D Heisenberg system [4.163] and can provide an alternative explanation for $M(H)$. The spin blocks fluctuate at a fast rate and disappear and reform in arbitrary locations. This is in contrast to the islands which are fixed by the steps or defects to one particular location. At any rate, we see that $M(H)$ curves are essential for understanding magnetism in UMS. Looking only at the remanent magnetization as many researchers have done, one would have concluded that 1 ML is not magnetic at 30 K. In the interpretation where 1 ML is superparamagnetic, the islands are fully magnetized up or down and localized on specific substrate sites. There is then no contradiction to the qualitative model of K_s due to *Néel* [4.164].

4.2.3.2 In-Plane Magnetic and Exchange Fields

In the next example, the external magnetic field is applied parallel to the sample surface. In this case, an awkward stray magnetic field is generated at the sample surface that severely affects the emerging low energy electrons. If the stray field is weak enough, in practice < 1 kA/m extending 1 mm from the sample surface, low energy cascade electrons emerging from the sample can still be focused adequately onto the entrance slit of the Mott polarimeter for measurement of P [4.165]. However, it is clear that only soft magnetic samples such as permalloy can be saturated in weak external fields. Any ultrathin sample deposited onto such a uniformly magnetized substrate will experience an exchange field H_{ex} due to the exchange interaction across the interface. The transfer of exchange fields solves the electron optical problem as the electrons are not Lorentz-deflected by H_{ex}, yet for achieving magnetic saturation, H_{ex} is as good as a real magnetic field. We will show below that H_{ex} ranging from 0–10^7 A/m may readily be generated with appropriate nominally nonmagnetic spacer layers between a magnetic substrate and magnetic overlayer. In fact, one of the most fascinating observations of recent years has been the oscillation of H_{ex} with the thickness of certain spacer layers such as Cr, Ru, Cu and others [4.166, 167].

Suppose an ultrathin ferromagnetic overlayer is prepared on top of a nominally nonmagnetic spacer layer such as Cr, and that this UMS sits on top of a thick soft magnetic substrate, for instance permalloy ($Ni_{78}Fe_{22}$). The substrate and overlayer are coupled by an exchange field H_{ex}. The fields acting in the overlayer are H_{ex}, the external field H and the anisotropy field H_A present in the overlayer. The magnetization M of the overlayer adjusts itself to a direction in which the sum of the resulting energies is zero. This situation has been numerically studied for the simplest possible model in which H_{ex} is assumed to be homogeneous over the whole interface[4.168, 169]. The hysteresis loop $M(H)$ of the overlayer is a replica of the substrate hysteresis loop if both H and H_A are much smaller than H_{ex}, but changes from regular to inverted when H_{ex} changes sign. Taking advantage of the short probing depth inherent in spin–polarized electron spectroscopy, *Donath* et al. [4.170] measured hysteresis loops of 1 nm thick sputter deposited overlayers coupled to bulk permalloy over Cr spacer layers in the thickness range $0 < x < 5$ nm. In this experiment, the low energy cascade electrons produced by a primary electron beam of 3 keV energy were collected into a beam and the degree P of spin polarization of the cascade electrons along the direction of the external field was measured. The loops of the overlayer determined by the cascade electrons were indeed replicas of the substrate loops as determined by the bulk sensitive magnetooptic Kerr effect except at certain thicknesses of the spacer layer x where they changed from regular to inverted indicating that the sign of the transferred exchange field oscillates in the case of sputtered polycrystalline NiFe/Cr/NiFe just as with Fe/Cr/Fe.

Insight into the critical magnetic properties of the overlayer and spacer layer is obtained when the loops are measured close to $H_{ex} = 0$, that is at the point where H_{ex} switches sign. This is realized experimentally by varying the thickness x of the spacer layer, but interestingly enough also by varying the temperature T. According to theoretical work for instance by *Edwards* et al. [4.171], one expects that mainly the amplitude of $H_{ex}(x)$ changes with T, but not the wavelength of the oscillations. Hence, at first there seems to be a severe contradiction to the theory. However, the change in sign of the experimental' H_{ex} is most likely due to the T dependent difference between exchange couplings in parts of the film with different thicknesses and/or defect structures.

The key to the understanding of the experiments [4.170] is the realization that it is not possible to have $H_{ex} = 0$ over the whole interface. There will always be some fluctuations of thickness and structure of the spacer layer creating patches of radius R where H_{ex} has a different sign compared to the rest of the film. In this situation, a magnetic domain will form in the patch, if the energy of a domain wall around the patch is smaller than the exchange energy gained by coupling to the substrate. The energy of a domain wall is $\sqrt{A_{11}K_s}$ per unit surface where A_{11} is the exchange stiffness on a path parallel to the surface and K_s the anisotropy of the overlayer. Hence, the surface energy of the patch is $= 2\pi R \cdot d\sqrt{A_{11}K_s}$ where d is the thickness of the overlayer. The exchange energy gained by coupling to the substrate is $R^2 \pi A_{ex}$, where A_{ex} is the coupling energy per unit interface area with the substrate. We obtain the following condition for domain formation in the patch:

$$R > 2d\sqrt{A_{11}K_s}/A_{ex}. \tag{4.11}$$

It is seen that the domains will "evaporate" as the size of the patches shrinks to a critical value. This is the two-dimensional analogue to the homogeneous nucleation of droplets in a gas. Typical values for the domain wall and exchange energy density close to the zero intercept of H_{ex} suggest that the limit occurs when the patch radius R is less than about 10 times the overlayer thickness, however A_{ex} critically depends on the structure of the spacer [4.172] and is therefore difficult to estimate. If no domains can be formed, *the ultrathin overlayer acts as one giant spin entity* in which all the spins remain coupled by A_{11}. However, if domains are formed, the entity is broken up into pieces, and the simplest possible model [4.168, 169] assuming a uniform average H_{ex} in the overlayer breaks down.

The experiments of *Donath* et al. [4.170] show that both possibilities can be realized in polycrystalline FeNi/Cr/FeNi. Figure 4.21 shows schematic hysteresis loops of the overlayer film for the case of $T = $ const, thickness x of spacer variable, and $x = $ const, T variable. Note that with the cascade electrons, one measures the overlayer magnetization only.

At $T = 90$, $H_{ex} > 0$ for $x = 0.3$ nm, and $H_{ex} < 0$ for $x = 1.2$ nm as the loop changes from regular to inverted. Close to $H_{ex} = 0$ at $x = 0.7$ nm, the loop is already inverted, but has a much smaller amplitude. This can only be understood by domain formation. Obviously, large enough patches in the film still have $H_{ex} > 0$, whereas for more than half of the interface $H_{ex} < 0$. The magnetization direction in each domain changes sign when the substrate switches, as we assumed for simplicity that the anisotropy of the overlayer $H_A \ll H_{ex}$, and $H \ll H_{ex}$. Considering the small period oscillations observed in [4.172] the breaking up of the overlayer film into large domains close to $H_{ex} = 0$ is expected with polycrystalline films. With $x = 1.2$ nm, the overlayer loop is inverted at $T = 90$ but regular at $T = 330$ K meaning that H_{ex} changed from a negative to a positive sign on increasing T. This phenomenon was observed only when the spacer layer was deposited at $T \cong 300$ K, but the transition H_{ex} is reversible; that is, irreversible processes such as interdiffusion cannot account for it. The

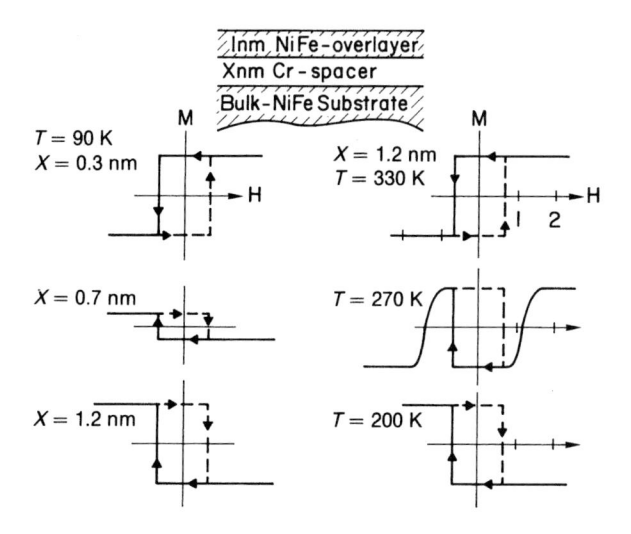

Fig. 4.21. A polycrystalline $Ni_{78}Fe_{22}$ film 1 nm thick is coupled over a Cr spacer layer of thickness x to bulk permalloy. The magnetization M of the overlayer film as it depends on the external magnetic field H and as measured with low energy cascade electrons is plotted schematically according to the results of [4.170]. The hysteresis loops $M(H)$ show the transition from positive to negative exchange coupling which occurs at $T = 90$ K, x variable and at $x = 1.2$ nm, T variable. The scale of the applied magnetic field H in kA/m is given in the loop at the upper right corner

transition occurs at $T \cong 270$ K where three switches of the overlayer magnetization M are observed when H varies from $+2$ kA/m to -2 kA/m. The field acting on M is $H' = H + H_{ex}$ as we assume $H_A = 0$ for simplicity. At large $H \geqslant +2$ kA/m, H' is positive and M is parallel to H. The first switching of M occurs when H' goes through zero yielding the result $H_{ex} \cong 1$ kA/m. The second switch of M occurs when H_{ex} changes sign as the substrate switches, and the third switch of M occurs again at $H' \leq 0$ but with $H_{ex} = +1$ kA/m because the substrate is already inverted. *Each of the switches of M occurs with full amplitude, indicating that the overlayer now acts as one giant spin entity.* According to 4.11 this means that the patches with reversed exchange coupling must be small, of the order of the thickness of 1 nm of the overlayer film. This, and the dependence of the phenomenon on the preparation conditions as reported in [4.170] suggests that ferromagnetic coupling across localized defects is responsible for it, and that this dominates in strength over the more delicate negative coupling at low T.

In conclusion, we see that important factors determining the magnitude and sign of H_{ex} are due to the chemistry, structure and the thickness of the nonmagnetic spacer layer. Most of the experiments dealing with magnetic exchange coupling across spacer layers use spectroscopies such as RHEED, Auger, etc. as evidence for continuous layer by layer growth though the lateral spatial resolution is insufficient as shown in the above example where defects with a small lateral extension of the order of the thickness of UMS are important. It appears that spin–polarized electron beam techniques can detect extremely small exchange couplings; in the example of Fig. 4.21, $H_{ex} = 1$ kA/m is readily determined amounting to the interface energy density of 1 µJ/m^2. Equation 4.11 suggests that the breaking of the giant spin ultrathin overlayer into domains should depend on its thickness d. Unique information on exchange interaction at defects can be gained from studying this and similar phenomena. It has long been anticipated that defects have a decisive influence on important magnetic properties such as coercivity. The small experimental coercivities cannot be explained by theory unless one assumes that rudimentary traces of domain walls or regions with negative exchange [4.173] exist somewhere in the material, e.g., at grain boundaries, interfaces or surfaces. The possibility is now at hand to design materials with "frustrated" patches for engineering the magnitude of the coercivity desired for the various applications.

4.2.4 Temperature Dependence of the Magnetization

The temperature dependence of the spontaneous magnetization $M_0(T)$ reveals the quantum mechanical nature of the magnetic interactions. Furthermore, the shape of $M_0(T)$, particularly close to the magnetic transition point T_C provides a critical test of various concepts on phase transitions. In the famous theory due to *Pierre Weiss* [4.188], one assumes that a mean field $H_m \propto M_0(T)$ exists, in which the magnetic moments are aligned against thermal agitation. The mean field theory is surprisingly successful in describing $M_0(T)$. It yields a universal, that is material independent temperature dependence of the reduced spontaneous magnetization $M_0(T^*)/M_0(0)$ with $T^* = T/T_C$. However, even in three dimensions, characteristic deviations from the mean field curve occur at low T^* due to spin wave excitations, and close to the transition point $T^* \cong 1$ due to spin–spin correlations. With the surface magnetization of bulk materials and with two-dimensional systems, these deviations are expected to become more prominent. In this section we will examine the contributions of magnetometry with spin–polarized electron beam techniques to the measurements of $M_0(T)$ in 2D and quasi-2D systems [4.174]. Conventional magnetometry on bulk magnets has already clearly shown that it is quite difficult to obtain the spontaneous magnetization $M_0(T)$ from the observed $M(H,T)$. It involves extrapolation of $M(H,T)$ to $H = 0$ at T = const. The resulting "$M_0(T)$" depends on preconceived ideas on how $M(H,T)$ should vary with H. Close to T_C, scaling theory postulates

$$M_0(T^*)/M(0) = (1 - T^*)^\beta, \tag{4.12}$$

where the critical exponent β is to be experimentally determined and compared to the theory. The uncertainty in the extrapolated $M_0(T^*)$ is particularly large close to T_C and translates into a large uncertainty of β and of T_C.

It should be noted that techniques measuring a scalar quantity do not yield the critical behavior of the spontaneous magnetization along a quantization axis. Examples of such scalar quantities include the hyperfine splitting of nuclear levels observed in, e.g., the Mössbauer effect, or the Zeeman splitting of electronic states observed in optical or electron spectroscopy. The scalar quantities depend on a spin–spin correlation function with a different T dependence compared to the one of the order parameter $M_0(T)$.

4.2.4.1 Two-Dimensional Ultrathin Films

Ultrathin polycrystalline NiFe films provide very simple and clear illustrations of the general features and problems encountered in the experimental determination of $M_0(T)$. Particularly simple is permalloy $Ni_{78}Fe_{22}$ in which the anisotropies are very small even with ultrathin samples. The films can be deposited by sputtering onto a thin Ta substrate, which in turn is located directly on the surface of a polished current carrying copper band producing the external magnetic field H [4.165]. A primary electron beam with an energy of 3 keV impinges on the permalloy film and the spin polarization $P(H,T)$ of the emerging low energy cascade electrons is measured to obtain information on $M(H,T)$.

Figure 4.22 shows, as an example, the results of measurements on a $Ni_{78}Fe_{22}$ sample with a thickness of 0.45 nm [4.165]. The thickness was determined by a quartz microbalance. As an insert, Fig. 3 also shows hysteresis loops at T = const. The spontaneous spin polarization $P_0(T)$ was obtained by linear extrapolation of $P(H,T)$ to $H = 0$. This extrapolation is quite obvious at low

T where magnetic saturation is clearly defined. However, as T_C is approached, the linear extrapolation becomes doubtful due to the curvature of $P(H,T)$. At $T^* \geq 0.9$, linear extrapolation is arbitrary, and the function $P(H,T)$ has to be known to obtain $P(0,T)$. It is also evident from Fig. 4.22 that the remanent polarization disappears at a much faster rate than the extrapolated saturation spin polarization $P_0(T)$. Close to T_C, $P(H,T)$ ressembles the paramagnetic Langevin function. These phenomena were observed already by *Weiss* and *Forrer* in 1926 [4.189] on bulk Ni, yet the parameter H/T in the Langevin function is ~ 100 times smaller in 2D films. Today, the explanation is clear: close to T_C, the magnetization breaks up into "spin blocks" rather than independent atoms. The magnetization of the spin blocks can be aligned in the external field. The amazing observation in Fig. 4.22 is how small the external field generating a sizeable magnetization is in 2D compared to 3D. This indicates that the spin blocks must be much larger in 2D. The question that remains in the 2D case is whether one sees the spin blocks, that is the inherent critical phenomena, or whether one simply has supermagnetic islands in which M_0 fluctuates according to 4.10. In the latter case, the apparent transition point T_0 should be identified with the blocking temperature T_B rather than with T_C. With supermagnetic islands, T_0 depends strongly on the concentration of impurities and the quality of the substrate, as T_B must depend critically on defects defining the size of the island. Again, the dependence of the critical behavior on material defects and purity has been studied in 3D much earlier [4.175].

However, the question is more important in 2D than in 3D, because the definition of a 3D superparamagnetic particle requires the unlikely existence of a closed surface of defects, whereas a 2D island only needs a closed line.

The experiments with 2D ferromagnets confirm extreme dependence of T_0 on impurities and the substrate preparation technique [4.165, 176]. Hence in many cases one would tend to identify the observed T_0 with T_B. However, theory gives no hint on how the true T_C should depend on the experimental conditions. Therefore, special experiments are necessary to decide whether the

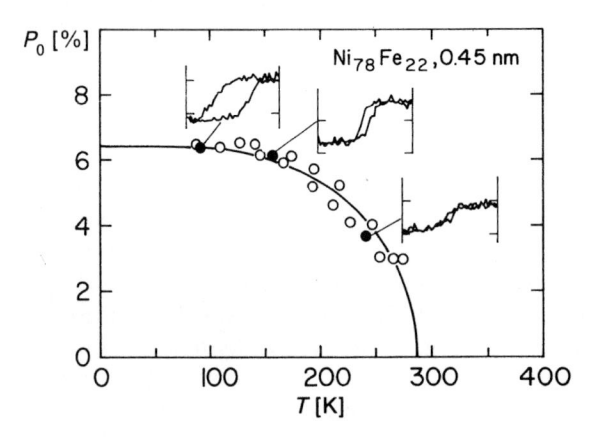

Fig. 4.22. Spin polarization $P_0(T)$ obtained by extrapolation of $P(H,T)$ to $H \to 0$ for a 0.45 nm thick NiFe film. The full curve is the Brillouin function for spin 1/2. In the hysteresis loops shown at $T = 91, 157,$ and 241 K, the difference between the adjacent marks on the ordinate indicates a 10% change of spin polarization and the applied field is swept from -2.0 to $+2.0$ kA/m. Data from [4.177, 165]

4.2 Probing Magnetic Properties with Spin-Polarized Electrons

observed T_0 is T_C of the 2D film or T_B of superparamagnetic islands. Before describing such experiments, we note that Fig. 4.22 also shows that the mean field curve for spin 1/2 reproduces the observed $P_0(T)$ within the experimental uncertainty if $T^* \leq 0.9$. This is in contrast to conventional spin wave theory which predicts a rapid linear decrease of $P_0(T)$ at low T with the low anisotropy material under consideration [4.177]. *Pini* et al. suggest that the interaction of spin waves can account for this failure as the interaction is more important in 2D than in 3D [4.178].

Kerkmann et al. have addressed the question on superparamagnetic islands versus the inherent spin fluctuations in an experiment [4.174]. They deposited ~ 1 monolayer of Co onto Cu(100) and measured the low energy cascade electron spin polarization excited from a spot of only ~ 20 nm diameter. *This high lateral resolution makes it possible to directly measure the spontaneous magnetization $M_0(T)$ within one domain without applying an external field.* Figure 4.23 shows that $M_0(T)$ decays fast with T and defines a sharp transition temperature T_0. To decide between $T_0 = T_C$ or $T_0 = T_B$ one uses the fact that superparamagnetic islands are anchored to the location of the defects and that their magnetization fluctuates slowly according to 4.10. The radius R of the islands or of the inherent "spin blocks" can be estimated from the field dependence $M(H,T)$ of the magnetization. Figure 4 shows that at $T = T_0$ there is a very large response to external fields as weak as $H = 8$ kA/m. From $R^2 \pi d M_0 H = kT_0$, one obtains

$$R \cong (kT_0/\pi dM_0 H)^{1/2}. \tag{4.13}$$

With 1 monolayer of Co, $d \cong 0.2 \times 10^{-9}$ m, $T_0 \cong 300$ K, $M_0 \cong 2$ V s/m². Hence at $H = 8$ kA/m, $R \cong 20$ nm. That is, the expected size of the islands is of the same order as the lateral resolution of the magnetization measurement. Therefore, fluctuations of the spontaneous magnetization or alterations of domains should become observable in this experiment if superparamagnetism is present. No such phenomena were seen within 1 K from T_C. Therefore, one can conclude that the inherent phase transition was observed with this Co/Cu(100) sample. In agreement with the general theory of 2D ferromagnets [4.163], it is the large spin blocks that generate the observed phenomenal response to external fields at $T \cong T_C$.

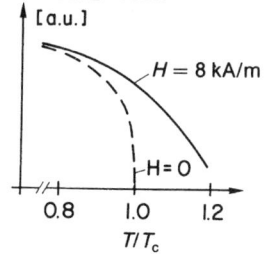

Fig. 4.23. In-plane magnetization of one layer of Co on Cu(100) withi one Weiss domain in zero applied field, measured via the low energy cascade spin polarization with a lateral resolution of 20 nm, from *Kerkmann* et al. [4.179]. The temperature dependence of the magnetization in an applied field of 8 kA/m is also shown

4.2.4.2 Quasi-Two-Dimensional Ferromagnetic Films

Considering the large response of 2D ferromagnets to external fields, it is of considerable interest to study particularly the paramagnetic region $T > T_C$ in all kinds of external disturbances. So far, the closest realization of such "quasi-2D" ferromagnets have been layered structures with small interplane coupling [4.180]. For instance, K_2CuF_4 crystallizes in magnetic layers with spin 1/2 and the dominant interaction is Heisenberg-like. The magnetic layers are separated by nonmagnetic spacers. The relative perpendicular exchange coupling $J_\perp/J$ is of the order of 10^{-3}, with J the average exchange within the layer. A larger spin asymmetry of approximately 1% exists in the coupling within the layers. Due to this asymmetry, the easy direction of the magnetization lies in the plane of the layers, yet no direction in that plane is preferred. This situation corresponds to the $X-Y$ model in a certain range of temperatures near T_C. Spontaneous magnetization does not appear in the $X-Y$ model, but rather infinite magnetic susceptibility. Hence any small external field or magnetic anisotropy in the plane of the film will produce a substantial magnetization. The Kosterlitz–Thouless transition point is expected to shift to higher temperatures as the fields or the anisotropies in the plane of the 2D film increase [4.181].

Ultrathin magnetic films coupled through nonmagnetic spacer layers to a 3D bulk ferromagnet make possible the engineering of quasi 2D-ferromagnets as will be shown below. The magnetization of the overlayer film can be measured separately from that of the bulk ferromagnet by virtue of the small probing depth of the spin polarized electron beam techniques.

Donath et al. [4.182] studied 2D polycrystalline Fe films exchange coupled through nominally nonmagnetic Ta layers of various thicknesses to bulk Fe. The exchange field H_{ex} transferred from the bulk through the Ta spacer layer into the overlayer film is the analoge of an external magnetic field applied parallel to the magnetic planes of the crystal. As long as one stays away from the Curie point of the bulk substrate, H_{ex} is constant in a small interval of temperatures. One magnetic layer in K_2CuF_4 corresponds to the ultrathin polycrystalline Fe film. This film is made up of many small crystallites which each have one or several easy directions of magnetization, but these directions are different in each crystallite if preferred crystallization directions and in-plane stress are absent. If the magnetic coupling between the crystallites is stronger than the individual anisotropies, their random orientation means that there is no preferred magnetization direction in the plane of the polycrystalline films, just as with K_2CuFe_4. Furthermore, the preferential direction of M is in the plane of the Fe film as well, rather than perpendicular to it because of the shape anisotropy. The ratio of the shape to the exchange energy is approximately 1% in Fe. Therefore, the analogy to K_2CuF_4 is quite complete.

Hirakawa and *Ubukoshi* [4.183] have measured $H(H_{ex}, T)$ with K_2CuF_4 where the rôle of the exchange field H_{ex} was taken by a real magnetic field applied parallel to the layers. The explanation in terms of the XY model [4.181] is convincing, in particular the shape of $M(H_{ex} = \text{const}, T)$ and the increase of

4.2 Probing Magnetic Properties with Spin-Polarized Electrons

the ordering temperature T_C with increasing H_{ex}. Although the magnetic transition is increasingly smeared out as H_{ex} increases, T_C is still clearly defined for instance by the point of inflection of the $M(H_{ex} = \text{const}, T)$ curve.

It is clear from the experiments described in this chapter that superparamagnetism is the real obstacle in observing 2D magnetism, yet the magnetic properties are also the best sensor of superparamagnetic behavior. This is illustrated in Fig. 4.24 where the spin polarization P of the low energy cascade electrons from a sputter-deposited Fe film of 0.5 nm average thickness is plotted versus the temperature. The film is deposited onto a Ta spacer of 0.7 nm thickness which in turn sits on a substrate assembly consisting of a thin Fe overlayer on a thicker permalloy substrate layer. The Fe substrate overlayer is magnetically saturated in the direction of the measurement of P by exchange coupling to the permalloy film. In this way, the bulk substrate has the electronic properties of Fe near the substrate surface, but also the desirable soft magnetic properties of permalloy. This experimental approach allows one to magnetically saturate the Fe substrate in weak external fields, but has no other critical importance.

The 0.5 nm thick polycrystalline Fe overlayer film is now a quasi-2D film because a large H_{ex} is transferred into it from the substrate through the Ta-spacer. Fig. 4.24 shows that P is nevertheless very low rising to only 1–2% as the sample is cooled to $T = 100$ K. This polarization did not change on annealing to $T = 450$ K. However, with submonolayer contamination of C and O, P increased dramatically. The increase of P occurred despite the well known fact that C and O attenuate the emission of spin-polarized electrons from the 3d band of Fe. Further adsorption of C and O did not affect $P(T)$ any more until severe contamination levels of the order of one monolayer or more were reached.

The key to understanding this phenomenon is that the shape of $P(H_{ex},T)$ also changed dramatically, namely from being more or less linear to a curve expected for $M(H_{ex},T)$. The curve fitted to the filled data points in Fig. 4.24

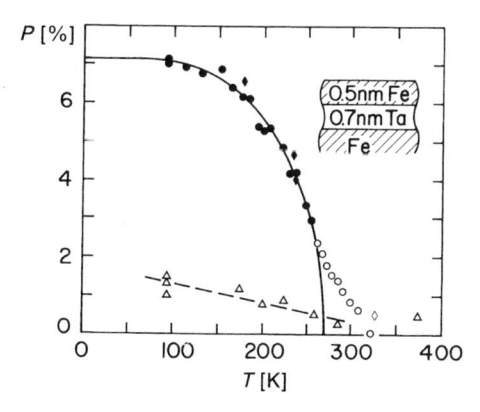

Fig. 4.24. $P(T)$ of the low energy cascade electrons of a polycrystalline Fe film 0.5 nm thick coupled to the bulk over a 0.7 nm thick Ta spacer. Triangles are observed immediately after deposition and after heating to 450 K, circles and diamonds after submonolayer contamination with C and O. The solid line is the Brillouin function for spin 1/2 in the absence of a transferred exchange field fitted to the filled symbols

taken with the "sub-monolayer contaminated" film is the mean field curve for spin 1/2 in $H_{ex} = 0$. We see that this curve does represent the data well except close to the transition point T_C where a tail occurs. This is due to the presence of H_{ex} transferred from the substrate and demonstrates that we now are dealing with a "quasi-2D" phase transition. The changes of P occurring during the conversion from the freshly deposited to the aged film are induced by the transition from superparamagnetism to ferromagnetism. In line with the findings of *Egelhoff* and *Steigerwald* [4.184], small amounts of contaminations such as C and O must have acted as surfactants allowing the Fe to lower its surface energy in order to spread out and gain surface energy of uncovered Ta. This is in agreement with the observation that annealing does not induce the transition from superparamagnetism to ferromagnetism and that T_C as far as one can tell does not change in that transition. Hence Fig. 4.24 demonstrates that it is possible, with the help of minute amounts of contamination of C and O acting as surfactants, to avoid the obstacle of superparamagnetism and sputter deposit ultrathin polycrystalline Fe-films on a Ta-substrate in a stable ferromagnetic state.

The next step is now to prepare the 2D films on a thinner Ta substrate in order to increase H_{ex} transferred from the substrate. It turns out that the changes of $P(H_{ex},T)$ occurring with the adsorption of residual gas molecules are less dramatic with a thinner Ta spacer, that is with increased H_{ex}. This is in agreement with the expectation that as H_{ex} increases, the superparamagnetic fluctuations must be increasingly suppressed. Finally, a stable $P(H_{ex},T)$ is reached in all cases, and is again well represented by the Brillouin function for spin 1/2 at low T/T_C. However, close to T_C a tail occurs which increases with decreasing thickness of the Ta spacer, that is with increasing H_{ex}, consistent with the phase transition in a quasi-2D system.

As the thickness x of the nonmagnetic Ta spacer decreases, the Fe overlayer couples more strongly to the 3D magnetic substrate and crossover from 2D magnetism to 3D magnetism occurs. Fig. 4.25 summarizes the results with Ta spacer thicknesses of $x = 0.3, 0.5$, and 0.7 nm thickness. The temperature is in reduced units $T = T/T_{CB}$ where $T_{CB} = 1043$ K is the Curie point of bulk Fe; the T dependence of the magnetization in bulk Fe is also given for comparison. The relative spin polarization $P(H_{ex},T)/P(H_{ex},0)$ for the quasi-2D films is plotted in the stable ferromagnetic state. The transition points T_C are increasingly smeared out as H_{ex} grows, yet T_C is still clearly defined. The solid lines are calculated mean field curves in an external magnetic field H_{ex}. They are not based on any theoretical model for quasi $-$ 2D films, yet they fit the experimental data well within the experimental uncertainty. From these fits one obtains $T_C = 421, 344, 270$ K, and $H_{ex} = 32, 16, 8$ MA/m for Ta spacer thicknesses of $x = 0.3, 0.5, 0.7$ nm respectively. One sees that the "ansatz"

$$H_{ex} = H_0 \, e^{-\alpha x} \tag{4.14}$$

describes the dependence of H_{ex} transferred through the Ta spacer of thickness x taking $H_0 = 80$ MA/m and $\alpha = 3.1$ nm^{-1}. This value of H_0 corresponds to the

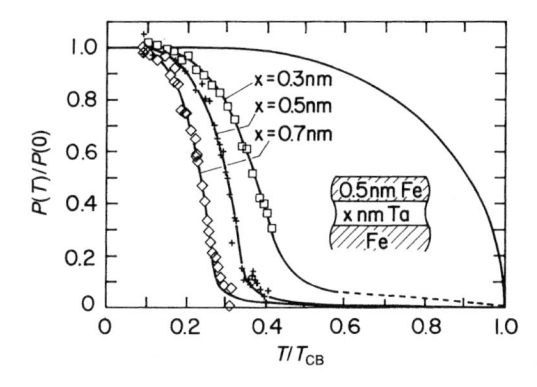

Fig. 4.25. Relative spin polarization $P(T')/P(0)$ of the low energy cascade electrons from the 0.5 nm polycrystalline Fe film in the ferromagnetic state for various thicknesses x of the Ta spacer. $T = T/T_{CB}$ where $T_{CB} = 1043$ K is the Curie point of bulk Fe. The solid curves are calculated from mean field theory for $s = 1/2$ in H_{ex} which is chosen to match the observations. The mean field curve for bulk Fe ($H_{ex} = 0$) is shown to illustrate the crossover from 2D to 3D magnetism

molecular field in bulk Fe reduced by ~ 30% to take into account the smaller number of nearest neighbors in the surface.

Theoretically, both a space anisotropy and a spin anisotropy lead to an increase of T_C in quasi-2D systems. In fact, the *dependence of T_C on the anisotropies is the typical signature of the 2D system* as it does not occur in the 3D system. Noting that H_{ex} = const in a temperature interval close to T_C of the overlayer, one is led to suppose that the observed shift of T_C is due to spin anisotropies yielding

$$T_C = C_1 T_{CB}/\ln(C_2 T_{CB}/H_{ex}), \qquad (4.15)$$

where C_1 and C_2 are constants and H_{ex} describes the strength of the anisotropy [4.181]. This general form of the dependence of T_C on anisotropies is found with many different 2D models and assumptions. Combining (4.14) and (4.15) yields

$$T_{CB}/T_C = A \cdot x + B. \qquad (4.16)$$

The experimental data are consistent with (4.16) as shown in [4.182].

4.2.5 Magnetism away from Equilibrium

When a magnetic field is applied suddenly to a magnetic material, the magnetization must change to reach a new equilibrium. This is a classical topic in magnetism, and the reversible and irreversible domain wall motions, Barkhausen jumps and processes of coherent rotation of magnetization are now being studied with UMS [4.185]. However, with the advent of intense pulsed laser beams, a more fundamental issue can also be addressed. In the focus of such a laser beam, large amounts of energy may be deposited in a solid in an extremely short time. The energy is initially generated in the form of electron–hole pair excitations, that is, the electron gas in the solid can be heated at a very fast rate. The hot electron gas generates phonons. The establishment of the equilibrium temperature between the electron gas and the phonons takes

~ 10^{-12} s. In a magnetic solid, the electrons and to a lesser extent also the phonons may generate spin waves (magnons), as well. It will be shown that the establishment of the magnetic equilibrium temperature takes as long as 10^{-10} s, the bottleneck being the transfer of the angular momentum to the lattice. Laser induced photoemission of electrons makes possible the measurement of the spin polarization of the electrons as it changes while the thermal equilibrium is established in a magnetic solid. Apart from its basic interest and novelty, this technique also has technological implications, for instance in thermomagnetic recording where the storage medium is heated by short laser pulses to induce magnetization reversal.

With UMS, ultrafast magnetometry should become a key issue in the future. First, it will make it possible to perform magnetic measurements at higher temperatures which can normally not be done because the layers interdiffuse. As diffusion is a slow process, fast heating at a rate of nanoseconds where electrons, phonons and magnons are in equilibrium combined with ultrafast measurement of the magnetization at the picosecond level will allow one to study the delicate UMS even at elevated temperatures. Second, the establishment of the magnon equilibrium temperature can now be determined directly. It depends on the mechanism by which the electron spin is coupled to the lattice, that is on the magnetic anisotropy. In order to depolarize the electrons or to change the direction of the spin polarization in a solid, one needs to transfer angular momentum to the solid. Hence the measurement of the spin lattice relaxation time provides basic information on the magnetic anisotropy which has its origin in the spin–orbit (l,s) interaction. With UMS, the (l,s) coupling can be very different from the bulk, particularly at the interfaces where the crystal symmetry is broken and the orbital moment is quenched to a lesser extent than in the bulk.

Upper and lower limits for the time needed to establish thermal equilibrium of the magnetization in a solid have been obtained with a simple experiment [4.186]. The sample, a polycrystalline piece of Fe, is magnetically saturated in an external field. The pulse of a laser is focused onto the surface of Fe in UHV. The laser pulse has two functions: (i) it heats the sample, and (ii) it induces the emission of photoelectrons from it. The power of the pulse can be large enough to melt the Fe. The melting is signaled by the onset of positive ion emission. The spin polarization *P* of the photoelectrons measures whether or not the Fe is still magnetized. Fig. 4.26 shows that *P* is reduced to zero on melting the Fe with a pulse duration of 20 ns. However, if the laser pulse is only 20 ps long, the spin polarization of Fe persists even in the liquid state. This clearly shows that the establishment of magnetic equilibrium takes longer than 20 ps and is faster than 20 ns. One can also show with this technique that the establishment of equilibrium between the heated electron gas and the phonons is much faster, as follows. For this experiment, the laser pulse has to be circularly polarized. Circularly polarized light can induce the emission of spin-polarized electrons, for instance in β-Sn. The polarization occurs only in a crystal and not in a liquid. The inset in Fig 4.26 shows that the polarization of the optically pumped photoelectrons from β-Sn is reduced to zero on melting even when the laser pulse is only

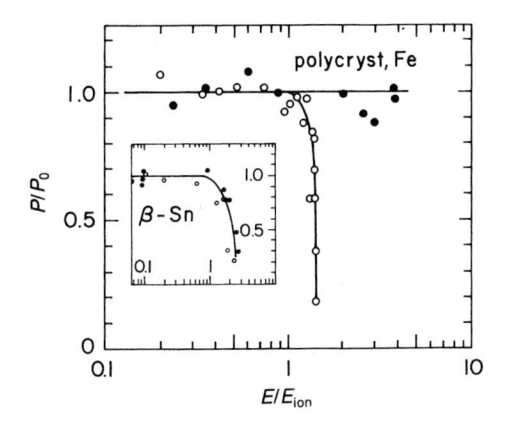

Fig. 4.26. Relative photoelectron spin polarization P/P_0 of pulsed laser excited photoelectrons. P_0 is the polarization of photoelectrons emitted at very low light intensity. The open circles are for a long laser pulse (20 ns duration) and the filled circles for a short laser pulse (30 p duration). The calibration of the laser pulses is in units of E_{ion} which is the energy at which positive ion emission sets in. With β-Sn, P is due to optical pumping with circularly polarized pulses, and there is no difference between short and long pulses. This proves that E_{ion} approximately indicates melting, and that the equilibrium between the hot electrons and the photons is established much faster than with magnons, from [4.186]

$\sim 10^{-11}$ s long. This proves that melting is properly indicated by the emission of positive ions.

In a more complete experiment, using a heating pulse with photon energy below the photoelectric threshold and a probing pulse with photon energy above photoelectric threshold, the time τ needed for establishment of the magnetization equilibrium can be explicitly determined. *Vaterlaus* et al. found for a Gd film on an Fe substrate $\tau = (1.0 \pm 0.8)10^{-10}$ s [4.187]. This is not the spin lattice relaxation time determined for instance in ferromagnetic resonance. The relaxation of the polarization of a *hot* spin polarized electron gas in equilibrium with a hot crystal lattice is described by τ. The technique can be applied to all kinds of materials. Due to the surface sensitivity of photoemission it is particularly suited to study the relaxation of the surface magnetization and of the magnetization in UMS.

References

Section 4.1

4.1 The basics of electron spectroscopies are described in textbooks on Surface Physics, e.g. G. Ertl J. Küppers: *Low Energy Electrons and Surface Chemistry* (VCH Verlagsgesellschaft, Weinheim 1985)

4.2 J. Kessler: *Polarized Electrons* (Springer, Berlin, Heidelberg 1985)

4.3   R. Feder (ed.): *Polarized Electrons in Surface Physics*, (World Scientific, Singapore 1985)
4.4   J. Kirschner: "*Polarized Electrons at Surfaces*" in *Springer Tracts in Mod. Physics*, Vol. 106 (Springer, Berlin, Heidelberg 1985)
4.5   C.J. Powell: J. Electron Spectrosc. Relat. Phenom. **47**, 197 (1988)
4.6   M.P. Seah, W.A. Dench: Surface and Interface Analysis, **1**, 2 (1979)
4.7   J.L. Erskine, E.A. Stern: Phys. Rev. Lett. **30**, 1329 (1973)
4.8   D.T. Pierce, H.C. Siegmann: Phys. Rev. **B9**, 4035 (1974)
4.9   D.P. Pappas, K.-P. Kämper, B.P. Miller, H. Hopster, D.E. Fowler, C.R. Brundle, A.C. Luntz, Z.-X. Shen: Phys. Rev. Lett. **66**, 504 (1991)
4.10 M.P. Gokhale, D.L. Mills: Phys. Rev. Lett. **66**, 2251 (1991)
4.11 D.L. Abraham, H. Hopster: Phys. Rev. Lett. **58**, 1352 (1987)
4.12 M. Donath, D. Scholl, H.C. Siegmann, E. Kay: Appl. Phys. **A52**, 206 (1991)
4.13 O. Paul, S. Toscano, K. Totland, M. Landolt: Surf. Sci. **251/252**, 27 (1991)
4.14 X. Zhang, H. Hsu, F.B. Dunning, G.K. Walters: Phys. Rev. B **44**, 9133 (1991)
4.15 O.M. Paul, Determination of energy-resolved inelastic mean free paths of electrons with spin-polarized secondary and Auger electron spectroscopy, Dissertation ETH No. 9210, Eidgenössische Technische Hochschule, Zürich (1990)
4.16 C.K. Sinclair: Proc. 8th Int. Symp. on High Energy Spin Physics, Minneapolis, 1988, AIP Conf. Proc. **187**, 1412 (1989)
4.17 E. Reichert: Proc. 9th Int. Symp. on High Energy Spin Physics, Bonn 1990, High Energy Spin Physics, ed. by K.-H. Althoff, W. Meyer, Vol. 1 (Springer, Berlin, Heidelberg 1991) pp. 303–317
4.18 J. Unguris, D.T. Pierce, R.J. Celotta: Rev. Sci. Instrum. **57** 1324 (1986)
4.19 L.G. Gray, M.W. Hart, F.B. Dunning, G.K. Walters: Rev. Sci. Instrum. **58**, 2195 (1987)
4.20 F.B. Dunning, L.G. Gray, J.M. Ratliff, F.-C. Tang, X. Zhang, G.K. Walters, Rev. Sci. Instrum. **58**, 1706 (1987)
4.21 F.-C. Tang, X. Zhang, F.B. Dunning, G.K. Walters, Rev. Sci. Instrum. **59**, 504 (1988)
4.22 J. Kirschner in [4.3] "Sources and Detectors for Polarized Electrons" pp. 245–286
4.23 J.J. McClelland, M.R. Scheinfein, D.T. Pierce: Rev. Sci. Instrum. **60**, 683 (1989)
4.24 J.J. McClelland, M.R. Scheinfein, D.T. Pierce: Rev. Sci. Instrum. **60**, 683 (1989)
4.25 D.P. Pappas, H. Hopster: Rev. Sci. Instrum. **60**, 3068 (1989)
4.26 D.M. Oro, W.H. Butler, F.-C. Tang, G.K. Walters, F.B. Dunning: Rev. Sci. Instrum. **62**, 667 (1991)
4.27 D. Tillmann, R. Thiel, E. Kisker: Z. Phys. B **77**, 1 (1989)
4.28 F.U. Hillebrecht, R. Jungblut, E. Kisker: Phys. Rev. Lett. **65**, 2450 (1990)
4.29 T.J. Gay, M.A. Khakoo, J.A. Brand, J.E. Furst, W.V. Meyer, W.M.K.P. Wijayaratna, F.B. Dunning: Rev. Sci. Instrum. **63**, 1 (1992)
4.30 M. Uhrig, A. Beck, J. Goeke, F. Eschen, M. Sohn, G.F. Hanne, K. Jost, J. Kessler: Rev. Sci. Instrum. **60**, 872 (1989)
4.31 J. Kirschner, R. Feder: Phys. Rev. Lett. **42**, 1008 (1979)
4.32 H. Hopster, D.L. Abraham: Rev. Sci. Instrum. **59**, 49 (1988)
4.33 D.T. Pierce, F. Meier, P. Zürcher: Appl. Phys. Lett. **26**, 670 (1975)
4.34 D.T. Pierce, F. Meier, P. Zürcher: Phys. Lett. A **51**, 465 (1975)
4.35 D.T. Pierce, R.J. Celotta, G.-C. Wang, W.N. Unertl, A. Galejs, C.E. Kuyatt, S.R. Mielczarek: Rev. Sci. Instrum. **51**, 478 (1980)
4.36 D. Conrath, T. Heindorff, A. Hermanni, N. Ludwig, E. Reichert: Appl. Phys. **20**, 155 (1979)
4.37 J. Kirschner, H.P. Oepen, H. Ibach: Appl. Phys. A **30**, 177 (1983)
4.38 S.F. Alvarado, F. Ciccacci, M. Campagna: Appl. Phys. Lett. **39**, 615 (1981)
4.39 F. Meier in Ref. [4.3]
4.40 H.-J. Drouhin, C. Hermann, G. Lampel: Phys. Rev. B **31**, 3872 (1985)
4.41 T. Maruyama, R. Prepost, E.L. Garwin, C.K. Sinclair, B. Dunham, S. Kalem: Appl. Phys. Lett. **55**, 1686 (1989)

References

4.42 F. Meier, A. Vaterlaus, F.P. Baumgartner, M. Lux-Steiner, G. Doell, E. Bucher: In *High Energy Spin Physics*, ed. by W. Meyer, E. Steffens, W. Thiel, Vol. 2 (Springer, Berlin, Heidelberg 1991) pp. 25–29
4.43 T. Maruyama, E.L. Garwin, R. Prepost, G.H. Zapalac, J.S. Smith, J.D. Walker: Phys. Rev. Lett. **66**, 376 (1991)
4.44 T. Nakanishi, H. Aoyagi, H. Horinaka, Y. Kamiya, T. Kato, S. Nakamura, T. Saki, M. Tsubata: Phys. Lett. A **158**, 345 (1991)
4.45 T. Omori, Y. Kurihara, T. Nakanishi, H. Aoyagi, T. Baba, T. Furuya, K. Itoga, M. Mizuta, S. Nakamura, Y. Takeuchi, M. Tsubata, M. Yoshioka: Phys. Rev. Lett. **67**, 3294 (1991)
4.46 G. Chrobok, M. Hofmann: Phys. Lett. A **59**, 257 (1976)
4.47 J. Unguris, D.T. Pierce, A. Galejs, R.J. Celotta: Phys. Rev. Lett. **49**, 72 (1982)
4.48 E. Kisker, W. Gudat, K. Schröder: Solid State Commun. **44**, 591 (1982)
4.49 H. Hopster, R. Raue, E. Kisker, G. Güntherodt, M. Campagna: Phys. Rev. Lett. **50**, 71 (1983)
4.50 A. Bringer, M. Campagna, R. Feder, W. Gudat, E. Kisker, E. Kuhlmann: Phys. Rev. Lett. **42**, 1705 (1979)
4.51 E. Tamura, R. Feder: Phys. Rev. Lett. **57**, 759 (1986)
4.52 R. Allenspach, Magnetic Characterization at Surfaces by Spin Polarized Electron Spectroscopies, Dissertation ETH No 7952, Eidgenössische Technische Hochschule, Zürich, 1985
4.53 J. Kirschner, "Spin-Polarized Secondary Electrons from Ferromagnets," in *Study of Surfaces and Interfaces by Electron Optical Techniques*, ed. by A. Howie and U. Valdre (Plenum, New York 1988) pp. 267–283
4.54 M. Taborelli, Magnetism of Epitaxial Thin Films and Single Crystal Surfaces Studied with Spin-Polarized Secondary Electrons, Dissertation, Eidgenössische Technische Hochschule, Zürich, 1988
4.55 H. Hopster: Phys. Rev. B **36**, 2325 (1987)
4.56 D.R. Penn, S.P. Apell, S.M. Girvin: Phys. Rev. B **32**, 7753 (1985)
4.57 J. Glazer, E. Tosatti: Solid State Commun. **52**, 905 (1984)
4.58 O.M. Paul, Determination of energy-resolved inelastic mean free paths of electrons with spin-polarized secondary and Auger electron spectroscopy, Dissertation ETH No. 9210, Eidgenössische Technische Hochschule, Zürich (1990) pp. 72–79
4.59 D.P. Pappas, K.-P. Kämper, H. Hopster: Phys. Rev. Lett. **64**, 3179 (1990)
4.60 D.P. Pappas, C.R. Brundle, H. Hopster: Phys. Rev. B **45**, 8169 (1992)
4.61 F.B. Dunning, G.K. Walters: "Elastic Spin-Polarized low Energy Electron Diffraction from Non-Magnetic Surfaces" in [4.3] pp.287–320
4.62 C.-G. Wang, B.I. Dunlap, R.J. Celotta, D.T. Pierce: Phys. Rev. Lett. **42**, 1349 (1979)
4.63 D.L. Abraham, H. Hopster: Phys. Rev. Lett. **59**, 2333 (1987)
4.64 R.J. Celotta, D.T. Pierce, G.-C. Wang, S.D. Bader, G.P. Felcher: Phys. Rev. Lett. **43**, 728 (1979)
4.65 S.F. Alvarado, R. Feder, H. Hopster, F. Ciccacci, H. Pleyer: Z. Phys. B **49**, 129 (1982)
4.66 D.T. Pierce, R.J. Celotta, J. Unguris, H.C. Siegmann: Phys. Rev. B **26**, 2566 (1982)
4.67 R. Feder: 'Principles and Theory of Electron Scattering and Photoemission' pp. 125–124 in ref. 4.3
4.68 J. Kirschner: Phys. Rev. B **30**, 415 (1984)
4.69 W. Dürr, M. Taborelli, O. Paul, R. Germar, W. Gudat, D. Pescia, M. Landolt: Phys. Rev. Lett. **62**, 206 (1989)
4.70 C. Carbone, S.F. Alvarado: Phys. Rev. B **36**, 2433 (1987)
4.71 S.F. Alvarado, C. Carbone: Physica B **149**, 43 (1988)
4.72 D. Pescia, D. Kerkmann, F. Schumann, W. Gudat: Z. Phys. B **78**, 475 (1990)
4.73 D. Kerkmann, D. Pescia, J.W. Krewer, E. Vescovo: Z. Phys. B **85**, 311 (1991)
4.74 E. Bauer: In *Chemistry and Physics of Solid Surfaces VIII*, ed. by R. Vanselow, R. Howe (Springer, Berlin, Heidelberg 1990)
4.75 M.S. Altman, H. Pinkvos, J. Hurst, H. Poppa, G. Marx, E. Bauer: Mat. Res. Soc. Symp. Proc., Vol. 232, 125 (1991)

4.76　R. Feder, S.F. Alvarado, E. Tamura, E. Kisker: Surf. Sci. **127**, 83 (1983)
4.77　S.J. Porter, J.A.D. Matthew: J. Phys.: Cond. Matter **1**, SB13 (1989)
4.78　G. Waller, U. Gradmann: Phys. Rev. B **26**, 6330 (1982)
4.79　E. Tamura, R. Feder, G. Waller, U. Gradmann: Phys. Stat. Solidi B **157**, 627 (1990)
4.80　A. Ormeci, B.M. Hall, D.L. Mills: Phys. Rev. B **42**, 4524 (1990)
4.81　A. Ormeci, B.M. Hall, D.L. Mills: Phys. Rev. B **44**, 12369 (1991)
4.82　R. Feder, H. Pleyer: Surf. Sci. **117**, 285 (1982)
4.83　S.F. Alvarado, M. Campagna, H. Hopster: Phys. Rev. Lett. **48**, 51 (1982)
4.84　S.F. Alvarado, M. Campagna, F. Ciccacci, H. Hopster: J. Appl. Phys. **53**, 7920 (1982)
4.85　B.H. Dauth, S.F. Alvarado, M. Campagna: Phys. Rev. Lett. **58**, 2118 (1987)
4.86　C. Rau, J. Magn. Magn. Mater. **31–34**, 874 (1983)
4.87　C. Rau, S. Eichner: Phys. Rev. B **34**, 6347 (1986)
4.88　D. Weller, S.F. Alvarado, W. Gudat, K. Schroder, M. Campagna: Phys. Rev. Lett. **54**, 1555 (1985)
4.89　D. Weller, S.F. Alvarado, M. Campagna: Physica **130B**, 72 (1985)
4.90　C.Rau, C. Jin, M. Robert: Phys. Lett. **138**, 334 (1989)
4.91　D. Weller, S.F. Alvarado: Phys. Rev. B **37**, 9911 (1988)
4.92　W. Weber, D. Kerkmann, D. Pescia, D.A. Wesner, G. Guntherodt: Phys. Rev. Lett. **65**, 2058 (1990)
4.93　H. Ibach, D.L. Mills: *Electron Energy Loss Spectroscopy and Surface Vibrations* (Academic Press, New York 1982)
4.94　H. Hopster: Phys. Rev. B **42**, 2540 (1990)
4.95　J. Kirschner, D. Rebenstorff, H. Ibach: Phys. Rev. Lett. **53**, 698 (1984)
4.96　H. Hopster, R. Raue, R. Clauberg: Phys. Rev. Lett. **53**, 695 (1984)
4.97　J. Kirschner: Phys. Rev. Lett. **55**, 973 (1985)
4.98　J. Kirschner, S. Suga, Surf. Sci. **178**, 327 (1986)
4.99　D. Venus, J. Kirschner: Phys. Rev. B **37**, 2199 (1988)
4.100　D.L. Abraham, H. Hopster: Phys. Rev. Lett. **62**, 1157 (1989)
4.101　D.R. Penn, P. Apell: Phys. Rev. B **38**, 5051 (1988)
4.102　H. Hopster, D.L. Abraham: Phys. Rev. B **40**, 7054 (1989)
4.103　J. Kirschner, J. Hartung: Vacuum **41**, 491 (1990)
4.104　J. Kirschner, E. Langenbach: Solid State Commun. **66**, 761 (1988)
4.105　K.-P. Kamper, D.L. Abraham, H. Hopster: Mater. Res. Soc. Symp. Proc., Vol. 231, 71 (1992)
4.106　K.-P. Kamper, D.L. Abraham, H. Hopster: Phys. Rev. B **45**, 14335 (1992)
4.107　*Photoemission in Solids I*, ed. by M. Cardona, L. Ley, Topics Appl. Phys., Vol. 26, (Springer, Berlin, Heidelberg 1978)
4.108　G. Busch, M. Campagna, P. Cotti, H.C. Siegmann: Phys. Rev. Lett. **22**, 597 (1969)
4.109　U. Bänninger, G. Busch, M. Campagna, H.C. Siegmann: Phys. Rev. Lett. **25**, 585 (1970)
4.110　S.F. Alvarado, W. Eib, F. Meier, H.C. Siegmann, P. Zürcher, "Photoemission of Spin-Polarized Electrons" In *Photoemission Photoemission and the Electronic Properties of Surfaces*, ed. by B. Feuerbacher, B. Fitton, R.F. Willis (Wiley, New York 1978) pp. 437–467
4.111　W. Eib, S.F. Alvarado: Phys. Rev. Lett. **37**, 444 (1976)
4.112　E. Kisker, W. Gudat, M. Campagna, E. Kuhlmann, H. Hopster, I.D. Moore: Phys. Rev. Lett. **43**, 966 (1979)
4.113　R. Clauberg, W. Gudat, E. Kisker, E. Kuhlmann: Z. Phys. B **42**, 47 (1981)
4.114　W. Gudat, E. Kisker, E. Kuhlmann, M. Campagna: Solid State Commun. **37**, 771 (1981)
4.115　R. Raue, H. Hopster, R. Clauberg: Phys. Rev. Lett. **50**, 1623 (1983)
4.116　B.T. Jonker, K.-H. Walker, E. Kisker, G.A. Prinz, C. Carbone: Phys. Rev. Lett. **57**, 142 (1986)
4.117　E. Kisker, K. Schröder, W. Gudat, M. Campagna: Phys. Rev. B **31**, 329 (1985)
4.118　D. Pescia, M. Stampanoni, G.L. Bona, A. Vaterlaus, R.F. Willis, F. Meier: Phys. Rev. Lett. **58**, 2126 (1987)
4.119　M. Stampanoni, A. Vaterlaus, M. Aeschlimann, F. Meier: Phys. Rev. Lett. **59**, 2483 (1987)
4.120　K.-P. Kämper, W. Schmitt, D.A. Wesner, G. Güntherodt: Appl. Phys. A **49**, 573 (1989)

4.121 C.M. Schneider, P. Schuster, M. Hammond, H. Ebert, J. Noffke, J. Kirschner: J. Phys.: Condens. Matter **3**, 4349 (1991)
4.122 W. Heinen, C. Carbone, T. Kachel, W. Gudat, J. Electron Spectrosc. Relat. Phenom. **51**, 701 (1990)
4.123 R. Rochow, C. Carbone, Th. Dodt, F.P. Johnen, E. Kisker: Phys. Rev. B **41**, 3426 (1990)
4.124 W. Clemens, T. Kachel, O. Rader, E. Vescovo, S. Blügel, C. Carbone, W. Eberhardt: Solid State Commun. **81**, 739 (1992)
4.125 K.-P. Kämper, W. Schmitt, G. Güntherodt, Phys. Rev. B **42**, 10696 (1990)
4.126 W. Weber, D.A. Wesner, G. Güntherodt, U. Linke: Phys. Rev. Lett. **66**, 942 (1991)
4.127 N.B. Brookes, Y. Chang, P.D. Johnson: Phys. Rev. Lett. **67**, 354 (1991)
4.128 B. Sincovic, P.D. Johnson, N.B. Brookes, A. Clark, N.V. Smmith: Phys. Rev. Lett. **65**, 1647 (1990)
4.129 C. Carbone, T. Kachel, R. Rochow, W. Gudat: Solid State Commun. **77**, 619 (1991)
4.130 R. Jungblut, C. Roth, F.U. Hillebrecht, E. Kisker: J. Appl. Phys. **70**, 5923 (1991)
4.131 C. Carbone, E. Kisker, Phys. B **36**, 1280 (1987)
4.132 C. Carbone, R. Rochow, L. Braichovic, R. Jungblut, T. Kachel, D. Tillmann, E. Kisker: Phys. Rev. B **41**, 3866 (1990)
4.133 M. Taborelli, R. Allenspach, G. Boffa, M. Landolt: Phys. Rev. Lett. **56**, 2869 (1986)
4.134 M. Taborelli, R. Allenspach, M. Landolt: Phys. Rev. B **34**, 6112 (1986)
4.135 O. Paul, S. Toscano, W. Hürsch, M. Landolt: J. Magn. Magn. Mater. **84**, L7 (1990)
4.136 M. Landolt: "Spin Polarized Secondary Electron Emission from Ferromagnets" In Ref. 4.3, pp. 385–421
4.137 J.B. Pendry: Phys. Rev. Lett. **45**, 1356 (1980)
4.138 J. Unguris, F. Seiler, R.J. Celotta, D.T. Pierce, P.D. Johnson, N.V. Smith: Phys. Rev. Lett. **49**, 1047 (1982)
4.139 H. Scheidt, M. Glöbl, V. Dose, J. Kirschner: Phys. Rev. Lett. **51**, 1688 (1983)
4.140 J.Kirschner, M. Glöbl, V. Dose, H. Scheidt: Phys. Rev. Lett. **53**, 612 (1984)
4.141 A. Seiler, C.S. Feigerle, J.L. Peña, R.J. Celotta, D.T. Pierce: Phys. Rev. B **32**, 7776 (1985)
4.142 C.S. Feigerle, A. Seiler: J.L. Peña, R.J. Celotta, D.T. Pierce: Phys. Rev. Lett. **56**, 2207 (1986)
4.143 M. Donath: Appl. Phys. A **49**, 351 (1989)
4.144 F.J. Himpsel: Phys. Rev. Lett. **67**, 2363 (1991)
4.145 C. Rau: J. Magn. Magnet. Mater. **30**, 141 (1982)
4.146 C. Rau, K. Waters, N. Chen: Phys. Rev. Lett. **64**, 1441 (1990)
4.147 M. Onellion, M.W. Hart, F.B. Dunning, G.K. Walters: Phys. Rev. Lett. **52**, 380 (1984)
4.148 R. Wiesendanger, H.J. Güntherodt, G. Güntherodt, R.J. Gambino, R. Ruf: Phys. Rev. Lett. **65**, 247 (1990)
4.149 R. Wiesendanger: Science **255**, 583 (1992)
4.150 S.F. Alvarado, P. Renaud: Phys. Rev. Lett. **68**, 1387 (1992)
4.151 G. Chrobok, M. Hofmann, G. Regenfuss, R. Sizmann: Phys. Rev. B **15**, 429 (1977)
4.152 M. Landolt, M. Campagna: Surf. Sci. **70**, 197 (1978)
4.153 E. Kisker, G. Baum, A.H. Mahan, W. Reith, B. Reihl: Phys. Rev. B **18**, 2258 (1978)

Section 4.2

4.154 G. Busch, M. Campagna, H.C. Siegmann: J. Appl. Phys. **41**, 1044 (1970)
4.155 H.C. Siegmann, F. Meier, M. Erbudak, M. Landolt: Adv. El. and El. Phys. **62**, 1 (1984)
4.156 *High Energy Spin Physics* ed. by W. Meyer, E. Steffens (Springer, Berlin, Heidelberg 1991)
4.157 M. Donath, D. Scholl, H.C. Siegmann, E. Kay: Appl. Phys. **A52**, 206 (1991)
4.158 O. Paul, S. Toscano, K. Totland, M. Landolt: Sur. Sci. **251**, 27 (1991)
4.159 M. Stampanoni: Appl. Phys. **A49**, 449 (1989), and Dissertation Nr. 8937, Eidgenössische Technische Hochschule, Zürich (1989)
4.160 J. Unguris, A. Seiler, R.J. Celotta, D.T. Pierce, P.D. Johnson, N.V. Smith: Phys. Rev. Lett. **49**, 1047 (1982)

4.161 M. Donath: Appl. Phys. **A49**, 351 (1989)
4.162 A.H. Morrish: *The Physical Principles of Magnetism*, (Wiley, New York 1965) p. 360
4.163 V. Pokrovskii: Adv. Phys **28**, 595 (1979)
4.164 L. Néel: J. Phys. Radium **15**, 225 (1954)
4.165 D. Mauri, D. Scholl, H.C. Siegmann, E. Kay. Appl. Phys. **A49**, 439 (1989)
4.166 D. Pescia, D. Kerkmann, F. Schuhmann, W. Gudat: Z. Phys. **B78**, 475 (1990)
4.167 S.S.P. Parkin, N. More, K.P. Roche: Phys. Rev. Lett. **64**, 2304 (1990)
4.168 H.C. Siegmann, P.S. Bagus, E. Kay: Z. Phys. **B69**, 485 (1988)
4.169 H.C. Siegmann, P.S. Bagus: Phys. Rev. B **38**, 10434 (1988)
4.170 M. Donath, D. Scholl, D. Mauri, E. Kay: Phys. Rev. B **43**, 13164 (1991)
4.171 D.M. Edwards, J. Mathon, R.B. Munitz, M.S. Phan: Phys. Rev. Lett. **67**, 493 (1991)
4.172 J. Unguris, R.J. Celotta, D.T. Pierce: Phys. Rev. Lett. **67**, 140 (1991)
4.173 Hinne Zijlstra: IEEE Trans. Magn. **15**, 1246 (1979)
4.174 H.C. Siegmann: Rev. Solid State Sci. **4**, 817 (1990)
4.175 Walther Gerlach: Sci. Rep. Tohoku Imp. Univ. **1**, 248 (1936)
4.176 W. Weber, D. Kerkmann, D. Pescia, D.A. Wesner, G. Güntherodt: Phys. Rev. Lett. **65**, 2058 (1990)
4.177 D. Mauri, D. Scholl, H.C. Siegman, E. Kay: Phys. Rev. Lett. **62**, 1900 (1989)
4.178 H.G. Pini, A. Rettori, D. Pescia, N. Majlis, S. Selzer: Phys. Rev. B **45**, 5037 (1992)
4.179 D. Kerkmann, D. Pescia, R. Allenspach: Phys. Rev. Lett. **68**, 686 (1992)
4.180 L.J. de Jongh, A.R. Miedema: *Experiments on Simple Magnetic Model Systems*, (Taylor & Francis, London 1974)
4.181 J.M. Kosterlitz, D.J. Thouless: Progress in Low Temperature Physics, Vol. VII B, (North Holland, Amsterdam 1978)
4.182 M. Donath, D. Scholl, H.C. Siegmann, E. Kay: Phys. Rev. B **43**, 3164 (1991)
4.183 K. Hirakawa, K. Ubukoshi: J. Phys. Soc. Jpn. **50**, 1909 (1981)
4.184 W.F. Egelhoff Jr., D.A. Steigerwald: J. Vac. Sci. Technol. **A7**, 2167 (1989)
4.185 J. Pommier, P. Meier, G. Pénisard, H.J. Ferré, P. Bruno, D. Renard: Phys. Rev. Lett. **65**, 2054 (1990)
4.186 A. Vaterlaus, D. Guarisco, M. Lutz, M. Aeschlimann, F. Meier: J. Appl. Phys. **67**, 5661 (1990)
4.187 A. Vaterlaus, T. Beutler, F. Meier: Phys. Rev. Lett. **67**, 3314 (1991)
4.188 P. Weiss: J. Phys. **6**, 667 (1907)
4.189 P. Weiss, R. Forrer: Ans. Phys. (Paris) **5**, 153 (1926)

5. Epitaxial Growth of Metallic Structures

The importance of characterizing the structure of ultrathin films derives from the extreme sensitivity of the magnetic properties to the film structure. Three powerful techniques which have played a key role in the determination of film structure are discussed in this chapter. In the first section *Arrott* describes the use of Reflection High Energy Electron Diffraction (RHEED) studies of epitaxial metal growth, a technique also widely used in semiconductor MBE. In the second section, *Egelhoff* discusses the use of angle-resolved Auger and photoelectron spectroscopies as powerful structural probes. The application of X-ray diffraction studies to metallic superlattices are discussed in the final section by *Clarke* and *Lamelas*. Increasingly, X-ray studies are being performed at powerful new synchrotron sources, not only for structural characterization but also for magnetic studies which exploit the polarization dependence of the magnetic X-ray scattering. However, the use of X-ray scattering in the laboratory is discussed here. The theory of RHEED and X-ray diffraction described here may usefully be read in conjunction with the discussion of the neutron reflection technique by *Bland* in Chap. 6.

5.1 Introduction to Reflection High Energy Electron Diffraction (RHEED)

A.S. ARROTT

5.1.1 Real Space and k-space

It is desirable to produce a picture in the mind's eye of what goes on when an electron leaves the RHEED gun, interacts with the surface of a crystal and then proceeds to excite a phosphor on the screen. Consider a television screen with an array of dots. Each dot is excited sequentially by electron beams to produce a picture. An alternative scheme would send the electron beams into an array of atoms. The beams would diffract simultaneously into many directions with varying intensity from dot to dot. To change the picture on the screen the atoms would have to be arranged. One could be watching the televised Olympic games or one could be visualizing the arrangement of atoms that is accounting for the action. Perhaps, one even could do both at the same time.

The RHEED screen is the window on the world of surface structure. Most of the information is processed using the pattern recognition ability of the mind and most of this is done in real time. Some patterns are recorded on video tape, some are stored digitally on computer disks, and some are photographed. Most recording techniques lose both the detail and the dynamic range available to the eye viewing the phosphor screen. Disappointment is the usual reaction to the final print of a RHEED screen pattern in the standard journals. A line scan across a portion of the screen or a scan of the intensity at a point with time is more apt to be satisfying in print.

The intensity on the RHEED screen reflects the interaction of the electron beam with the sample surface. Actually it is not just the surface, but the atoms in the surface region. It has been suggested that this region be called the *selvedge* in analogy with the edge of a bolt of cloth. It is a fair analogy, but it will be resisted here. The word *surface* means as deep into the crystal as probed by a given beam under given circumstances. A 10 keV beam can penetrate through 20 nm and still retain a small elastic component. At grazing incidence the beam is refracted down into the crystal at a small but finite angle, typically 0.02 rad, penetrating to less than 1 nm, where penetration is defined as a fall off of $1/e$ in intensity of the elastic component of the beam. Under certain conditions of diffraction this penetration is confined to the upper one or two layers.

Each picture on the RHEED screen is caused by a particular arrangement of atoms. One learns to recognize the picture, but one has to be told the arrangement responsible for the image. For several reasons, one can not deduce full knowledge of the arrangement from the picture. A given arrangement of atoms produces different pictures depending on its orientation with respect to the beam. One can turn the sample while watching the RHEED screen. It is fun to see the diffraction pattern change. One develops hand-eye coordination. One can steer the sample into a desired orientation. There is a pattern of intersecting arcs called Kikuchi lines which appears with enhanced intensity and beautiful symmetry for major crystallographic orientations. One does not have to understand Kikuchi lines to use them or to avoid their interference with particular observations by changing orientations.

As one turns the crystal or directs the beam to other parts of the sample the RHEED screen patterns change in both anticipated and unanticipated ways. "What was that?" is often heard. A small crystallite sitting on a surface can be the source of amusement as its own diffraction pattern appears in forward transmission.

The first aim of the experimentalist is to produce a 'good' pattern. This requires the extremely important steps of surface preparation. The meaning of 'good' has changed somewhat in recent years as better surfaces have been prepared. In the beginning sharp spots meant small crystallite impurities on the surface or asperities of the surface. Broad spots meant that one was seeing the periodicities of the surface, but with penetration of the beam into the sides of an undulating surface. Sharp streaks meant 'good' flat surfaces, with the streaks being associated with the diffraction rods of the reciprocal lattice of a two-dimensional lattice. At one time sharp streaks were the most welcome sign to the experimentalist.

But the diffraction rods of a two-dimensional lattice intersecting the sphere of reflection create points, not streaks, except near evanescence. Now it is well appreciated that it is sharp spots that indicate a truly good surface, as long as they are not coming from crystallities or asperities. It is easy to recognize the differences between the two types of sharp spots once both have been seen.

5.1.1.1 The Power of RHEED

RHEED is an important tool in surface science, for the following reasons. The power of RHEED is that it produces pictures directly related to atomic arrangements. It is observed in real time. The geometry of the RHEED beam at glancing angles is convenient. It does not interfere with the growth process or with other monitoring devices. It is inexpensive on the scale of the cost of surface science laboratories.

Typical types of information obtained in real time are:

(1) the quality of surface preparation;
(2) the orientation of the crystal, or a selected grain of a *mosaic* crystal;
(3) the crystal structure and morphology of the starting surface and of the overlayers during and after growth;
(4) the number of atomic layers deposited in a *layer-by-layer* growth.

The first two work primarily on pattern rercognition. The third is pattern recognition based on understanding of diffraction followed by quantitative analysis of diffraction angles and intensities. The fourth is best observed on a chart recording of the intensity of the specular reflection. One counts atomic layers by the periodic variations in the intensity of the reflection. The brain is better at reading intensity versus time from a chart than from the memory of the visual cortex. (Though it is surprising how with enough practice one can do well looking for maxima and minima in time of the intensity on a screen.) The periodicity of the *RHEED oscillations* is closely related to the time to deposit one layer, but questions about phase and intensity of the oscillations require pictures of the atomic arrangements during growth plus the application of diffraction theory.

5.1.1.2 Electron-Surface-Interactions

In RHEED, electrons leave the source one at a time, but are recorded collectively over the integration time of the phosphor. The beam intensity is $\sim 1\,\mu A$ or 10^{13} electrons per second. The electron travels close to the speed of light spending no more than 10^{-14} s interacting with the crystal. At 10 keV the electron has the energy to cause lots of excitement in the crystal, but spends insufficient time to provide much impulse. A typical encounter of a 10 keV electron with the atomic potential results in deflections of less than 50 mrad (milliradians). There are many processes by which the electron gives energy to the crystal. These result in heating through phonon and plasmon oscillations, the generation of photons and X-rays, and the emission of electrons in chain reactions. Each of these things is a field of study and each is utilized in surface science. In the case of Auger spectroscopy an electron enters the crystal and at a later time of the order of 10^{-15} s an electron leaves the crystal. In between these two events the paths of the electrons are obscured by some 10^3 theoretical papers.

Elastic Scattering. It is assumed in RHEED that there is a magic fraction of electrons that interact with the crystal without changing the state of the crystal in any way, except to impart some momentum to the center of mass of the whole. The scattering is then *absolutely elastic*. This fraction can be described by a wave function $\psi(r)$ that is a macroscopic thermodynamic quantity which obeys a one-body Schrödinger equation in which the crystal is represented by another macroscopic thermodynamic quantity $V(r)$ called the *optical potential* [5.1]. The world of RHEED then divides nicely into discussions of what is the optical potential that goes into the Schrödinger equation and what is the wave function that results. The arrangement of atoms and their scattering power all go into the optical potential, which is complex. The imaginary part of the potential accounts for all of the electrons removed from the category of being absolutely elastically scattered. It also takes into account the effects of those inelastic processes where the electron gives energy to the lattice and then recovers it before leaving the crystal, like a neutron diving into a nucleus, pulling out of the dive, and escaping before being captured. Such events do change the phases of the waves.

In RHEED there are many quasi-elastic events that mimic elastic scattering. An electron can lose energy to a long wavelength plasmon but still interfere with itself to produce a diffraction pattern. Kikuchi patterns come from electrons that have lost some energy while channeling into the crystal. Most discussions of electron diffraction start from the one-body Schrödinger equation. *Sears* [5.1] in his treatise on Neutron Optics, Sect. 2.8, demonstrates that this is an *exact* description of the absolutely elastic scattering. Whether or not there is such a thing as absolutely elastic scattering for electrons, it is certainly the best available starting point.

5.1.1.3 Theoretical Problems

The construction of the optical potential is model building. The solution of the one-body Schrödinger equation is diffraction theory. It is the role of theory to

5.1 Introduction to Reflection High Energy Electron Diffraction (RHEED)

do both. The physics should be in the model building, but this is not quite the case. The problem is that the diffraction theory is too difficult for most models. So approximations are made in the theory of diffraction and limitations are placed on the complexity the models. It takes physical insight to know what approximations to make and even more to know how good they are. This leads to a vocabulary that describes the approximations in terms of physical pictures of the waves and the *Fourier components* of the potential, which are being included and which are being left out. By using the one-body Schrödinger equation one is already leaving out all the inelastic processes. Yet some of the approximate methods used for solving the elastic equation can be adapted to treat inelastic processes using the physical reasoning that went into making the mathematical approximations.

Needless to say this gives rise to some confusion on the part of the experimentalist who wants to look at the RHEED screen and have a picture of what the atoms are doing without distinguishing between *elastic coherent* and *inelastic coherent* scattering or between *kinematic* theory using the *first Born approximation* and *n-beam dynamical* theory, not to mention *evanescence, surface resonance,* or *plasmons.*

What the experimentalist might appreciate is a computer program in which he inputs the positions of all the atoms in the surface region, with whatever morphology might happen during deposition. The output of the program is the pattern on a monitor which is then manipulated by a joy stick that allows the orientation of the theoretical crystal to be changed with respect to the beam. That program does not exist. What do exist are a few scanning tunneling microscope STM pictures and RHEED patterns of the same surface. Someday there will be video disks on which one can see the atomic arrangements and the RHEED patterns side by side for sufficient number of samples that the experimentalist will be able to exercise the brain's pattern recognition ability to the point of seeing the atoms move in the mind's eye while the changing patterns are viewed on the screen.

5.1.1.4 An Introduction to the Literature

There are many excellent texts on diffraction and on electron diffraction both at high energies as in transmission electron microscopy [5.2a–e] and at low energies as in low energy electron diffraction (LEED) [5.3a–b].There are experts in the field of RHEED who could write similar books on RHEED but, unfortunately, they have not. The need for such a book was recognized by *Larsen* and *Dobson* who arranged the 1987 workshop on *RHEED and Reflection Electron Imaging of Surfaces* [5.4]. The proceedings serve as the current foundation of the field. One purpose of this chapter is to make explicit some of the things that at such workshops it is assumed "everybody knows".

The 1987 workshop brought together most of the major players at that time. The authors and titles of 20 of the chapters are listed in the reference section with comments [5.4 a–t]. Their references will indicate which papers they recommend. In his paper on surface structural determinations, *Beeby* [5.4b] reflects on the recent emergence of RHEED as a distinct part of electron diffraction. Two other workshops resulting in books in which RHEED is an important consideration were *Thin Film Growth Techniques for Low-Dimensional Structures* [5.5] and *Kinetics of Ordering and Growth at Surfaces* [5.6]. Selected chapters from these books are listed in the reference section [5.5a–f] and [5.6a–g].

RHEED is a new field, but it has a 65 year history going back to the seminal experiments that showed the wave-like nature of electrons [5.7]. In his classic review of 1969, *Bauer* refers to the energy range from 10 to 40 keV as RED [5.7c]. RHEED as a title of a field of study first appears in the late 1960s at the same time as the emergence of Molecular Beam Epitaxy, MBE. MBE and RHEED have grown together [5.8], both feeding on the availability of ultra-high vacuum systems.

The potential of RHEED was recognized from the beginning of electron diffraction, but commercial electron microscopes have not been compatible with ultra-high vacuum. An ultra-high vaccum microscope is a major facility to which one might add a MBE machine. By contrast, RHEED is an accessory to a MBE machine. A description of the design considerations that went into a state-of-the-art RHEED system developed at the Philips Research Laboratory, where phosphor screens and electron guns are a specialty, has been given by *van Gorkum* et al.[5.9a]. Other papers on the instrumentation of RHEED are given as references[5.9b–f].

In 1972 the work of *Menadue* and *Colella*[5.10a–c] established the experimental and theoretical foundations of modern RHEED in papers that are well worth reading today. A series of experimental and interpretive papers on the subject of RHEED intensity oscillations by the Philips group [5.11a–h] and by *Cohen* and his students [5.12a–i] were important to the growth of RHEED in the 1980s. *Ichimiya*, not present at the 1987 workshop, has continued to contribute to the foundations of RHEED through the last decade [5.13a–g]. References [5.4–13] should be sufficient for an efficient assault on the RHEED literature, except for the works of *Tong* [5.14] who has recently applied his expertise in LEED to RHEED.

Physical science articles since 1989 are cited on CD ROMs. One can type *RHEED* and find all the abstracts of all articles that use it in the title, abstract or key word list. It takes an hour to find and print them all out. The interested reader should be able to do this at his local library.

5.1.2 The Surface as a Diffraction Grating

To introduce the geometry and apparatus of RHEED the sample will be considered both as a simple reflector [5.15] and as a diffraction grating [5.16]. The geometry of RHEED as a simple reflector is shown in Fig. 5.1. The incoming beam, 0.3 mm across with a 0.0125 nm wavelength, is represented by

5.1 Introduction to Reflection High Energy Electron Diffraction (RHEED)

a plane wave of wave vector k_0. The vector g_n perpendicular to the reflecting surface represents the momentum imparted to the wave by the crystal. The final wave vector k_f is determined by the point where the Ewald sphere of vectors allowed by the conservation of energy is pierced by g_n. For a finite reflector there are small transverse components to g_n. The possible k_f are the intersections with the Ewald sphere of a finite rod with g_n in its center. Many of the figures, including Fig. 5.1, combine both k-space and real space. Diffraction demands that one think in these complementary spaces.

In reciprocal space the line perpendicular to the surface from the tip of the incoming wave vector is called the specular diffraction rod. In real space the through beam can be made visible on the screen by deliberately letting some of the incoming beam miss the reflector. The reflector casts a shadow on the screen in the sense that any diffuse scattering, from a less than perfect reflector, occurs only above y_2 measured from the position of the through beam. For a simple reflector the amplitude of the scattered wave would decrease with increasing θ (and g_n). The ratio of the reflected amplitude to the amplitude of the incoming wave goes to -1 in the limit of glancing angle.

An iron whisker (0 0 1) surface, which is about as perfect as one could desire, acts for a 10 keV electron beam like a 700 line per cm optical grating does for

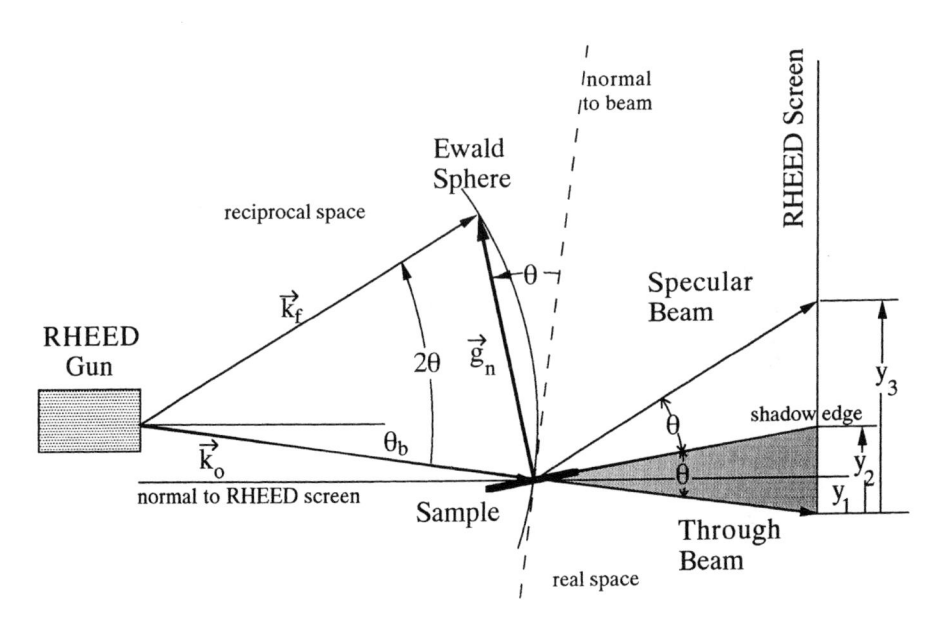

Fig. 5.1. The geometry of RHEED treating the sample as a simple reflector. The incoming beam makes an angle θ_b with respect to the normal to the RHEED screen and an angle $\pi/2 - \theta$ with respect to the normal to the surface of the reflector. In k-space, or reciprocal space, on the left, the initial beam has momentum k_0 and the reflected beam has momentum k_f with $k_f = k_0$. The Ewald sphere is the locus of all wave vectors of magnitude k_0. The simple reflector can transfer momentum only perpendicular to its surface. The momentum transferred is g_n

a HeNe laser. For the ~10 keV beam the wavelength is ~0.0125 nm and the spacing of the atoms on the (0 0 1) surface is 0.2866 nm, or a ratio of 23 to 1. For the laser the wavelength is 632 nm and the grating spacing is 1.5 μm.

5.1.2.1 Diffraction Angles

Grating Lines Perpendicular to the Beam. An optical grating can be described as a series of line scatterers. Let the incident beam be perpendicular to the lines. Each line is the source of a wave, the phase of which is set by the incoming wave. This kinematic model, with which the reader is likely to be familiar, yields the grating formula

$$k_0 \cos \theta_n = k_0 \cos \theta_0 + nG, \tag{5.1}$$

where $k_0 = 2\pi/\lambda$ is the magnitude of wave vector of the incoming wave and also of the outgoing waves if the scattering conserves energy; θ_0 is the angle of grazing incidence; θ_n is the angle of the nth order diffraction beam; and $G = 2\pi/a_0$ where a_0 is the periodicity of the grating. Angles are generally measured from the normal in optics. The angle of grazing incidence is more natural for RHEED and is used here. This case, with the beam perpendicular to the lines, is considered first to show why it can often be ignored in RHEED.

The nature of the problem of diffraction can be illustrated using the grating formula. There is a particularly useful description of the conservation of momentum and energy which is found using the geometrical construction of the sphere of reflection, often called the Ewald sphere. The incoming wave vector and the outgoing wave vectors have the same magnitude to conserve energy, therefore the tips of all possible vectors involved in reflection must lie on a sphere of radius k_0. This is not quite the case for there are imaginary solutions to the wave equation that turn the sphere into the planet Saturn with a ring of imaginary roots outside of the sphere in the equatorial plane. These are called evanescent waves, and they are important in RHEED. The sphere of reflection for the case of the beam perpendicular to the scattering lines is shown in Fig. 5.2.

The origin of k-space is the center of the sphere. The end of the incoming wave vector is the origin for the k-space description of the grating. The lines through the tips of the grating vectors are called the diffraction rods. They are perpendicular to the plane of the grating. The intersections with the sphere for positive k_z, below the equator of the sphere, are transmitted beams if the grating is not perfectly reflecting or fully absorbing. Scattering to the evanescent wave k_e gains momentum G from the grating. The evanescence wave is not seen, but it can be diffracted again by the crystal. The real part of the z component of momentum is zero; the imaginary part gives an exponential decrease in the wave amplitude with $-z$, the normal direction.

A point to note about Fig. 5.2 is that, for small angles of incidence, the reflection is at a small angle only for the specular beam. The beam at $-G$ is at a sufficiently large angle that the required momentum transfer in the z direction

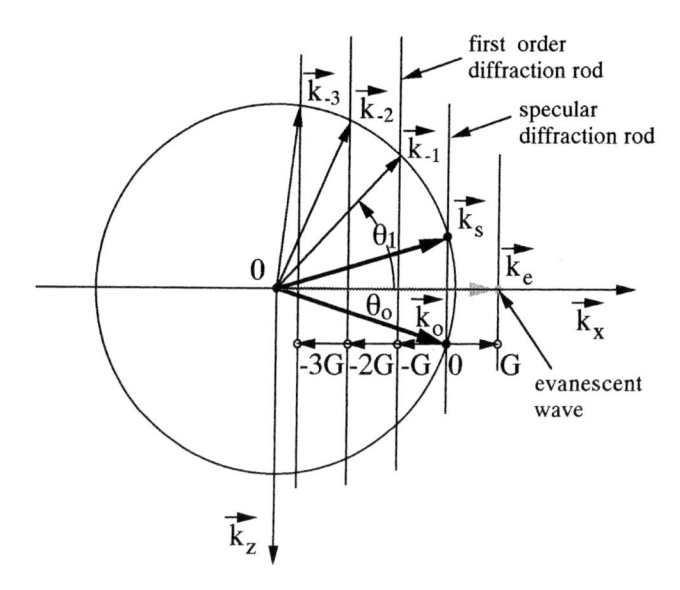

5.1 Introduction to Reflection High Energy Electron Diffraction (RHEED)

Fig. 5.2. The sphere of reflection in k-space for a grating with periodicity $a_0 = 2\pi/G$. The incoming wave vector k_0 is at the grazing angle θ_0. Scattering to the specularly reflected beam k_s imparts momentum $2k_0 \sin\theta_0$ to the grating in the normal direction. Scattering to the first order diffraction imparts momentum $k_0 \sin\theta_0 + k_0 \sin\theta_1$ in the normal direction and momentum G in the plane of the surface in the direction of the beam

is much larger than for the specular beam. The Fourier transform of a step function, representing a surface discontinuity, has components that decrease as $1/\delta k_z$ where δk_z is the momentum transfer normal to the surface. The intensity of reflection decreases rapidly with momentum transfer normal to the surface.

The angle of the first order diffraction is given for small angles by

$$\theta_n = \sqrt{\theta_0^2 - \frac{2nG}{k_0}}, \tag{5.2}$$

which for $G = -2\pi/(0.2866 \text{ nm}) \cong -2.2 \text{ Å}^{-1}$, $k_0 = 2\pi/(0.0124 \text{ nm}) \cong 50 \text{ Å}^{-1}$, and $\theta_0 = 0.02$ (in radians) gives $\theta_1 = 0.3$, for which the reflection would be down by a factor of eight in amplitude if the scattering from the lines were isotropic, which it is not for 10 keV electrons. Arguing from the uncertainty principle using $\Delta p \Delta x \sim h$ with Δx of the order of the Bohr radius puts most of the scattering of a 10 keV electron well within 0.1 rad. Note also that the angle in (5.2) becomes imaginary with increasing n for positive G. The imaginary angle corresponds to evanescence. Note that in diffraction the negative values of G are usually of interest.

The intensity of scattering with a change in the x component of the momentum (along the beam) is negligible in RHEED if the scatterers have the periodicity of the lattice. For surface features such as steps or reconstructions of long period, as the famous 7×7 structure of Si, the momentum transfers in the

x direction are small enough that the scattering is within the scattering cone, 0.1 rad, of a 10 keV electron. Then one considers reflections outside of the zero order in k_x. Indeed, the sensitivity of RHEED to such features is much amplified when they are in the direction of the beam.

Grating lines along the beam. The more interesting case in RHEED is when the grating is turned 90 degrees to put the periodicity perpendicular to the plane of the grating normal and the incoming beam. In this case the action is in k_y with k_x remaining unchanged in reflection. The description of reflection from a grating becomes three-dimensional with the beam in the x-z plane being deflected into the $\pm y$ directions as shown in Fig. 5.3.

The vector labeled *transmitted* becomes the incoming momentum and the other vectors to the circle are the diffracted wave vectors in the k-space interpretation. The scattering vectors are $\pm G\hat{y}$ augmented by the momentum transfer in the direction normal to the plane of the grating, indicated by the diffraction rods through the tips of $\pm G\hat{y}$. There are additional Fourier components of the interaction potential of the grating given by $\pm nG\hat{y}$. For the case drawn here these lie well enough outside of the sphere of reflection that their evanescent waves are highly damped and weakly excited.

For the drawing in Fig. 5.3 the spacing of the scattering lines and the wave vector of the incoming beam have been chosen such that there are only the

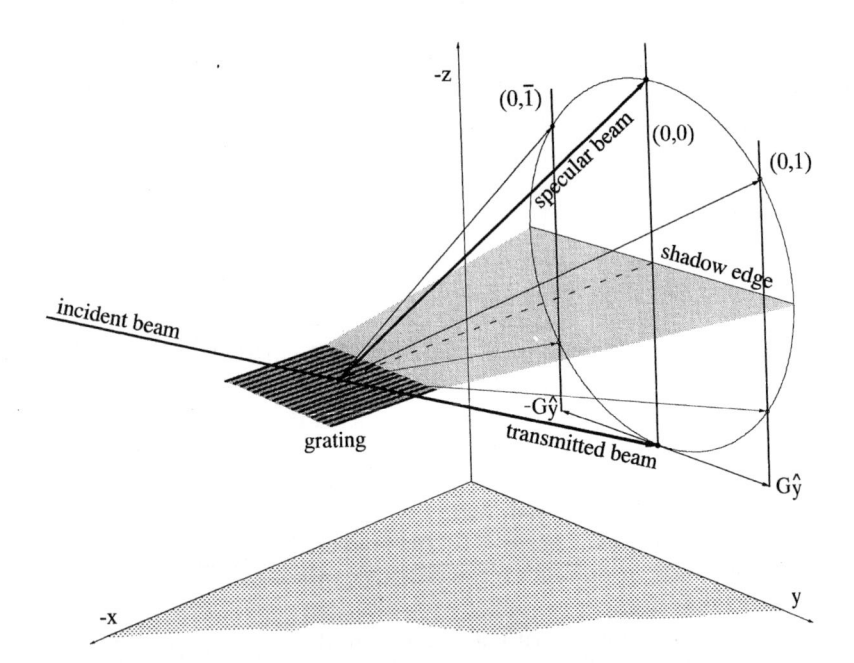

Fig. 5.3. Reflection from a grating periodic in the y direction with the beam in the x–z plane. The drawing has a double meaning. It can be interpreted in real space with the y–z wall representing the RHEED screen or it can be interpreted in k-space in which case the circle is on the sphere of reflections

5.1 Introduction to Reflection High Energy Electron Diffraction (RHEED)

specular beam and two diffracted beams. If the grating spacing were increased the vectors G would become shorter and more spots would be seen coming first from $\pm 2G$ and then $\pm 3G$. If every other scattering line were removed from the grating, the vector labeled G would become $2G$ with a new vector of half the length appearing. The specular beam would remain as it was. When G is just equal to $k_0 \sin \theta$, the radius of the circle, the vertical diffraction rod through the tip of G would be tangent to the circle and to the sphere of reflection. In this case the rod cuts the sphere of reflection in a streak and there is an evanescent component to the diffraction. The strictly evanescent wave is confined to the grating if it is perfect.

For a finite, perfectly spaced grating the diameter of the diffraction rod is $1/n$ of the length of G, where n is the number of lines in the total grating. By measuring the length of the streak near evanescence it is possible in principle to count the total number of lines in the grating. But generally the incoming beam has a range of angles sufficiently wide to obscure the effect of the finite size of the grating.

Azimuthal Dependence of the Grating Pattern. Next consider the effect of rotating the grating so that the beam is no longer quite parallel to the scattering lines. This is illustrated in Fig. 5.4. The grating is rotated through an azimuthal angle ϕ. The specular spot remains where it was. The spots to the right move down and disappear at the shadow edge. The spots to the left move up with new ones appearing from the shadow edge. If the grating is rotated by $\pi/2$ one returns to the geometry of Fig. 5.2. The loci of the diffraction spots on rotation through π are shown in Fig. 5.5. The intensity of the diffraction spots decreases as the required momentum transfer increases.

5.1.2.2 Intensities

The diffraction rods correspond to Fourier components of the interaction potential of the grating. Modulating the spacing of the scattering lines produces new periodicities. Reconstructing the actual arrangement of the scattering lines from the positions of the diffraction rods can only be done by guess work. What is needed is a theory of the intensities [5.16]. The simplest theory would make the intensities proportional to the squares of the Fourier components of the interaction potential. This is essentially what is done in kinematic theory. It is a good guide and often the only model that is calculable once the spacing of the scattering lines becomes complicated. It is most useful because it is a simple matter to write a program for a personal computer that Fourier analyses a one-dimensional distribution of scattering lines.

The First Born Approximation. The kinematic treatment is also called the first Born approximation. From a rigorous start with the one particle Schrödinger equation one makes an approximation where the incoming wave is unaffected by its interaction with the optical potential. The outgoing wave is generated by

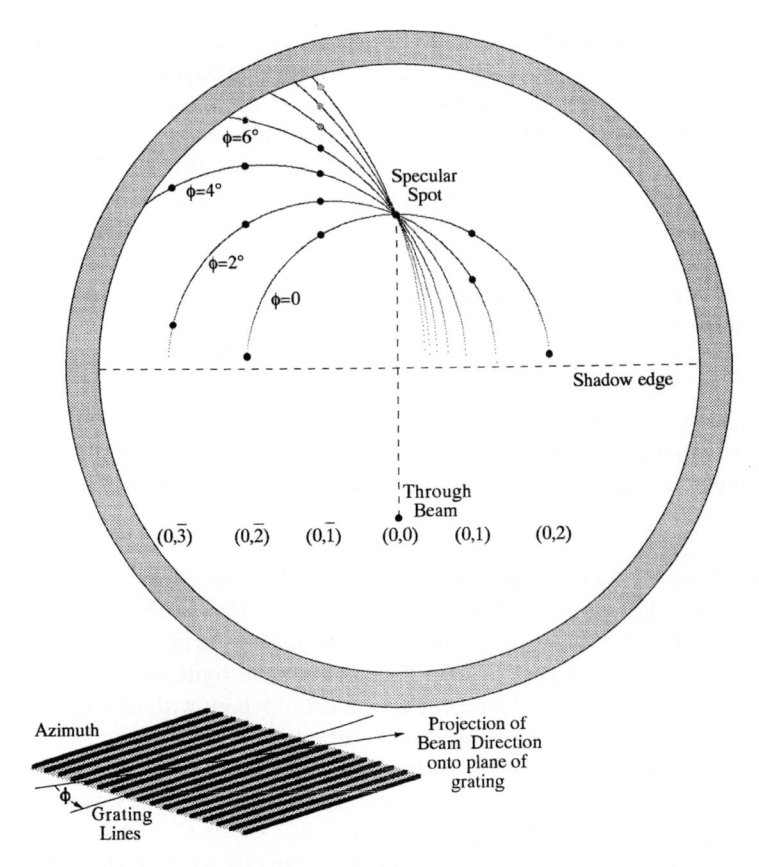

Fig. 5.4. Effect on the RHEED screen pattern of small azimuthal rotations of the diffraction grating in the geometry of Fig. 5.3. The spots on the left move away from the shadow edge. Those on the right move toward and disappear on reaching the shadow edge. The circular arcs through the specular spot are the intersection with the sphere of reflection with a plane that is perpendicular to the grating lines and passes through the origin and the specular spot

the incoming wave, but the outgoing wave is not considered to generate any additional waves, nor to feed back into the incoming wave. Scattering is from one plane wave state to another. As a result the problem reduces to that of each outgoing wave k_f and the incoming wave k_i appearing in a matrix element for scattering into that direction. The kernel has the form

$$e^{-ik_f \cdot r} V(r) e^{ik_i \cdot r} = V(r) e^{i\kappa_s \cdot r}. \tag{5.3}$$

This is a Fourier analysis of the potential $V(r)$ for the scattering vector

$$\kappa_s = k_f - k_i. \tag{5.4}$$

The matrix is integrated over the volume of the scatterer to find the scattering amplitude. The intensity is then proportional to the square of the scattering

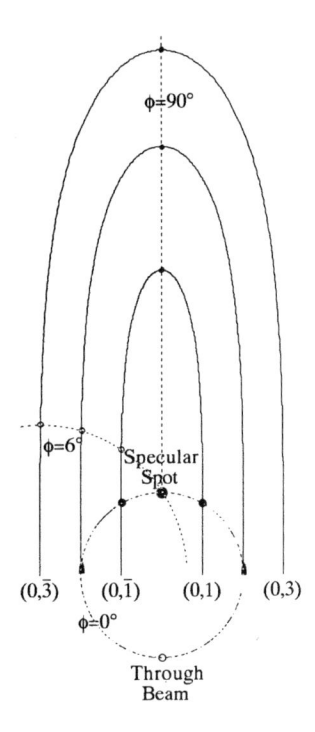

5.1 Introduction to Reflection High Energy Electron Diffraction (RHEED)

Fig. 5.5. The locus of spots on the sphere of reflection as a grating is rotated through 180 degrees. The geometry of Fig. 5.2 determines the position of the spots at $\phi = 90°$. In Fig. 5.3 $\phi = 0$

amplitude if the incoming beam can be treated as a superposition of independent plane waves. If that is the case, the scattered intensity is integrated over all of the incoming beams. If the incoming beam is a converging spherical wave, the amplitudes of the plane wave components of that wave are added and then squared to get the intensity.

In fact the waves in the region of interaction are not simple plane waves, but waves very much modified by the potential with which they interact. Schrödinger's equation with the optical potential is solved for the interacting waves in dynamical theory. The restriction to simple plane waves in the crystal puts limitations on the validity of the kinematic theory. Nevertheless the kinematic approximation has some validity and it is often the best that one can do. Almost all of our understanding of defects in surfaces is derived from kinematic calculations.

From Potentials to Charge Densities. In the first Born approximation it is possible to replace the potentials by charge densities [5.17] using Poisson's equation that relates the Laplacian of the potential to the charge density:

$$\nabla^2 \phi(r) = -\frac{\rho}{\varepsilon_0}, \qquad V(r) = -|e|\phi(r). \tag{5.5}$$

The scattering amplitude for a momentum change κ_s can be expressed as the Laplacian of the phase factor of the scattering amplitude:

$$F(\kappa) = \int_{-\infty}^{\infty} V(r) e^{i(\kappa_s \cdot r)} \, dr = -\frac{1}{\kappa_s^2} \int_{-\infty}^{\infty} V(r) \nabla^2 e^{i(\kappa_s \cdot r)} \, dr. \tag{5.6}$$

Integration by parts then permits the replacement of potentials by charge densities:

$$F(\kappa) = \frac{1}{\kappa_s^2} \int_{-\infty}^{\infty} e^{i(\kappa_s \cdot r)} \nabla^2 V(r) \, dr = \frac{e}{\varepsilon_0 \kappa_s^2} \int_{-\infty}^{\infty} e^{i(\kappa_s \cdot r)} \rho(r) \, dr. \tag{5.7}$$

Thus the scattering amplitude is the Fourier transform of the charge density. The scattering intensity is proportional to $F^*(\kappa)F(\kappa)$. The integral in (5.7) is broken into a sum over all atom sites and integrals over the charge density of each atom site. If the atoms are all the same, the sum becomes $g(\kappa)$, the structure factor of the system of atoms, and the integral becomes $f(\kappa)$, the form factor of the atom. If the atoms are all the same, the scattered intensity is proportional to the Fourier transform of the atom–atom pair correlation function.

For weak interactions with a bcc cubic lattice in a rectangular mosaic block with N_x, N_y, and N_z cubic unit cells, the structure factor is

$$g(\kappa) = \frac{\sin(\tfrac{1}{2} N_x a \kappa_x)}{\sin(\tfrac{1}{2} a \kappa_x)} \frac{\sin(\tfrac{1}{2} N_y a \kappa_y)}{\sin(\tfrac{1}{2} a \kappa_y)} \frac{\sin(\tfrac{1}{2} N_z a \kappa_z)}{\sin(\tfrac{1}{2} a \kappa_z)} \left\{ 2 \cos\left[\frac{a(\kappa_x + \kappa_y + \kappa_z)}{4}\right] \right\}, \tag{5.8}$$

where a is the lattice parameter of the cubic unit cell. The first three terms result from sums over the cubic units. The last term comes from the sum over the positions of the atoms (two for bcc) in the unit cell. The reciprocal lattice points (rlp's) are the points in momentum space where the three denominators vanish while the final factor does not.

The simplest assumption is that the atoms are point scatterers. In that case the form factor would be independent of momentum transfer. If the scattering takes place over some volume of the atom, the form factor would fall of with $|\kappa|$, somewhat like

$$f(\kappa) = \frac{2}{r_1 \kappa} \sin\left(\frac{r_1 \kappa}{2}\right), \tag{5.9}$$

or

$$f(\kappa) = \frac{1}{1 + (r_1 \kappa)^2}. \tag{5.10}$$

In one case the form factor is oscillatory and in the other it is monotonically decreasing with the scattering vector. The calculation of form factors for electrons interacting with atoms is a major quantum mechanical undertaking.

The kinematic picture is effective in describing X-ray diffraction from individual mosaic blocks of crystalline materials. It does not work well for electron

5.1 Introduction to Reflection High Energy Electron Diffraction (RHEED)

diffraction because the modification of the incoming waves by the strong scattering invalidates equating the scattering amplitude to the Fourier analysis of charge density. The strong interaction can make the first atomic layer more important than the underlying layers.

Kinematic theory applied to RHEED looks like diffraction from a grating if it is assumed that only the topmost atoms contribute to the structure factor. This assumption is hard to justify within the assumptions of kinematic theory itself. The best argument comes from dynamic theory, where it is seen that, under strongly diffracting conditions, the elastically scattered electron interacts primarily with the electrons on the surface side of the topmost atoms. Usually this will account for the wave vectors of the elastically diffracted electrons in RHEED. Calculation of intensities from the kinematic approximation are usually not successful, but some aspects of line profiles can be treated.

The connection between a surface and the grating of scattering lines is simply to consider lines of atoms parallel or close to parallel to the beam as the scattering lines giving rise to diffraction perpendicular to the beam. The charge density modulation along the lines can be ignored because these give rise to the diffraction, discussed for Fig. 5.2, which can be neglected because the scattering angles are large.

5.1.2.3 Singular Surfaces

An optical grating is made by starting with a flat surface and then modulating it to produce the lines. An atomic surface comes modulated; the trick is to get it atomically flat. An atomically flat surface is called a *singular* surface. {Experimentalists use a looser definition of singular [5.18]}. *The structure factor for the topmost atoms of a singular* (0 0 1) *surface is that of a square array.*

$$g(\kappa) = \frac{\sin(N_x a\kappa_x/2)}{\sin(a\kappa_x/2)} \frac{\sin(N_y a\kappa_y/2)}{\sin(a\kappa_y/2)}, \qquad (5.11)$$

where N_x and N_y are the number of atoms coherently sampled by the diffracting beam in the x and y directions, respectively. This peaks for $a\kappa_x = n\pi$ and $a\kappa_y = m\pi$ where n and m are integers. Because $g(\kappa)$ is independent of k_z, the structure factor for the topmost atoms of a singular surface is a set of rods in the z direction, originating on a square array of reciprocal lattice points. The positions of the spots in the diffraction pattern are determined by the square array of uniform diffraction rods with widths $\Delta k_x = 1/N_x a$ and $\Delta k_y = 1/N_y a$ intersecting the Ewald sphere of reflection. Experimentally one could define a singular surface as one that is flat enough that it is the coherence of the beam and not the terrace sizes that determines the line widths.

For a bcc lattice, the square array has the spacing of the bcc unit cell lattice parameter. For a fcc lattice, the square array has the lattice spacing of the fcc nearest neighbor distance which is the fcc lattice parameter divided by $\sqrt{2}$. In both cases the diffraction pattern does not reflect the underlying fcc or bcc structure of a singular surface, unless one knows the square lattice parameter of bcc Ni should be close to that of bulk Fe while that of fcc Ni should be close to that of bulk Ni. The distinction between the diffraction patterns would appear in the modulation of the structure factor in the z direction.

In fact, the underlying layers do affect the intensities of RHEED. The diffraction rods are not uniform, e.g., the specular intensity depends on κ_z. In particular, along each diffraction rod there are peaks in intensity which correspond to the wave vectors of the reciprocal lattice of the bulk lattice. (The positions of these peaks must be corrected for refraction at the surface as discussed below for dynamical theory). In terms of kinematic theory the experimental observations are somewhere in between diffraction from point scatterers at the surface, that produce uniform rods, and diffraction from the bulk, that produces intensity only when reciprocal lattice points lie on the sphere of reflection. This can be simulated by considering the diffraction to take place from a bulk lattice in which σ, the effective crosssection of the atoms for scattering, decreases with distance from the surface. If the $\sigma_n = \sigma_0 \exp(-n/M)$, and the lattice is taken to be bcc, the square of the structure factor becomes

$$g^2(\kappa) = 2e^{-(N+1)/M} \frac{\cosh(N/M) - \cos(Na\kappa_z)}{\cosh(1/M) - \cos(a\kappa_z)}$$

$$\times \frac{1 - \cos(Na\kappa_x)}{1 - \cos(a\kappa_x)} \frac{1 - \cos(Na\kappa_y)}{1 - \cos(a\kappa_y)}$$

$$\times \left\{ \cosh\left(\frac{1}{2M}\right) - \cos\left[\frac{1}{2}a(\kappa_x + \kappa_y + \kappa_z)\right] \right\}. \tag{5.12}$$

The dependence of $g^2(0, 0, \kappa_z)$, normalized to 1 for $\kappa = 0$, upon M, the number of layers for the attenuation by $1/e$, is shown in Fig. 5.6. As M decreases, the bulk reciprocal lattice points extend in the direction perpendicular to the surface. In the limit where scattering is just from the atoms at the surface, the extensions coalesce into uniform rods.

5.1.2.4 Vicinal Surfaces

A surface that is cut at a small angle to a low index plane is called *vicinal*. A vicinal surface is still flat if the average number of steps per unit length is constant and the variation in the density of the steps is much less than the average density of steps. The normal to the low index plane makes a small angle with respect to the normal to the surface. If the terraces are perfectly uniform, the unit cell extends from one step to the next.

For the vicinal surface it is convenient to consider the topmost rows of atoms as scattering lines, taking into account their z coordinate as well as their y coordinate, while maintaining the approximation that each line acts as a wave source whose phase is determined by the incoming wave. This treatment will

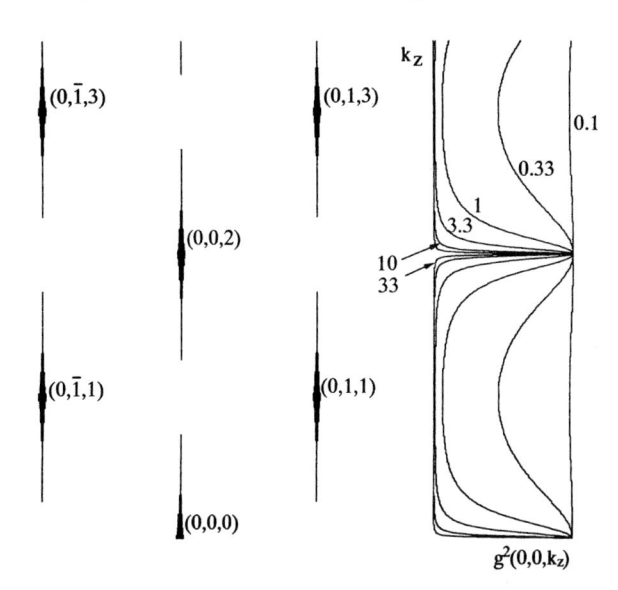

5.1 Introduction to Reflection High Energy Electron Diffraction (RHEED)

Fig. 5.6. The structure factor $|g|$ for a bcc lattice where the scattering cross section σ_n is assumed without justification to decrease exponentially with layer number from the surface. The reciprocal lattice is shown with the horizontal width of the lines proportional to $\log(|g|)$ for the case where $\sigma_n = \exp(-n/M)$ with $M = 3.3$. The dependence of g^2 on κ_z and M (from 0.1 to 33) is shown on the right for $\kappa_y = \kappa_x = 0$

introduce a κ_z dependence to the structure factor, even though all of the atoms below the surface are neglected. Such an approximation also neglects the effects arising from the fact that the incoming wave has to go through some surface atoms to reach others in the surface and that the outgoing beams go through other atoms on the way out. This effect is also there in the case of the singular surface where all the atoms are affected equally. For the stepped surface, atoms near the steps are shielded differently than those in the middle of the terraces.

Diffraction from a Vicinal Surface. Diffraction from a stepped surface is of general interest in RHEED. It serves as a prototype of methods used to extract information on defect structures. Consider reflection from a crystal cut at a small angle α with respect to a low index plane, for example the (0 0 1) plane of bcc iron. A distinction is made between the case where the step direction is along the direction of the beam as in Fig. 5.7 and the case where the step direction is perpendicular to the beam as in Fig. 5.9.

For the step direction along the beam, a further distinction is made between looking down the steps and looking up the steps. The model of the (0 0 1) surface of bcc Fe with steps is shown at the bottom of Fig. 5.7. The structure factor, found by summing over the atomic sites, is a product of the structure factor for a set of N steps spaced by D and the structure factor for the n atoms with spacing a on each terrace:

$$g(k_x, k_z) = \frac{\sin\left(\frac{ND}{2}k_x\right) \sin\left(\frac{an}{2}(k_x \cos\theta + k_z \sin\theta)\right)}{\sin\left(\frac{D}{2}k_x\right) \sin\left(\frac{a}{2}(k_x \cos\theta + k_z \sin\theta)\right)}, \tag{5.13}$$

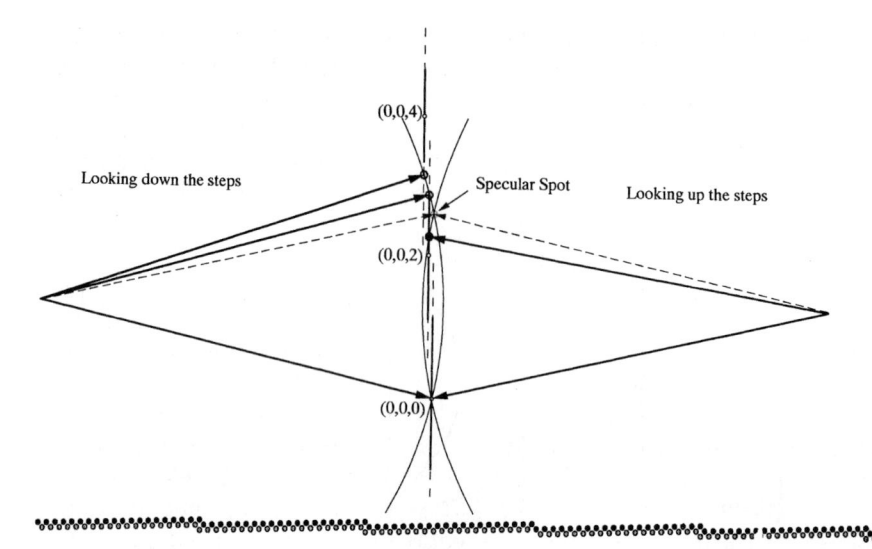

Fig. 5.7. Diffraction from a vicinal surface with the beam looking down the steps on the left and looking up the steps on the right. The surface atoms shown as solid spheres at the bottom of the diagram are considered as the source of diffraction. The diffraction rods perpendicular to the mean surface are shown as lines varying from solid to dotted to convey the z dependence of the structure factor for the stepped surface as given by (5.10). A line which would pass through the points labeled $(0,0,0)$, $(0,0,2)$, $(0,0,4)$ would be perpendicular to the atomic planes

where

$$\frac{D}{a} = \sqrt{(n-0.5)^2 + 0.5^2} \quad \text{and} \quad \sin\theta = \pm\frac{a}{2D}. \tag{5.14}$$

The peaks in the structure factor are centered on the reciprocal lattice positions (rpl's) of the bcc lattice $\{(h,k,l)$ with the restriction that $h+k+l$ is even$\}$. The rpl's are rotated with respect to the normal by the angle θ which is positive for looking down the steps and negative for looking up the steps. If the surface atoms were point scatterers, the intensity would be given by the square of the structure factor. The structure factor is sharp in the κ_x direction, parallel to the surface, but extends in the κ_z direction, normal to the surface. For $\kappa_x = 0$ the structure factor peaks for any integer multiple of $2\pi/(a\sin\theta)$, but, unless the integer is zero, this is a large scattering angle and the intensity is not observable. The structure factor for the specular reflection reaches its first zero for $\kappa_z \cong 2(2\pi/a)$, which is the position of the $(0,0,2)$ rpl. With increasing κ_z, $g(0,0,\kappa_z)$ oscillates with decreasing amplitude. Similarly, for $\kappa_x = -2\pi/D$, the maximum in κ_z occurs for $\kappa_z \cong 2(2\pi/a)$ and the first zeros about this maximum are for $\kappa_z \cong 0$ and $\kappa_z \cong 4(2\pi/a)$. Note that the distance to the first zero does not depend on the vicinal angle. These facts are illustrated in Fig. 5.7 where the solid lines represent values of κ_x and κ_z for which the square of the structure factor is > 0.5. The dashed lines end at the first zeros. As the vicinal angle is decreased, the centers of the lines rotate toward the specular rod. The lengths of the streaks do not change as they coalesce into the specular rod. The specular spot is beyond the first zero for the case illustrated. It will not be as strong as the intensities where the rods through the $(0,0,2)$ and $(0,0,4)$ rlp's intersect the sphere of reflection. When looking down the steps, the strong intensities are at angles greater than the specular angle. The opposite is the case when looking up the steps.

In Fig. 5.7 the angle of the incoming beam has been chosen so that the line of $(0,0,2n)$ rlp's intersects the sphere of reflection between the $(0,0,2)$ and the $(0,0,4)$ for the case of looking down the steps. This is near the kinematic *anti-Bragg position*. Near the anti-Bragg position the steps give rise to a pair of spots, one from the rod through the $(0,0,2)$ and the other from the rod through the $(0,0,4)$. The truly specular reflection from the vicinal surface would be weak at the angle of incidence shown. If the drawing were made near a Bragg position where the sphere of reflection passed through the $(0,0,2)$ rlp, the effects of the steps would still be weakly present, in principle, as a pair of weak spots on either side of the strong $(0,0,2)$ reflection. They would be the specular reflection and intersection of the rod from the $(0,0,4)$. The intensity of the strong reflection, in practice, would mask the two weak reflections. When the beam comes from the left, the diffraction rods pierce the sphere of reflection above the specular spot. When the beam comes from the right the intersections lie below the specular spot. The separation of the spots depends on the angle of incidence, the bulk lattice spacing in the z direction and the vicinal angle. For the sake of illustration the energy of the beam was decreased from 10 keV to 0.6 keV. The positions of the diffraction spots for 10 keV are shown in Fig. 5.8.

Figure 5.7 is not to scale for a 10 keV beam. A lower energy beam (~ 0.6 keV) was used to make the drawing clear. The resolution possible at 10 keV for steps is illustrated in Fig. 5.8 for the (0 0 1) face of bcc Fe with 50 atoms per terrace, looking along the step direction. At the second anti-Bragg condition the separation of the two spots where the $(0,0,2)$ and the $(0,0,4)$ rod intersect the sphere of reflection is close to 0.01 rad. In this case, where the vicinal angle is comparable to the Bragg angle, one can double the sensitivity by going to the first anti-Bragg condition to measure the difference between the specular spot and the rod from the $(0,0,2)$. In the limit of vicinal angles small compared to the Bragg angle, the angular separation is just twice the vicinal angle. By line width analysis of the specular spot, *Pukite* [5.19] determined that some crystals were flat to 0.3 µm. The resolution of *Pukite's* modified commercial 10 keV RHEED system for average terrace size was judged to be 0.8 µm, corresponding to 0.2 mm on the RHEED screen.

As seen by comparing Fig. 5.9 with Figs 5.7, 8, the sensitivity to the vicinal angle is much less for looking across the steps. The spacing of the vertical lines is the same in both cases, but the slope of the tangent to the sphere of reflection amplifies the effect for the beam parallel to the step direction. There is approximately a factor of 50 more sensitivity looking along the steps for a 10 keV beam. The same 0.2 mm on the RHEED screen corresponds to an average terrace size

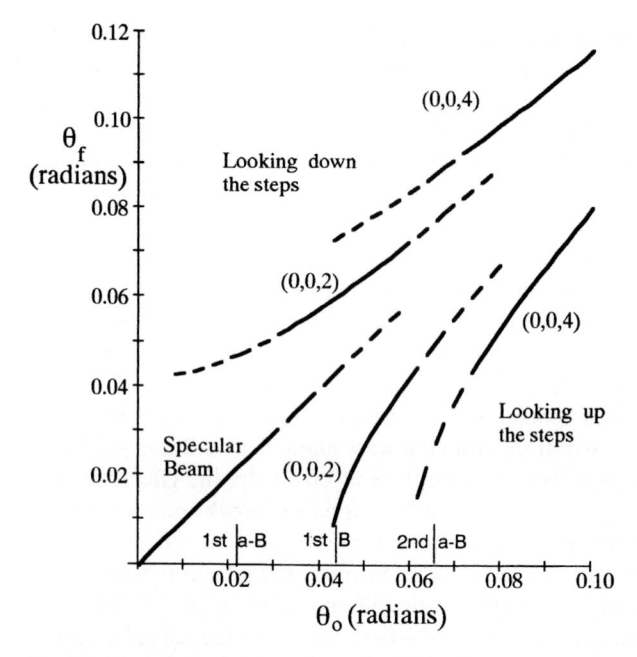

Fig. 5.8. The relation between the angle of diffraction and the angle of incidence for reflection from a stepped surface with the beam along the direction of the steps for a 10 keV beam, the (0, 0, 1) surface of bcc Fe, and 50 atoms per terrace. For the specular spot the two angles are the same by definition. Looking down the steps the (0, 0, 2) and (0, 0, 4) rods pierce the sphere of reflection at increased angle. Looking up the steps these rods pierce the sphere at decreased angle or not at all if the angle of incidence is small. The intensity of the (0, 0, 2) is a maximum in the region of the first Bragg position, as calculated naively

of 18 nm for the beam across the steps. The coherence of the beam is much less perpendicular to the direction of the beam than parallel, both from the geometry and the focussing of the beam.

The mean spacing of steps is generally known from X-rays during the preparation of a substrate. Of more interest is the distribution of steps and such questions as the straightness of the step edges. The statistics of steps can be investigated with RHEED. When diffraction amplitudes are squared to obtain the reflected intensity, phase information is lost. In kinematic theory the intensity is related to the pair correlation function. It is the pair correlation of step edges that is obtained by analysis of line shapes in experiments looking up and down the steps. The subject of distributions of steps can be pursued by starting with [5.20].

5.1.2.5 Spot Profile Analysis

The concept of the pair correlation function is used to interpret line widths and line shapes in general. An example is the analysis of one atom high islands that

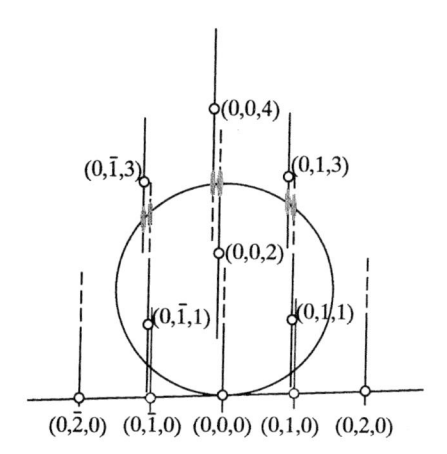

5.1 Introduction to Reflection High Energy Electron Diffraction (RHEED)

Fig. 5.9. The RHEED screen for diffraction from a vicinal surface with the beam looking across the direction of the steps. The diffraction rods are perpendicular to the average surface. A line from the (0, 0, 0) to the (0, 0, 4) reciprocal lattice points is perpendicular to the atomic planes. The separation of the diffraction rods is the same as in Fig. 5.7, but the separation of the pairs of diffraction spots is much less. The lightly presented (0, 1, 0) rlp is shown because it is not likely that there will be complete cancellation of the contributions from adjacent bcc steps

form during the growth of a fractional layer on top of a well formed surface layer. The line width is a measure of the mean separation between islands, that is, the island–island correlation length. For example, a line width (fwhm) of 1 mm perpendicular to the line joining the through beam and the specular spot on the 10 keV RHEED screen corresponds to an island-island correlation length of 7 nm. The measurement of such correlation is of sufficient interest that *Henzler* [5.12a–b] developed a LEED system specifically for spot profile analysis. This is a built-in feature of RHEED. The effect of islands on a surface is to widen the diffraction rods. When the correlation length is short enough the diffraction rods become wide enough to cut the sphere of reflection over extended regions which produce streaks on the RHEED screen, as illustrated in Fig. 5.10. This source of RHEED streaks is characterized by an obvious increase in streak width. The drawing is for a 10 keV electron beam with an angle of incidence of 50 mrad on the $(0,0,1)$ surface of bcc Fe, viewed from slightly above the shadow edge at a small azimuthal angle. Again mixing real space and momentum space this also illustrates the RHEED screen and beams in the laboratory. If the correlation between the islands is sharp in momentum space, the rods becomes hollow cylinders and the contours labeled a appear as split lines on the RHEED screen.

When the surface is undulating, the beam enters the sides of the hills, goes through them, and diffracts. The rlp's individually broaden because of finite size effects. Fuzzy ellipsoids about each rlp then intersect the Ewald sphere if the rlp

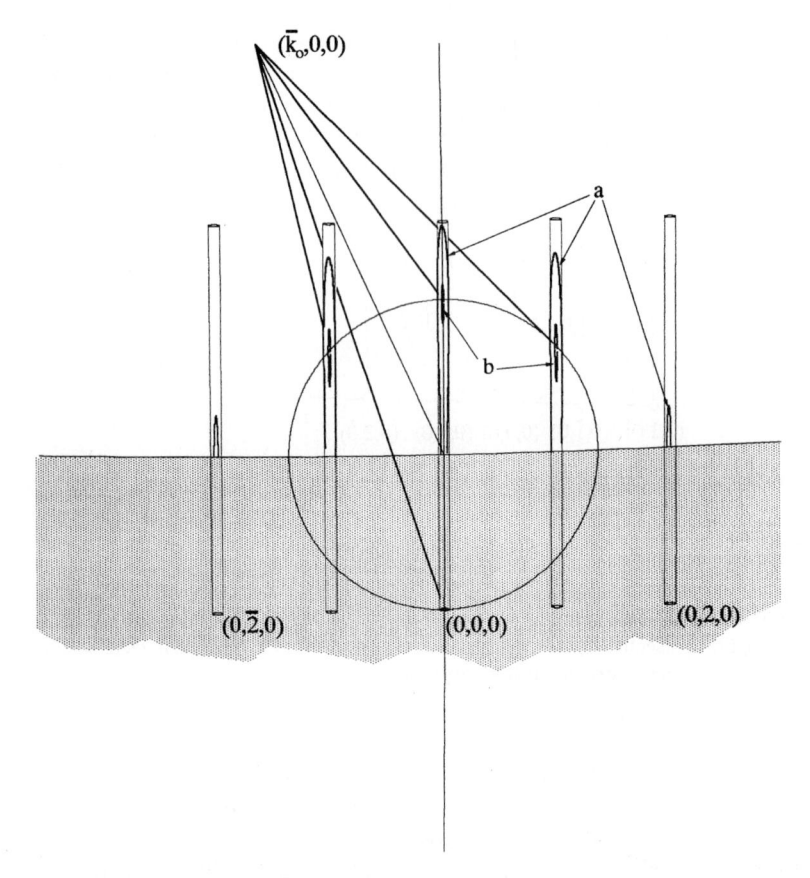

Fig. 5.10. Fat rods cutting the sphere of reflection. When one-atom-high islands form during the growth of a partial layer on top of a well formed substrate, the reciprocal lattice rods centered on g_n take on sideways components from the island–island correlation function. The cylinders about the rods represent the level of 10% of the maximum structure factor squared. The rods intersect the sphere of reflection in the contours labeled a. The contours labeled b are for rods that are five times sharper. The ordinarily evanescent $\pm (0, 2, 0)$ appears as a real streak as the rod extends back to intersect the sphere in case a but not in case b

is sufficiently close to the sphere. This results in broad fuzzy spots resembling the projections of the rlp's on the sphere of reflection.

5.1.2.6 RHEED Streaks

A mosaic crystal may have a statistical distribution of axes which can be described as a Gaussian distribution of the $(0, 0, 1)$ axes with x and y components, of the $(0, 1, 0)$ axes with z and x components and of the $(1, 0, 0)$ axes with

5.1 Introduction to Reflection High Energy Electron Diffraction (RHEED)

y and z components. For a nominally $(0, 0, 1)$ surface these can be viewed as a tilt forward and backwards with respect to the incoming beam in the x direction, rotations about the x axis that produces y components of the $(0, 0, 1)$ and rotations about the z axis that are equivalent to a change in azimuth. If the mosaic crystal is a smooth flat cut, each mosaic block will have the same surface normal, but a different vicinal angle α. The distribution of vicinal step directions will include up and down the steps as discussed in relation to Fig. 5.7, across the steps as illustrated in Fig. 5.9, and everything in between. The specular spot does not change its position with the vicinal angle of a mosaic block. The big effect comes from the position in k-space of the $(0, 0, 2)$ rlp and its accompanying diffraction rod. As the $(0, 0, 2)$ rlp is tilted away from the specular rod in the direction of the incoming beam, the angle of the outgoing wave changes rapidly as the beam looks up the steps. The outgoing angles for a given incoming angle as a function of the vicinal angle α are shown in Fig. 5.11 for the diffraction rod through the $(0, 0, 2)$ rlp. A reciprocal lattice rod that is close to intersecting the equator of the sphere of reflection will show a similar amplification of the effect of the mosaic.

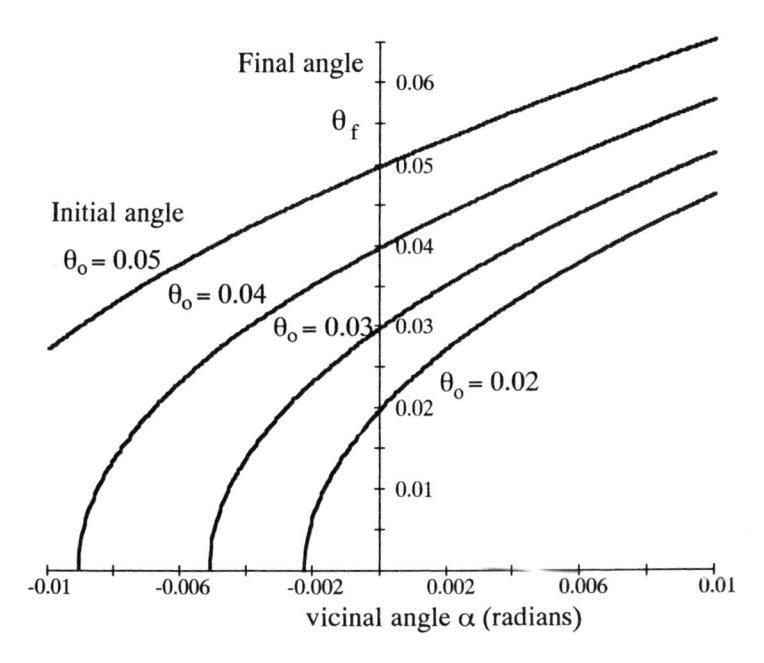

Fig. 5.11. The dependence of final angle θ_f {for the diffraction rod through the (0, 0, 2) rlp} on vicinal angle α for a range of incoming angles θ_0. A mosaic spread of 2 mrad can cause a streak through the specular spot on the RHEED screen with an angular range of as much as 20 mrad. A mosaic crystal can produce RHEED streaks that are extended in the κ_z direction but sharp in the κ_y direction

5.1.3 Waves Inside a Slab

The discussion so far has avoided solving Schödinger's equation. Before proceeding further with the discussion of RHEED, it may be worthwhile considering the waves inside of the crystal that accompany reflection from the surface. The full problem has been discussed many times in the literature starting with Bethe[5.7b] in 1928. There is a small group of theorists that actually calculate RHEED intensities by choosing a suitable optical potential and using the power of modern computers to solve the problem numerically. These programs have concentrated on accounting for the intensity of the specular spot as a function of the incoming angle θ_0 of perfect crystals. There is very little data which directly compares experiments and theory. The experiment has not yet been done on Fe whiskers which present the best metallic surfaces and are of interest in the field of magnetism.

Consideration is given to the elementary aspects of reflection from a perfect crystal by dividing the problem into four parts. The first is a discussion of reflection from a slab with a uniform complex optical potential. The second is reflection from a slab with a periodic variation of the optical potential in the direction of the surface normal. In the first two cases the beam is in the x, z plane. The third case is reflection from a slab with a periodic variation of the optical potential in the direction perpendicular to the x, z plane with the beam satisfying the Bragg condition. The final case is reflection from a slab with periodicity in both the y and z directions. Then some consideration is given to how calculations are actually done, dropping the restriction of periodic potentials.

5.1.3.1 Reflection From a Uniform Slab with Absorption

The one-body Schrödinger equation is particularly transparent for reflection from a parallel-sided slab with a uniform optical potential [5.15]. The entrance surface is at $z = 0$ with the wave approaching in the positive z direction and reflecting into the negative z direction. The back surface is at $z = d$. The wave equation outside has solutions that are plane waves that satisfy translationally invariant boundary conditions. There is an incoming plane wave and a reflected wave. The wave vector of the reflected wave maintains the components of the incoming wave vector parallel to the surface. The perpendicular component must have the same magnitude as the incoming wave for elastic scattering. On the far side of the parallel-sided slab the transmitted wave has the same wave vector as the incident wave, These three waves have amplitudes $1, r,$ and t with wave vectors in the x, z plane

$$\mathbf{k}_0 = \mathbf{k}_t = k_z \hat{z} + k_x \hat{x}, \qquad \mathbf{k}_r = -k_z \hat{z} + k_x \hat{x}. \tag{5.15}$$

The wave function for $z < 0$ is

$$\phi_0(z, x) = e^{i(k_z z + k_x x)} + r e^{i(-k_z z + k_x x)}. \tag{5.16}$$

5.1 Introduction to Reflection High Energy Electron Diffraction (RHEED)

The energy is $E_0 = (\hbar k_0)^2/2m$ outside and inside of the slab. The optical potential energy V inside of the slab is complex. It is expressed in reduced units as

$$\frac{2m}{\hbar^2} V = \frac{2me}{\hbar^2} \phi_0 = -|v_0| - i|v_i|. \tag{5.17}$$

The units of the reduced potential energy are those of k_0^2. The introduction of the absolute value signs is to avoid any confusion about the fact that the electron wants to be in the slab and the complex part of the potential energy must be negative to act as an absorber. The potential energy that goes into Schrödinger's equation is negative. The electron gains kinetic energy on entering the surface. The magnitude of the wave vector increases; the wavelength decreases. While watching a RHEED screen it may be helpful to remember the analogy of light going from air to water. For RHEED the vacuum is the air and the sample is the water. The reflectivity goes to one in the limit of zero grazing angle. Critical angles appear only when the electron is going from the sample to the vacuum. The wave vectors inside of the slab have the same k_x as outside. The z components of the wave vectors inside, $\pm q$, are complex,

$$q = \sqrt{k_z^2 + |v_0| + i|v_i|} = q_r + i\eta \tag{5.18}$$

with real part q_r and a positive imaginary part η. The symbol q is used for the z components inside of the slab because there will be several z components once periodic potentials are introduced, and one needs the room for subscripts. The real and complex parts are given by:

$$q_r = \sqrt{\frac{k_z^2 + v_0 + \sqrt{(v_0 + k_z^2)^2 + v_i^2}}{2}},$$

$$\eta = \sqrt{\frac{-(k_z^2 + v_0) + \sqrt{(v_0 + k_z^2)^2 + v_i^2}}{2}}. \tag{5.19}$$

The real part of the wave vectors external and internal to the slab are given geometrically, as shown in Fig. 5.12, by constructing the sphere of reflection for waves outside and inside the slab. The sphere inside has a larger radius given by the real part of $\sqrt{k_0^2 + |v_0| + i|v_i|}$. As v_0 is typically 3 Å^{-2} and k_0^2 is typically 2500 Å^{-2}, it is necessary to distort the scale to illustrate the Ewald construction.

The wave function inside has two independent solutions to the second order differential equation

$$\Psi(z, x) = \psi_+ e^{i(qz + k_x x)} + \psi_- e^{i(-qz + k_x x)}, \tag{5.20}$$

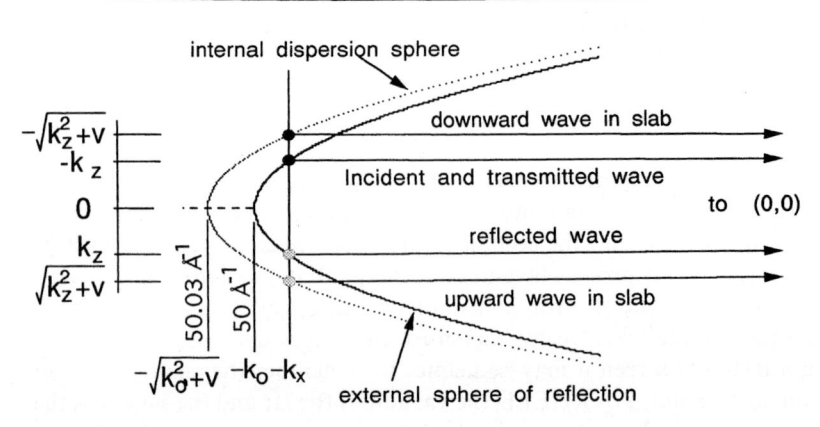

Fig. 5.12. Geometric representation of the relation between internal (downward and upward) and external (incident, reflected and transmitted) wave vectors. The external and internal spheres of reflection are shown on a distorted 50:1 scale to visualize the very small difference between the diameters of the two spheres. The points on the spheres of reflection are the tails of the wave vectors with their tips at the origin of k-space (0, 0), far off the page. The points shown correspond to an incident angle $\theta = 0.035$ rad, for $k_0 = 50$ Å^{-1} and $v = 3$ Å^{-2}. If the wave were generated inside the slab for k_x between k_0 and $\sqrt{k_0^2 + v}$, the external waves would be evanescent on both sides of the slab.

where the amplitudes

$$\psi_+ = \frac{2k_z(q + k_z)}{(q + k_z)^2 - (k_z - q)^2 e^{-2\,d\eta + 2i\,dq_r}},$$

$$\psi_- = \frac{2k_z(k_z - q)}{(k_z - q)^2 - (k_z + q)^2 e^{2\,d\eta - 2i\,dq_r}} \qquad (5.21)$$

are found by matching the wave functions and their derivatives on the two surfaces:

$$1 + r = \psi_+ + \psi_-, \qquad t\,e^{idk_z} = \psi_+ e^{iqd} + \psi_- e^{-iqd},$$

$$k_z(1 - r) = q(\psi_+ - \psi_-), \qquad t\,k_z e^{idk_z} = q\psi_+ e^{iqd} - q\psi_- e^{-iqd}. \qquad (5.22)$$

The wave vector in the exponential has been expressed in terms of its real and imaginary parts to emphasize which terms dominate when the attenuation $d\eta$ is large.

The general structure of dynamical theory is already here in the description reflection. The solution of Schrödinger's equation is in terms of allowed wave vectors inside of the slab. They are those for which

$$q^2 + k_x^2 + k_y^2 = k_0^2 + |v_0| + i|v_i|. \qquad (5.23)$$

The real part of this is a sphere in momentum space. Inside of the slab it is called the dispersion surface. To satisfy the boundary conditions the tangential components are picked by the incoming wave. The q's are found by the intersec-

5.1 Introduction to Reflection High Energy Electron Diffraction (RHEED)

tion with the dispersion surface of a vertical line (normal to the surface) through the tail of the incoming wave vector drawn as in Fig. 5.12.

The reflection from the back surface is negligible if the attenuation is large. Then

$$\psi_+ = \frac{2k_z}{q + k_z}, \quad \psi_- = 0. \tag{5.24}$$

The reflection amplitude is

$$r = \frac{k_z - q}{k_z + q}, \tag{5.25}$$

which goes to -1 as the incoming angle and k_z go to zero.

The reflectivity is

$$R = r^*r = \frac{(k_z - q_r)^2 + \eta^2}{(k_z + q_r)^2 + \eta^2}. \tag{5.26}$$

For large k_z, $q_r - k_2$ goes to $v_0/2k_z$ and the reflectivity falls as θ_0^4. The electron density inside is

$$\Psi(z)\Psi^*(z) = \frac{4k_z^2}{(q_r + k_z)^2 + \eta^2} e^{-2z\eta}. \tag{5.27}$$

At the surface this goes to zero for small angles and to one for q_r large compared to η. For values suitable for bcc iron with a 10 keV electron beam $k_0 = 50\,\text{Å}^{-1}$, $|v_0| = 3\,\text{Å}^{-2}$ and $|v_i| = 0.5\,\text{Å}^{-2}$. For an incoming angle $\theta_0 = 50$ mrad the reflectivity is less than 0.01 and the penetration of the beam is 1 nm for a decrease to $1/e$.

The above result was obtained by taking the limit of large attenuation after solving the boundary conditions on both surfaces. The same result applies if the back wave is neglected from the beginning and only the boundary conditions for the front surface are applied. Then there are only the one internal wave amplitude and the amplitude of the external reflected wave to be determined.

For reflection from a slab thin enough for the back wave to be important, the result with a complex potential is too complex to contemplate analytically, but it is easily obtained using complex arithmetic in FORTRAN. The reflectivity from a slab without attenuation is

$$R = \frac{|v_0|^2 \sin^2\left(d\sqrt{|v_0| + k_z^2}\right)}{|v_0|^2 \sin^2\left(d\sqrt{|v_0| + k_z^2}\right) + 4k_z^2(k_z^2 + |v_0|)}. \tag{5.28}$$

which shows that the reflectivity oscillates with thickness of the layer and with the angle of incidence. The reflectivity with attenuation for a slab 2.866 Å thick, corresponding to 2 ML of iron, is compared with reflection from a thick slab of

204   5. Epitaxial Growth of Metallic Structures

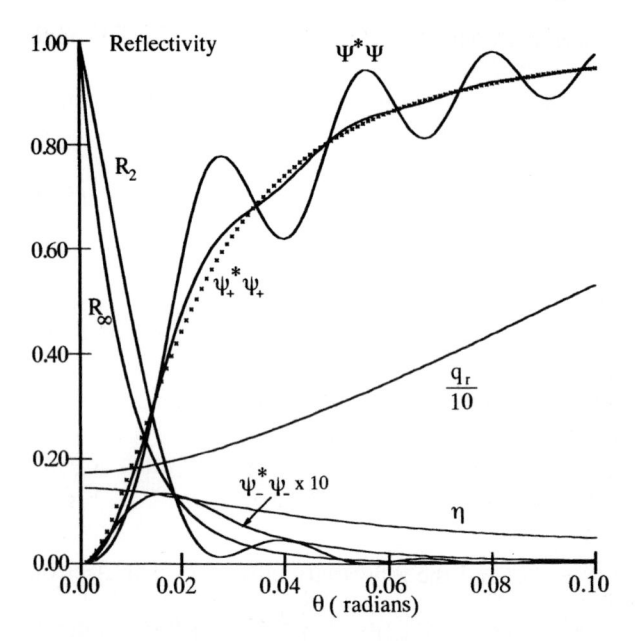

Fig. 5.13. The reflectivity from a parallel-sided slab with a uniform optical potential, $v = 3 \text{ Å}^{-2}$, approximating the average potential in bcc iron, as a function of the angle of grazing incidence θ of a 10 keV electron beam. The curve labeled R_∞ is for a slab of thickness greater than 5 nm. The curve labeled R_2 is for a slab 0.2866 nm thick. The dotted curve is $\psi_+^* \psi_+$ at the surface for the thick slab. The adjacent, slightly oscillating curve is $\psi_+^* \psi_+$ for the thin slab

iron in Fig. 5.13. The reflectivity for the thin slab oscillates about the reflectivity for the thick slab. The much smaller value of $\psi_-^* \psi_-$ had to be multiplied by 10 to make it visible, but the total wave $\Psi^* \Psi$ exhibits strong oscillations from the beating of the much smaller amplitude wave against the larger amplitude wave. The real q_r and the imaginary η parts of the wave vector change with angle as shown, where the units are Å^{-1}. The attenuation in the limit of small angles has the intensity dropping to 1/e by 7 Å. The oscillation with angle in the reflected intensity for two layers is stronger for the square well potential than it would be for a more smoothly varying potential.

5.1.3.2 Potentials Periodic in z

For a sinusoidal potential with period a_0 and wave vector $G = 2\pi/a_0$ in the z direction the reduced potential energy is

$$v(z) = -|v_0| - i|v_i| + (v_{rG} + iv_{iG})(e^{iGz} + e^{-iGz}). \tag{5.29}$$

A warning from the beginning is that the restriction $2|v_{iG}| \leq |v_i|$ must be applied or else the imaginary potential will create electrons. The sinusoidal potential turns Schrödinger's equation into Mathieu's equation [5.22] after eliminating the exp $\{i(k_x x + k_y y)\}$ dependence on x and y:

$$-\left(\frac{\partial \partial}{\partial z \partial z} \Psi(z)\right) - (v - 2v_G \cos(Gz)) \Psi(z) = k_z^2 \Psi(z), \tag{5.30}$$

where

$$v = |v_0| + i|v_i| \quad \text{and} \quad v_G = v_{rG} + iv_{iG}. \tag{5.31}$$

The two solutions of this equation are one wave that propagates into the crystal from the front surface and one that propagates from the back surface on reflection. The latter will not be important for RHEED.

To a good approximation the solution is a plane wave with wave vector q given by (5.18) if q is *not close* to $-G/2$. The meaning of *not close* is that $q + G/2$ is large compared to $\sqrt{|v_G|}$. In that case the problem is only slightly modified from the treatment with a uniform optical potential. Actually there is some mixing of all waves with wave vector $q \pm nG$. The mixing to $q + G$ and $q - G$ is of order v_G^2. For a sinusoidal potential the mixing to $q + nG$ and $q - nG$ is of the order v_G^{2n}. A real potential will have Fourier components v_{nG} for which the coupling from q to $q + nG$ will be of the order v_{nG}^2. The ratio of the wave components comes directly from the differential equation. The overall amplitude of the solution is determined from the boundary conditions at the front surface which also determine the amplitude of the external reflected wave.

If q is close to $-G/2$, the wave is, to a good approximation, a coupled pair of waves with wave vectors $q = -G/2$ and $q + G = G/2$ which form a damped standing wave in z. The value of q changes with the incoming wave vector. As before, k_x and k_y will be the same inside and outside of the slab. The value of q is found by the intersection of a vertical line through the tail of the incoming wave vector and the dispersion surface generated by Mathieu's equation. To a first approximation this surface is a series of spheres in k-space of radius $\sqrt{k_0^2 + |v_0|}$ separated by $G\hat{z}$. To a second approximation the periodic potential modifies the spheres in the regions where they would cross. The dispersion surface consists of coaxial corrugated cylinder, as depicted in Fig. 5.14. The thick solid lines show the sphere of reflection outside of the slab with the horizontal portion representing evanescent waves outside of the crystal. The large black dot represents a wave vector of an external wave. It generates an internal wave which is a coupling of the series of plane waves originating on the dispersion surface at the open circles along the dashed line, which is the locus of all vectors with the same k_x and k_y as the incoming beam. The unmarked crossings of the dispersion surface along the dashed line are the wave vectors of the reflected wave from the back surface which is being ignored. The compound wave has components both away from the front surface and back toward it, but the net flow is inward, except when the incoming wave excites waves in the gap represented by the horizontal line linking the outer cylinder and the inner cylinder of the dispersion surface. The horizontal line is the real part of q in the

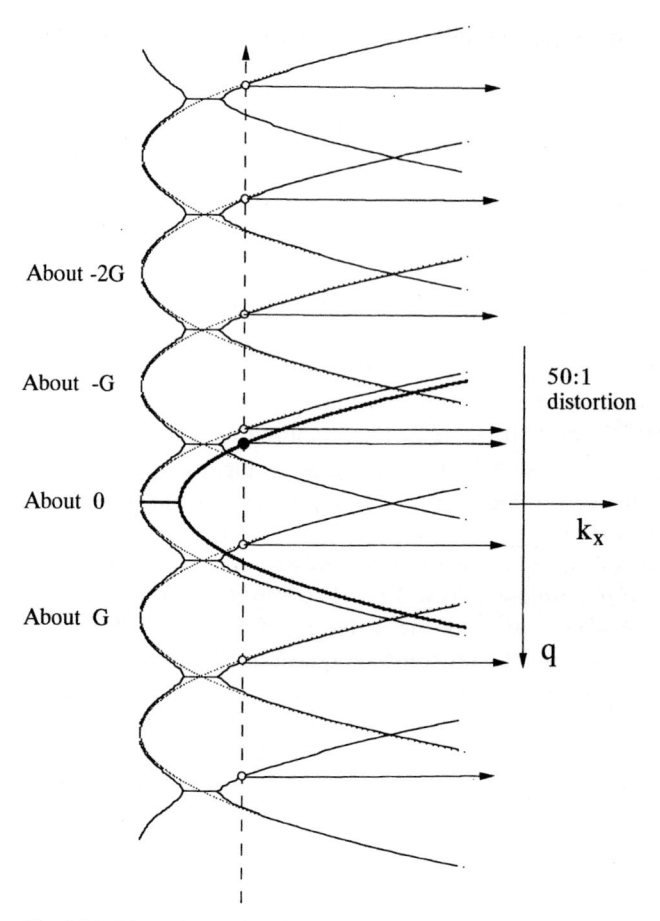

Fig. 5.14. Dispersion surface for a sinusoidal potential in the z direction for potentials suitable for bcc Fe for a 10 keV beam are shown with a distortion of 50 to 1, stretched along the k_x axis as in (Fig. 5.12). The unperturbed spheres of radius $\sqrt{k_0^2 + |v|}$ about each of the reciprocal lattice points $\pm nG\hat{z}$ are shown as dotted lines. The perturbation by the potential $2v\hat{G}\cos Gz$ is shown as solid lines. The perturbation breaks the dispersion surface into concentric corrugated cylinders

gap. For the values chosen to draw the dispersion surface, the lower end of the gap is just accessible with a finite k_z for the incoming wave. If there were no imaginary part to the potential, the reflectivity would go to one in the gap. In the gaps the waves are stationary in the z direction, propagating in the plane of the periodic potential. There is also a gap where the next neighbor spheres appear to cross, but for a sinusoidal potential it is not visible on this scale. It becomes important for a real potential with a v_{2G} Fourier component when $q = -G$, as shown in Fig. 5.15. An alternative diagram would show just the principle q with vectors fanning out to each of the reciprocal lattice points $(0.0, \pm nG)$ which are the centers of the unperturbed spheres.

5.1 Introduction to Reflection High Energy Electron Diffraction (RHEED)

The z component of the fundamental wave vector for small k_z is given by

$$q = \pm \sqrt{\frac{G^2}{4} + k_z^2 + v_0 + iv_i - \sqrt{G^2(k_z^2 + v_0 + iv_i) + v_G^2}} - \frac{G}{2}, \tag{5.32}$$

where the $+$ sign is for above the gap and the $-$ sign is for below the gap. The dependence of q on $\theta = k_z/k_0$ is shown in Fig. 5.15. The curve is computed in sections by changing the G in (5.32) to nG for the nth gap.

The spatial dependence of the wave function is

$$\Psi(x, y, z) = \Psi(z) e^{i(k_x x + k_y y)}, \tag{5.33}$$

where $\Psi(z)$ is a sum of coupled plane waves of the form

$$\Psi(z) = \sum_{n=-\infty}^{\infty} c_n e^{i(q+nG)z} = \cdots + c_0 e^{iqz} + c_1 e^{i(q+G)z} + \cdots \tag{5.34}$$

The ratio of c_0 to c_1 (and all the rest of the c_n's) is found from the differential equation by the same process that is used to generate the equation for q. Substituting the expression for $\Psi(z)$ into the differential equation leads to an infinite set of coupled algebraic equations. Under the approximation that at most two of the coefficients are large for any choice of k_z, only two of the equations survive

$$-c_{-1} v_G - c_1 v_G = c_0 (k_z^2 + v_0 - q^2) \tag{5.35}$$

and

$$-c_0 v_G - c_2 v_G = c_1 (k_z^2 + v_0 - (q+G)^2). \tag{5.36}$$

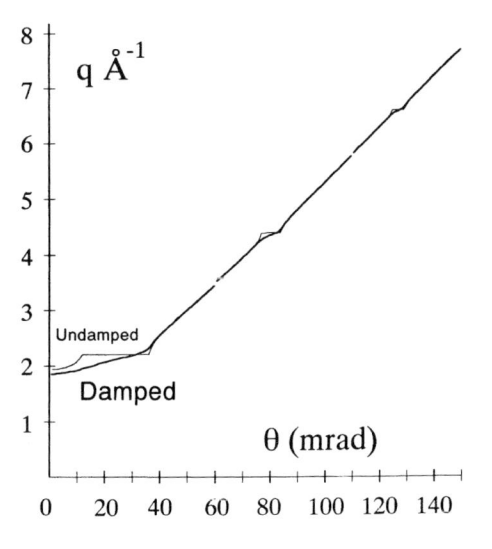

Fig. 5.15. The calculated dependence of the principle wave vector inside the crystal on the angle of incidence θ for reflection from a potential periodic in z with (heavy line) and without (light line) damping from the imaginary part of the optical potential. The calculation takes into account only one other wave coupled to the primary wave. The choice of the important second wave is changed from section to section of the calculated curves

Neglecting the terms c_{-1} and c_2 gives a pair of coupled linear homogeneous equations that have a solution only if the relation between q and k_z is as given in (5.32). The ratio of the coefficients is

$$\frac{c_1}{c_0} = -\frac{k_z^2 + v_0 - q^2}{v_G}. \tag{5.37}$$

The value of the c's and the reflection amplitude are found from the boundary conditions

$$c_0 q + (q + G)c_1 = k_z(1 - r), \qquad c_0 + c_1 = 1 + r. \tag{5.38}$$

These give the reflection amplitude in terms of k_z,

$$r = \frac{-2k_z(k_z^2 + v_0 - q^2) + 2k_z v_G}{-(q + G + k_z)(k_z^2 + v_0 - q^2) + v_G(q + k_z)} - 1, \tag{5.39}$$

which reaches large values when $k_z \cong q \cong -G/2$.

The reflectivity, $R = r^*r$ as a function of the angle of incidence is shown in Fig. 5.16. This curve is calculated in three sections, changing the choice of which two coefficients are important for each diffraction peak. When the real part of $q = -G/2$, the reflectivity is large, actually total if the imaginary part of the optical potential were neglected. The two parts of the wave have equal |amplitude|and form a standing wave in the z direction with a definite phase relation with respect to the potential. The wave is attenuated by $\sim 1/e$ for each period of the potential as shown in Fig. 5.17. This is consistent with systematic studies of RHEED intensities which show that the influence of a substrate on the diffraction pattern of an overlayer changes very little after two monolayers, for example, Au on Ag (5.23). To a first approximation, when the diffraction condition is fulfilled for a periodic potential in the z direction, the wave propagates above the center of the first plane of atoms.

If there were a step up somewhere on the surface, there would be a transition region. The beam just above the centers of the first layer before the step would

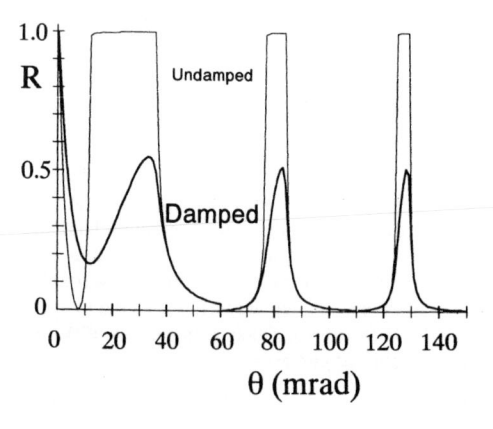

Fig. 5.16. The calculated dependence of the intensity of the reflected wave outside of the crystal on the angle of incidence θ for reflection from a potential periodic in z with (heavy line) and without (light line) damping from the imaginary part of the optical potential as in Fig. 5.15

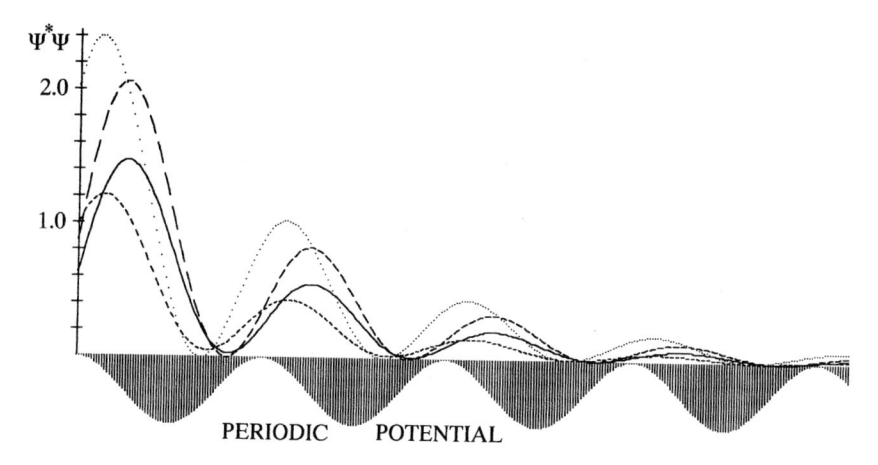

Fig. 5.17. The calculated dependence of the internal wave intensity inside of the crystal distance from the surface for the same two choices of damping as used in Figs. 5.15, 16. The periodic potential is shown at the bottom. The waves are phase locked to the potential, but the phase changes with angle of incidence. The dotted curve is for no damping at an angle of incidence of 0.020 rad. The curve with short dashes is for the same angle of incidence with damping corresponding to Fe. The solid curve and the curve with the long dashes are for an angle of incidence of 0.030 rad with and without damping, respectively

penetrate into the region between the first and second layer after the step. The steady state should be recovered with a characteristic length given by $k_0/v_i \cong 100$ Å. This is an indication of why one must be skeptical in applying the kinematic model.

It should be noted that for the singular slab there is only the specularly reflected beam outside of the crystal. This was also the case in considering reflection from a plane of atoms. The components $nG\hat{z}$, did not show up in the reflection until the vicinal surface was considered. The same is true for dynamical theory. The components with $q + nG$ are present inside of the crystal, but they and the reflected wave all are generated by the incoming wave. For a vicinal surface the dynamical theory will have each of these components generating waves outside of the crystal which in turn generate waves inside that use the same set of internal wave vectors but in different combinations [5.24]. Then half of the wave amplitudes appear as independent variables to be found by matching boundary conditions with the set of external waves. This is better understood, perhaps, after looking at the case of a potential periodic in the y direction.

5.1.3.3 Potentials Periodic in y

When the Bragg condition is fulfilled for a potential periodic in the y direction, the beam continues to travel down into the crystal while forming a standing wave in the direction of the periodic potential. The periodicity in the y direction is not particularly interesting unless it is accompanied by a periodicity in the z direction. This case is discussed in Sect. 5.1.3.4. The main purpose of studying the periodicity in the y direction is to show how the boundary conditions differ when there are extra external waves generated. To scatter strongly the crystal should be rotated azimuthally by ϕ where $2k_x \sin \phi = G$.

The geometry for the Bragg condition is shown in Fig 5.18. The k_x direction bisects the angle of rotation. The wave vector of the periodic potential is $\boldsymbol{G} = -G\hat{y}$. The incident beam has wave vector.

$$\boldsymbol{k}_0 = k_z \hat{z} + k_x \hat{x} + G/2\, \hat{y}. \tag{5.40}$$

In addition to the specularly reflected beam of amplitude r and wave vector

$$\boldsymbol{k}_r = -k_z \hat{z} + k_x \hat{x} + G/2\, \hat{y}. \tag{5.41}$$

there is a diffracted beam of amplitude g and wave vector

$$\boldsymbol{k}_g = -k_z \hat{z} + k_x \hat{x} - G/2\, \hat{y}. \tag{5.42}$$

The external wave function for $z < 0$ is

$$\Phi(x,y,z) = e^{ik_x x}\left(e^{i((G/2)y+k_z z)} + r e^{i((G/2)y - k_z z)} + g e^{i(-(G/2)y - zk_z)}\right). \tag{5.43}$$

Schrödinger's equation inside the crystal is

$$\nabla^2 \Psi(x,y,z) + (v + 2v_G \cos(Gy) + k_0^2)\, \Psi(x,y,z) = 0. \tag{5.44}$$

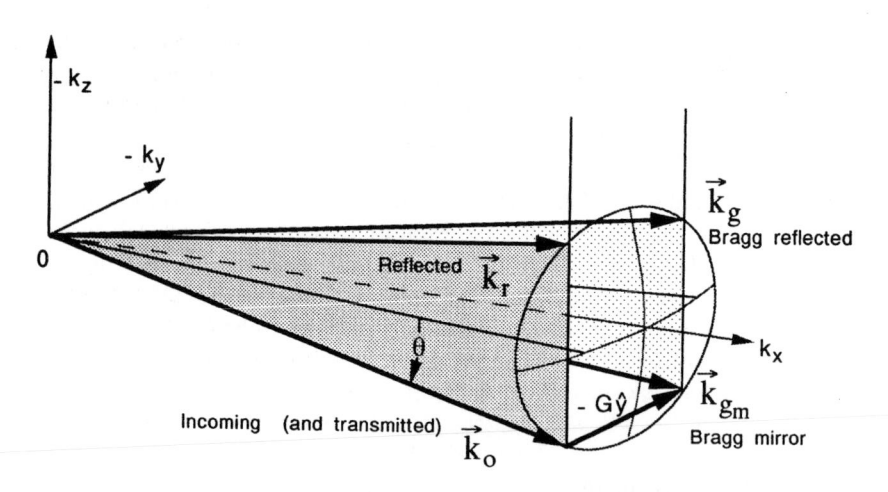

Fig. 5.18. The external wave vectors for a periodic potential with wave vector G in the y direction. The k_x direction is chosen along the bisector of $-G\hat{y}$. The incident beam has $k_y = G/2$ to satisfy the Bragg condition for all angles of incidence θ. The x–z planes have constant phase for the periodic potential. The internal waves are pairs of plane waves which propagate in the x–z plane while creating standing waves in the y direction

5.1 Introduction to Reflection High Energy Electron Diffraction (RHEED)

One looks for solutions of the form

$$\Psi(x, y, z) = \sum_{n=-\infty}^{\infty} c_n e^{i(G(n - 1/2)y + xk_x + qz)}. \tag{5.45}$$

The coefficients are related by an infinite set of algebraic equations

$$c_n(q^2 + k_x^2 + G^2(n - \tfrac{1}{2})^2 - v - k_0^2) - c_{n-1}v_G - c_{n+1}v_G = 0. \tag{5.46}$$

The important two coefficients are for $n = 0$ and $n = 1$ for which

$$c_0(q^2 + k_x^2 + G^2(\tfrac{1}{2})^2 - v - k_0^2) - c_{-1}v_G - c_1 v_G = 0 \tag{5.47}$$

and

$$c_1(q^2 + k_x^2 + G^2(\tfrac{1}{2})^2 - v - k_0^2) - c_0 v_G - c_2 v_G = 0. \tag{5.48}$$

Neglecting the effects of c_{-1} and c_2 produces a pair of homogeneous equations which yield the ratio c_1/c_0 and the condition that determines q. The two roots of q that give waves propagating into the crystal are

$$q_A = \sqrt{k_z^2 + v + v_G}, \qquad q_B = \sqrt{k_z^2 + v - v_G} \tag{5.49}$$

for which the ratios of the coefficients are

$$c_1^A = c_0^A, \qquad c_1^B = -c_0^B. \tag{5.50}$$

The two parts of each wave are strongly coupled. They each form standing waves in the direction of the wave vector of the periodic potential while being transmitted into the crystal. The differential equation now appears to have four solutions, two traveling in the positive direction, as given above, and two that are reflected from the back surface, which are being ignored. How is it that now there are four solutions to one second order differential equation? The answer is that there is now more than the one external beam to excite the internal waves. The incident beam and the reflected beam are one wave. Each of the other external waves is an additional wave. The incident beam and the reflected beam excite all the harmonics of one fundamental in-going wave. Half of the harmonics then reflect and transmit at the front surface. Each of the reflected beams excites a set of harmonics as required by the differential equation, but with different ratios of amplitudes. The result of all this is that one can then treat the waves q_A and q_B as independent but strongly coupled, (q_A to $q_A + G$) and (q_B to $q_B + G$). These are standing waves in the y direction of the periodic potential propagating parallel to the planes of constant potential:

$$\Psi(x, y, z) = 2e^{ixk_x}\left(c_0^A e^{izq_A}\cos\frac{Gy}{2} - ic_0^B e^{izq_B}\sin\frac{Gy}{2}\right). \tag{5.51}$$

This wave must match at the boundary with the wave outside. Because there are two different projections of the waves in the plane of the surface, there are four boundary conditions to determine r, g, c_0^A and c_0^B.

$$r = \frac{k_z^2 - q_A q_B}{(k_z + q_A)(k_z + q_B)}, \qquad g = \frac{2k_z(q_B - q_A)}{(k_z + q_A)(k_z + q_B)}, \tag{5.52}$$

$$c_0^B = \frac{2k_z}{k_z + q_B}, \qquad c_0^A = \frac{2k_z}{k_z + q_A}. \tag{5.53}$$

The reflected amplitude drops monotonically with angle of incidence. The diffracted amplitude builds up for small angles because there is no diffraction if there is no penetration of the beam. The reflectivity goes to one at small-angles because there is no wave inside of the crystal. The amplitudes and intensities of the reflected and diffracted waves are represented in Fig. 5.19.

5.1.3.4 Potentials Periodic in y and z

For a $(0, 0, 1)$ surface of a bcc lattice with the beam in the x direction, the most important rlp's are $(0, \pm 1, 1)$ and $(0, 0, 2)$. The $(0, 0, 2)$ corresponds to the periodicity of the atomic planes in the z direction. In treating the periodicity in the z direction in Sect. 5.1.3.2, the effects of the periodicity in the in-plane directions were ignored. No specifications of the direction of the beam with respect to periodicities in x or y were made. If the beam is directed along the $(1, 0, 0)$ and the angle of incidence is increased to the point where the Bragg condition is fulfilled for the $(0, 0, 2)$ with the beam in the x–z plane, then the Bragg condition is also fulfilled for the $(0, \pm 1, 1)$ without azimuthal rotation of

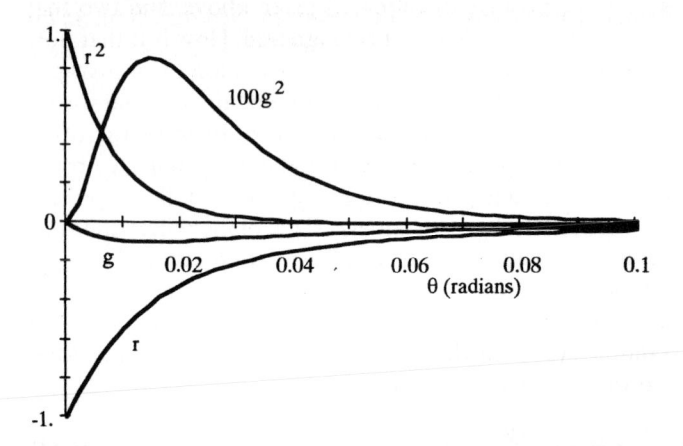

Fig. 5.19. The dependence on angle of incidence θ of the reflectivity from a periodic potential parallel to the surface at the azimuth for the Bragg condition. The amplitude and intensity are shown for the specular spot, r and r^2, and for the diffracted beam, g and g^2. The intensity of the diffracted beam is multiplied by 100

5.1 Introduction to Reflection High Energy Electron Diffraction (RHEED)

the crystal. The sphere of the dispersion surface about the origin and the sphere about the (0, 1, 1) intersect in the (0, −1, 1) plane which intersects the x–z plane. In this case there are four strong internal beams at the same time of which three are independent and two of these are related by symmetry. There are three strong external beams in addition to the incident beam, providing three pairs of boundary conditions at the entrance surface. The problem is treated as above except there are now more equations, but the symmetry helps. There are three fundamental values for the z component of the internal wave vector for a given incident beam which are found from the solution of an equation which is cubic in q^2. Each of these generates a coupled group of four waves for which the ratio of the coefficients is known. The complex wave vectors are found by numerical tracking, starting well away from a diffraction peak where they can be written down by inspection of the unperturbed spheres about each of the rlp's of each of the important Fourier components of the potential.

5.1.3.5 Arbitrary z Variations of the Potential

An important application of LEED has been the study of the relaxation of the surface, in particular a difference in the interlayer spacing between the surface layer and the penultimate layer compared to the lattice parameter of the bulk. Reflection from a stepped surface can reveal the height difference between adjacent terraces, but that will correspond to the bulk lattice parameter because both of the observed layers will have the same relaxation. One needs to apply dynamical theory without invoking a periodic potential in the z direction. This is done by treating the surface as if it were made of a series of very thin subatomic sheets parallel to the surface. The transmitted and reflected beams are calculated for each sheet, keeping track of all the beams in each direction. So far this method has been applied to only a few cases for RHEED [5.25 a–h]. These are all for structures with a small unit cell. With a larger unit cell the method can be applied to a stepped surface [5.26].

Slice Method. The principle of the slice method can be illustrated for the case of translational invariance in the aforementioned sheets as in optics [5.15]. In each of N slices the potential is treated as constant, varying in z from slice to slice. The wave vectors q_n are then known in terms of the potentials. The waves in two adjacent regions are

$$\Psi_n = F_n e^{iq_n z} + B_n e^{-iq_n z} \tag{5.54}$$

and

$$\Psi_{n-1} = F_{n-1} e^{iq_{n-1} z} + B_{n-1} e^{-iq_{n-1} z}, \tag{5.55}$$

where the coefficients F_n and B_n are for the amplitudes of the forward and backward waves in each region. There is no backward wave in the final region. Matching the waves and their z derivatives at the interface z_n between the two regions produces a matrix transformation between the coefficients

$$\begin{bmatrix} F_n \\ B_n \end{bmatrix} = \frac{1}{2q_n} \begin{bmatrix} (q_{n-1} + q_n)e^{iz_n(q_{n-1} - q_n)} & (-q_{n-1} + q_n)e^{-iz_n(q_{n-1} + q_n)} \\ (-q_{n-1} + q_n)e^{iz_n(q_{n-1} + q_n)} & (q_{n-1} + q_n)e^{-iz_n(q_{n-1} - q_n)} \end{bmatrix} \begin{bmatrix} F_{n-1} \\ B_{n-1} \end{bmatrix}. \tag{5.56}$$

This can be written symbolically as

$$\begin{bmatrix} F_n \\ B_n \end{bmatrix} = [M_n] \begin{bmatrix} F_{n-1} \\ B_{n-1} \end{bmatrix}. \tag{5.57}$$

The first surface is at $n = 1$. The region $n = 0$ has the incoming wave of amplitude $F_0 = 1$ and the reflected amplitude $B_0 = r$, which is what is to be calculated. The final region beyond z_{N+1} has $F_{N+1} = t$ and $B_{N+1} = 0$. The full relation is

$$\begin{bmatrix} t \\ 0 \end{bmatrix} = [M_{N+1}][M_N][M_{N-1}] \cdots [M_2][M_1] \begin{bmatrix} 1 \\ r \end{bmatrix} = \begin{bmatrix} M_{11} & M_{12} \\ M_{21} & M_{22} \end{bmatrix} \begin{bmatrix} 1 \\ r \end{bmatrix}. \tag{5.58}$$

To find out the wave at the surface it is necessary to follow the beam to the backside of the crystal or well into the crystal to where t can be taken as negligible, which is determined by the matrix $[M_n]$ becoming a unit matrix. The reflection amplitude is given by

$$r = -\frac{M_{21}}{M_{22}}. \tag{5.59}$$

This was for a wave with translational invariance parallel to the surface with just two Fourier components q and $-q$ feeding back on one another. The principle is the same if each slice has potentials periodic in y or x. The vectors have more components and the matrices become larger. The practical considerations in carrying out the procedure are the concern of the specialists. A stepped surface has been treated by using a large unit cell [5.26].

For treating reflection electron microscopy, *Cowley* has developed the perpendicular slice method. The slices perpendicular to the surface include the vacuum above the surface [5.27 a–f]. In this case the beam is scattered into the forward direction only. The beams can be followed slice by slice without having to work all the way to the end to find out what happens at the beginning. The emergence of an evanescent wave from a step down has been treated using the perpendicular slice method. This appears to be the most promising theoretical method for treating surface imperfections.

Differential Equation Method. The discrete matrix method can be made continuous. Then the elements satisfy non-linear first order differential equations which are solved numerically by taking finite steps. This sounds like going in a circle, but there are well developed numerical methods for differential equations. This approach has been applied to RHEED by *Meyer-Ehmsen* [5.28]. If the *Meyer-Ehmsen* formalism were applied to the simple case of a single periodicity in the y direction with an arbitrary potential in the z direction, the matrices would be three by three, resulting in nine simultaneous first order equations. The wave function $\Psi(x, y, z)$ is periodic

5.1 Introduction to Reflection High Energy Electron Diffraction (RHEED)

in y for each value of z:

$$\Psi(x, y, z) = e^{i(k_x x + k_y y)} [\psi_1(z) e^{iGy} + \psi_0(z) + \psi_{-1}(z) e^{-iGy}]. \tag{5.60}$$

The *Bethe* approach produces three simultaneous equations

$$\begin{bmatrix} \dfrac{\partial \partial}{\partial z \partial z} \psi_1(z) \\ \dfrac{\partial \partial}{\partial z \partial z} \psi_0(z) \\ \dfrac{\partial \partial}{\partial z \partial z} \psi_{-1}(z) \end{bmatrix} = \left(-\dfrac{v(z)}{8} \begin{bmatrix} 2 & -1 & 0 \\ -1 & 2 & -1 \\ 0 & -1 & 2 \end{bmatrix} \right.$$

$$\left. + \begin{bmatrix} iK_1 & 0 & 0 \\ 0 & iK_0 & 0 \\ 0 & 0 & iK_{-1} \end{bmatrix}^2 \right) \begin{bmatrix} \psi_1(z) \\ \psi_0(z) \\ \psi_{-1}(z) \end{bmatrix}. \tag{5.61}$$

where

$$K_1 = k_0^2 - (G_y + k_y)^2 - k_x^2, \qquad K_0 = k_0^2 - k_y^2 - k_x^2,$$
$$K_{-1} = k_0^2 - (-G_y + k_y)^2 - k_x^2. \tag{5.62}$$

The functions $\Psi_G(z)$ are described as modulated plane waves in the forward and backward directions:

$$\psi_G(z) = f_G(z) e^{iK_G z} + b_G(z) e^{-iK_G z} \tag{5.63}$$

to arrive at the definition of the reflection matrix $[R]$ connecting two vectors that combine the functions and their z derivatives:

$$\left[\psi_G(z) - \frac{1}{iK_G} \frac{\partial}{\partial z} \psi_G(z) \right] = [R] \left[\psi_G(z) + \frac{1}{iK_G} \frac{\partial}{\partial z} \psi_G(z) \right]. \tag{5.64}$$

The reflection matrix satisfies the differential equation

$$[R'] = -[iK][R] - [R][iK] - ([R] + 1)[2iK]^{-1}[v(z)]([R] + 1). \tag{5.65}$$

where $[R']$ contains the z derivatives of the matrix elements R_{ij}. The matrices in $[K]$ are diagonal. The calculation appears tractable using standard numerical recipes applied to complex numbers. These recipes are presumably sophisticated enough to take care of problems of convergence that defeat naive attempts at solutions.

5.1.4 Applications of RHEED to the Study of Growth

5.1.4.1 Structure Determinations

For several decades it has been anticipated that RHEED would eventually compete with LEED in the determination of surface structure. This has not happened. For a single setting of the crystal with respect to the beam, LEED produces a pattern giving the full symmetry of the surface. To obtain the same information from RHEED the sample must be rotated about its surface normal and the information from successive screen images processed. The rotation of the sample is complicated because most RHEED systems do not place the sample on a proper goniometer which leaves one point on the sample invariant with respect to the beam. Even on a proper goniometer, if the sample surface is vicinal, the step effects discussed and illustrated in Figs. 5.6–8 make it difficult to track a reflection spot. Quantitative aspects of LEED depend on tracking the intensity along a particular diffraction rod as the energy of the incoming electron is varied. For RHEED the energy is fixed. The sample must be rotated mechanically to change in the angle of incidence. For LEED changing the energy is controlled electronically. In both cases the position of the diffracted beam on the screen changes as the diffraction rod is scanned. This calls for special data acquisition and analysis systems that have been developed commercially, continually improving as technology progresses, over the past 30 yr, for LEED, but not yet for RHEED. RHEED instruments have been modified to scan diffraction rods [5.29].

5.1.4.2 RHEED Oscillations

A special power of RHEED is its convenience for obtaining information in real time during epitaxial growth. The quality of the surface can be judged by observing the shapes of spots. A change in crystal structure, e.g., a reconstruction of a surface, is immediately detected. If the reconstruction has been previously studied, its pattern is also immediately recognized. Perhaps the most popular use of RHEED is to follow the growth by monitoring the intensity of the specular reflection. For many growths all of the intensities including the diffuse inelastic background oscillate with time during deposition. Color enhanced intensity maps of RHEED are reproduced in [5.11f]. RHEED oscillations were intensively studied in the growth of semiconductors by MBE [5.11, 12]. When MBE techniques were employed for the growth of metals, RHEED oscillations were also observed [5.30]. RHEED oscillations give convincing evidence that one can grow in a mode that can be called *layer-by-layer*. The term is used loosely to describe the situation where principally one layer is growing at a time with not too many holes in the layer below and not too many atoms on the layer above.

The absence of RHEED oscillations can be *good or bad*; *good* if the growth is by rapid diffusion to steps which then sweep across the surface maintaining what looks to RHEED as a steady state and *bad* if the atoms stick where they hit in such as way as to produce statistical hills. The former is associated with growth at elevated temperatures. Low temperature growth does not necessarily produce bad growths.

When Fe grows on Fe whiskers [5.6e], there are no oscillations above 650 K. Between 580 K and 620 K the oscillations have large intensity and contrast,

5.1 Introduction to Reflection High Energy Electron Diffraction (RHEED)

closely resembling ideal oscillations. Ideal oscillations have the shape of a series of parabolas centered about a minimum that is down to the background on each cycle of growth. The intersections of the parabolas at the maxima are sharp. The oscillations can be damped even at 620 K by increasing the flux from 1 ML per minute to 5 ML per minute, but they recover completely if the flux is reduced. The surface produced when Fe grows on an Fe whisker at 600 K is as good as any produced by sputtering and annealing at 1000 K, which is the temperature at which whiskers are originally grown.

For the homoepitaxial growth of Fe below 500 K, one-atom-high islands with highly correlated distances between centers are deduced from the RHEED pattern which is explained by a diffraction rod which is a cylindrical shell intersecting the Ewald sphere, as discussed in the caption of Fig. 5.10. With increasing temperature the hollow diffraction rod narrows as the correlation length increases continuously, starting from 30 atoms at 300 K. At 100 K the correlation length is still about 30 atoms but the width of distribution is much less well defined, that is, the wall of the cylindrical shell of the diffraction rod becomes less sharp.

There are two schools of thought on the subject of RHEED oscillations. The problem is to describe diffraction from a half filled surface. The interference school [5.12, 31] says that the intensity decreases because the path length from the RHEED source to the RHEED screen is different reflecting from the tops of the filled layer of atoms compared to reflecting from the tops of the growing layer. The channeling school [5.11, 32] says that the intensity drops because the electrons easily enter the crystal through step edges and then get scattered in many directions. Those that work with extremely good surfaces such as Fe whiskers or singular cuts of Si observe the interference effects. Those who deal every day with surfaces that have high densities of steps find evidence for channeling. As in all good arguments, both are right to some degree. Fe whiskers exhibit oscillations in intensity at the anti-Bragg condition with minima as small as one percent of the peak intensities during homoepitaxial growth at temperatures from 300 to 650 K. But for growth at 100 K the oscillations look more like GaAs with ratios of peak to minimum no more than 1.2. These growths produce spectacular Kikuchi patterns [5.33], indicating the role of channelling in addition to the interference effects.

One can use RHEED oscillations to determine the number of monolayers deposited. For the first several layers there can be phase shifts before a steady state is reached, but then the time for deposition of one monolayer can be measured to an accuracy of better than two percent. This can be used to calibrate deposition monitors placed away from the position of the sample. The total deposition can be found from the total time. However, the deposition of the material equivalent to 10 layers does not mean that there are ten full layers. If the RHEED oscillations are strong there should be a good correlation between the peak of the oscillation and the filling of the top layer. It is speculation to make quantitative statements about just how good are the interfaces from

RHEED data alone. Magnetic studies of the interactions between ferromagnetic layers separated by non-magnetic layers [5.34] provide strong evidence that good surfaces are obtained.

5.1.4.3 The Funnelling Down Model

It seems well established that the diffusion of metal atoms is not playing an important role in growth at 100 K [5.35]. The observation of RHEED oscillations at low temperatures has been explained by *Evans* [5.36]. To form rather smooth surfaces it is not necessary that atoms migrate. Consider the (0 0 1) surface for a bcc or fcc lattice. The atoms can stick where they hit without causing excessive hill formation if they are required to stick in four-fold hollow sites and not on the sides of the pyramids that form. This is called the funnelling down model. Presumably the atoms retain sufficient momentum gained from the cohesive energy on approaching the surface to carry them to the four-fold site without being caught on the sides of pyramids. There is always a four-fold hollow site at the base of the pyramid. Regions between the highest peaks must be filled before a four-fold hollow site forms at the top of a pyramid.

As an example, let there be N atom sites on the surface. If n_{i-1} sites are occupied in the $i-1$ layer, there will be $(n_{i-1}/N)^4$ four-fold hollow sites for deposition of the ith atomic layer. If n_i of these are already occupied, the probability of finding a suitable hollow site to form the next layer is $(n_{i-1}/N)^4 - n_i/N$. Let the fractional coverage of a given layer be $c_i(t) \equiv n_i/N$. The development of the coverages of the layers depends on the deposition rate D and is governed by a set of simultaneous finite difference equations.

$$c_i(t + \delta t) = c_i(t) + D\,\delta t \frac{c_{i-1}^4(t) - c_i(t)}{\sum_{n=1}^{N}(c_{i-1}^4(t) - c_i(t))}, \quad (5.66)$$

where the time step is δt, N is greater than the index of the topmost occupied layer, the denominator is a normalization factor that insures that each arriving atom lands on one of the layers, and $c_0 = 1$. These equations give the results shown in Fig. 5.20 for the coverage of each layer with time. Note that two layers dominate the growth process at any particular time. In perfect layer-by-layer growth one layer would finish before the next starts. In rougher growth there would be an increase in the number of active layers with time. In the kinematic model of diffraction from the topmost atoms for the condition of interference between alternate layers, the RHEED intensity for the funnelling down model varies with time as shown also in Fig. 5.20. In perfect growth, the oscillations would be parabolic. For rougher growth the RHEED intensity would fall monotonically with time. After the deposition of sufficient material to complete four layers, $t = 4\tau$, there are still two percent holes in the third layer and 20%, occupation of the fifth layer. The fourth peak of the RHEED intensity calcu-

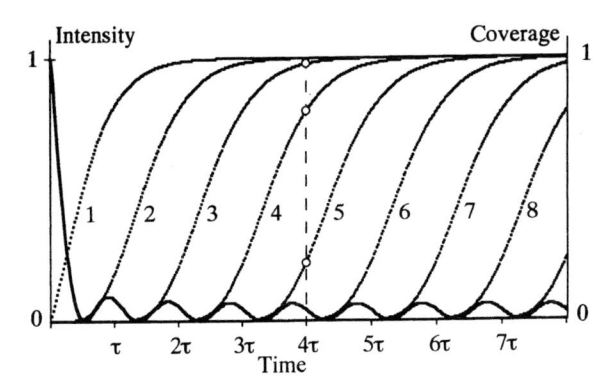

5.1 Introduction to Reflection High Energy Electron Diffraction (RHEED) 219

Fig. 5.20. The time dependence of the development of coverage for the funnelling down model. The successive fractional coverage of successive atomic layers (1 to 8) is calculated using (5.66) starting from a flat substrate, $n = 0$. The RHEED intensity is calculated kinematically at the anti-Bragg condition

lation occurs before $t = 4\tau$ and well before the fourth layer is completed. (This depends on the model of growth.) The behavior should be somewhat better than this to be classified as good layer-by-layer growth. The appearance of strong RHEED oscillations would be the signal of good layer-by-layer growth.

The funnelling down model may be taken as the dividing line between good and not so good layer-by-layer growth. In perfect layer-by-layer growth one layer would finish before the next starts. Interfaces that conform to the funnelling down model are still quite respectable for many purposes. but they would leave something to be desired in the study of how the magnetic interaction between two ferromagnetic layers depends on an intervening non-magnetic layer.

These ideas concerning the interpretation of RHEED oscillations receive support from magnetic studies of exchange interactions between two ferromagnetic layers separated by a non magnetic interlayer. *Slonczewski* has given a mechanism for the observed biquadratic exchange coupling [5.37]. It depends on the existence of extended regions where the thickness of the non-magnetic layer differs by one or two atoms from the thickness of adjacent regions. The mean size of the regions should be comparable to the magnetic exchange length. The effect should be decreased if the regions are too small or too large. The equations given above for the funnelling down model say nothing about the size of these regions, only that they exist. Calculations allowing for diffusion show that the surface has one-atom-high islands whose mean separation is determined by the temperature of growth. Growing at low temperatures produces islands that are on the small side. Growing at higher temperatures, where diffusion is significant, produces larger islands. The observed narrowing of RHEED intensity profiles with increasing growth temperature supports this conclusion. As the growth temperature and the island sizes increase, the biquadratic exchange effects are also found to increase [5.38]. If the growth were to approach perfect layer-by-layer, the biquadratic exchange should again decrease.

Models of growth with diffusion are discussed in the literature along with conjectures about the accompanying RHEED patterns. In the next few years there will be many studies of growth using RHEED, LEED and STM on the same samples [5.39]. These will serve as an analog computer to provide solutions to problems of surface diffraction from non-periodic systems. Then the pattern recognition power of the brain will be able to conclude with justifiable confidence what the atoms are doing as the RHEED patterns change on the screen.

5.2 X-Ray Photoelectron and Auger Electron Forward Scattering: A Structural Diagnostic for Epitaxial Thin Films

W.F. EGELHOFF, Jr.

The technique of X-ray photoelectron and Auger electron forward scattering has developed in the past few years into one of the most valuable structural diagnostics for investigating the early stages of epitaxial thin-film growth. A review is given here of this technique with special emphasis on its application to epitaxial films of interest to the thin-film magnetism community. The aim of this review is provide an easily readable, self-contained account directed primarily at experimentalists, with the hope of making forward scattering accessible to new or occasional users. Some familarity with XPS and Auger spectroscopy is assumed, as is an elementary knowledge of quantum mechanics. A basic description of the essential electron-atom scattering phenomena is presented in terms of simple semiclassical models. These models contain valuable insights that are not readily apparent in a completely quantum mechanical treatment. The forward-scattering technique is based on the observation that when photoelectrons or Auger electrons with a kinetic energy of several hundred eV or more are emitted from a crystal surface, enhanced intensities are observed along the internuclear axes connecting the emitting atom with its nearest and next-nearest neighbors. This effect is a result of the forward scattering, or forward focussing, which occurs when the outgoing electron wave is deflected from its initial trajectory by the potential of an atom or atoms) overlying the emitter. Since electrons emitted by atoms in the top atomic layer at the surface do not exhibit such forward-scattering enhancements, this effect is an excellent diagnostic of whether or not a particular atom is in the top layer. For an atom below the top layer, forward scattering is an excellent diagnostic of the local structural environment around the emitting atom. As a probe of short-range order, this technique has important applications in thin film magnetism. It can provide important insights into structural factors that influence magnetic properties such as epitaxial growth morphology, surface segregation, interdiffusion at interfaces, stacking faults, and lattice expansions and contractions.

5.2.1 Introduction

The field of thin film magnetism seems perennially awash with controversy. This unfortunate state of affairs has two main causes. First, magnetic properties are notoriously structure-sensitive, with small structural rearrangements of atoms often producing major changes in magnetic properties. Second, the structure of thin films, at the atomic level, is often complex, quite sensitive to preparation conditions, and sometimes not easily reproduced from one laboratory to another. What this situation clearly calls for is an improvement in structural characterization techniques. The purpose of this review is to describe one of the newer techniques for surface and interface structural analysis and to give examples of the kind of insights it can provide.

The technique of X-ray photoelectron and Auger-electron forward scattering has been developed during the past several years. It is based on the strong forward scattering, or forward focussing, which occurs when X-ray photoelectrons (XPS) or Auger electrons emitted with kinetic energies of several hundred eV, or more, by near-surface atoms are scattered by overlying lattice atoms. This scattering process produces enhanced intensities in the directions connecting the emitting atom with its overlying nearest and next-nearest neighbor atoms. Since core-level and Auger peaks are element-specific, these observations constitute a probe of short-range order around a particular element.

This probe is very effective for studying surface alloying or interdiffusion, surface segregation, epitaxial growth modes, and lattice expansions and contractions. The type of information provided (usually, nearest-neighbor directions) is an excellent complement to the information provided by extended X-ray absorption fine structure (EXAFS) measurements (usually, nearest-neighbor distances). A major advantage of XPS and Auger forward scattering over many structural diagnostics is the speed with which measurements can be made. An angular scan can often be recorded in a few seconds. This allows many surface structural-rearrangement processes, such as interdiffusion of surface segregation, to be monitored almost continuously.

Photoelectron diffraction is a larger field of which XPS forward scattering is a subset [5.40–54]. The present review concentrates on this subset because the ease with which forward-scattering experiments can be implemented makes them useful to a large community. In the broader sense, photoelectron diffraction studies of epitaxial films usually must be accompanied by complex multiple scattering calculations to model the data, or by holographic inversion techniques [5.55–78] that are still under development. Either approach represents a major undertaking for a group new to those methods. In contrast, the relative simplicity of XPS and Auger forward-scattering experiments make them much more accessible, especially for the many groups in the field of thin film magnetism which already possess a spectrometer for XPS or Auger measurements. The simplicity of the elementary scattering processes involved and the unusually direct information content of the data would seem to warrant a wider usage of XPS and Auger forward scattering by the thin-film magnetism community. Although there are a number of review articles on the broad topics of photoelectron and Auger electron diffraction [5.40–54], so far there has been only one that concentrates primarily on XPS and Auger forward scattering [5.54]. The purpose of the present review is to expand and update that earlier review and to concentrate on applications of interest to the thin-film magnetism community. Readers interested in XPS and Auger diffraction in the broader sense as well as those interested in semiconductor epitaxy should consult excellent reviews by *Fadley* [5.52] and by *Chambers* [5.53].

The diffraction of photoelectrons by a crystal lattice was first reported by *Siegbahn* et al. in 1970 [5.79, 80]. They found, in XPS studies of NaCl(100), that the core-level peaks and the Auger peaks exhibited enhanced intensities along the principle crystal axes $\langle 001 \rangle$ and $\langle 011 \rangle$. In the early literature, these enhancements were usually thought to be analogous to the Kikuchi bands [5.81] found in transmission electron diffraction [5.79, 80, 82–110]. Later it became clear that at typical XPS and Auger kinetic energies of several hundred eV or more, forward scattering in the top few atomic layers at the surface is a better description of the phenomenon [5.111–127]. At typical XPS and Auger energies Kikuchi bands, which originate from deeper than the top few layers, are a weak effect and play a only minor role in the observed intensity distributions [5.128].

The earliest application of XPS forward scattering to surface structure occurred in the late 1970s in the study of molecules adsorbed on surfaces [5.129,130]. This application has turned out to be particularly useful for determining adsorption sites, determining whether a diatomic molecule bonds to a surface in a vertical orientation or is tilted, and also for determining the amplitude of "wagging" vibrations [5.129–152]. This material is outside the scope of the present review but has been covered in other recent reviews [5.52, 54, 150].

In 1984, a very large increase began in the number of papers appearing on the topic of XPS and Auger forward scattering [5.45–50, 52–54, 111–294]. This increase followed in the wake of the recognition of the valuable information which the technique provides on the near-surface structure of single crystals and epitaxial films [5.111–120]. Extensive theoretical work [5.58, 59, 100, 116, 120 –134,138, 139, 142, 153-194] on the underlying physics has progressed so well that there now seems to be no major unresolved issue. The stage would seem to be set for a much wider usage of XPS and Auger forward scattering by the hundreds of instruments, worldwide, capable of making such measurements. To facilitate this wider usage, the present review gives a general overview of the basic electron-atom scattering processes involved and a selection of experimental examples. This material is intended serve as a guide to experimentalists interested in making use of this powerful, relatively new technique. This guide should make XPS and Auger forward scattering readily accessible to a wide range of experimentalists, especially to new or occasional users, and in many cases, permit them to gain important insights into the structure of their epitaxial films without having to resort to complicated theoretical calculations for the interpretation to their data. Of course, theoretical calculations to model the data will always be needed if the maximum possible amount of information is to be extracted from any given data set. However, since full multiple-scattering calculations are quite tedious [5.41, 50, 59, 85, 152, 153, 155, 157, 163, 164, 169, 173–178, 182, 183, 188–194, 211–218], it would seem worthwhile to present an overview of what can be done without resorting to them.

5.2.2 The Basics of Electron–Atom Scattering

One of the most important developments in our understanding of the XPS and Auger forward-scattering phenomenon came when *Tong* and coworkers [5.162, 163] pointed out how much of the basic physics can be described using easily understood classical and semiclassical models. Figure 5.21 presents their their semiclassical picture of forward scattering. A large ring of solid angle of the outgoing electron wave is deflected into the forward direction by the attractive atomic potential. The part of the wave scattered into the forward direction is in-phase all around the illustrated ring so that a constructive interference occurs in the forward direction. Tong et al. used the term "forward focussing" for this effect [5.155, 158, 162, 163]. Given the general acceptance of the validity of this model, a brief discussion of quantum-mechanical and semiclassical treatments of electron–atom scattering is worthwhile.

5.2.2.1 Quantum and Semiclassical Models

A frequent observation for scattering problems in physics is that when the wavelength of the scattering particle is shorter than the effective range of the scattering potential, a semiclassical description of the scattering event can often provide much physical insight without a great loss of accuracy [5.295–301]. The basic ideas, which may be traced back to *Dirac's* book [5.295], are associated with the concept that a particle may be represented by a wave packet whose size is of the order of the de Broglie wavelength of the particle. If the wave packet is smaller than the effective range of the scattering potential (i.e. if the interaction time is short enough so that spreading of the wave packet due to the uncertainty principle is small during the interaction time), then the propagation of the wave packet along all possible classical scattering trajectories adequately describes the quantum-mechanical scattering event [5.295–301].

XPS and Auger forward scattering generally turn out to be borderline cases for meeting these criteria. This situation means that a simple semiclassical calculation cannot provide numerical results of extreme accuracy, but will nevertheless give a qualitatively correct picture of the scattering event. Semiclassical descriptions derive their popularity from the fact that, in problems

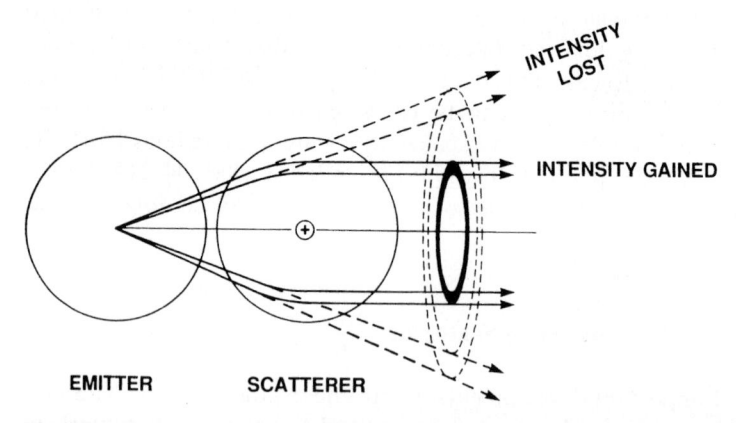

Fig. 5.21. A semiclassical description of forward scattering. The attractive potential of a scattering atom deflects a ring of solid angle into the forward direction. Since this part of the scattered wave is in-phase around the cylindrically symmetric axis, a constructive interference occurs and produces an enhanced intensity along the internuclear axis. [5.163]

for which they are reasonably accurate, they usually provide clear, easily understood visual images and a qualitative, intuitive grasp of the essential scattering processes [5.296–300]. In a full quantum-mechanical solution, such insights are often obscured or lost in complexity.

Figure 5.22 may help provide a better understanding of the rationale for Fig. 5.21 Figure 5.22 presents a sample of classical trajectories simulating scattering in a Cu lattice by a realistic Cu potential [5.302,303] of electrons at kinetic energies of 317 eV (corresponding to XPS of the Cu $2p_{3/2}$ core level using MgK$_\alpha$ X-rays), 917 eV (the Cu CVV or $L_3M_{4,5}M_{4,5}$ Auger line), and 10^5 eV (as in transmission electron microscopy). Some general points to take note of in Fig. 5.22 are:

1) The closer an electron comes to the core of the atom the larger the angle through which it is scattered.
2) The higher the kinetic energy of the electron, the closer it must come to the core of the atom in order to be scattered through a given angle.
3) Point (2) generally means that, with increasing electron kinetic energy, a smaller fraction of solid angle of the outgoing electron wave will be available for deflection into the forward direction.
4) For electrons well above 1 keV, only a small volume near the core of the scattering atom exhibits a potential with a sufficiently rapid spatial variation ($F = -dV/dr$) to scatter the electron significantly.
5) Point (4) means that most of a crystal lattice will appear to high energy electrons as empty space, and the distance between elastic scattering events will be very long compared to the lattice constant.

The electron wavelengths are 0.7, 0.4 and 0.04 Å for the energies of 317, 917 and 10^5 eV, respectively. Upon close examination of the trajectories in Fig. 5.22

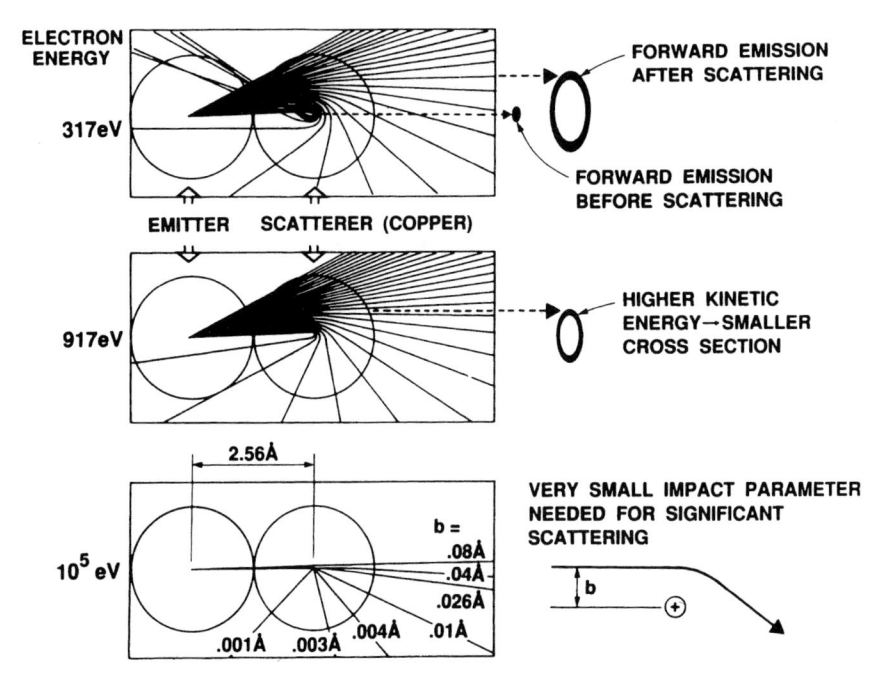

5.2 X-Ray Photoelectron and Auger Electron Forward Scattering

Fig. 5.22. Several illustrations of how, with increasing electron kinetic energy, the electron must pass steadily closer to the nucleus to be scattered significantly. The trajectories are classical calculations for a realistic Cu potential. [5.264]

it is apparent that the effective range of the potential (as approximated by the impact parameter required for forward scattering) is larger than the electron wavelength in each case. Thus, this criterion for using a semiclassical description is approximately met.

To give an idea of the extent of wave-packet spreading, a 917 eV electron described by a minimum-uncertainty wave packet of width 0.4 Å will spread by only 11% during the time it takes it to travel a distance equal to the diameter of a Cu atom [5.304].

The situation is actually more favorable than these simple comparisons would indicate since, deep inside the atom, the electron has been accelerated by falling into the potential well and actually has a shorter wavelength than the values given above. A more rigorous criterion which takes this effect into account is [5.301]:

$$\frac{4\pi k}{\lambda dk/dx} \gg 1,$$

which is to say that the fractional change in the momentum of the electron is small over a distance equal to the electron wavelength. The left side of this inequality is ~ 6 for scattering into the forward direction at energies of 317 and

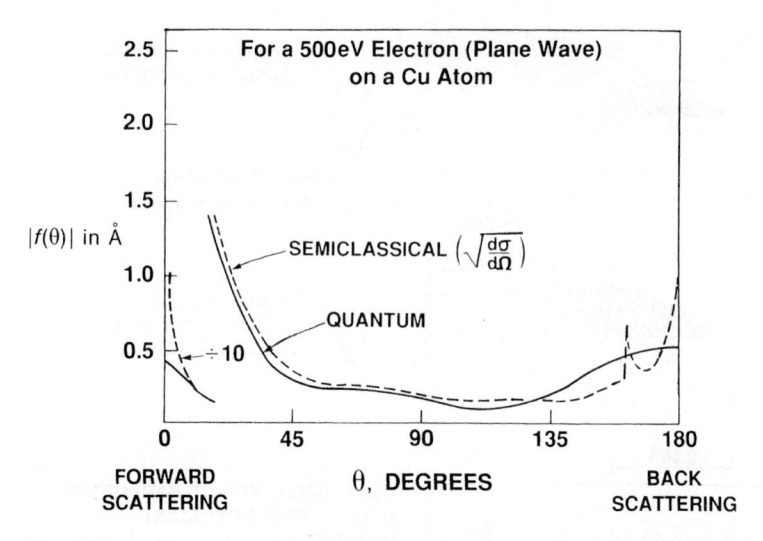

Fig. 5.23. An illustration of how quantum and semiclassical calculations given similar results for the amplitude of the scattered wave for a 500 eV electron plane wave incident on a Cu atom. [5.306,309]

917 eV [5.303]. The implication given here (by the value of ~ 6) is one of guarded optimism, which is borne out by the comparison given in Fig. 5.23. Figure 5.23 shows the quantum and semiclassical scattering amplitudes for a 500 eV plane wave incident on a Cu atom. The quantum mechanical calculation was made by the standard methods described in Refs. [5.305–308]. The semiclassical calculation is done at the simple WKB-level [5.309], in which the trajectories are classical (as where the ones in Fig. 5.22) and the phase of the wave is computed as the time integral of the Lagrangian (i.e. the action integral along the path) [5.298,299]. In such a model, the wave function for each trajectory may be written [5.298]

$$\psi(x) = \left(\frac{p_0}{p(x)}\right)^{1/2} \exp(\pm i\phi(x)/h),$$

where x is the path of the trajectory, p_0 and $p(x)$ are an initial momentum and the momentum at x so that the probability density is inversely proportional to the velocity, and $\phi(x)$ is the phase along the path x [5.299], computed as

$$\phi(x) = \int_0^{t(x)} L\,dt = \int_0^x p(x)dx = \int_0^x \frac{h}{\lambda(x)}dx,$$

where L is the Lagrangian, $T - V$, and $\lambda(x)$ is the electron wavelength. In the final form, it becomes clear that the action integral is simply the accumulated number of electron wavelengths along the path. In the usual semiclassical manner [5.297], the scattering amplitude corresponds to the square root of the

5.2 X-Ray Photoelectron and Auger Electron Forward Scattering

differential scattering cross section, $\sqrt{(d\sigma/d\Omega)}$, where,

$$\frac{d\sigma}{d\Omega} = \frac{b}{\sin\theta} \cdot \left(\frac{d\theta}{db}\right)^{-1}, \tag{5.67}$$

which may be viewed as relating areal cross-section $d\sigma = 2\pi b\, db$ to a solid angle $d\Omega = 2\pi \sin\theta\, d\theta$, where θ is the scattering angle and b is the impact parameter [5.298]. In this way the observable, intensity, is the square of either the semiclassical or the quantum mechanical scattering amplitudes. In the semiclassical treatment the phase is important only when a given scattering angle can result from more than one value of b. In such cases, all trajectories leading to a given scattering angle must be summed to get the semiclassical scattering amplitude according to [5.297]

$$f(\theta) = \sum_j \sqrt{(d\sigma/d\Omega)_j} \exp(i\phi_j). \tag{5.68}$$

The mechanism by which different values of b can give the same value of θ may be understood in the context of the 317 eV trajectories of Fig. 5.22. As b decreases, θ increases, eventually reaching 180°, and as b continues to decrease θ goes, as it were, "beyond" 180° looping partially around behind the atom. The trajectories that loop around the atom can lead to pronounced effects, and it is worthwhile to consider them somewhat further. Note first, that the illustrated trajectories are only one slice of a family of trajectories that, like a solid of rotation, has cylindrical symmetry about the internuclear axis. The trajectories that go "beyond" 180° have an *effective* θ which is less than 180°. The effective θ of such a trajectory can coincide with the θ of another trajectory (having a larger value of b) that loops around the opposite side of the scattering atom but does not go "beyond 180°. Thus, two trajectories with different values of b, and thus different phases, can have the same observed θ, and this situation calls for the use of Eq. (5.68).

In some cases the phase difference between two such trajectories is large enough to produce destructive interference. Examples of this effect may be observed in the local minima exhibited in the curves of Fig. 5.24. These minima are sometimes called generalized Ramsauer–Townsend minima [5.51,311]. Such minima are especially prominent at low electron kinetic energies, which tend to produce trajectories exhibiting the looping-around or quasi-orbiting behavior [5.312]. At large kinetic energies, this behavior tends to be suppressed [5.312]. Compare, for example, the 317 and 917 eV results in Fig. 5.22. What is happening here is that, at higher energies, the scattering is approaching the Coulomb limit of 180°. The Coulomb limit is reached when all significant scattering occurs well inside all screening electrons so that the electron experiences a Z/r potential, where Z is the nuclear charge. For lower kinetic energies, the electron does not have to come so close to the nucleus to be scattered through large angles, and here the potential Z_{eff}/r varies faster than $1/r$ since Z_{eff} increases with decreasing r as the electron penetrates the

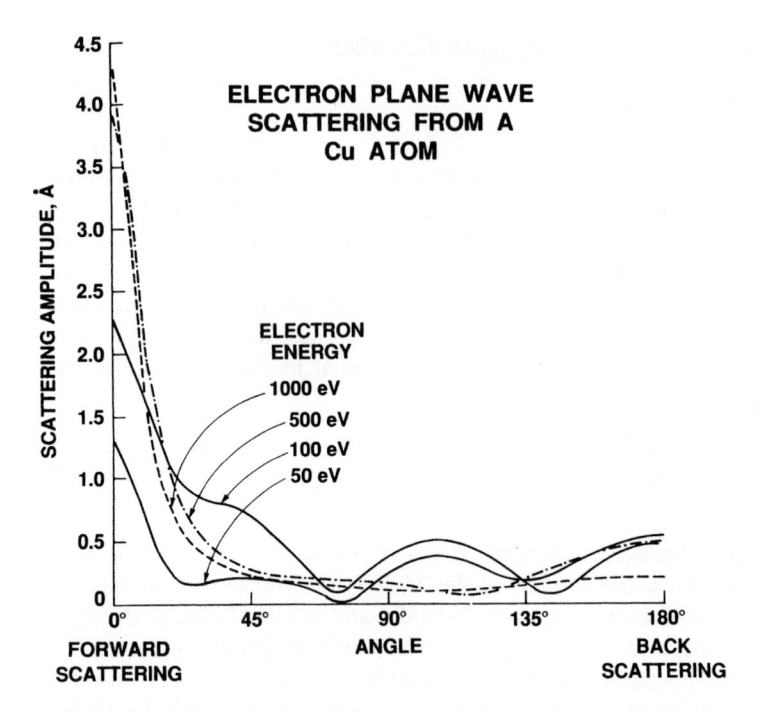

Fig. 5.24. Quantum mechanical calculations of the scattering amplitude for an electron plane wave incident on a Cu atom at different kinetic energies. [5.306, 308]

screening-electron shells [5.54]. This faster-than-$1/r$ effect gives the looping-around behavior [5.312].

This discussion of Z_{eff} might seem to suggest that large-Z atoms such as Au might exhibit much stronger forward scattering than a smaller-Z atom such as Ni. However, Ni and Au actually exhibit similar forward-scattering strengths in XPS [5.224]. The reason for this result is that the many deep core-level electrons in Au screen the nucleus, reducing Z_{eff} at the distances of b which scatter electrons with typical XPS and Auger energies into the forward direction. At much higher electron kinetic energies, the difference between Ni and Au would become apparent, but at typical XPS and Auger energies all the transition metals are rather similar in their strength of forward scattering. Studies of adsorbed diatomic molecules demonstrate that scattering by O or N atoms produces forward-scattering peaks of modest but unable strength. No observation of XPS or Auger forward scattering by H atoms have been reported.

The similarity in forward-scattering strengths exhibited by the transition metals means that it is essentially impossible to identify, by element, at atom responsible for the scattering. The element emitting the electrons is of course known from the kinetic energy of the electrons.

5.2 X-Ray Photoelectron and Auger Electron Forward Scattering

The agreement between quantum and semiclassical models in Fig. 5.23 is encouraging. Apart from three semiclassical singularities, which are smoothed out in a quantum treatment by wave packet spreading, the agreement is quite good. The singularities at 0° and 180° are due to sin θ in Eq. (5.67) and the one at 160° is due to a zero in the deflection function $d\theta/db$ (a rainbow effect), which occurs when the looping around effect reaches a maximum.

The usefulness of semiclassical models of electron–atom scattering has been demonstrated before in other areas such as low-energy electron diffraction (LEED) [5.310], gas-phase electron–atom scattering (both elastic and inelastic) [5.313–321], EXAFS [5.322], electron microscopy [5.323–328], and inelastic-loss processes in solids [5.325–328]. The most relevant of these are the semiclassical LEED calculations of *Jauho* et al. [5.310], who used a simple WKB-level semiclassical approach to computing LEED I–V curves for Ni(1 0 0) and Ni(1 1 0) from 20–200 eV. They found that the agreement with experiment, while only fair at the lower end of this energy range, was nearly as good as that of full quantum calculations at the high end of this energy range. Furthermore, they pointed out that the agreement should improve steadily above 200 eV since higher electron kinetic energy (and hence shorter wavelength) should increase the accuracy of semiclassical models. These results indicate an excellent prognosis for the use of semiclassical concepts to provide an intuitive understanding of XPS and Auger forward-scattering phenomena.

Figure 5.25 presents one of the simplest examples available to illustrate the concept of Fig. 5.21. A half-monolayer of Ni atoms was deposited on Cu(1 0 0) at cryogenic temperatures to prevent agglomeration. Since the Ni atoms are present on the surface as adatoms, no forward scattering is observed at 45° in the Ni $2p_{3/2}$ photoelectrons ($E_k = 397$ eV), and the angular distribution looks much like the instrument response function. However, when a thin epitaxial film of Cu is deposited over the Ni, the Ni $2p_{3/2}$ photoelectrons forward scatter off the Cu atoms and an enhanced intensity is observed along the Ni–Cu bond axis.

In Fig. 5.25 some of the useful insights provided by forward scattering are immediately apparent:

1) The angle at which the forward-scattering peak appears can be used to identify by emitter-scatter axis, or bond direction.
2) In epitaxial films, agglomeration or interdiffusion can be ruled out when the forward-scattering peak is absent as in Fig. 5.25a and surface segregation of the Ni can be ruled out when the forward-scattering peak is present as in Fig. 5.25b.
3) It is interesting to note that the forward-scattering effect is so strong here that the Ni photoelectron intensity is actually increased at 45° by the Cu overlayer. Thus, the usual effect of the attenuation of XPS and Auger intensities by overlayer films is overridden in this case by forward scattering.

Although forward scattering generally accounts for the most prominent peaks in angular distributions of the XPS and Auger emission from crystalline surfaces for kinetic energies of several hundred eV, it should be emphasized that

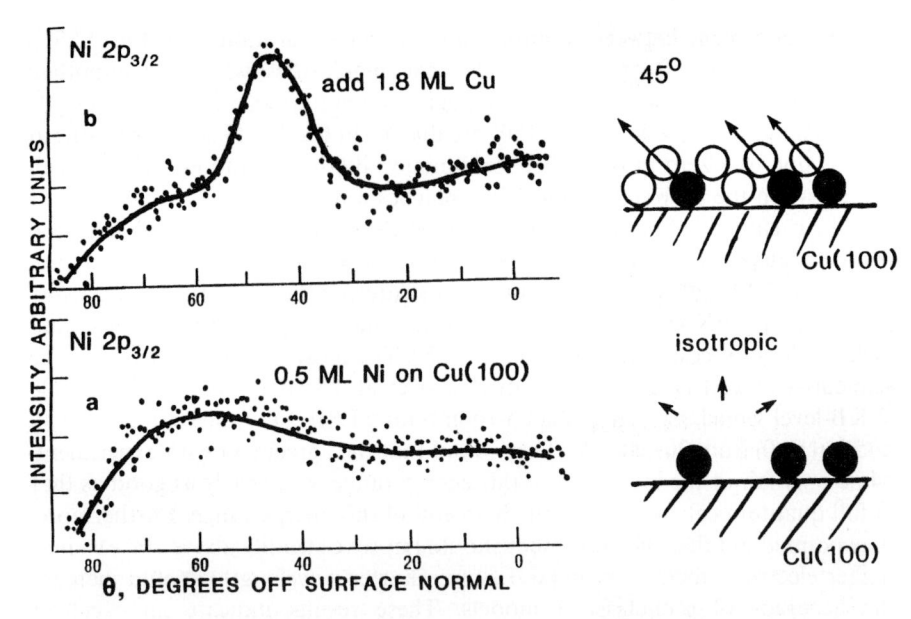

Fig. 5.25. An illustration of XPS forward scattering. In (**a**), the Ni $2p_{3/2}$ emission is approximately isotropic since the Ni atoms are at the surface. In (**b**), an epitaxial film of Cu produces a strong forward-scattering peak along the Ni–Cu axis. Note that the noise level in this early example is unusually high. Modern instruments generally yield noise levels comparable to the solid lines. [5.113]

other, generally weaker, peaks appear and should not be mistaken for forward-scattering features. Among the most common of these additional features are the ones arising from first-order constructive interferences [5.48, 53, 54, 116, 129, 133, 140, 155, 157, 160, 167, 170]. Figure 5.26a presents a semiclassical illustration of the distinction between zeroth order and first order effects, and Fig. 5.26b presents results of a quantum-mechanical calculation of the effects.

Several points worth noting about first-order constructive interferences are:

1) They take the form of a ring of intensity around the forward-scattering direction.
2) These rings are sometimes distinct enough to be clearly discernable in the [5.71, 72, 329], but more often it is difficult to pick out anything resembling a complete ring [5.57, 59, 60, 62, 69, 70, 83, 102, 192, 279].
3) First-order constructive interferences are generally weaker than forward scattering peaks because the ring distributes the intensity over a larger solid angle.
4) An exception to point (3) can occur when two first-order constructive interferences overlap and interfere constructively, a situation which occurs with many fcc lattices in a direction midway between the nearest and

5.2 X-Ray Photoelectron and Auger Electron Forward Scattering

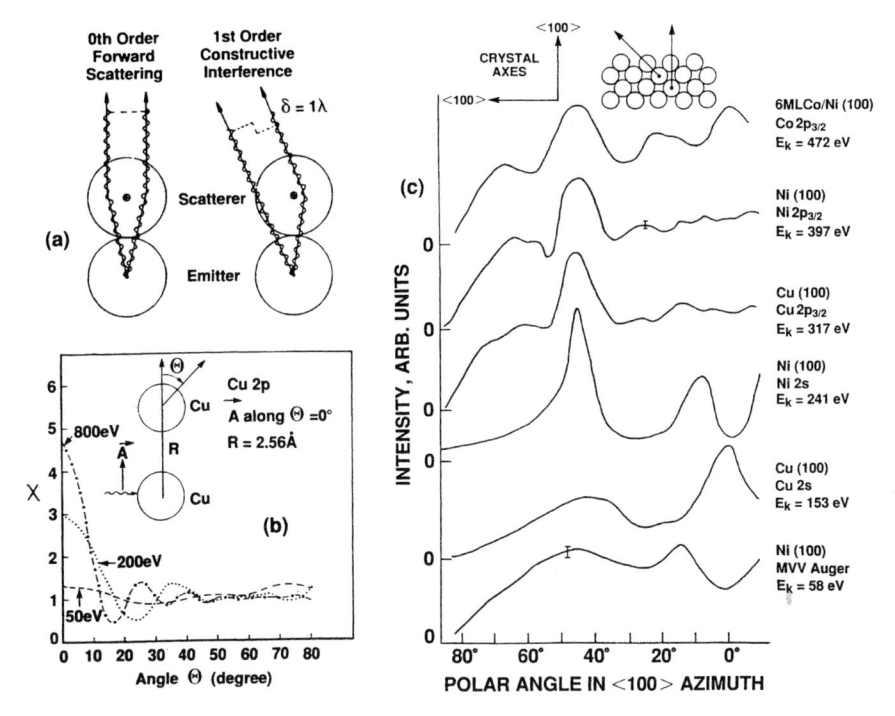

Fig. 5.26. An illustration, (a) of the distinction between zeroth-order forward scattering and first-order constructive interference effects [5.224]. The calculated relative intensity of these effects is illustrated in (b) [5.155]. The kinetic-energy dependence of the forward-scattering peaks is illustrated in (c) using the indicated core levels and Auger lines from isomorphic fcc (1 0 0) crystal surfaces [using Cu(1 0 0), Ni(1 0 0), and a 6 ML epitaxial film of Co deposited on Ni(1 0 0)]. The inset at the upper right is a profile of the surface in the azimuth in which the data is recorded, ⟨0 0 1⟩, indicating the origin of the 0° and 45° peaks. [5.116]

next-nearest neighbor axes [5.48, 53, 54, 62, 116, 155, 157, 160, 167]. Some examples of this effect are pointed out in Sect. 5.2.3.

5) As seen in Fig. 5.26b, the first-order constructive interferences shift to smaller angle with increasing electron energy (shorter wavelength).
6) First-order constructive interferences shift to smaller angles with increasing internuclear distance, and these angles are a primary source of information in photoelectron holography [5.57, 71].

Before leaving the subject of theoretical models, it is worth pointing out that a purely classical model of Auger angular distributions has been promoted recently by *Frank* et al. [5,329–334]. This model stands in stark contrast to all previous work in suggesting that the angular dependence of Auger intensities can be understood on the basis that atoms cast shadows. This model predicts internuclear axes that are exactly opposite to those inferred from forward scattering, and thus leads to totally different conclusions about surface crystal structure. This atoms-cast shadows model has, to date, apparently attracted no support in the literature and has been widely dismissed as erroneous, both on grounds of physical theory and on grounds that the atomic structures it predicts are crystallographically unreasonable [5.62, 180, 194, 254–258, 278, 279].

5.2.2.2 Kinetic Energy Dependence

It has been widely recognized that electron kinetic energies of at least a few hundred eV are required for forward-scattering peaks to be a reliable source of structural information [5.47–49, 52–54, 62, 111–120, 155–165, 190]. Since some XPS and Auger data only comply marginally with this criterion (and some not at all), it is important to have a good understanding of the inherent limitations in the use of forward-scattering peaks.

Figure 5.26c illustrates how the forward-scattering peaks evolve with increasing kinetic energy for a series of data taken on samples having a common crystal structure (fcc[1 0 0] as illustrated in the inset). The forward-scattering peak at 45° corresponding to the nearest-neighbor axis appears to become a reliable structural diagnostic above 241 eV, while the part of the forward-scattering peak at 0° corresponding to the next-nearest-neighbor axis only appears at 472 eV.

The general shape of the curves changes little from the 472 eV data in Fig. 5.26c upon increasing the kinetic energy. A family of curves in Ref. [5.62] for Ag(1 0 0) in the same azimuth as Fig. 5.26c shows that the same general shape persists up to 1477 eV. Other work on Pd(1 1 1), using Ti-Kα X-rays and going up to electron kinetic energies as high as 4170 eV, has demonstrated that with increasing energy forward-scattering peaks gradually become more narrow (as expected on the basis of Fig. 5.26b) and less prominent (as expected on the basis of Fig. 5.22) [5.335].

These trends continue in the limit of extremely large electron kinetic energy. Forward-scattering peaks become more narrow but remain an observable feature of the data. For example, such peaks are clearly observed in the angular distribution of 250 to 500 keV electrons emitted in the radioactive decay of ^{110}Ag nucleii in an Ag(1 0 0) crystal [5.336]. The full width at half maximum is 3° at 250 keV and 2° at 500 keV, and in both cases the forward-scattering peak is about 25% over background.

In Fig. 5.26c, the difference at 0° between the 241 eV data and the 153 eV data illustrates the risks associated with using forward-scattering interpretations at too low a kinetic energy. While one might think the 0° peak at 153 eV was a reliable forward-scattering feature, the 241 eV data, with a dip at 0° and a peak at 10°, shows such an interpretation is entirely misleading. To be reliable, the kinetic energy must be large enough for the peaks to have converged on their high-energy limit. The high-energy convergence occurs at a lower kinetic energy for nearest-neighbor forward scattering (45° in Fig. 5.26c) than for next-nearest-neighbor forward scattering (0° in Fig. 5.26c). One reason for this

5.2 X-Ray Photoelectron and Auger Electron Forward Scattering

effect is that the next-nearest-neighbor atom, being further away, intercepts a smaller fraction of the total emitted electron wave and thus can redirect less of it into the forward direction. This means that forward scattering in inherently weaker the greater the emitter–scatterer distance and can more easily be masked by other diffraction processes. One such diffraction process likely to be important in eliminating the 0° peak at 241 eV is the scattering into the 0° direction by the four overlying nearest-neighbor atoms (two are illustrated and two are not: one is above the page and one below). When the emitted wave goes out to these nearest neighbors and scatters through 45° into the 0° direction, the phase of this contribution will differ from that of the forward-scattered wave from the next-nearest neighbor atom. There is also a path-length difference. These factors mean that the nearest-neighbor contribution may well interfere destructively with the next-nearest neighbor contribution. In fact, it appears in the data of Fig. 5.26c that the interference at 0° goes destructive, constructive, and destructive at 58, 153 and 241 eV, respectively, and it is likely that phase and path-length difference effects are be involved.

The interference effects just described become less important at higher kinetic energy because the scattering amplitude at 45° relative to that at 0° falls off quickly with increasing kinetic energy, as may be seen in Fig. 5.24. This fall off in the large-angle scattering relative to forward scattering is of central importance for the usefulness of forward scattering above a few hundred eV kinetic energy.

Figure 5.27 gives a schematic illustration of the factors associated with this falloff. First, at relatively low energies, such as 50 eV, trajectories passing through even the outer regions of the atom are scattered through significant angles. Second, as noted earlier, in the outer to middle regions of the atom the potential changes faster than $1/r$ so that a small change in impact parameter gives a large change in scattering angle. Third, also as noted earlier, low energies

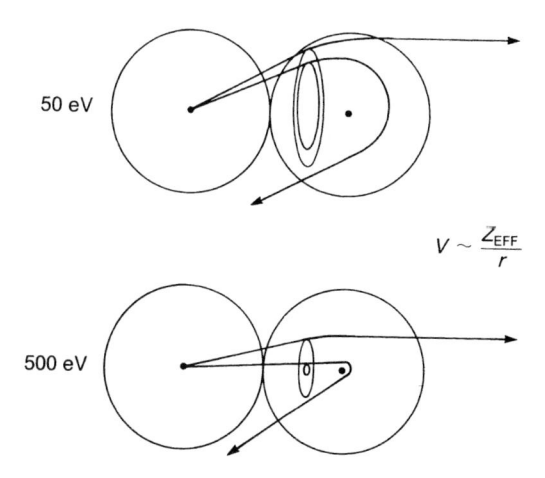

$$V \sim \frac{Z_{EFF}}{r}$$

Fig. 5.27. An illustration of why forward scattering dominates at high electron kinetic energies. At low energies, large impact parameters (where the potential varies faster than $1/r$) produce scattering into a wide range of angles. In contrast, at high energies only very small impact parameters produce large-angle scattering so that the cross-section for large-angle scattering is smaller that for forward scattering [5.54]

tend to favor quasi-orbiting trajectories. These factors mean that at low energies the scattering is more isotropic than at several hundred eV. At 500 eV, for example, the impact parameter for large-angle scattering is smaller than for forward scattering; hence the cross-section is smaller, and this reduction is reflected in the scattering amplitudes shown in Fig. 5.24.

Figure 5.28 illustrates schematically an important additional reason why large-angle scattering events become relatively unimportant at large kinetic energies. There are many electron paths that can lead to a given emission angle, such as those leading to 45°, as illustrated in Fig. 5.28. The forward-scattering paths have the advantage of cylindrical symmetry, as illustrated by the ring labeled 1, which means they all come out in-phase [5.120]. Viewed in terms of amplitudes added in the complex plane, shown in Fig. 5.28, this in-phase amplitude constitutes a vector of large magnitude at given phase angle, ϕ. The many other paths involving larger scattering angles that also lead to 45° emission, as illustrated by the ones labeled 2 and 3, will generally contribute smaller vectors at random phase angles. The vectors are smaller since the cross section for large-angle scattering is small above a few hundred eV: i.e. the electron must come close to the core of the atom to be so strongly scattered. The phase angles tend to be random since they are a result of the path-length differences, which are considerably larger than the electron wavelength (e.g., $\lambda = 0.55$ Å at 500 eV whereas $d_{Cu-Cu} = 2.56$ Å in Cu), and the phase shifts due to scattering, which vary strongly with scattering angle. Thus, the forward-scattering contribution to the sum of amplitude vectors in generally the

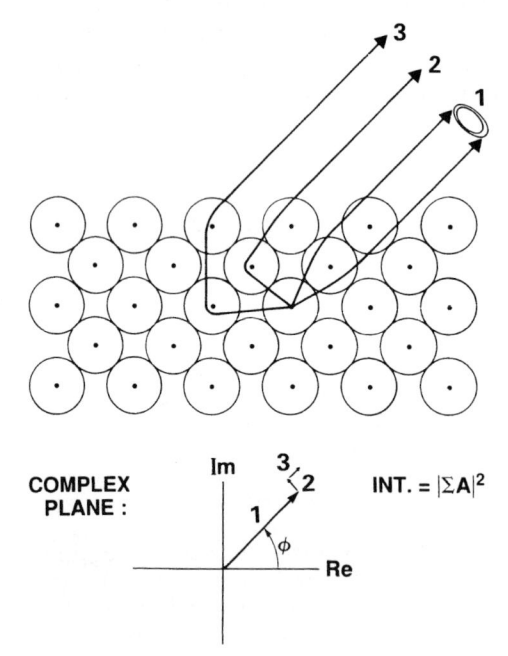

Fig. 5.28. An illustration of why, at high electron kinetic energies, forward scattering is usually the dominant contribution to the emission intensity along rows of atoms. The cylindrical symmetry of the row yields a large amplitude at one phase angle, whereas other trajectories, usually involving larger scattering angles, contribute a small amplitude at a different phase angle. [5.54]

5.2 X-Ray Photoelectron and Auger Electron Forward Scattering 235

dominant contribution, with the numerous other paths contributing only a weakly-perturbing random walk in that neighborhood, as Fig. 5.28 indicates.

Figure 5.28 also makes it easy to understand why photoelectron diffraction is so complicated at low kinetic energies. If the vectors 2 and 3 were comparable in magnitude to vector 1, as might be the case at say 50 eV, then 2 and 3 together with the many other vectors like them would dominate. Moreover, since the magnitudes and phase angles will change rapidly with angle of observation, a complex angular distribution (with no sample interpretation) will result.

Figure 5.29 provides further useful insights into a different type of kinetic energy dependence in forward scattering. Here a broad energy scan of the Ni(1 0 0) emission has been partitioned into the parts of the spectrum that exhibit and the parts that do not exhibit forward scattering. There are several important points to note in Fig. 5.29:

1) Forward scattering is apparent for several tens of eV in the tail of inelastic electrons associated with each elastic peak [5.184,186,189,190]. This result is understandable since each inelastic-loss event typically deflects the electron trajectory a few degrees (at these energies) and the typical energy loss is only 10–20 eV [5.325,337].
2) One hundred eV or more below the elastic peaks, the inelastic electron background is nearly isotropic due to the randomizing effect of multiple inelastic losses [5.184,186,189,190].

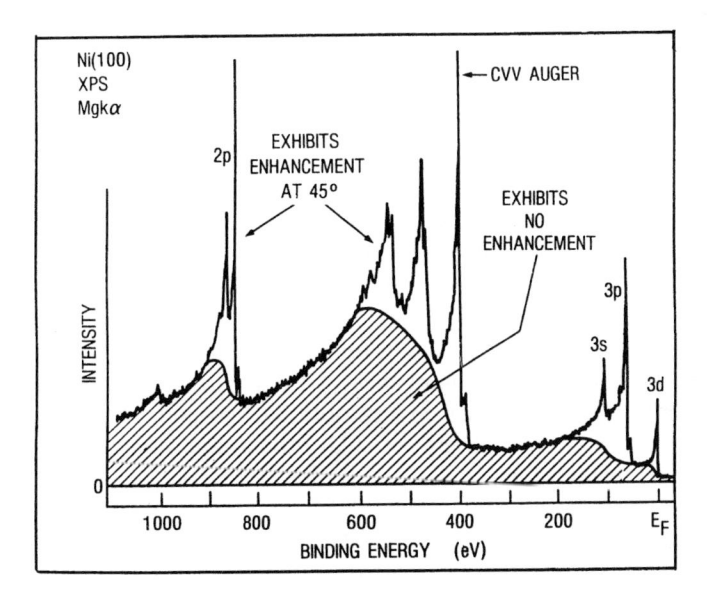

Fig. 5.29. An illustration of the breakdown of the entire spectrum of Ni(1 0 0) into parts enhanced (white) and not enhanced (cross-hatched) at 45° due to forward scattering [5.111]. The enhanced parts are largely elastic photoelectron emission while the parts not enhanced are largely inelastic secondary electron emission (see text for details)

3) To get an accurate measure of elastic peak intensity versus angle the background underlying the peak needs to be subtracted.
4) The simplest and easiest way to accomplish this subtraction is by iteratively measuring the peak intensity and then the background intensity at a kinetic energy a few eV higher than the peak and taking the difference. Such data can be recorded by slow rotation of the sample angle to accumulate a plot.
5) Great care must be exercised in making this subtraction whenever a peak of interest from an epitaxial film happens to fall at an energy for which the substrate signal exhibits forward scattering.
6) As an example of point (5), the 3p core level of Cu falls between the 3s and 3p core levels of Ni, a region which Fig. 5.29 indicates will exhibit forward scattering. Compounding the problem is the slope in the data between the Ni 3s and 3p. These factors mean that careful adjustments must be made in analysing the Cu 3p core-level data when epitaxial films of Cu are grown on Ni(1 0 0).

Finally, on the topic of the kinetic energy dependence, it should be pointed out that a small angular correction is needed to make forward-scattering peaks agree with internuclear axes, particularly at lower kinetic energies. This is due to refraction of the electron at the surface as it escapes from the inner potential of the solid [5.45, 88]. The correction may be calculated from the assumptions that the total electron kinetic energy is reduced by the value of the inner potential as it crosses the surface and that the electron momentum parallel to the surface is conserved. For a typical inner potential of 15 eV and a 1000 eV electron emitted at 45°, this correction is less than 0.5° [5.45,88].

5.2.2.3 The Role of Multiple Scattering

In the earliest forward-scattering studies of epitaxial films, it was found that the atoms contributing to the forward-scattering peaks were quite near the surface, and that the deeper atoms made a somewhat more isotropic contribution to the observed intensity [5.111,112]. This effect is evident in Fig. 5.30 in which the forward-scattering peak at 45° almost reaches its full intensity after deposition of only 2 ML Cu on Ni(1 0 0). The predominant effect of thicker Cu films is for the forward-scattering peak to ride up higher on a smooth background. Therefore, in the thicker films, the deeper layers of Cu atoms are contributing a more-or-less isotropic background.

Tong et al. were the first to explain this effect, pointing our that it is a consequence of multiple forward scattering [5.158,164]. Tong et al. used the term defocusing to describe the way in which multiple forward scattering disperses a forward-scattering peak [5.158,164]. Figure 5.31 presents the essence of the argument by Tong et al. Illustrated in Fig. 5.31a is simple forward scattering due to a single-scattering event, as discussed earlier in connection with Fig. 5.21. In Fig. 5.31b a trajectory which is scattered into the forward

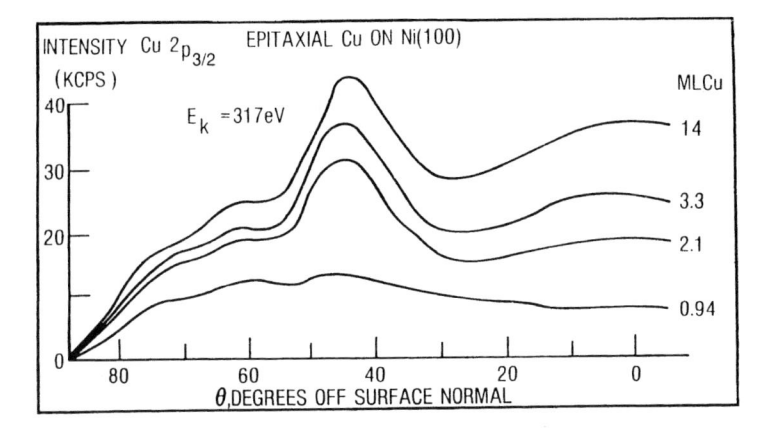

5.2 X-Ray Photoelectron and Auger Electron Forward Scattering

Fig. 5.30. An illustration of forward scattering in the Cu $2p_{3/2}$ core-level intensity for epitaxial Cu on Ni(1 0 0) as a function of Cu thickness in monolayers (ML). The data are recorded in the $\langle 0\,0\,1 \rangle$ azimuth. [5.111]

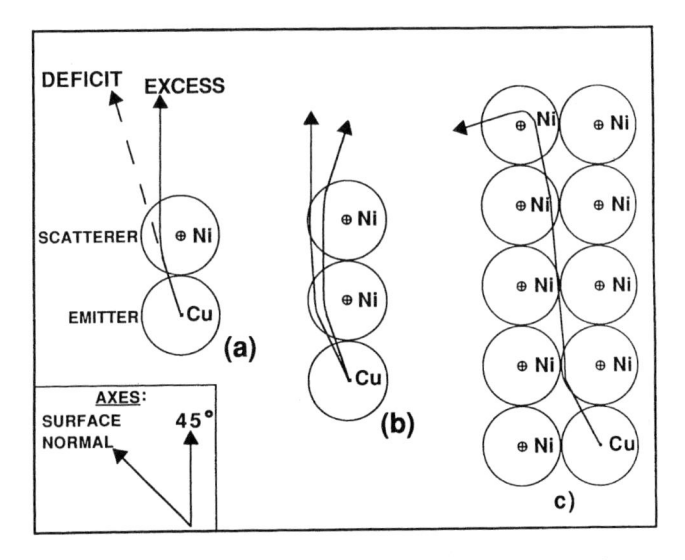

Fig. 5.31. A schematic illustration of important multiple-scattering processes in electron emission. The first one or two scattering events tend to be forward focussing and subsequent events tend to be defocusing. [5.120]

direction by the first scattering atom is seen to be scattered out of the forward direction by a second scattering atom. However, two scattering events can still give forward scattering for trajectories initially slightly further off axis. As the number of scattering atoms above the emitter increases, as Fig. 5.31c illustrates, the initial conditions for forward scattering become increasingly restrictive.

Trajectories tend to wander off into adjacent rows of atoms and get scattered away, diminishing the intensity of the forward-scattering peak and narrowing the width of the intensity that remains [5.120,191,192]. These classical effects are enhanced if the quantum mechanical spreading of the wavepacket is taken into account.

Figure 5.32 presents theoretical results which contrast multiple-scattering and single-scattering calculations in a situation in which defocussing is important. A single scattering calculation is a far easier method for modeling a given dataset. One simply allows an electron wave, Ψ_0, from the emitter to arrive unscattered at each lattice atom, scatter off that atom generating a scattered wave, Ψ_s, and then allow Ψ_0 and all the Ψ_s waves to propagate with no further scattering. The sum of the amplitudes at infinity is squared to get the intensity,

$$I(\theta) = |\Psi_0 + \Sigma_s \Psi_s|^2 = \left| \frac{\exp(-ikr)}{r} + f_s(\theta) \frac{\exp[ik(r-r_s)]}{r-r_s} \right|^2, \tag{5.69}$$

where Ψ_0 has been represented as a spherical wave, k is the electron wavevector, r_s is the position of the scattering atom s, and $f_s(\theta)$ is the complex scattering amplitude for atom s. Modeling data with a method such as this is attractive and has been widely practiced because the calculations are so simple to do. In many simple cases useful insights can be obtained from such calculations [5.45, 48, 49, 52, 53, 68, 100, 101, 109, 116, 126–134, 142, 143, 145, 147, 149, 156, 157, 159–161, 165–172, 181, 183–190, 195–209, 219–222, 231, 232, 234, 240–246, 248–255, 273, 275]. They are particularly useful for distinguishing between true forward-scattering features and first-order constructive interferences. However, Fig. 5.32 illustrates how wrong such calculations can be in cases in which multiple scattering is important. The failure of the single scattering model to include events of the type illustrated in Figs. 5.31b and 5.31c produces a drastic disparity with the multiple scattering model after only a few multiple forward-scattering events.

The major implications of Fig. 5.32 are that single-scattering calculations are generally useful only as a qualitative guide and that for highly accurate modeling of experimental data, to extract the maximum possible amount of structural information, multiple-scattering calculations are essential for all but the simplest systems (e.g., the amplitude of wagging vibrations in adsorbed diatomic molecules [5.129–33, 139, 148, 149].)

It is usually true that the higher the electron kinetic energy the larger the number of multiple forward-scattering events that the required to defocus a forward-scattering peak. For example, in Ref. [5.120] it was found that approximately twice as many atoms were required for defocusing a forward-scattering peak of 917 eV electrons as for 317 eV electrons. It also appears that for relatively light atoms, such as Al, two to three times as many of them are required for defocusing than for heavier atoms such as Cu [5.183–191, 193]

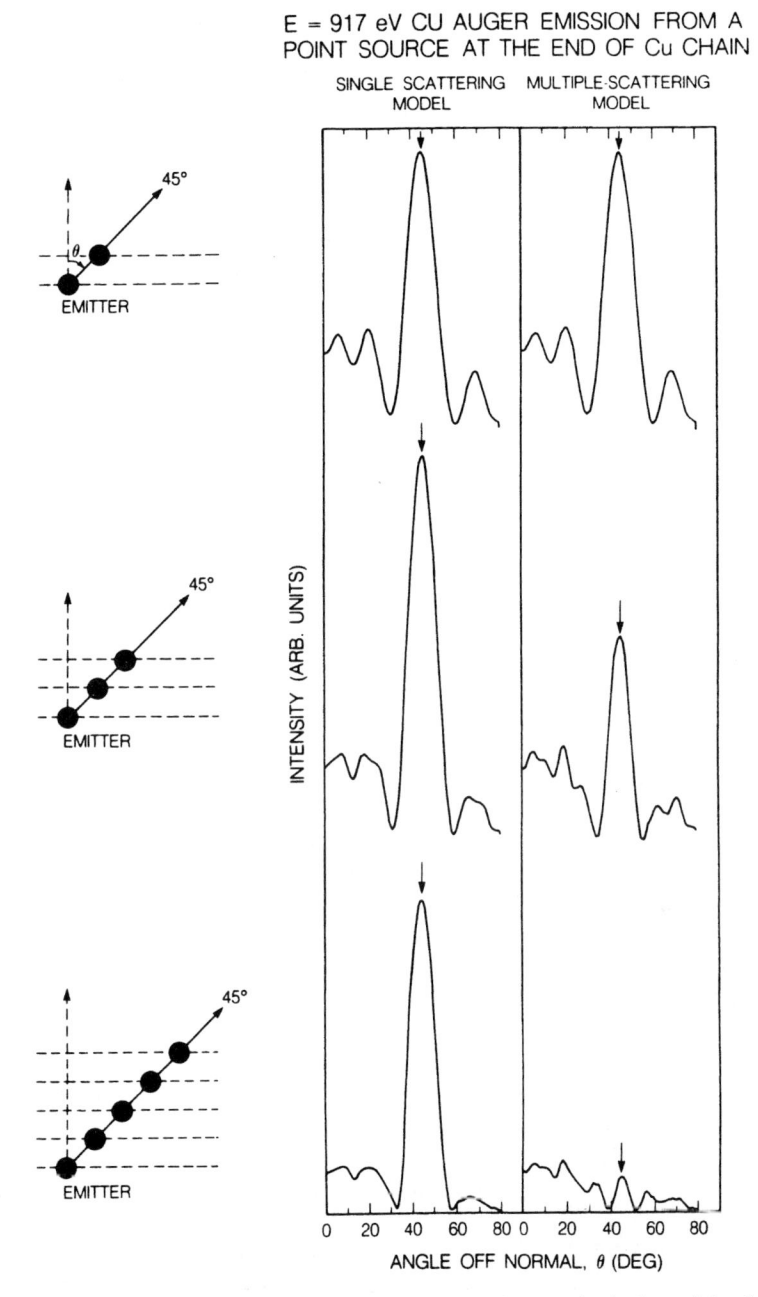

5.2 X-Ray Photoelectron and Auger Electron Forward Scattering

Fig. 5.32. An illustration of the importance of multiple scattering in determining forward-scattering peak intensities. The polar-angle intensity distribution is calculated with single scattering and multiple scattering for Cu Auger emission at 917 eV from a point source at the end of row of Cu atoms. The left-hand diagrams illustrate the geometry used for each calculation. Vertical arrows in the two right-hand panels indicate the forward-scattering, or forward-focusing, peaks. [5.177]

Figure 5.33a presents an experimental illustration of defocusing [5.120]. The angular dependence of Cu Auger peak at $E_k = 917$ eV is presented for a Cu monolayer on Ni(1 0 0) with various thicknesses of overlying Ni. Fig. 5.33b illustrates the arrangement of atoms and the crystal axes for these sandwich structures. For 0 ML of overlying Ni, i.e. a bare Cu ML, the angular distribution is quite featureless. The apparent peak at 75° is largely an artifact of a steep instrumental cutoff around 80°, but is also partly due to a weak first-order constructive interference associated with forward scattering in the plane of the surface ($\theta = 90°$). Upon depositing 2 ML Ni on the Cu, a strong forward-scattering peak appears at 45° and a weaker one at 0°. These peaks sharpen up, for reasons discussed just above, upon increasing the Ni thickness to 4 ML. The peak observed at 20° is due to the coincidence of two first-order constructive interference peaks. It happens in this particular case that first-order

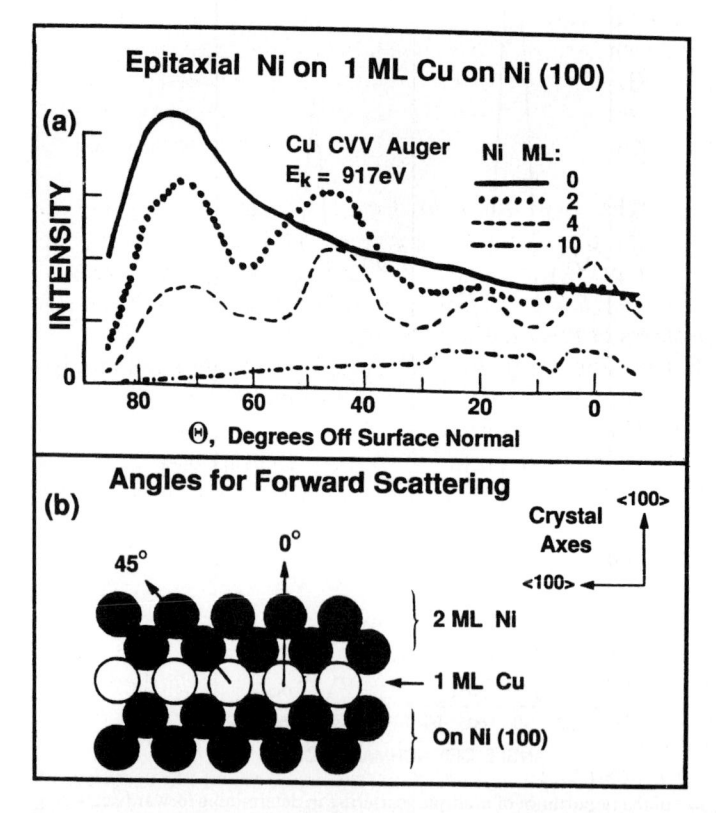

Fig. 5.33. The polar-angle dependence of the intensity in the $\langle 0 0 1 \rangle$ azimuth of (**a**) the core-valence-valence (CVV or $L_3 M_{4,5} M_{4,5}$) Auger emission for 1 ML of epitaxial Cu on Ni(1 0 0) and for subsequent deposition of epitaxial Ni overlayers on the 1 ML Cu. A typical structure is illustrated in (**b**). [5.120]

peaks associated with both the 0° and the 45° axes happen to fall near 20° and interfere constructively. Since the amplitudes add and are squared, the intensity is greater than would be expected on the basis of Fig. 5.26b.

In interpreting data one must always be on the look out for effects such as that due to overlapping first-order constructive interferences. Fortunately, prediction of approximately where such interferences will occur is a back-of-the-envelope calculation using Eq. (5.69) and the $f(\theta)$ values of Ref. [5.305–307]. Estimating the position of the first-order constructive interferences associated with strong forward-scattering peaks is always a good precaution to take for any angular scan which has not previously been interpreted. This precaution will prevent mistaking peaks such as the one at 20° in Fig. 5.33a for forward-scattering peaks.

In Fig. 5.33a, a major change occurs upon increasing the Ni thickness to 10 ML. The forward-scattering peak at 45° has vanished, and the one at 0° has diminished by half. The difference in behavior between these two peaks is an excellent illustration of defocusing. One might have thought that the peaks would behave similarly; however, the key here is that along the 45° trajectory there is a Ni atom for each ML while along the 0° trajectory there is only one Ni atom for every other ML. Thus, the 10 atoms in the 45° trajectory are sufficient for the build-up of a forward-scattered feature and its subsequent complete defocusing, but the 5 atoms in the 0° trajectory are not sufficient to complete the defocusing.

The patterns of behavior evident in Fig. 5.33b are quite useful for characterizing the depth of origin of features in XPS and Auger angular distributions. For example, if an element diffuses up through an overlayer towards the surface, then the features in Fig. 5.33b develop and can be used to characterize the progress of this diffusion [51.14].

Figure 5.34a presents another interesting example of the effects of multiple scattering and defocusing [5.338]. Here a 4 ML Fe film on Ag(1 0 0) is compared to a 4 ML Ni film on Cu(1 0 0). There is a striking difference in the extent of defocusing by the 4 ML film of the 45° forward-scattering peaks from the substrate. In the Ag case, the 45° peak is greatly diminished while in the Cu case it is still strong. In both cases, the 0° peak is strong.

The explanation of the results in Fig. 5.34a is given in Fig. 5.34b. When the electron wave forward-scattered into the 45° direction leaves the Ag(1 0 0) and enters the Fe film, it finds Fe atoms directly in its path. These atoms disperse or defocus the forward scattering much more effectively than do overlayer atoms aligned with substrate atoms. This effect is easily understood on the basis of simple semiclassical scattering events which, as illustrated in Fig. 5.34b, remove intensity from the beam.

In the case of Ni on Cu(1 0 0), alignment of atoms occurs since both metals are face-centered cubic(fcc) and the lattice constants differ by only 2.6%. The Ni atoms, being aligned with the Cu atoms, give weaker defocusing of the 45° forward-scattering peak in the Cu emission than do the Fe atoms on Ag(1 0 0), and the peak is still prominent in Fig. 5.34a.

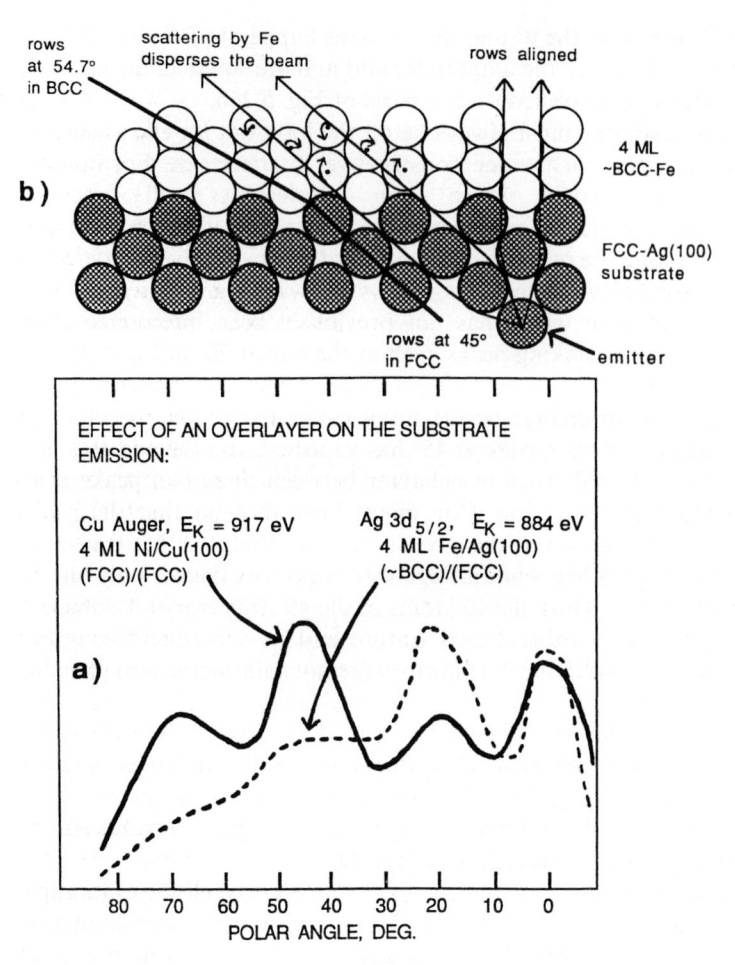

Fig. 5.34. An illustration of how the alignment of rows of atoms affects the defocussing process. A 4 ML film of Fe on Ag(1 0 0) allows a strong forward-scattering peak in the Ag emission to escape along aligned rows at 0° but not along misaligned rows at 45°. For 4 ML Ni on Cu(1 0 0), the rows are aligned along both 0° and 45°, and strong forward-scattering peaks are seen at both angles in the Cu emission. [5.338]

Along the 0° direction, the overlayer atoms are aligned with the substrate in both cases and the forward-scattering peaks are observed in both angular distributions. Thus, the intensity of a forward-scattering peak can be markedly affected by whether or not there are strong scattering atoms blocking the exit channels. One possible use for this effect would be in helping to locate the binding site of an adsorbed monolayer.

5.2.2.4 The Role of Angular Momentum

In the foregoing discussion, the initial outgoing XPS or Auger electron has been discussed as if it were an outgoing wave of spherical symmetry with no angular momentum, e.g., an $L = 0$ or s-wave. In reality, it is well known that this is often not the case. For energies well above the photoemission threshold, dipole selection rules generally favor $\Delta L = +1$, so that in XPS on, say, a 2p core level, the initial outgoing wave would be predominantly an $L = 2$ spherical harmonic, i.e., d-wave. In Auger emission, the outgoing wave is often a superposition of several different L states, with those from $L = 0$ up to about $L = 3$ being the most common.

It might be expected that these angular momentum states would play an important role in determining the observed angular dependence of XPS and Auger peaks. However, it was pointed out rather early in studies of forward scattering that the dominant angular anisotropies seemed to depend more on the electron kinetic energy than on the state of origin [5.111, 112, 156]. It was found in data for Ni(1 0 0) and Cu(1 0 0) that the $L_3M_{4,5}M_{4,5}$ Auger electrons and the 3s, 3p, and 3d X-ray photoelectrons (all of which has kinetic energies on the order of 1000 eV) exhibited qualitatively similar angular distributions, and it was concluded that in all cases the outgoing electron wave forward scattered in a similar manner [5.111,112]. Subsequent work has borne out these conclusions, and has shown that, while L states can make striking differences in angular distributions at electron kinetic energies below 100 eV [5.179, 180, 193, 278, 282, 283] their main effect at ~ 1000 eV is to change the relative intensities of the forward-scattering peaks [5.62, 126, 178, 180, 190, 193, 205, 260].

Figure 5.35 provides a simplified, semiclassical illustration of what occurs in these two energy regimes. The energies are chosen to represent two Auger emission peaks of Cu, the $L_3M_{4,5}M_{4,5}$ peak at $E_k = 917$ eV and the $M_{2,3}M_{4,5}M_{4,5}$ peak at $E_k = 57$ eV. The L values of 1 and 3 are chosen to illustrate the effect of different angular momentum in the outgoing electron. The classical description of angular momentum in Fig. 5.35 has the appearance of a central circle with tangential trajectories radiating out from it. (By analogy, note that such trajectories may be generated by swinging a ball on a string and releasing it, a process in which angular momentum is conserved.) In Fig. 5.35 the circumference of each central circle is simply L times the wavelength of the electron (i.e., zero radial velocity). By simple kinematics, an electron leaving the circle and propagating away on a tangential trajectory conserves angular momentum. Superimposing the whole family of such trajectories (wave packets) would provide a semiclassical representation of the electron wave propagating away from the atom and rotating at the same time to maintain constant angular momentum.

The purpose of Fig. 5.35 is to illustrate how angular momentum modifies the scattering picture presented in Fig. 5.22, which represented the emitted electron as an $L = 0$ or s-wave state. This modification is best visualized by noting in Fig.

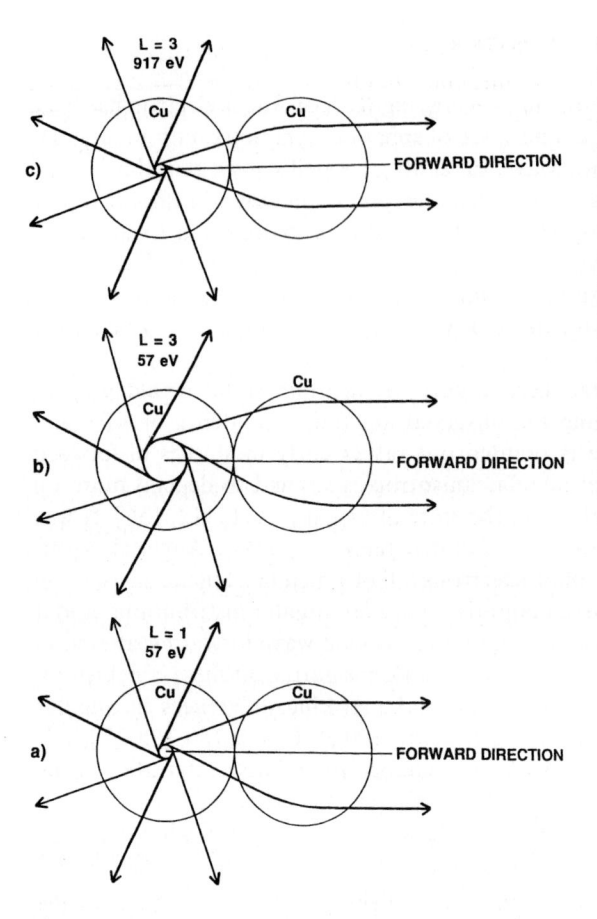

Fig. 5.35. Schematic electron trajectories illustrating the effect of angular momentum on forward scattering. For low angular momentum and low kinetic energy, (**a**), little evidence of any effect is apparent. For high angular momentum and low energy, (**b**), an asymmetry is introduced into the forward scattered wave which means that the wave will not be completely in-phase around the cylindrical axis (cf. Fig. 5.21) and the forward-scattering intensity will be reduced. For high angular momentum and high energy, (**c**), little evidence of any effect is apparent

5.35 the following. The part of the emitted wave that enters a neighboring scattering atom behaves *somewhat* as if it originated off-center from the emitting atom. This may not make much difference to the scattering in a case such as Fig. 5.35a in which the effective "offset" is small; however one could readily imagine the effect to be important in Fig. 5.35b. Here, the two illustrated scattering paths leading to forward scattering are clearly inequivalent in that the upper path scatters through a smaller angle than the lower one. This difference means that the upper one has a significantly smaller phase shift upon scattering, and the constructive interference in the forward direction (discussed in connection with Figs. 5.21 and 5.22) will be reduced or even replaced by destructive interference [5.180, 193, 278, 282, 283]. To visualize destructive interference of this type, it may be helpful, as a rough analogy, to consider Fig. 5.26a and imagine trajectories (intermediate between the two cases illustrated) for which $\delta = 1/2$.

Quantum mechanical calculations bear out the general picture of the quenching of forward scattering at low E_k for increasing L. These calculations

5.2 X-Ray Photoelectron and Auger Electron Forward Scattering

indicate that, for the case of 100 eV electrons and a Cu–Cu emitter–scatterer, the forward scattering is essentially quenched upon going from $L = 2$ to $L = 3$ [5.180, 193]. It appears [5.282, 283, 339–342] that such destructive interference for large L values at low kinetic energies is one reason why reduced intensities are sometimes observed along internuclear axes [5.329–334].

The main purpose of Fig. 5.35 is to provide a qualitative illustration of how angular momentum is primarily important at low kinetic energies and at L values of 3 or more. Even in Fig. 5.35a, where $L = 1$ at 57 eV, very little asymmetry is apparent in the scattering. Likewise in Fig. 5.35c at 917 eV, the "offset" due to $L = 3$ appears to be very small and, qualitatively, no great difference in the scattering would be expected from the $L = 0$ case illustrated in Fig. 5.21. A rough numerical estimate can be made by noting that in this energy range for first-row transition metal atoms the scattering phase shift for electrons of several hundred eV increases roughly 3% of π for every degree of scattering [5.305–307]. The "offset" in Fig. 5.35c subtends about 4° at the scattering atom center. Thus, one might estimate a difference in phase shifts between the two trajectories of only 12% of π. This would reduce the strength of the constructive interference only slightly (a phase shift of π would be destructive), relative to the $L = 0$ case, and this rough estimate is borne out by quantum mechanical calculations [5.180, 193].

With increasing kinetic energy, the result of Fig. 5.35b converges rather quickly on the result of Fig. 5.35c. This convergence occurs because the electron wavelength varies as the inverse square root of the kinetic energy (in Å, $\lambda = 12.263/\sqrt{E}$, when E is in eV). For example, the central circle in Fig. 5.35b will shrink by half when the kinetic energy increases by only $\sqrt{2}$ to 81 eV. Thus, for kinetic energies above a few hundred eV, $L = 3$ and $L = 0$ states give similar results [5.180, 193].

A few points of clarification about Fig. 5.35 may help those interested in pursuing it in more depth. First, the effect of the atomic potential of the Cu emitter atom is neglected for clarity. (It would have the effect of shrinking the central circle and causing the trajectories inside the atom to spiral.) All that really matters is that the linear trajectories outside the atom (beyond the influence of the atomic potential) correctly characterize the angular momentum state. Neglect of the potential makes the construction in Fig. 5.35 particularly simple. Second, the scattering trajectories presented are illustrative, not calculated. Finally, Fig. 5.35 only illustrates one slice of a family of trajectories that are cylindrically symmetrical about the emitter–scatterer axis, in the same way Fig. 5.26a relates to Fig. 5.21.

The above discussion of angular momentum states might lead one to think that an $L = 1$ wave would resemble a p orbital or that an $L = 2$ wave would resemble a d orbital, and that such a symmetry might somehow be superimposed on the angular dependence. However, in general this is not the case. Such orbitals represent only one of the magnetic quantum number subshells (e.g., for $L = 1$, $m_l = 0, \pm 1$). The actual outgoing wave is a superposition of these subshells and is usually more isotropic than any one of them.

For photoelectrons, the angular distribution of the actual outgoing wave can be characterized by a single parameter, β, using an equation, valid within the dipole approximation, that has along been known from atomic physics to be

$$d\sigma/d\Omega = (\sigma_T/4\pi)[1 - 0.5\beta P_2(\cos\theta)], \tag{5.70}$$

where $d\sigma/d\Omega$ is the differential cross section, σ_T is the total cross section, β is the asymmetry parameter, $P_2(\cos\theta) = 0.5(3\cos^2\theta - 1)$, and θ is the angle between the photon and photoelectron propagation directions [5.340, 343, 344]. For photoelectrons of several hundred eV kinetic energy, that originate from a filled core level, β tends to be slightly over 1.0 [5.340]; the resulting outgoing wave is typically in the form of an oblate spheriod with an approximately 2:1 ratio of width to height. If the X-ray source and analyser are fixed and the sample is rotated to record and angular distribution, then the initial oblate spheriod is also fixed. If forward scattering dominates, the effect of this departure from sphericity is minimal. If the X-ray source and the sample were fixed and the analyser rotated then a cosine-like modulation would be superimposed on the data. However, this modulation would only appear to influence the shape of the background, not the forward-scattered peaks which are relatively narrow.

For Auger emission, there is no simple equation analogous to Eq. (5.70). However, the outgoing electron wave is generally not highly anisotropic [5.348,349]. Since the decay process involves two electrons, the outgoing wave is generally a superposition of a number of subshells, and this has a randomizing effect [5.348,349]. Therefore, the implicit assumption in the present discussion that the outgoing Auger electron wave is initially spherical is probably quite adequate. Note that this assumption does not necessarily mean that $L = 0$. For example, it is well-established that for Cu the $M_{2,3}M_{4,5}M_{4,5}$(57 eV) Auger emission is dominated by $L = 3$ [5.179, 282, 283, 341, 342].

The foregoing discussion blurs somewhat the distinction between plane-wave phase shifts and spherical-wave phase shifts in electron–atom scattering. Since this topic is much discussed in the photoelectron diffraction literature, a word of clarification may be in order here. Plane-wave phase shifts can often be used even though the actual wave incident on the scattering atom is spherical. A spherical wave can only be approximated as a plane wave incident on the scattering atom if either the internuclear distance is large or if the important scattering occurs near the core of the scattering atom. Otherwise, the spherical-wave trajectory has to be scattered through a larger angle to achieve the same post-scattering direction as a plane-wave trajectory. This approximation gives rise to the much-discussed spherical-wave or curved-wave corrections to the phase shifts in photoelectron diffraction [5.129–133, 138, 162, 164, 165, 169, 170, 211–218, 350–355c]. As might be expected from viewing Fig. 5.27 these corrections are quite important at lower energies, as in Fig. 5.27 at 50 eV where important scattering occurs in the fringes of the scattering atom. The corrections may be of slight importance for forward scattering at a few hundred eV, as in Fig. 5.27 at 500 eV, but are of little importance for large-angle scattering above a few hundred eV since the scattering volume is so small that there is little

5.2 X-Ray Photoelectron and Auger Electron Forward Scattering

curvature present in the wavefront for that region. Thus, the plane wave calculations of Figs. 5.23 and 5.24 are quite reasonable as a basis for a qualitative discussion of forward scattering at a few hundred eV and above, and the main effect of spherical-wave corrections is to reduce slightly the magnitude of the enhancement in the forward direction [5.129–133, 138, 162, 164, 166, 169, 170, 211–218, 350–355c].

5.2.2.5 Analogies in Elastic Backscattering

The forward-scattering effects discussed in this review are not, of course, limited to XPS and Auger experiments but are quite general features of electron–atom scattering. One example of this generality is the way an electron beam incident along a row of atoms tends to be "focussed" onto the core of near-surface atoms. Figure 5.36 illustrates how this process may be understood as the application of reciprocity to forward scattering. The cross-section for electron impact on the core may even be higher for subsurface atoms than for surface atoms. This principle is just the time reversed analog of the point, made in connection with Fig. 5.33, that in the forward-scattering direction the Auger emission intensity of a subsurface atom can exceed that of a surface atom. This focussing of an incident beam on the core is important since small-impact-parameter collisions of this type dominate large-angle elastic scattering (or backscattering) and core ionization with its consequences, X-ray or Auger emission. Thus, when an incident electron beam of a few keV or more is incident along a row of atoms, local maxima are observed in the *total* yield of elastically backscattered electrons, Auger electrons, and in X-ray emission [5.356–373]. These maxima are a consequence of incident electrons tending to focus on the atomic cores more efficiently at these incident angles than would be case for random angles, before they dissipate their energy through inelastic scattering processes involving predominantly the valence electrons [5.120].

Another interesting example of forward scattering appears in the angular distribution of electrons elastically backscattered from crystals. Perhaps the

CORE ⟶ FORWARD SCATTERED
CORE ⟵ INCIDENT BEAM

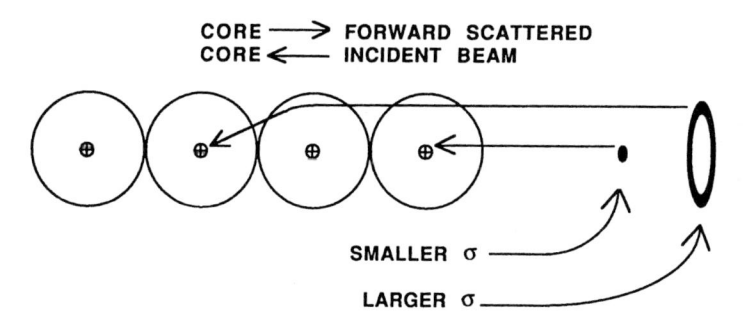

SMALLER σ

LARGER σ

Fig. 5.36. An application of reciprocity to forward scattering. An electron beam incident along a row of atoms tends to be focused on the core of atoms in subsurface layers. [5.120]

most familiar case of elastic backscattering is that of LEED, a technique is which intense emission directions are normally associated with diffraction from a lattice and not with short-range forward scattering. It is thus of interest to see how, with increasing beam energy, the dominant effect goes over from the usual LEED pattern to forward scattering. Figure 5.37 presents such data for Cu(1 0 0) [5.292, 293]. The angular distribution of the 917 eV Auger emission is illustrated for $\langle 1\,0\,0\rangle$ azimuth, and it exhibits the usual forward-scattering peaks at 0° and 45°. Also illustrated are the angular distributions for the elastically scattered intensity at several beam energies. The beam is incident at an angle 35° out of the plane for which the data are plotted. At 200 and 400 eV, the intensity maxima can be indexed according to the usual LEED criteria, although at 400 eV a shoulder is developing at 45°. By 900 eV, no remnant of the LEED pattern is evident, and the angular distribution strongly resembles the Auger result as forward scattering becomes dominant in the backscattered intensity.

The suppression of the LEED pattern at energies greater than 400 eV in Fig. 5.37 may be understood as a Debye–Waller effect, in which the electron wavelength becomes comparable to the amplitude of lattice vibrations so that lattice atoms scatter incoherently (out-of-phase) [5.117]. This situation means each atom acts *as if* it were an independent scatterer [5.117]. Furthermore, when

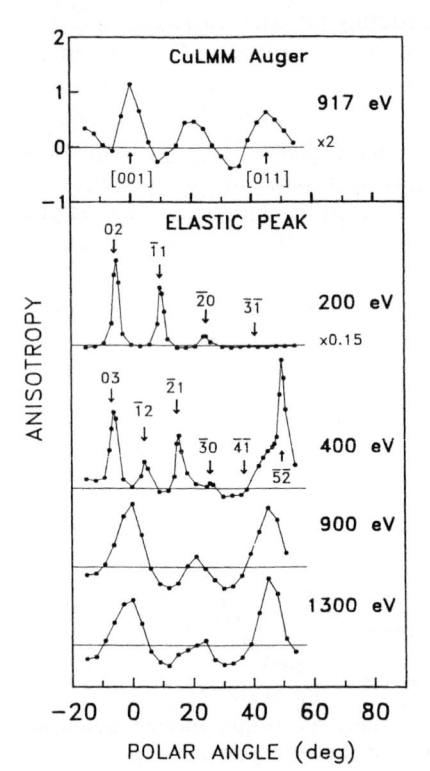

Fig. 5.37. Illustrations of the angular distribution of the Cu CVV Auger emission from Cu(1 0 0) in the $\langle 0\,0\,1\rangle$ azimuth and of the elastic backscattering as a function of beam energy, illustrating the cross-over form the LEED regime to the forward-scattering regime. [5.292]

5.2 X-Ray Photoelectron and Auger Electron Forward Scattering

an incident electron scatters from an atom, the backscattered part of the electron wave is roughly isotropic if the energy is large enough (as in the 1000 eV curve of Fig. 5.24) [5.117]. This isotropic backscattered wave then behaves in a manner similar to an XPS or Auger electron: it undergoes forward scattering on its way out of the lattice [5.117]. Thus incoherence transforms a probe of long-range order (LEED) into a probe of short-range order (forward-scattering peaks in the elastic backscattered distribution) [5.117]. Similar forward scattering also occurs in the *inelastic* backscattered distribution above a few hundred eV [5.383,384], (which can be incoherent even without considering the Debye–Waller factor).

Although this type of forward scattering lacks the elemental specificity of XPS or Auger signals, it is nevertheless useful in structural analyses for which elemental specificity is unnecessary [5.164,170,171,176,377]. One attractive feature about forward-scattering peaks of this type is that they may be observed in a standard LEED instrument simply by raising the beam voltage to ≥ 1 keV.

In the present context of electron backscattering, a few comments are appropriate about the so-called Kikuchi patterns sometimes observed in LEED instruments and in 2-D displays of XPS and Auger data [5.379–385]. These patterns generally occur at electron kinetic energies of 1 keV or more, although in some cases traces of them can be observed at energies as low as 500 eV [5.380]. These patterns take the form of fuzzy bands of enhanced intensity connecting low-index crystal directions. They may be understood as examples of classical electron channeling between planes of atoms. As noted in connection with Fig. 5.31c, a high-kinetic-energy electron wavepacket can travel in one direction for a short distance between rows of atoms before wavepacket spreading leads to large-angle scattering. This effect produces enhanced intensities radiating out from between the principal crystal planes. However, at kinetic energies around 500 eV to 1 keV, these so-called Kikuchi bands are generally weaker than the forward-scattering peaks and are often so weak that lines to guide the eye have to be sketched on top of the data to make the bands recognizable [5.380–385]. Moreover, these bands are generally interrupted near the forward-scattering peaks by the destructive interference that occurs at an angle about halfway out to the first-order constructive interference (e.g., imagine $\delta = 1/2\lambda$ in Fig. 5.26a). Therefore, in this energy regime it seems most appropriate to view the angular distribution of the intensity as being dominated by forward scattering from nearest and next-nearest neighbor atoms, with lesser contributions being made by first-order constructive interferences and by the short-range classical analog of channeling.

At energies above several keV, true Kikuchi patterns begin to develop as the elastic mean-free-path of electrons becomes long enough (i.e., as the cross-section for elastic scattering becomes small enough) so that the standard two-beam approximation for dynamical diffraction becomes valid [5.386]. The propagating beam can then develop nodes on the planes of atoms with antinodes between the planes of atoms, thus rendering the channeling inherently quantum mechanical. The result is that very distinct Kikuchi bands become apparent, each with a width corresponding to twice the reciprocal lattice vector of the channeling planes. The bands then have sharp edges, which no longer need to be sketched in for clarity [5.386].

In concluding this section on the basics of electron-atom scattering, it should be re-emphasised that the goal here has been to concentrate on the dominant features in the XPS, Auger, and backscattered angular distributions and to explain how they can often be understood in terms of simple semiclassical models. However, it should be acknowledged that there are many details in the fine structure of such angular distributions that cannot be explained in terms of easily understood elementary phenomena. The attempt here has been to concentrate on features that can be easily interpreted. To understand the fine structure or to extract the maximum possible amount of structural information from any given data set, it is essential that theoretical modeling of the data be carried out with full multiple-scattering calculations and variation of the structural parameters to optimize agreement between theory and experiment [5.155, 158, 163, 173–177, 183, 184, 186, 188, 191, 192, 280]. A particularly good example of this approach is presented in Ref. [5.177].

5.2.3 Experimental Problems of Current Interest

The foregoing section has attempted to provide a basic introduction to the most important aspects of XPS and Auger forward scattering. The following sections provide an overview of some of the more noteworthy examples of how this technique can be used to solve or at least gain insight into surface structural problems of current interest.

For the sake of simplicity, all of the following examples pertain to single crystals or epitaxial films. However, this does not mean that XPS and Auger forward scattering is of no value in polycrystalline or non-crystalline materials. Indeed, whenever a solid has a tendency for internuclear or bond axes to have a preferrential alignment [5.387], forward-scattering enhancements should be observed, and there are some recent data to substantiate this expectation [5.388,389]. Therefore, it would seem that forward scattering has much untapped potential in the area of partially ordered systems [5.387].

5.2.3.1 *Evaluating Layer-by-Layer Growth*

One of the most frequently asked questions about epitaxial growth is whether or not it is layer-by-layer. This issue is particularly important for ultrathin-film magnetism for two reasons. First, it is widely recognized that structural imperfections could be expected to alter magnetic properties, and second, in magnetic studies the films of interest are often high-surface-free-energy transition metals (e.g., Fe, Co, or Ni) deposited on low-surface-free-energy noble-metal substrates (e.g., Cu, Ag, or Au). Bauer was the first to point out that, in such systems, the equilibrium growth morphology is governed by the balance of surface and interface free energies [5.390, 391]. Layer-by-layer growth is expected if the inequality

$$\sigma_d + \sigma_i - \sigma_s < 0, \tag{5.71}$$

is satisfied, where σ_d is the surface free energy of the deposited film, σ_i is the interfacial free energy of the film-substrate interface, and σ_s is the surface free energy of the clean substrate (where the surface free energies are all positive quantities). Basically, layer-by-layer growth is expected if it produces the lowest free energy of the system. If strain is important in the system, a strain term should be added to the left side of Eq. (5.71). If an increasing strain term for the growing film causes the left side of (5.71) to exceed zero at some thickness, Stranski–Krastanov growth may be expected to result, for which layer-by-layer growth ceases and 3D clusters or islands form. If the left side of (5.71) exceeds zero from the beginning, Volmer–Weber growth may be expected, for which the film does not wet the substrate and growth starts as 3D clusters or islands.

The available values of the surface free energies for the transition metals are generally significantly larger than those for the noble metals [5.392–394]. It is difficult to find values for interfacial free energies, but they may be estimated from the heats of formation of the alloys [5.395]. The heat of formation for a transition-metal-noble-metal alloy is usually weakly endothermic, and σ_i is then predicted to be a small positive quantity [5.236]. This means that the left side of (5.71) exceeds zero for Fe, Co, or Ni on Cu, Ag or Au, even from the beginning, and the equilibrium growth morphology should not be layer-by-layer.

A deposited thickness equivalent to 1 ML is probably the best case for evaluating whether or not layer-by-layer growth occurs. If it does, the substrate should be completely covered and no second-layer formation should be present in the deposited film. XPS and Auger forward scattering are among the best techniques available for assessing whether or not a deposited monolayer lies flat without either any second-layer formation or interdiffusion. The reason for this is simply that if all atoms are present in the top layer, no forward-scattering peaks are observed and the angular distribution is approximately isotropic.

Figure 5.38 presents an illustration of how such an assessment may be made [5.228]. The data for a 1 ML film of Cu on Ni(100) are presented to illustrate the point that, when a monolayer is predicted by (5.71) to wet the surface, no forward-scattering peak is observed. For a 2 ML Cu film, the double-layer structure gives the expected forward-scattering peak at 45°. These spectra provide examples against which the cases of 0.1 and 1 ML Fe on Cu(1 0 0) may be judged. Since a prominent peak is observed in both Fe cases at 45°, the implication is clear that layer-by-layer growth is not occurring in this system [5.396–401]. This conclusion has been substantiated by other techniques, which have demonstrated that in this system 1 ML Fe deposited on Cu(1 0 0) agglomerates and intermixes [5.399–401].

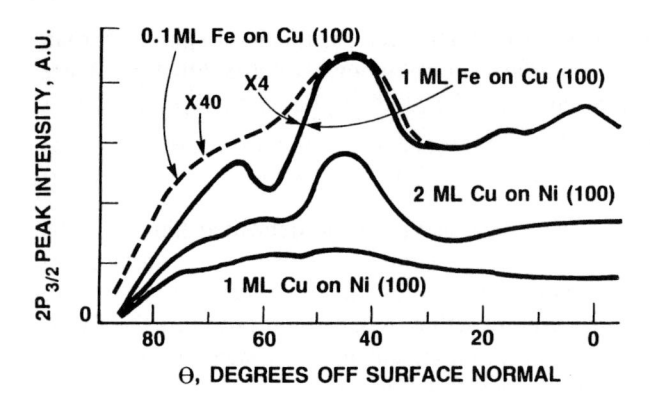

Fig. 5.38. The angular distribution of the XPS intensity in the $\langle 0\,0\,1 \rangle$ azimuth for the $2p_{3/2}$ peak of the deposited element. The strong forward-scattering feature at 45° demonstrates that 0.1 and 1.0 ML Fe deposited on Cu(1 0 0) at 300 K do not lie flat on the surface as does 1 ML Cu on Ni(1 0 0)

Similar XPS forward-scattering data have been obtained for the growth of Co on Cu(1 0 0) [5.260]. The results have been analysed with model calculations to extract the average Co island thickness and the fractional coverage of the Cu(1 0 0) as a function of the amount of Co deposited [5.260]. Figures 5.39 and 5.40 present plots of these data and the results of the analysis [5.260]. Presentation in the form of Fig. 5.40 is particularly useful for assessing the extent of Co agglomeration. For example, after deposition of 1 ML of Co, the average Co island thickness is 1.6 ML, and the Cu(1 0 0) surface is 65% covered (i.e. the surface is 35% exposed Cu, 35% a Co double layer (or intermixed Cu on Co), and 30% a Co monolayer). Recent determinations of the amount of exposed Cu by a different technique [5.404] yielded very similar results, and tend to confirm this general picture of Co growth on Cu(1 0 0) at 300 K [5.405].

Other demonstrations by forward scattering of 1 ML transition-metal films which do not lie flat on the surfaces of noble metal substrates include Fe on Cu(1 0 0) [5.189], Fe on Cu(1 1 1), (1 1 0), and (1 0 0) [5.405], Cr on Ag (1 0 0) [5.229, 231], and Mn on Ag(1 0 0) and Cu(1 0 0) [5.251]. There is one report of a counter-example in which 1 ML Cr appears to lie flat on Ag(1 0 0) at 440 K [5.268]. It is not clear why this result is so inconsistent with Eq. (5.71). There is one forward-scattering study reporting a 1 ML transition-metal film, Fe, lying flat on a semiconductor surface, ZnSe [5.266, 267].

5.2.3.2 Lattice Expansions and Contractions

In recent years, the thin film magnetism community has shown a great deal of interest in using epitaxy to produce novel crystal structures [5.406]. For example, it is now possible to produce, by epitaxial growth on Cu(1 0 0)

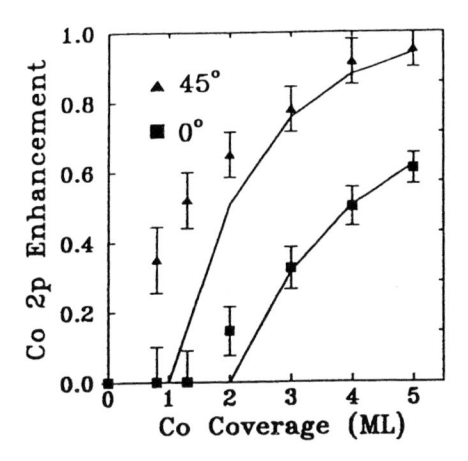

5.2 X-Ray Photoelectron and Auger Electron Forward Scattering   253

Fig. 5.39. The forward-scattering enhancements from Co films on Cu(1 0 0) at 0° and 45° as a function of Co thickness. The solid lines are model calculations for an ideal layer-by-layer growth mode. Deviations from this growth mode are seen below 2 ML [5.260]

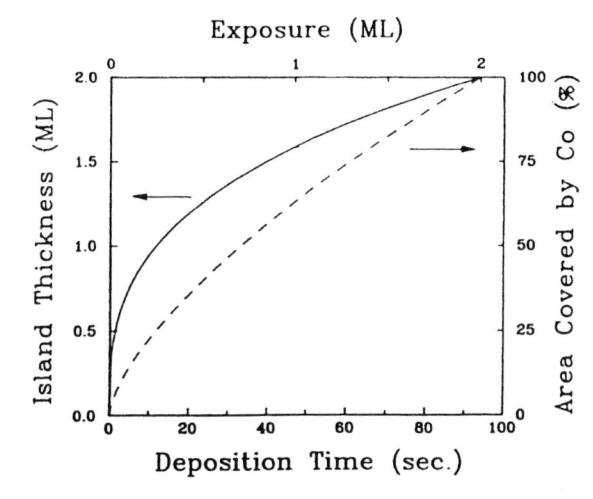

Fig. 5.40. A plot of the average island thickness (solid line) and the percentage of the Cu(1 0 0) surface covered by Co (dashed line) versus deposited Co thickness. These results are derived from the data of Fig. 5.39 [5.260]

substates, reasonably stable Fe or Co films of a few ML thick in structures that are very nearly fcc [5.407]. Such films often have very interesting magnetic properties [5.408]. Novel crystal structures provide for the field of magnetism a new dimension to explore, and work in this area is already beginning to deepen our understanding of magnetism [5.409].

Lattice expansions and contractions in epitaxial films are often a key ingredient in producing novel crystal structures. The reason for this is that a very thin epitaxial film is often strained to match the substrate psudomorphically. The in-plane strain generally produces an expansion or contraction in the vertical atomic layer spacing because such an effect reduces the change in atomic volume. Forward scattering is an excellent approach for observing such changes in the lattice spacing, and if LEED or RHEED is also available so that the in-plane spacing can be observed, then the crystal structure of the film can often be determined.

Figure 5.41 presents an example of a contraction in layer spacing for epitaxial Mn films on Ag(1 0 0) [5.251]. It is easy to visualize what occurs here. Since Mn is a smaller atom than Ag, it would be expected that when a Mn film is stretched in-plane, as it were, to match the larger Ag lattice, a contraction in layer spacing would occur, according to Poisson's ratio, to reduce the change in Mn atomic volume. In this case, LEED indeed shows the Mn(1 0 0) surface is matched in-plane with the Ag(1 0 0). The forward-scattering peak in the Mn emission at 48° or 51° indicates that the Mn layer spacing is substantially less than that of the underlying Ag lattice (which exhibits a peak at 45°). The nearest-neighbor forward-scattering peak appears at 54.7° for a bcc lattice, so the Mn is intermediate between fcc and bcc. This result establishes that the crystal structure of the Mn film is body-centered tetragonal [5.251].

The changes in Mn layer spacing with Mn thickness indicated by the forward-scattering peaks in Fig. 5.41 are in remarkably good agreement (within a few percent) with corresponding determinations for (1 0 0) MnAg superlattices using X-ray diffraction [5.410]. This agreement suggests that these changes are a property of the Mn on Ag(1 0 0) and not affected by the presence of Ag(1 0 0) on both sides of the Mn in the superlattices.

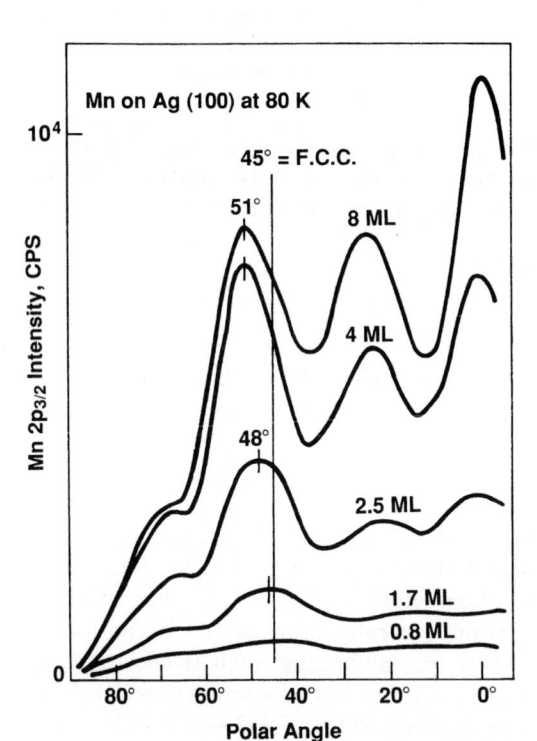

Fig. 5.41. The angle distribution of the XPS intensity in the $\langle 0 0 1 \rangle$ azimuth for the Mn $2p_{3/2}$ peak for various thicknesses of Mn on Ag(1 0 0). The peaks at 0° and $\sim 50°$ are due to forward scattering, and the peak at $\sim 25°$ is a first-order constructive interference. [5.251]

5.2 X-Ray Photoelectron and Auger Electron Forward Scattering   255

Note here, that the peaks in Fig. 5.41 at about 25° are not forward-scattering features but are predominantly first-order constructive interferences, as discussed in Sect. 5.2.2a.

An interesting contrast with the Mn on Ag(1 0 0) case is the case of Mn films on Cu(1 0 0) [5.251]. Copper is smaller atom than Mn so the Mn film may be imagined to be compressed in-plane to match the Cu(1 0 0). The expected outward expansion in Mn layer spacing is indeed manifested in the forward scattering with the peak appearing at 34° to 40° depending on the Mn film thickness.

A case in which a bcc, or very nearly bcc, structure does occur is Fe on Ag(1 0 0). For Fe films thicker than about 8 ML, a forward-scattering peak is observed almost exactly at 54.7° [5.251] This occurrence may be understood either as a consequence of the stretching and contracting described above or as a consequence of the near-perfect lattice match between bcc-Fe(1 0 0) and fcc-Ag(1 0 0).

In systems such as those discussed here, a 1° change in forward-scattering peak angle corresponds to an ~ 3.5% change in layer spacing. Such a change can readily be determined by careful angle-resolved measurements, particularly if the forward-scattering peaks of the substrate are used as a reference. It may be noted that since the forward-scattering peaks originate in the top few layers at the surface, the layer spacing provided is, inevitably, an average over these several layers.

5.2.3.3 Surface Segregation

Surface segregation can be a very important phenomenon in epitaxial growth of transition-metal films on noble-metal substrates, particularly at elevated substrate temperatures. For example, it was mentioned above in Sect. 5.2.3a that Fe and Co deposited on Cu(1 0 0) around 300 K tend to agglomerate into double layers. However, for deposition at somewhat higher temperatures (e.g. 400 K), Cu readily segregates onto the Fe or Co [5.227, 405, 411–414]. In fact, when 1 ML of Fe or Co is deposited at 450 K, the surface layer that is observed is almost entirely Cu. For thicker films, this segregating Cu gradually gets left behind in the growing Fe or Co lattice, and a dilute alloy with Cu is formed.

Although it is often thought that alloys such as CoCu will not form naturally due to the heat of alloying being rather endothermic, it should be noted that such restrictions do not necessarily apply at surfaces. Surface alloying during epitaxy differs from bulk alloying due to the influence of surface-free-energy terms such as those in Eq. (5.71). In many cases, thermodynamics favors surface segregation of low-surface-free-energy substrate atoms. If kinetic limitations cause these atoms to be left behind gradually in the growing film, a metastable alloy often results [5.227].

When an alloy forms in this manner, RHEED oscillations may well be observed due to the layer-by-layer growth of the alloy [5.416–419]. It is important, then, not to interpret RHEED oscillations as layer-by-layer growth of the pure element, with the implication that the film-substrate interface is atomically sharp.

As might be expected, such segregation effects can have a pronounced effect on magnetic properties [5.411–414]. It is thus of much interest to know when segregation has occurred and to have an understanding of the kinetics of the process. For a simple diagnostic of whether or not segregation has occurred, the most easily applied technique is probably not forward scattering but CO titration or H_2 titration. The titration methods provide an indication of the composition of the top atomic layer at the surface [5.227,413].

Forward scattering is, however, an excellent method for following the kinetics of surface segregation in specially prepared prototype samples, and the insights provided can serve as a base of knowledge for understanding the general characteristics of surface segregation kinetics. Figure 5.42a illustrates a type of epitaxial sandwich structure that is particularly well-suited to observing the kinetics of surface segregation in transition-metal-on-noble-metal systems [5.235]. The experiment is performed by rocking the sample back and forth (at

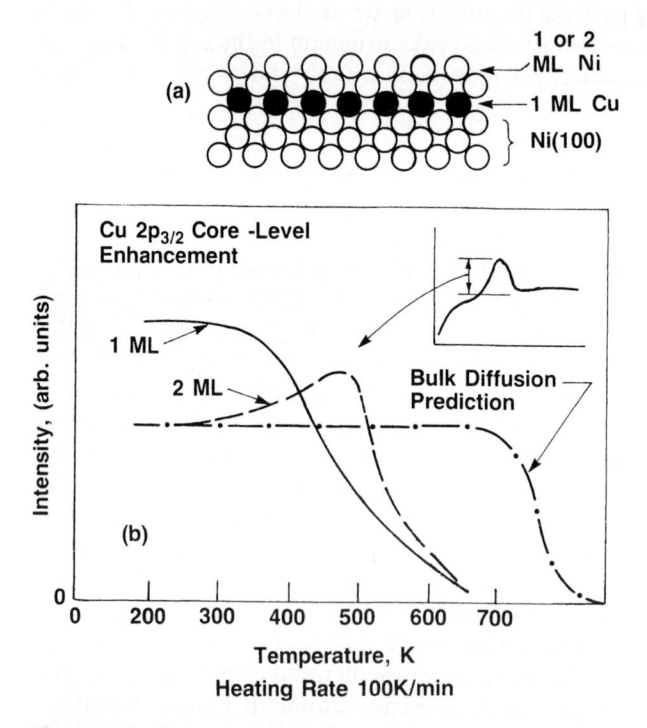

Fig. 5.42. An illustration of how the kinetics of surface segregation may be followed with forward scattering. Two Ni-Cu-Ni(100) sandwich structures were prepared as illustrated in (**a**), and the forward-scattering intensity in the Cu $2p_{3/2}$ peak is plotted in (**b**) as a function of temperature as the samples are heated. The dot-dashed line indicates the prediction made using the known bulk diffusion parameters for Cu in Ni [5.235]

5.2 X-Ray Photoelectron and Auger Electron Forward Scattering

$\sim$ 10 cycles/min.) through the forward-scattering peak in the Cu emission at 45° while the sample is heated. The segregation of Cu to a Ni surface is well-known to be an exothermic process, and as Fig. 5.42b illustrates, the forward-scattering peak disappears when the Cu segregates.

When the starting sample has 1 ML of Ni over the Cu monolayer, the segregation is seen to begin around 300 K and to be a rapid process at 450 K [5.235]. Diffusion by the usual bulk vacancy mechanism is clearly not involved here. The parameters for bulk diffusion of Cu in Ni are well-known and predict that each Cu atom will undergo a lattice-site hop by the vacancy mechanism only once in every 10^7 y at 450 K. Only at 877 K does this rate reach 1 hop/s [5.235] Clearly, a lower activation energy mechanism is required, and a combination of different surface diffusion processes seems to be the answer [5.235].

The usefulness of applying forward scattering to the study of segregation as in Fig. 5.42 may be illustrated by noting that, for angles around the forward-scattering peak (45°), the Cu intensity, counterintuitively, actually goes down as Cu segregates to the surface (analogous data may be seen in Fig. 5.25, but note the reversed roles of Cu and Ni). Thus, simple electron-escape-depth arguments often seen in the literature concerning the use of XPS and Auger signals for studies of surface segregation may be misleading, and forward scattering is seen to be a more refined probe.

When the starting sample has 2 ML of Ni over the Cu monolayer, the forward-scattering peak initially increases, as the Cu begins to move from the third layer into the second layer, but then decreases as the Cu gets into the top layer. In both the 1 and 2 ML Ni cases, the segregation is nearly complete at 650 K, a temperature still far too low for the bulk-vacancy-diffusion mechanism to apply ($\sim$ 1 hop in 5 d). Moreover, in both cases, a simple rate equation involving one activation energy cannot explain the shape of the curves in Fig. 5.42b. Additional data on this system (in which the sample temperature is raised quickly to a value which is then held constant so that isothermal rates of segregation can be measured) also show that no simple rate equation describes the process. Instead, it appears that a variety of processes or mechanisms is involved, with a range of activation energies. These data suggests that, at each stage, when the temperature is raised quickly to a value which is then held constant, new mechanisms become active, run to near-completion, and then the temperature must be raised again to allow new mechanisms to turn on (see Ref. [5.235] for details and possible mechanisms). Thus, the segregation mechanisms are very complex, and the mechanisms which operate at 300 K are very different from the ones that operate at 600 K. Therefore, in epitaxial growth whenever segregation of substrate atoms is a possibility, some determination of its extent is an important aspect of characterizing the imperfections in the epitaxial film.

Figure 5.42b also illustrates the segregation rate predicted by the known bulk diffusion parameters. It was found that a Ni overlayer of 10 ML was required to produce segregation at this rate. The important implication of these results is that the top several layers at a surface can exhibit surface segregation at temperatures well below the onset of bulk diffusion. These results also indicate why it has been found that, for thicknesses greater that a few ML, high-quality epitaxial films can often be produced by deposition at a low substrate temperature and subsequent gentle annealing [5.227] intermixing and segregation are suppressed during deposition at the low temperature, and during gentle annealing, they are suppressed due to bulk diffusion being negligible.

5.2.3.4 Interdiffusion at Interfaces

Forward scattering, being essentially a diagnostic of short-range order around the emitting atom, is an excellent probe for studying interdiffusion at interfaces. The concept of interdiffusion at interfaces during epitaxial growth may seem to have an extensive overlap with the surface segregation phenomena discussed in the previous section. It is indeed true that there is much overlap, but there is at least one way in which a clear distinction can be drawn. When a deposited atom finds its way into a lattice site previously occupied by a substrate atom or into an interstitial site of the substrate, then it seems fair to term the process interdiffusion.

One mechanism by which such interdiffusion can occur is "exchange diffusion", a process which was received much notoriety lately [5.401b,420–433]. "Exchange diffusion" is a process in which surface diffusion proceeds by the concerted motion of an adatom rolling down into the surface as a surface atom rolls up to form a new adatom. This place-exchange phenomenon has recently been found to occur for a variety of metals on fcc(1 0 0), fcc(1 1 0), and bcc(1 1 0) surfaces [5.401b,420–433]. Perhaps the most surprising result is the low temperatures at which exchange can occur. For Pt adatoms on Ni(1 1 0) the process is observable at 105 K [5.427], and for Pt adatoms on Pt(1 0 0) it is observable at 175 K [5.425]. For metals less refractory than Pt, it very likely occurs at even lower temperatures.

Molecular-dynamics simulations and potential-energy-surface calculations support the experimental observations [5.422, 424, 426, 430, 432]. To cite just one example, the exchange process for the Ni on Au(1 0 0) system is predicted to occur within picoseconds after deposition, even at 100 K! [5.430]. Clearly, this is a phenomenon that needs to be considered more seriously in magnetic thin film studies.

One system, important for surface magnetism, in which exchange diffusion appears to occur is Fe on Ag(1 0 0) [5.433]. It has become apparent in recent years that something very strange indeed is occurring during the growth of the first ML of Fe on Ag(1 0 0) [5.4]. For example, the RHEED oscillations (at glancing beam incidence) for growth at 300 K exhibit peaks at 0.17, 0.58, and 1.2 ML before settling down to regular peaks at 2, 3, 4, 5 ML, etc. [5.436, 437].

XPS forward-scattering data on this system provide a partial insight into what is going on. Figure 5.43 shows that, for Fe films of 0.5 ML and 1.0 ML

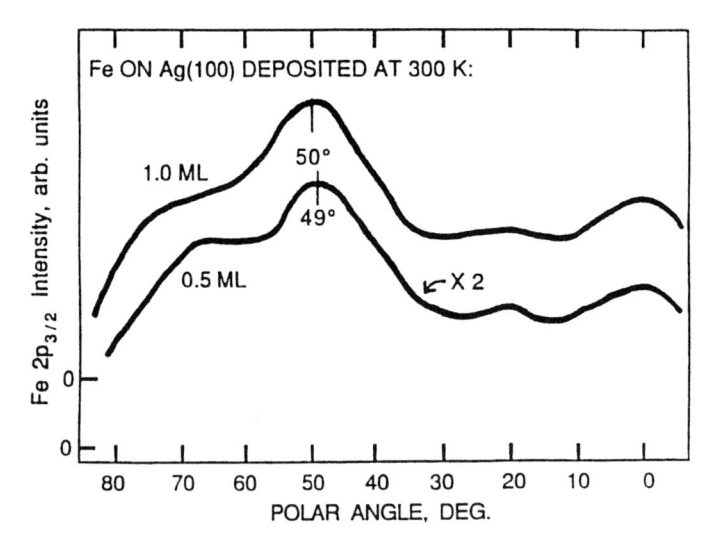

Fig. 5.43. The angular distribution of XPS intensity in the $\langle 0\,0\,1 \rangle$ azimuth for the Fe $2p_{3/2}$ peak for 0.5 and 1.0 ML Fe deposited on Ag(1 0 0) at 300 K. The forward-scattering peaks at 0° and ~ 50° indicates that the Fe neither wets Ag(1 0 0) no agglomerates into bcc-Fe islands. [5.443]

thickness, a peak is observed around 49°–50° and another at 0°. With increasing Fe thickness, the 50° peak shifts continuously to 54.7° by a deposition of about 8 ML, indicating that, within the XPS probing depth of several ML, the structure is essentially bcc-Fe. This result is consistent with the widely accepted view that, on Ag(1 0 0), Fe grows in a bcc structure [5.438]. However, in the earliest stages of growth, Fig. 5.43 shows that the forward-scattering peak is almost exactly half-way between the fcc (45°) and bcc (54.7°) angles. This result suggests that the Fe is not agglomerating into multilayer Fe islands (which should be bcc). Instead, it appears that the Fe exchanges with the Ag to produce a structure that is intermixed in the top few layers and is thus intermediate between fcc and bcc, as would be expected if approximate conservation of the Ag and Fe atomic volumes is maintained. The thermodynamic driving force for this exchange, which lets Fe go in and Ag come out, would clearly be the much lower surface free energy of Ag. This effect is easily large enough to overcome the endothermic heat of alloying of Fe and Ag, and the available thermodynamic values suggest the net driving force for the exchange is ~ 0.5 eV/atom [5.433].

Evidence is beginning to accumulate that exchange diffusion might be a common phenomenon in magnetic thin films on noble-metal substrates. For example, there is recent evidence from STM that exchange diffusion occurs at 300 K during deposition of Fe on Cu(1 0 0) [5.401b,439]. However, in the case of atoms of nearly the same size, forward scattering cannot provide insights into whether or not exchange diffusion has occurred.

5.2.3.5 Stacking Faults

On fcc(1 1 1) surfaces, two types of 3-fold coordination sites are present, those that continue the **abcabc** stacking sequence of the fcc lattice and those that do not. When an epitaxial film nucleates at the sites that do not maintain the proper stacking, the result is termed a stacking fault. In many cases, an adatom experiences only a very small binding energy difference between the two types of sites [5.440, 441]. When epitaxy proceeds by islands nucleating on flat terraces, stacking fault sites may thus be expected to be common. When the deposited atoms are mobile enough to reach steps and growth proceeds outward from steps, the fcc stacking may be expected to be maintained.

The available evidence suggests that stacking faults are indeed a common occurrence in epitaxy on fcc(1 1 1) surfaces [5.442–448]. Their occurrence is likely to be important for magnetic properties A simple ball model of an fcc(1 1 1) surface can be used to demonstrate that very severe lattice distortions will occur around the grain boundaries separating regions of stacking-fault growth from regions of fcc growth. Therefore, in order to characterize more fully the structural imperfections in such epitaxial films, it would be desirable to have an *in situ* diagnostic for the occurrence of stacking faults.

Forward scattering is very useful for identifying the presence of stacking faults. Figure 5.44 illustrates the use of forward scattering for this purpose in the case of a 3.7 ML Fe film deposited at 300 K on Cu(1 1 1) [5.338]. The data are taken in the ⟨1 1 2⟩ azimuth so that the Cu Auger angular distribution from

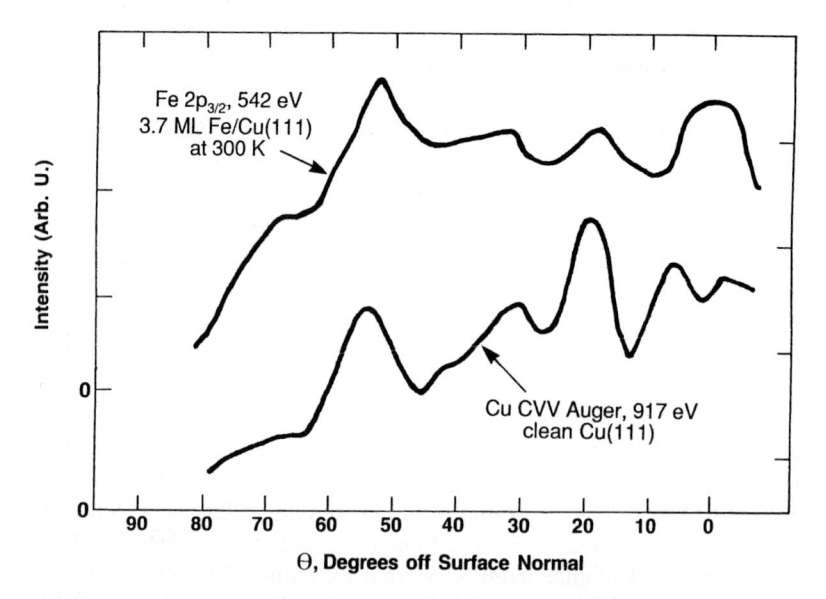

Fig. 5.44. An illustration of how forward scattering can establish the presence of stacking faults. The angular distribution of the CVV Auger emission from clean Cu(1 1 1) and of the Fe $2p_{3/2}$ peak intensity for 3.7 ML Fe deposited on Cu(1 1 1) are presented in the ⟨1 1 2⟩ azimuth

clean Cu(1 1 1) exhibits next-nearest neighbor and next-next-nearest neighbor forward scattering at 54.7° and 19.5°, respectively. No forward-scattering peak is apparent at 0° because the **abcabc** stacking sequence means that emitter-scatterer distance (**a** − **a**) is too long. However, in the Fe $2p_{3/2}$ emission from the 3.7 ML Fe film, a pronounced peak is apparent at 0°. This observation almost certainly means that stacking faults are present in the Fe lattice to give some **abab** stacking, and the resulting proximity of next-nearest neighbors (a–a) along the 0° direction produces a forward-scattering peak.

If growth consists of a mixture of stacking faults and fcc stacking, other forward-scattering peaks are predicted at 19.5°, 35.4°, and 54.7°. Since there is some evidence for each of the peaks in Fig. 5.44, it seems likely that such a mixture is indeed occurring during growth (the slight shifts in peak positions indicate a small expansion of the Fe layer spacing). Further support for the presence of stacking faults is found in the fact that the LEED pattern of the surface is 6-fold symmetric, unlike the 3-fold symmetric pattern of the Cu(1 1 1) substrate.

It may be of interest to contrast this growth of Fe with the growth of Cu on Cu(1 1 1). When a 4.5 ML Cu film was deposited on Cu(1 1 1) at 80 K, the Auger angular distribution was indistinguishable from that of the initial surface (as in Fig. 5.44). The absence of any evidence for stacking faults suggests that Cu on Cu(1 1 1) may be mobile enough even at 80 K to reach (and grow out from) step sites to preserve the fcc stacking. Support for this interpretation is given by the LEED pattern which, though fuzzy, was 3-fold symmetric [5.338].

5.2.3.6 Surfactants in Epitaxy

In the few years since it was first suggested and demonstrated that adsorbates might be used to deliberately modify epitaxial growth in a favorable manner [5.227], there has been rapid growth in the level of interest in this idea [5.449–459]. The basic idea is that an absorbate can be expected to modify the balance of surface and interface free energies which govern the equilibrium growth mode Eq. (5.71).

Figure 5.45 illustrates the first known example of an adsorbate being used to modify an epitaxial growth mode [5.227]. As discussed earlier, in Sect. 5.2.3a, high-surface-free-energy metals like Fe or Ni tend to agglomerate and intermix when they are deposited around room temperature on low-surface-free-energy metals like Cu. The free energy of the system is lowered by minimizing the amount of Fe or Ni exposed. However, an adsorbate such as oxygen forms considerably stronger bonds with Fe or Ni than with Cu. This difference is the key to modifying growth. Figure 5.45 illustrates how, with 0.5 ML of O present, the most stable state of a 1 ML film of Fe or Ni on Cu(1 0 0) is to have the Fe or Ni lie flat on the Cu(1 0 0) with the oxygen on top. This arrangement is favored because it maximizes the number of strong Fe–O or Ni–O bonds, and leaves oxygen, which has no broken or dangling bonds, on the surface.

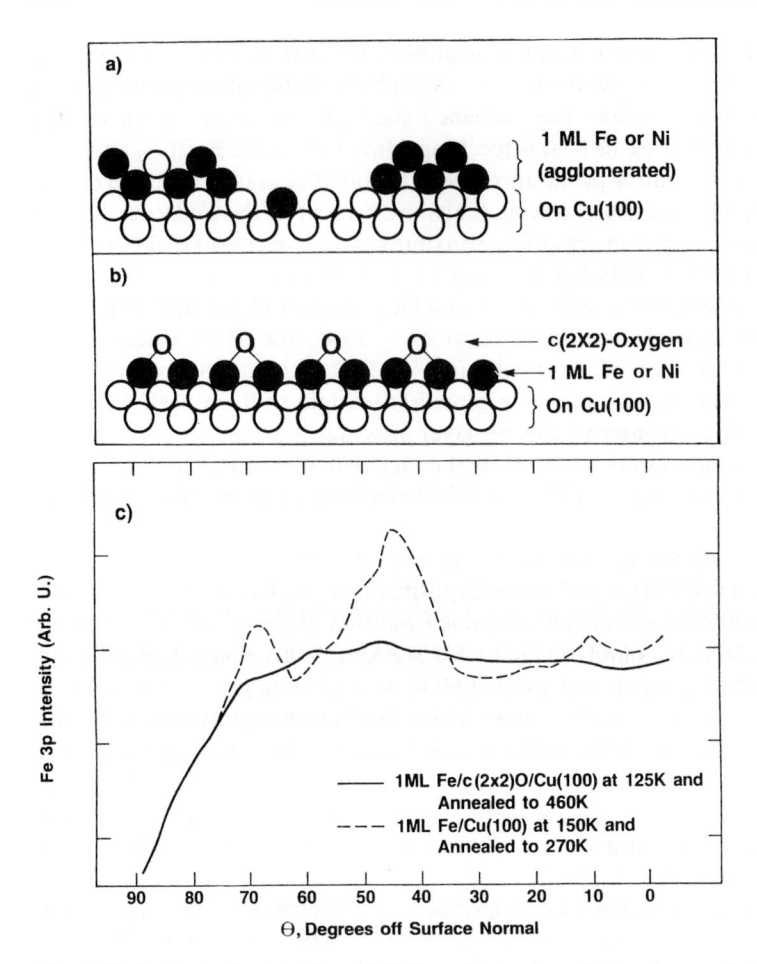

Fig. 5.45. An illustration (a) of how 1 ML Fe agglomerates and intermixes on Cu(1 0 0) at 300 K, (b) how oxygen acts as a surfactant to suppress this behavior, and (c) the XPS data for Fe on Cu(1 0 0) supporting this interpretation. [5.227, 401b, 439]

The XPS forward-scattering data provide the evidence supporting the schematic illustration of Fig. 5.45. A pronounced 45° forward-scattering peak is observed in the Fe emission for even the best-prepared 1 ML Fe film (deposited at low temperature and gently annealed). The driving force for agglomeration is strong. However, with 0.5 ML O present, the system can be annealed at 460 K to allow surface diffusion to order the surface well [and produce sharp LEED spots with a $c(2 \times 2)$ pattern] without any evidence of a forward-scattering peak at 45°. Since oxygen resides rather deep in the 4-fold coordination site [5.460], and is a weak scatterer, it is expected to produce a weak forward-scattering peak near grazing emission (not observed in Fig. 5.45).

Very recent STM studies have suggested that oxygen does indeed tend to suppress the agglomeration of Fe on Cu(1 0 0) [5.461]. Thus, it would seem that

5.2 X-Ray Photoelectron and Auger Electron Forward Scattering

surfactants will be of increasing importance is attempts to gain improved control over epitaxial growth in the future. For example, it could be that an adsorbate such as oxygen would provide a means for suppressing the intermixing by exchange diffusion discussed in the previous section. Although there is no confirmation of this suggestion, recent data do show that the presence of oxygen in the Fe on Ag(1 0 0) system does produce striking changes in the structure of the film and in its magnetic properties [5.462].

One important property that surfactants should have in order to be useful is that they should readily segregate to the surface of a growing film. The initial work on this topic did indeed show that many small, light-atom adsorbates have a remarkable tendency to "float out" to the surface during growth. For adsorbates such as oxygen on metals such as Cu or Ni, this tendency is strong even at cryogenic substrate temperatures! [5.236].

5.2.4 Conclusions

The major conclusions of this review may be summarized as follows:

1) Angle-resolved XPS and Auger emission from single crystals and epitaxial films generally exhibit strong forward-scattering features.
2) These features are caused by the emitted electron being strongly scattered by the potential of atoms surrounding the emitter. Forward scattering (or forward focussing) dominates and produces beams of enhanced intensity along internuclear axes (or bond directions). Next-nearest neighbor directions also produce forward scattering, but this is a weaker effect due to the greater distance involved.
3) The minimum kinetic energy that the electron must have for nearest-neighbor forward scattering to the reliably observable is about a few hundred eV.
4) The strength of forward scattering increases with atomic number of the scattering atom. For electron kinetic energies of several hundred eV, the effect is very strong for transition-metal atoms and a weak but detectable effect for C, N or O. The effect is expected to be almost negligible for H.
5) A good, qualitative understanding of the basic scattering phenomena underlying the data can be provided by simple semiclassical models.
6) Much useful structural information can be obtained directly from the data without need of theoretical analysis, deconvolution, or other processing. However, theoretical modeling is necessary to extract the maximum possible amount of structural information from any given data set.
7) In studies of epitaxial growth, XPS and Auger forward scattering often provide the best method for determining whether or not layer-by-layer growth occurs.
8) Surface segregation and interdiffusion at interfaces can be observed with particular clarity by XPS and Auger forward scattering.

9) XPS and Auger forward scattering are very good techniques for observing stacking faults, and the presence of expanded or contracted layer spacings in epitaxial films.
10) One of the most attractive features of XPS and Auger forward scattering is its ease of use. Valuable structural information can often be obtained with no more effort than rotating a sample and recording a peak intensity.

Acknowledgements. The author wishes to express his gratitude for numerous stimulating discussions, which have contributed much to the content of the present review, with S.A. Chambers, C.S. Fadley, S.Y. Tong, B.P. Tonner, H.L. Davis, M.A. Van Hove, and C.J. Powell.

5.3 X-Ray Studies of Ultrathin Magnetic Structures

R. CLARKE and F.J. LAMELAS

> *"... and even in gold they are; for we do find seeds of them, by our fire, and gold in them; and can produce the species of each metal more perfect thence, than nature doth in earth."*
>
> Ben Jonson, *The Alchemist II.i*

We describe the application of X-ray diffuse scattering techniques to layered ultrathin magnetic structures. We focus attention on three distinct aspects of the structure: layering (composition modulation), in-plane epitaxy, and atomic stacking sequences, each of which has important implications for the magnetic behavior. Deviations from perfect crystallinity, such as interface spreading, stacking faults, and layer thickness fluctuations, are a central issue in these systems and must be included in any useful description of the structure. Various techniques for modeling imperfect ultrathin structures are discussed, including both analytical approaches and numerical simulations. We illustrate the intricate dependence of magnetic behavior on epitaxial structure with recent results on Co-based superlattices grown by molecular beam epitaxy. In this context we stress the role of epitaxial structure in establishing spin anisotropies, and in stabilizing novel metastable phases of magnetic elements.

5.3.1 Introduction

The interdependence of structure and physical behavior is nowhere more emphatically demonstrated than in the field of magnetism [5.463]. This is especially true in ultrathin magnetic materials where relatively subtle structural modifications at the level of the atomic lattice can, intentionally or unintentionally, dominate the magnetic characteristics.

Given the rather primitive state of our current understanding of microscopic growth mechanisms, very often what is grown is not precisely what was intended. In the best tradition of empirical science, new phenomena have been discovered in the course of unsuccessful attempts to achieve highly perfect structures. For example, one of the most spectacular properties of ultrathin magnetic structures, giant magnetoresistance, is enhanced when some degree of interface disorder is present [5.464]. Regardless of whether such departures from crystalline perfection lead to desirable or interesting physical properties, it is essential to obtain as complete a picture as possible of the atomic arrangement, including the nature of the interfaces, their strain distribution, and their epitaxial relationships.

Of all the structural characterization techniques X-ray scattering has served not only as the 'workhorse' for routine analysis, but has also been developed to a point where it now provides the most sensitive and highest resolution data on the interface structure at the heart of the physics of ultrathin magnetic layers. New directions based on synchrotron radiation promise even more powerful applications for X-ray techniques. An excellent example is the use of the spin dependent scattering cross section of X-rays as a probe of magnetic structure.

Clearly no one technique is sufficient for a full structural determination and a number of complementary approaches have been brought to bear on the problem. RHEED (Reflection High Energy Electron Diffraction) as a monitor of the surface structure during growth, and HRTEM (High Resolution Transmission Electron Microscopy) for studies of dislocations and other faults, are particularly noteworthy in this context.

In this chapter we will describe X-ray diffuse scattering methods as they are applied to studies of ultrathin magnetic structures. We will emphasize the diverse types of information that X-ray scattering can provide with appropriate modeling. Taking examples from our work on Molecular Beam Epitaxy samples of Co-Au and Co-Cu superlattices [5.465, 466] we will also illustrate the relevance of this structural information to the unusual features of magnetism in ultrathin magnetic structures. Of particular interest are the large perpendicular anisotropy and the stabilization of novel epitaxial phases with unique magnetic properties.

5.3.2 Overview of the Problem

At one level X-ray diffraction is regarded as a routine structural characterization method for analyzing artificially prepared superlattices, since basic features of the scattering (e.g., the positions of the X-ray peaks) can be interpreted by inspection. However, a more complete analysis of X-ray data calls for the comparison of measured intensities with those of a realistic scattering model. This is especially true in the case of metallic superlattices which are usually far from perfect and require that the scattering model include various types of structural disorder. It is remarkable how much detailed information can be extracted from such models. For the purposes of X-ray analysis it is useful to separate the various aspects of an epitaxial layer structure into several distinct components:

Layering, refers to the chemical composition of the individual layers, their thicknesses and periodicity (if layers are repeated regularly as in a superlattice), and the abruptness of the interfaces between the different layers. The key question here is what type of roughness is present: steps and terraces, or continuous interdiffusion?

In-plane epitaxy describes the crystallographic relationships of succeeding layers of the structure, both with respect to the substrate and with respect to each other. Important issues to be taken into account here include relative mismatch of the lattices corresponding to the different layers, the strains arising in response to this mismatch, and the reduction in coherence that might result from strain-relieving dislocations at the epitaxial interface. The influence of surfaces and interfaces on magnetic anisotropy is a key issue in these materials [5.467].

Stacking refers to the relative crystallographic arrangement of neighboring atomic planes. A simple example of particular relevance to transition metal magnetic structures is the stacking of close-packed planes of atoms in either the hexagonal close-packed (hcp) sequence: ABABAB..., or in a face centered cubic (fcc) sequence: ABCABC.... Examples of each of these sequences are found in ultrathin magnetic structures and, as we shall see in the examples given below, the magnetic properties are strongly influenced by the stacking arrangement via the symmetry of the local ordering. An important point here is that the internal energies of the different stacking sequences are very similar and so stacking faults are to be expected. This particular defect has been linked by some authors to a possible mechanism for perpendicular anisotropy [5.468].

5.3.3 X-Ray Diffuse Scattering

Before introducing the various types of X-ray measurements which probe the three aspects of the structure outlined above it may be helpful first to briefly describe some of the instrumentation that we use to perform these scans.

5.3.3.1 X-Ray Instrumentation

Most of the X-ray measurements described here are performed using a 12 kW Rigaku rotating anode X-ray generator. It is advantageous to use a Mo anode which produces relatively energetic X-rays (17.5 keV; $\lambda \approx 0.71$ Å). This results in a penetrating X-ray beam which is useful in performing scans in transmission geometry, and also when the sample is mounted in a partially absorbing enclosure such as a cryostat or an annealing furnace.

The X-ray beam is monochromatized using a flat slab of Highly Oriented Pyrolytic Graphite. This monochromator has a rather large angular divergence ($\approx 0.2°$) due to its mosaic structure, but it has high integrated reflectivity resulting in an X-ray probe with high incident flux. For higher resolution measurements we use a Ge(1 1 1) monochromator but at the cost of substantially reduced incident flux.

The X-ray beam from the monochromator strikes the sample located at the center of a four-circle diffractometer (Fig. 5.46). We have found this to be an extremely versatile instrument for investigating thin film samples. In this geometry, the scattering plane, which contains the incident and scattered beams, is horizontal. The detector angle with respect to the incident beam is denoted 2θ by convention. The axis of the detector motion is fixed and vertical. The three remaining diffractometer circles are employed in aligning the sample with respect to the incident beam. The ω circle has a fixed vertical axis. The χ circle is mounted on the ω circle and has a horizontal axis which rotates with ω. Lastly, the ϕ circle is mounted in the χ circle. We will not deal at length with the various computer controlled rotations which these various axes undergo in the scans. We note only that it is possible to orient the sample arbitrarily with respect to the incoming beam.

5.3.3.2 Types of X-Ray Diffractometer Scans

The most commonly employed X-ray technique for layered materials is the *perpendicular* scan where intensities are measured along the specular rod. During this scan (Fig. 5.47a) the scattering vector is perpendicular to the superlattice layers and serves as a probe of layer structure in the growth direction. At relatively large wavevectors, on the order of the reciprocal lattice spacings of the bulk lattice, this scan is sensitive to a combination of the local atomic ordering and the elemental modulation throughout the superlattice. At small wavevectors, of order $2\pi/\lambda_{SL}$ (where λ_{SL} is the superlattice period), the scattering is sensitive to modulation in elemental composition only.

In a second type of scan, the *in-plane* scan, the scattering vector lies in the plane of the superlattice (Fig. 5.47b). With the sample-plane vertical, the χ circle of the diffractometer can be used to measure the X-ray scattering as a function of azimuthal angle. This scan serves as a probe of atomic arrangements within the growth plane and is useful, for example, in determining epitaxial relationships between the film and the substrate and in estimating in-plane coherence lengths. This type of scan is often performed in a glancing angle geometry at incidence angles below the critical angle for total external reflection of X-rays. In the glancing incidence geometry the penetration depth of the X-ray beam at typical ($\lambda \approx 1$ Å) energies is $\lesssim 100$ Å. Such a penetration depth may be favorable in certain applications but in our case we are interested in sampling the entire superlattice and the substrate. In addition, the glancing incidence geometry is favored by the high flux available at X-ray synchrotrons, given the small

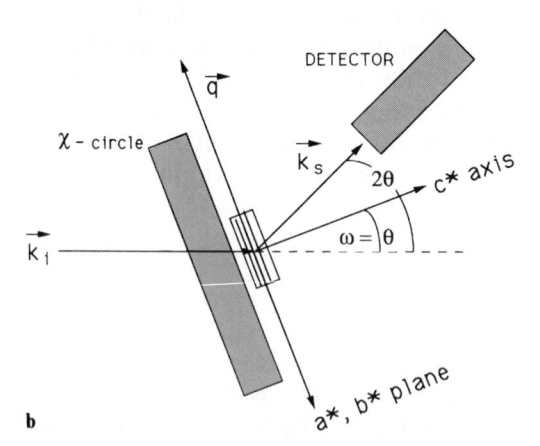

Fig. 5.46. (a) Four-circle diffractometer. (1) Detector 2θ-circle; (2) ω-circle; (3) χ circle; (4) ϕ circle; (5,6,7,8) stepping motors; (9) sample support; (10) incident beam X–Y slit. (b) Top view showing scattering geometry for in-plane scans.

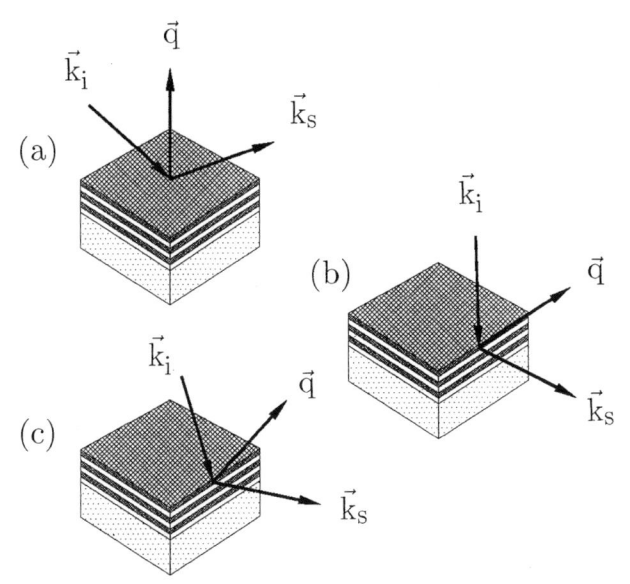

Fig. 5.47. Types of diffractometer scans for X-ray diffuse scattering measurements on ultrathin layered structures: (a) perpendicular scan (b) in-plane scan (c) c^* scan

scattering volume. In contrast, the in-plane scans which we will describe were all performed in transmission geometry (Fig. 5.47b). In this geometry the entire superlattice and the substrate are sampled by the beam, the initial sample alignment is straightforward, and scattering intensities with a rotating anode source are acceptable.

The third important type of X-ray scan for our purposes is the c^* scan (Fig. 5.47c), where the scattering vector has both in-plane and out-of-plane components. This scan, which is performed in transmission geometry along a non-specular rod, is sensitive to the lateral stacking coherence of the atomic layers. The trajectory of this scan in reciprocal space is shown in Fig. 5.48

5.3.4 Modeling Ultrathin Layered Structures

5.3.4.1 Layering Characterization

In what follows we will use a one-dimensional kinematical treatment to calculate X-ray scattering profiles from a stack of ultrathin layers. A note of caution is worthwhile here. The kinematical approximation assumes that contributions from multiple scattering are negligible, it neglects interference between scattered and incident radiation, and assumes that the reduction in intensity of the scattered wave inside the crystal lattice (due to extinction and absorption) is small. For metallic superlattices, which tend to be far from perfect crystals, these

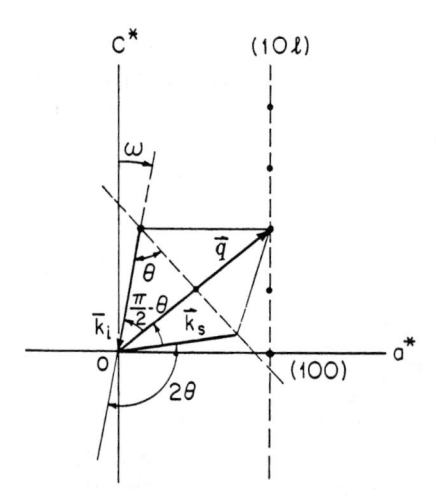

Fig. 5.48. Geometry of c^* scan. In this example, the tip of the diffraction vector, q, moves along the (1 0 l) reciprocal space direction

approximations are well obeyed except at very low diffraction angles close to the critical angle where the total external reflection of the X-ray beam occurs. In this case it is necessary to perform a more detailed calculation of the X-ray reflectivity [5.469], one which includes dynamical scattering effects (see following Section).

In the kinematic approximation one may calculate the specular scattering amplitude from a layer structure as

$$A(q) = \sum_{n=1}^{M} f_n e^{iqr_n} \tag{5.72}$$

where the sum is over each of the M monolayers in the superlattice, f_n is the layer scattering factor (atomic scattering factor multiplied by the atom density), q is the scattering vector amplitude, and r_n is the position of the nth monolayer.

Segmüller and *Blakeslee* [5.470] have obtained a closed-form expression for an ideal superlattice containing N_L bilayers of materials A and B with lattice spacings d_A and d_B and layer scattering factors f_A and f_B:

$$A(q) = \frac{\sin[\pi q(N_A + N_B)N_L]}{\sin[\pi q(N_A + N_B)]} \left[f_A \frac{\sin(\pi q N_A d_A/d)}{\sin(\pi q d_A/d)} \right.$$
$$\left. + f_B \exp[i\pi q(N_A + N_B)] \frac{\sin(\pi q N_B d_B/d)}{\sin(\pi q d_B/d)} \right], \tag{5.73}$$

where the weighted average of the lattice parameters, d, is specified by $N_A d_A + N_B d_B = (N_A + N_B)d$. The scattering intensity is obtained by squaring this amplitude.

As we shall see, real samples often deviate significantly from the ideal case. Nonetheless the model is useful as a guide. Plots of the scattering intensities derived from this model reveal a number of general features. First of all, the spacing of the superlattice satellites is inversely proportional to the superlattice periodicity, λ_{SL} (Fig. 5.49a,b). Secondly, an asymmetry in the superlattice satellite intensities will result when the atomic spacings d_A and d_B are not equal [Fig. 5.49c,d]. If the heavier element has a larger plane spacing, the lower angle satellites of the bulk peak will be more intense and vice versa. In fact the entire sequence of satellite intensties will in general be non-uniform and will be sensitive, at fixed period λ_{SL}, to variations in individual layer thicknesses $N_A d_A$ and $N_B d_B$. Sets of satellite intensities will vanish when the layer thicknesses $N_A d_A$ and $N_B d_B$ have integer ratios. For example, if $N_A d_A = N_B d_B$ the even order superlattice satellites will vanish, if $N_A d_A = (1/2) N_B d_B$ or $N_A d_A = 2 N_B d_B$ every third satellite will vanish, and so on.

The simplest deviation from an ideal superlattice that one might consider is the occurrence of diffusion at the interface between the A and B layers. A damping of the higher order satellites is to be expected since these correspond to the higher order Fourier coefficients that are necessary in reproducing compositionally abrupt profiles. In the limiting case of a sinusoidal composition profile only the first order satellite intensities will be non-vanishing. On the other hand, since interfacial diffusion does not diminish the long-range order of the superlattice, it does not lead to peak broadening. *McWhan* et al. [5.471] have developed

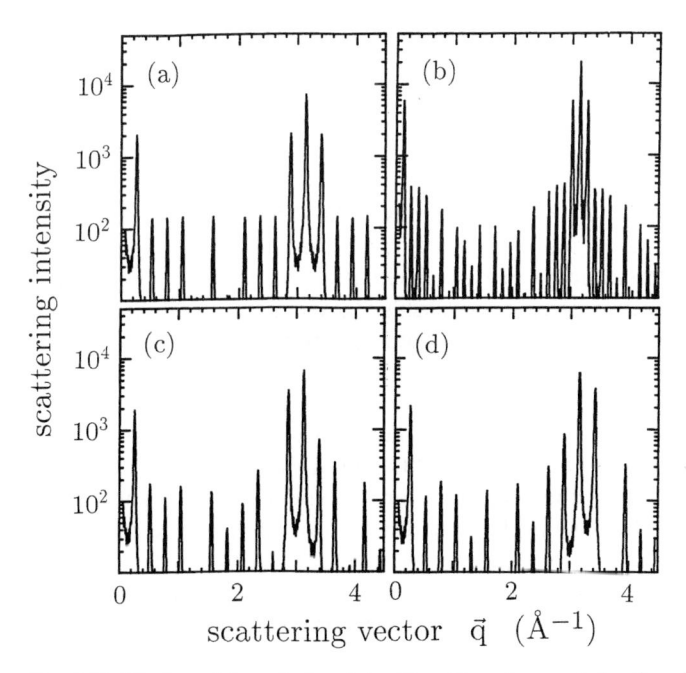

Fig. 5.49. Ideal model scattering intensities. Except as noted otherwise, the parameters are $d_A = d_B = 2.0$ Å, $f_A = 0, f_B = 1, N_A = 5, N_B = 7, N_L$ = total number of bilayers = 10. (a) parameters as noted above. (b) Halving of satellite spacing when superlattice period is doubled by setting $N_A = 10, N_B = 14$. (c), (d) Asymmetry of high-angle peaks induced by lattice spacing difference in A and B layers: (c) $d_A = 1.95, d_B = 2.05$; (d) $d_A = 2.05, d_B = 1.95$

a closed-form expression for satellite intensities for the case of trapezoidal composition profiles.

A second type of imperfection is the occurrence of irregularities in the superlattice layer thickness. These might arise, for example, in the growth of multilayers while the deposition rates are fluctuating. In the case of a crystallographically ordered superlattice, one may consider *discrete* fluctuations in layer thickness. In the case when the superlattice has amorphous or randomly oriented polycrystalline layers, which includes a large fraction of samples being prepared currently, the layer thickness fluctuations can be *continuous* rather than discrete. Both of these types of thickness fluctuations destroy the long-range order of the superlattice, resulting in a finite structural coherence length and diffraction peak broadening. However, the effects turn out to be less severe in the case of discrete fluctuations, as one might expect.

Sevenhans et al. [5.472] have derived an analytical expression for the scattering intensity from an amorphous-crystalline superlattice with a continuous Gaussian distribution of amorphous layer thicknesses. They obtain the intensity:

$$I(q) = \langle A(q)A^*(q) \rangle$$

$$= f^2(q) \frac{\sin^2(Nqd/2)}{\sin^2(qd/2)} \left[M + 2 \sum_{j=1}^{M-1} (M-j) e^{-q^2 j/4c^2} \cos[qj(Nd+a)] \right], \quad (5.74)$$

where f is the scattering factor of the crystalline layers, N is the number of atomic planes in a crystalline layer, d is the plane spacing in the crystalline layer, M is the number of bilayers, a is the average thickness of the amorphous layers and c^{-1} is the width of the amorphous layer thickness distribution. The scattering factor of the amorphous layers is set to zero. When the amorphous layers which separate successive crystalline layers have thicknesses that fluctuate on the scale of the atomic plane spacing, phase coherence is lost across the crystalline layers. The high angle diffraction peaks are substantially broadened as the coherence length drops to the thickness of an individual crystalline layer. An interesting sidelight of this work is that averaging is performed on the scattering intensities and not the amplitudes. This correctly reproduces the diffuse scattering.

Clemens and *Gay* [5.473] have extended the analysis of cumulative disorder to the case of discrete fluctuations in layer thickness, using a generalized Patterson function method. As in the previous example they consider layers of an optically dense material separated by layers with zero scattering factor. Their results in the case of continuous disorder are identical to those of *Sevenhans* et al.

In a recent article *Locquet* et al. [5.474] combine features of these earlier models and derive an expression for the scattering intensity which simultaneously includes both continuous and discrete fluctuations in layer thickness. They apply their model to the crystalline-crystalline case (in which both constituents of the superlattice are crystalline). They find that discrete fluctuations can indeed broaden high-angle peaks in the limit of large amplitude fluctuations and large lattice mismatch between layers. At low angles they find that continuous fluctuations have little effect, but peak broadening can occur in the presence of discrete fluctuations of sufficient magnitude. Therefore, they recommend a fitting scheme where the low-angle profiles are fitted by adjusting the width of discrete layer thickness fluctuations, and then high-angle fitting is carried out with the continuous fluctuation width as an adjustable parameter. These observations are relevant to the Co–Au case [5.465] where the lattice mismatch is quite large ($\sim 13\%$).

5.3.4.2 Computer Simulations of the Scattering Intensity

As modeling of superlattices becomes more involved, simultaneously including structural disorder of various kinds, the derivation of closed-form expressions for the scattering intensity at some point becomes intractable. We have recently described a numerical approach to modeling the intensities obtained in a perpendicular scan [5.475]. In our model the calculation of the scattering intensity is carried out by solving Eq. 5.72 numerically. We begin with an ideal superlattice of a given chemical composition (set of f_A, f_B) and lattice parameters d_A, d_B [Fig. 5.50a]. Next we introduce disorder in the layer thicknesses [Fig. 5.50b] using a random number generator to choose N_A and N_B from a parabolic layer-thickness distribution. Diffusion effects [Fig. 5.50c] are now introduced by allowing the scattering factors and layer spacings to vary exponentially according to the distance r', nearest interface. For example, within the A layer

$$f_n = f_A + \tfrac{1}{2}(f_B - f_A)\exp(-2r'/t_d), \tag{5.75}$$

where t_d is the interfacial diffusion width. The atomic plane spacing at the AB interface is assumed to be $\tfrac{1}{2}(d_A + d_B)$.

Once the values of d_n and f_n are determined for each atomic plane the scattering amplitude is calculated using (5.72) and squared to give the scattering intensity. Lastly, this intensity is multiplied by the Lorentz and polarization factors, and instrumental broadening is introduced by convoluting with a Gaussian resolution function.

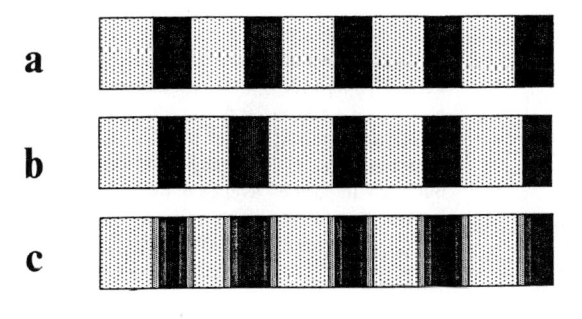

Fig. 5.50. Illustration of superlattice disorder introduced into numerical simulation. **a** Perfect superlattice **b** Layer thickness disorder **c** Simultaneous layer thickness disorder and interfacial diffusion

Layer thickness fluctuations destroy the superlattice periodicity. Higher-order superlattice peaks may appear split and/or shifted, as shown in Fig. 5.51b. During a particular execution of the program, the fine structure in the scattering intensity is sensitive to the actual sequence of numbers of atomic planes N_A^i and N_B^i that is used in the scattering calculation. N_A^i and N_B^i in turn are determined by the sequence of values returned by the random number generator. On the other hand, a real sample is composed of many local domains within the area sampled by the X-ray beam, each of which contains a different sequence of layer thicknesses. Thus the measured superlattice peaks are broadened rather than split into a set of sharp peaks. We simulate this effect in our calculation by averaging intensities over structures which are generated by different sequences of random numbers [Fig. 5.51c]. An example of the numerical simulation of Co–Au superlattices is shown in Fig. 5.52 for both low angle and high angle satellites. Typical values of t_d extracted from these fits are 1.3–1.5 monolayers. Fluctuation amplitudes for the layer thicknesses are ~ 2 monolayers and lateral coherence lengths of 0.5 to 2 μm are inferred from the widths of low-angle rocking curves. From these numbers we concluded that our Co–Au superlattices have essentially atomically abrupt interfaces. In this connection, we have found that spin echo NMR[5.476], a technique that is very sensitive to the local atomic environment, can better distinguish between step-like interfaces and interdiffusion.

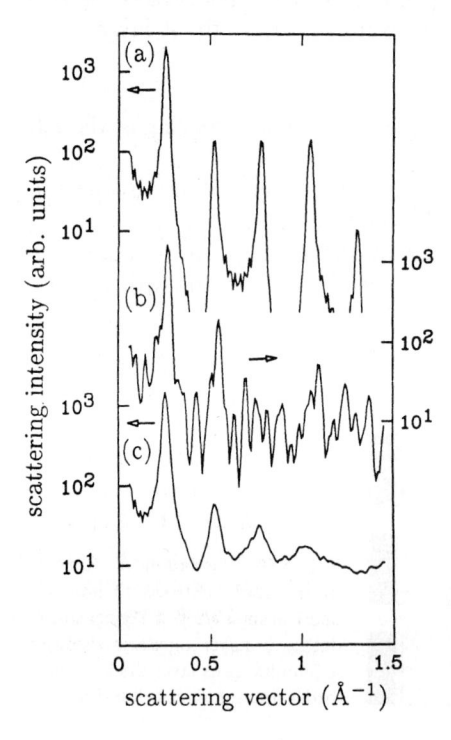

Fig. 5.51. Effect of layer thickness disorder on low-angle superlattice scattering intensities. (a) Perfect superlattice with parameters $N_A = 5$, $N_B = 7$, $d_A = d_B = 2.0$ Å, $f_A = 0$, $f_B = 1$. (b) Same parameters as (a), but with layer thickness fluctuations of ± 1 monolayer. (c) Same as (b), but with intensities averaged over 100 calculations. $N_L = 10$ for all simulations

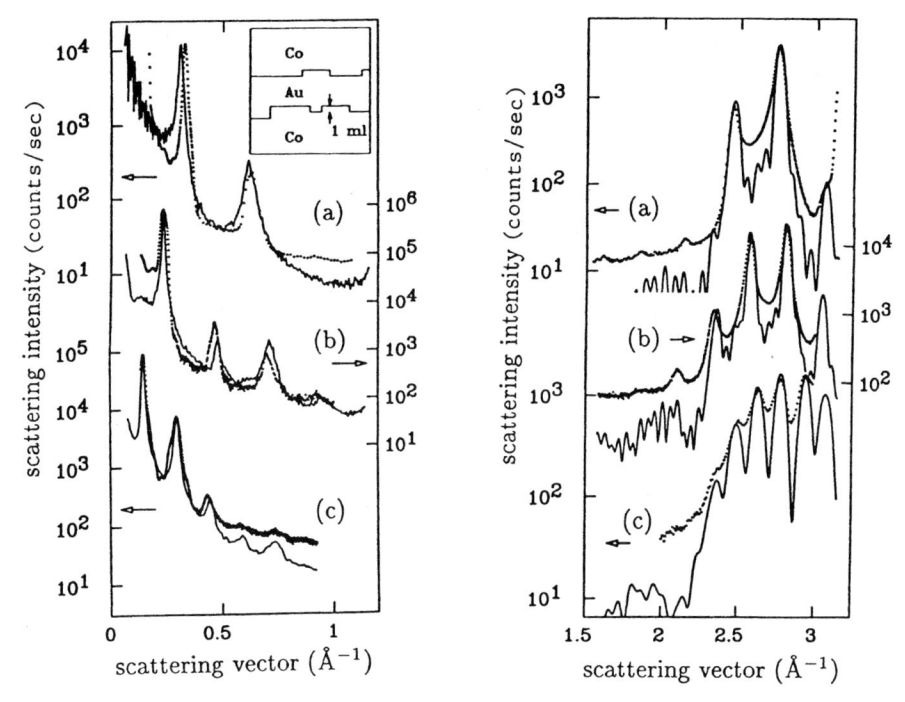

Fig. 5.52. Measured (points) and calculated (solid curves) low-angle and high-angle scattering intensities for three Co–Au superlattices. **a** 5 Å Co – 16 Å Au; **b** 10 Å Co – 16 Å Au; **c** 30 Å Co – 16 Å Au; Inset: Interface profile inferred from simulations

Small Angle Scattering. In the previous section we showed how the extended length scales associated with multilayers and superlattices give rise to diffraction features at relatively small Bragg angles. While we have discussed the interpretation of the basic features of the small angle data using a purely kinematical scattering theory, a more rigorous approach may be taken by using an optical model. In this section we consider some special considerations relevant to small angle X-ray measurements of thin film samples and their interpretation.

When the angle of incidence for X-rays, θ, is less than a few degrees, there can be significant effects due to refraction. This is most clearly manifested in the total external reflection of X-rays at a critical angle, θ_c, given by $\theta_c = \sqrt{2\delta}$, where δ is the decrement in the complex refractive index $\tilde{n} = 1 - \delta - i\beta$. Typically $\delta \sim 10^{-5}$ and $\theta_c \sim 0.2°$. As θ approaches θ_c, the intensity of X-rays scattered from the multilayer depends strongly on the optical constants of the constituent layers, and thus on the relative phases of the X-ray beams propagating in the various layers of the sample. The resulting interferences lead to a pronounced modulation of the low-angle reflectivity function from which one can extract information about the thicknesses of the layers, composition gradients, and interface roughening.

Miceli et al. [5.477] have pointed out that refraction effects can noticably shift the low-angle $0\,0\,l$ Bragg peaks away from the positions given by the simple kinematical expression $\lambda = 2d \sin \theta$ (see Fig 5.52a). The effect is to underestimate, by a few percent, the multilayer periodicity based on a simple measurement of the $0\,0\,l$ peak positions. Dynamical effects can also lead, in general, to asymmetries in the low-angle peak shapes. In order to analyze the low-angle X-ray data correctly it is necessary to include these optical effects explicitly in a calculation of the reflectivity as a function of the angle of incidence. A powerful technique has been developed for multilayers by *Barbee* et al. [5.469]. The calculation uses an iterative Fresnel approach, where the reflection coefficients are determined at each interface in the sample. For example, the reflection coefficient at the interface between layers j and $j+1$ is given by

$$R_{j,j+1} = a_j^4 \left[\frac{R_{j+1,j+2} + \mathscr{F}_{j,j+1}}{R_{j+1,j+2} \mathscr{F}_{j,j+1} + 1} \right], \tag{5.76}$$

where

$$\mathscr{F}_{j,j+1}^{\sigma} = \left[\frac{E_j^R}{E_j} \right]^{\sigma} = \frac{g_j - g_{j+1}}{g_j + g_{j+1}} \tag{5.77}$$

and

$$\mathscr{F}_{j,j+1}^{\pi} = \left[\frac{E_j^R}{E_j} \right]^{\pi} = \frac{g_j/\tilde{n}_j^2 - g_{j+1}/\tilde{n}_{j+1}^2}{g_j/\tilde{n}_j^2 + g_{j+1}/\tilde{n}_{j+1}^2} \tag{5.78}$$

are the Fresnel coefficients for reflection of the σ and π components of polarization from this interface. Here $g_j = (\tilde{n}_j^2 - \cos^2 \theta)^{1/2}$ and a_j is an amplitude factor, $e^{-i\pi/\lambda g_j d_j}$. The computation of the reflectivity starts at the substrate and works recursively back to the surface. The resulting reflectivity is related to the intensity by $I(\theta)/I_0 = R_{12}^2$.

An example of the use of low-angle X-ray reflectivity measurements for the analysis of ultrathin magnetic layers is shown in Fig. 5.53. Here we compare data on one of our MBE-grown cobalt–copper superlattices, measured using a high resolution X-ray diffractometer on a synchrotron radiation source, with reflectivity curves calculated using the optical model described above. The calculated reflectivity curve shown in Fig. 5.53a is for an ideal stack of 26 bilayers of Co (14.0 Å)/Cu (11.4 Å) grown on Ge-buffered (1 1 0) GaAs. The model includes the additional buffer layers of bcc-Co and Au required to initiate coherent growth of the superlattice in a (1 1 1) orientation; also included is a (15 Å) Au cap layer. The prominent peak at ~ 0.25 Å^{-1} corresponds to the first ($l = 1$) superlattice Bragg peak. The series of regularly spaced fringes arises from the finite thickness of the thin film superlattice. These fringes are sensitive to roughness on a scale shorter than the longitudinal coherence length of the X-ray beam (about 1000 Å in our measurements).

In order to achieve a reasonable fit to the data [Fig. 5.53b] we found it necessary to introduce *steps* at the substrate interface. The effect of the steps,

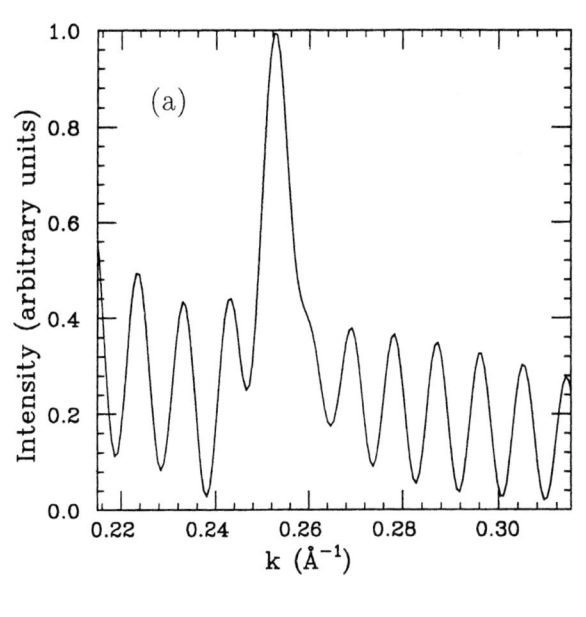

Fig. 5.53. Low-angle X-ray reflectivity from a cobalt-copper superlattice [Co (14.0 Å)–Cu (11.4 Å)]$_{26}$. **a** calculated reflectivity for ideal case with atomically abrupt interfaces. **b** comparison of measured reflectivity (circles) with calculated curve including interfaces steps

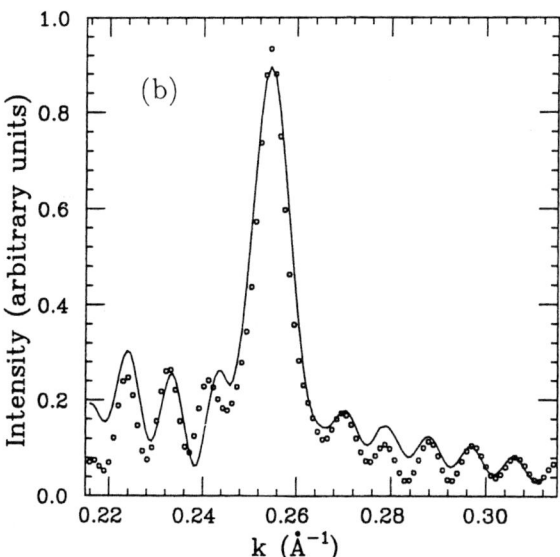

which we estimate to be ~ 2 ML in height, is to reduce the amplitude of the finite thickness fringes by introducing phase shifts between rays reflected from different terraces. The origin of the interface steps is in a small amount of miscut (typically $\sim 0.1°$) of the GaAs substrate which produces a stepped 'vicinal' surface. While the terraced interface model fits our data quite well there are significant discrepancies especially in the phase of the interference fringes immediately below the $l = 1$ Bragg peak. This may be due to anomalous scattering contributions arising from an absorption edge in the Cu layers close to the energy of the X-ray beam (~ 8 keV). These are not included in the model calculation at this point.

The measured intensity shown in Fig. 5.53b includes only the specular component of the reflectivity. A broad diffuse component, due to short-range correlations of the interface steps from layer to layer, has been subtracted from the raw data. Analysis of the diffuse scattering in the terms of interface roughness remains a significant challenge for the detailed characterization of the interfaces in multilayer samples. However, as this example illustrates, there is a great deal of useful information to be gained from the specular reflectivity.

5.3.4.3 In-Plane Scans

In order to perform in-plane measurements in transimission geometry on samples grown on GaAs the substrates are mechanically thinned down to ~ 100 μm. This is not necessary for samples grown on less absorbing substrates such as sapphire or Si. A substrate X-ray peak is used to align the sample on the diffractometer. Successive radial scans in which the scattering vector lies in the plane of the film [Fig. 5.54] were performed. After each radial scan the χ circle rotates the sample through an increment of typically 1° and the next radial in-plane scan is performed. In this way it is possible to plot out a contour map of the in-plane scattering intensity. An example is shown in Fig. 5.55.

Figure 5.55 is a contour plot of the measured in-plane scattering intensity for a Co–Au superlattice with nominal layer thicknesses of 20 Å for Co and 16 Å for Au. The presence of the Au $2\bar{2}0$ peak and the Co $11\bar{2}0$ peak shows that both Co and Au layers are oriented within the film plane. The metal peak positions relative to those of the GaAs substrate peaks confirm the epitaxial relationships given earlier by RHEED: Au $[20\bar{2}] \parallel$ GaAs $[001]$ and Co $[11\bar{2}0] \parallel$ Au $[2\bar{2}0]$. The splitting of the Co and Au peaks shows that the in-plane structure is incoherent from layer to layer. That is, even though there is sufficient coherence at the interface to orient each layer with respect to the previous one, the average

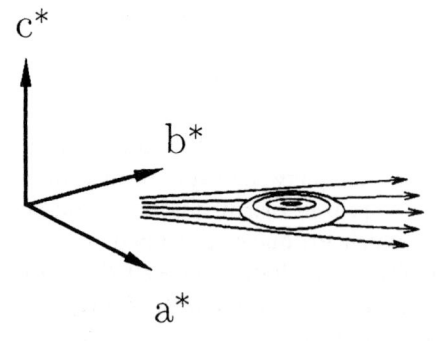

Fig. 5.54. Series of in-plane radial scans in the $a^\star b^\star$ reciprocal lattice plane, passing through a diffuse in-plane peak. Each arrow marks the trajectory of the tip of the scattering vector during a single in-plane scan

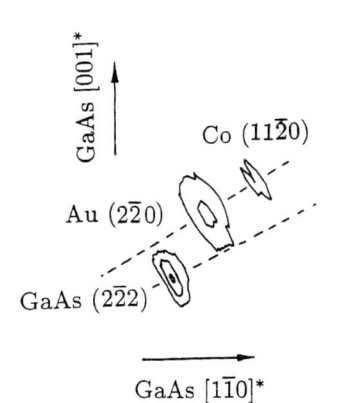

Fig. 5.55. In-plane scattering intensity contour plot for Co–Au superlattice with 20 Å Co – 16 Å Au bilayers. The three intensity contours are drawn at 50, 25 and 10 counts s^{-1}. The background intensity is 8 counts s^{-1}. The region shown lies far from the origin, which is located at the convergence of the two dashed lines

lattice parameters within the layers have relaxed towards their bulk values. We will discuss this peak splitting in greater detail below.

In contrast, if the interface lattice mismatch is not too big ($\lesssim 2\%$) the interface can have a coherent epitaxial structure. Fig. 5.56 is a scattering intensity contour plot for a sample with nominal layer thickness of 40 Å Co and 20 Å Cu. Our conclusions with respect to orientation of the layers within the film plane are identical to those in the Co–Au case. However, there is no evidence of any peak splitting at the cubic $2\bar{2}0$ position. This indicates that the entire superlattice structure occurs with the same in-plane lattice spacing. Note that for the case of the Co–Cu superlattice the mismatch is considerably smaller ($\sim 2\%$) than for Co–Au ($\sim 13\%$).

Once scattering intensity contour maps such as the ones shown in Figs. 5.55 and 5.56 have fixed the orientation of the superlattice layers we can turn to a more detailed investigation of the in-plane scattering. For example, we are interested in the strains in the superlattice layers, i.e. the shift in the measured peak positions relative to their bulk values. In the case of Co–Au superlattices, we obtain these strains from in-plane radial scans passing through the Au $2\bar{2}0$ and Co $11\bar{2}0$ peaks (scans along the upper dashed line in Fig. 5.55). Examples of such scans, in the case of superlattices with varying Co layer thicknesses, are plotted in Fig. 5.57.

If we examine the trend in peak positions in Fig. 5.57 by plotting the corresponding lattice strains (relative to the bulk) we see that the Co layer strain relaxes towards zero with increasing Co layer thickness, while the strain present in the Au layers remains essentially constant (see Fig. 5.58). We can also obtain an in-plane coherence length from the radial scans. The width of the Au $2\bar{2}0$ peak in Fig. 5.57 is nearly constant at ~ 0.14 Å^{-1}, giving a coherence length of $2\pi/0.14 = 45$ Å. This corresponds to the approximate distance between interface misfit dislocations resulting from the large mismatch between Co and Au.

The measurements of in-plane strain in ultrathin magnetic layers have been very useful in exploring a possible origin of perpendicular magnetic anisotropy.

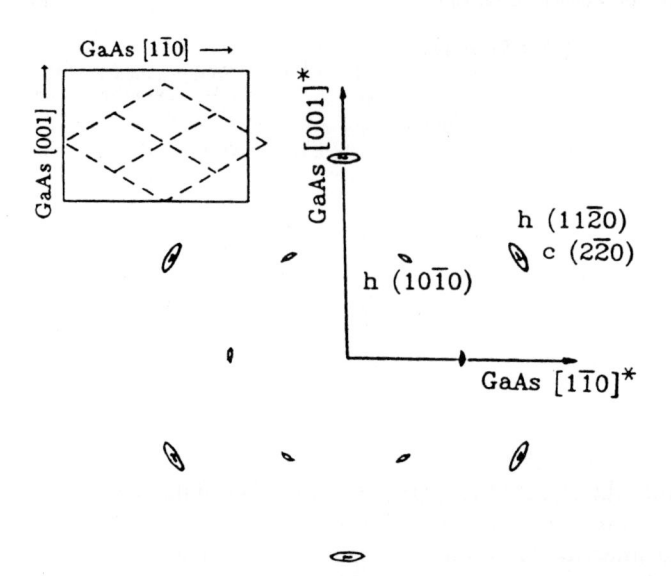

Fig. 5.56. In-plane scattering intensity contour plot of a 40 Å Co – 20 Å Cu superlattice. The two intensity contours are drawn at 130 and 65 counts s^{-1}. The background is approximately 60 counts s^{-1}. Inset: Epitaxial relationship (in direct space) between the unit cells of the metal superlattice (dashed line) and the GaAs substrate (solid lines)

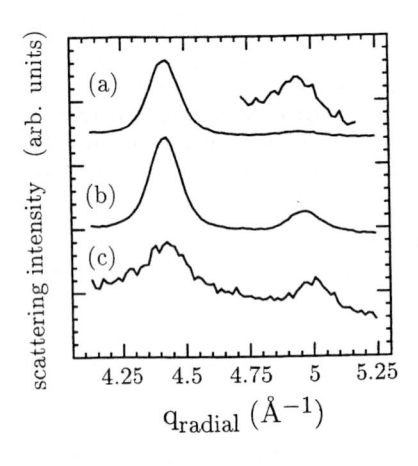

Fig. 5.57. In-plane radial scans through the Au $2\bar{2}0$ and Co $11\bar{2}0$ peaks for a series of Co–Au superlattices. The Au thickness was 16 Å in all three samples. **a** 10 Å Co layers, the scale near the Co peak is expanded to show the peak position. **b** 20 Å Co layers **c** 30 Å Co layers

The strain contributes a magnetoelastic component to the anisotropy energy which for Co can be large enough in some circumstances to counteract the shape anisotropy. If the Co is under tensile stress, as in the case of Co–Au and Co–Cu superlattices, the magnetoelastic energy favors a perpendicular spin orientation. In Fig. 5.59 we illustrate the crossover to a perpendicular easy axis which occurs in Co–Cu superlattices, possibly as a result of the build up in tensile strain when the Co layers are thinner than ~ 10 Å [5.478]. These findings illustrate a strong

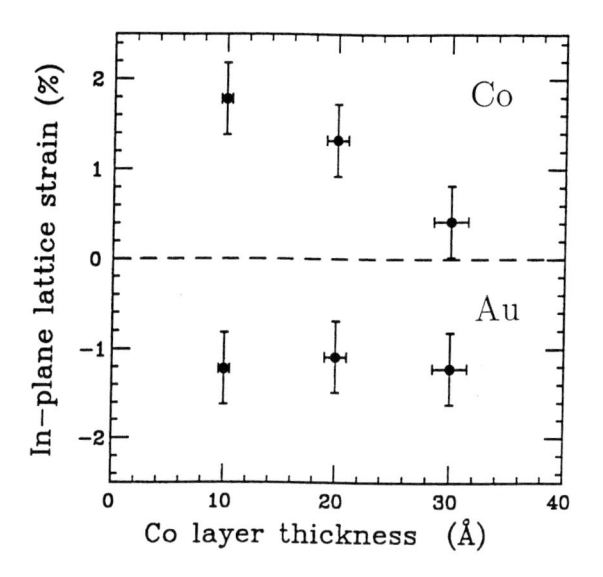

Fig. 5.58. In-plane lattice strains vs. Co layer thickness for a series of Co–Au superlattices, with a constant Au layer thickness of 16 Å. The points on this plot are derived by comparing the peak positions in Fig 5.57 with those of bulk Co $(11\bar{2}0)$ and Au $(2\bar{2}0)$ peaks. The Co layer strains are tensile (upper points) while those of the Au layers are compressive.

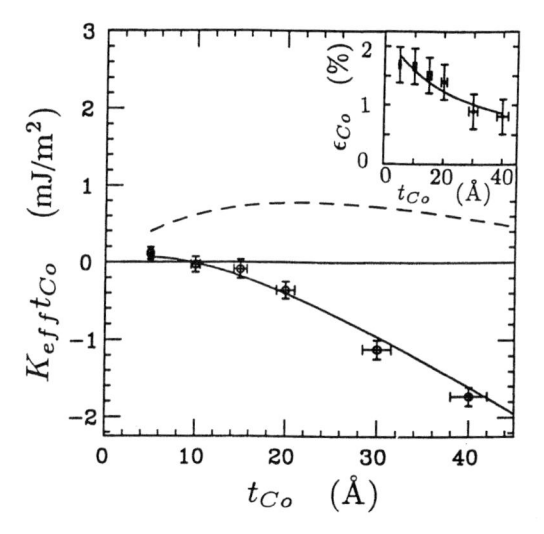

Fig. 5.59. Effective magnetic anisotropy, K_{eff}, for a series of Co–Cu superlattices in which the Co has fcc stacking. The solid curve is a theoretical fit including magnetoelastic contributions. The measured Co epitaxial strains are shown inset. The dashed curve is a calculation using magnetocrystalline anisotropy and magnetoelastic constants of hcp Co [5.478]

link between anisotropy and epitaxial structure. Theoretical studies are called for to look into this question more deeply, specifically the effects of strain on spin-orbit contributions in the 3d valence bands [5.479].

Fig. 5.59 also illustrates another interesting structural phenomenon specific to epitaxial systems: the stabilization of a metastable phase as a result of interfacial matching, in this case the fcc phase of Co. The dashed line depicts the expected anisotropy based on the normal hcp bulk structure. The magnetocrystalline anisotropy is very much suppressed in the fcc phase and the perpendicular anisotropy is considerably weakened compared to the case of hcp symmetry.

5.3.4.4 Stacking Structure

Many of the ultrathin magnetic structures of current interest are formed by the stacking of layers of close-packed metal atoms. At typical deposition rates (0.1–1.0 Å s^{-1}) and substrate temperatures (50–200 °C), the growth of such structures cannot be considered an equilibrium process. As a result, faulting of the stacking sequence is highly likely during growth. Moreover, the presence of interfaces in the structure can strongly influence the stacking sequences, for example through epitaxial strains, or *via* effects arising from the electronic band structure. That is, apart from producing faults in a superlattice, such effects during an eptaxial growth process can induce the formation of stacking sequences corresponding to metastable phases.

The simplest example of two phases which differ only in the stacking sequence is that of hcp (ABAB...) vs. fcc (ABCABC...) stacking. Using hexagonal notation, the difference between these two phases emerges in the sequence of diffraction intensities along the $\overline{hkh + kl}$ rods in reciprocal space when $h - k \neq 3m$, e.g., along $10\bar{1}l$. (Diffraction intensities where $h - k = 3m$, such as along $00l$, are unaffected by hcp-fcc transformations.) For the hexagonal case, reciprocal lattice points along $10\bar{1}l$ are separated by $2\pi/2d$; in the cubic case the separation is $2\pi/3d$, where d is the close-packed layer spacing. For either phase the signature of stacking faults is a characteristic streaking of the diffraction features along the $10\bar{1}l$ rods. (See for example, the texts of *Warren* and *Guinier* [5.480]/[5.481]). This subject has been treated extensively in the metallurgical literature concerned with stacking transitions in the bulk. In metallic superlattices, X-ray scattering studies of the layer stacking have been made in Ti–Ag [5.482], Ru–Ir [5.483] and Co–Cu [5.466, 484]. We note that local hexagonal and cubic stacking symmetries in Co layers can also be distinguished using nuclear magnetic resonance techniques, as demonstrated in several Co–Cu measurements [5.476, 485, 486].

We have analyzed diffuse X-ray scattering data from several ultrathin magnetic structures by comparing the measured intensities with those given by *Jagodzinski's* two-parameter model [5.487] for growth faulting. This model may be described by a growth rule for close-packed structures, where the probability of cubic (c) or hexagonal (h) site occupancy by a given layer depends on the previous three layers. Specifically, if the previous three layers are stacked according to the hcp sequence (e.g., ABA), the succeeding layer will occupy the fcc (C) sites with probability α and the hcp (B) sites with a probability $1 - \alpha$. That is, $P(h \to c) = \alpha$ and $P(h \to h) = 1 - \alpha$. Similarly $P(c \to c) = \beta$ and $P(c \to h) = 1 - \beta$. A perfect fcc structure corresponds to $\alpha, \beta = 1$ and an hcp structure corresponds to $\alpha, \beta = 0$.

Holloway has presented an analytical solution of the scattering intensity along $10\bar{1}l$ in the *Jagodzinski* model [5.488]. In Fig. 5.60 we plot calculated intensities generated using this solution along with the measured intensities for a Co–Cu superlattice with 10 Å Co layers. The two peaks in this plot arise from the cubic (fcc) stacking symmetry in this sample, while those corresponding to

5.3 X-Ray Studies of Ultrathin Magnetic Structures

hcp Co (which would occur at $q_c* = 0$ and $q_c* = \pm 1.54 \text{ Å}^{-1}$) are entirely absent. We find that the measured profiles correspond to $\alpha = 0.99$ and $\beta = 0.90$. This value of β implies that stacking faults in the predominantly cubic structure will occur with a 10% probability; the value of α indicates that whenever a stacking fault does occur the succeeding layer will occupy fcc sites with a probability of 99%. That is, there are almost no hcp regions in the crystal other than those associated with isolated faults. As an example of a structure corresponding to the calculated intensities of Fig. 5.60, the following is a 70-layer sequence generated using $\alpha = 0.99$ and $\beta = 0.90$:

... cchccccchccccccccccccccccccchccccccccchcccchchccccccccccccccccchcc cccc ...

The structure described by Fig. 5.60 is near the cubic limit, in that it is essentially a cubic structure containing growth faults. However, in the general case, more complicated structures may arise in a superlattice. For example, a Co–Cu superlattice in the thick-layer regime may contain a stacking sequence which looks something like

... cchccccchcccccccccccccccccccchccccchchhchhhhchchhhhhccchccccchccc ccccchc ...

In this case clusters of layers with correlated hcp stacking occur within the (thick) Co layers. In order to apply the *Jagodzinski* model to such a structure it is necessary to treat the system as separate regions (i.e. Co and Cu layers) described by different values of the parameters α and β. Given that the stacking correlations in such a structure are rather short-ranged, due to the prevalence of random faults, a valid approach to modeling the $1\,0\,\bar{1}\,l$ scattering intensity is to sum the intensities from two components with different parameters, weighted by

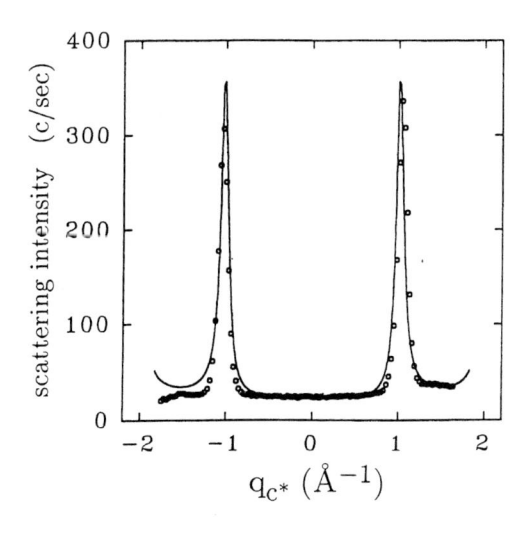

Fig. 5.60. X-ray scattering intensities along $(1\,0\,\bar{1}\,l)$ for Co–Cu superlattice with 10 Å Co layers and 20 Å Cu layers. The solid line is a fit to the model described in the text.

the thickness of the layers. That is,

$$I = t_{Co}I(\alpha_{Co}, \beta_{Co}) + t_{Cu}I(\alpha_{Cu}, \beta_{Cu})$$

Another approach is to employ analytic solutions to more sophisticated models, such as the three-parameter model of *Sebastian* and *Krishna* [5.489], which has been developed in the context of hcp-fcc phase transformations in the bulk. In this model the parameters describing the stacking sequence are: γ, the probability of random growth faults in an hcp phase: α, the probability of nucleation of fcc growth sites; and β, the probability of continued fcc growth at the nucleation sites. This model therefore lends itself well to the quantitative analysis of stacking transitions where *correlated* regions of one type of stacking (e.g., fcc) coexist with regions of another (e.g., hcp). This is often found to be the case in metallic superlattices grown in the (1 1 1) orientation.

5.3.5 Future Directions

We have emphasized how important it is to compare measured X-ray intensity profiles with those calculated from models of the structure. As modelling of artificial structures becomes more sophisticated, representing defects in a more realistic way, the need for numerical simulations will continue to expand. At present the quality of most metallic heterostructures does not warrant the use of scattering approaches beyond the kinematical approximation. As growth mechanisms become better understood and more control over the growth parameters is achieved, the crystalline perfection will improve to the point where dynamical scattering methods are necessary. While a few examples of metallic superlattices exist (e.g., Dy–Y [5.490, 491]) where the crystallographic quality approaches that of semiconductor heterostructures, the field is currently far from this ideal.

At the point where dynamical diffraction approaches become worthwhile for these systems, it should be relatively straightforward to implement techniques which are already in routine use for the analysis of epitaxial semiconductor materials. For example, application of the Tagaki–Taupin formalism [5.492] should be particularly useful because the equations describing the reflectivity from epitaxial layers incorporate non-uniform strains in a convenient way. In this connection it is worth noting that our group has recently developed new real-time dynamical X-ray scattering methods for studies of annealing in semiconductor heterostructures [5.493, 494].

In the context of epitaxial strain, current detailed work on ultrathin layers such as that described here suggests that defects intrinsic to the growth may be responsible for significant lattice distortions. In particular, there have been several recent reports [5.475, 495] indicating that the expansion of the lattice parameter in the growth direction may be a common feature of current growth techniques. The consequences for magnetic behavior have yet to be explored systematically.

In the future we will undoubtedly see much more structural work on these systems being performed at synchrotron radiation sources, especially measurements which require tunable X-ray beams (e.g., anomalous scattering performed near an absorption edge). In situ synchrotron measurements, performed in a time-resolved mode (for example during growth) will also find increasing application in studies of the interface structure.

Acknowledgements. We gratefully acknowledge the contributions of W. Vavra, S. Elagoz, D. Barlett, C.H. Lee and H. He to molecular beam epitaxy growth of samples described in this work. Thanks also to S. Elagoz for help with low-angle X-ray reflectivity analysis. This work was supported in part by ONR Grant N00014-92-J-1335. The synchrotron studies were performed at the National Synchrotron Light Source on the AT&T Bell Laboratories beam line X-16B. The NSLS is supported by the U.S. DOE, Basic Energy Sciences, Materials and Chemical Sciences under Contract No. DE-AC02-76CA0016.

References

Section 5.1

5.1 V.F. Sears: *Neutron optics* (Oxford University Press, New York, 1989) pp. 49–57. This section may be considered required reading for students of diffraction

5.2a J.M. Cowley: *Diffraction physics* (North Holland Publishing, Amsterdam, 1975) This is the basic text for graduate students

5.2b *High-Resolution Transmission Electron Microscopy and Associated Techniques* ed. by P.R. Buseck, J.M. Cowley, L. Eyring (Oxford University Press, 1988)
Chap. 3. J.M. Cowley: "Elastic Scattering of Electrons by Crystals;"
Chap. 4. J.M. Cowley: "Elastic-Scattering Theory;" and
Chap. 13. K. Yagi: "Surfaces."

5.2c P.J. Goodhew, and F.J. Humphreys: *Electron Microscopy and Analysis* (Taylor & Francis, London, 1988)

5.2d J.C.H. Spence: *Experimental High-Resolution Electron Microscopy*, (Oxford University Press, 1988). Chap. 5 on High-Resolution Images of Periodic Specimens is of particular interest. Earlier chapters give a Bibliography of books that Spence has found useful

5.2e G. Thomas, M.J. Goringe: *Transmission Electron Microscopy of Materials*: (John Wiley & Sons, New York, 1979). Applies dynamical theory to defects, nice pictures

5.3a M.A. Van Hove, W.H. Weinberg, C.-M. Chan: *LEED, Experiment, Theory and Surface Structure Determination* (Springer Berlin, Heidelberg 1986)

5.3b K. Heinz; Structural Analysis of Surfaces by LEED, Prog. in Surface Sci. 27, 239–326 (1988). With 309 references and 25 figures

5.4 *Reflection High-Energy Electron Diffraction and Reflection Electron Imaging of Surfaces*, ed. by P.K. Larsen, P.J. Dobson (Plenum Press, New York (1988)) Selected chapters are listed below:

5.4a S. Ino: "Experimental Overview of Surface Structure Determination by RHEED." Using a 40 keV and clever optics, *Ino* produces spectacular RHEED patterns that should be in an art gallery

5.4b J.L. Beeby: "Structural Determination Using RHEED." Highly recommended review

5.4c M.G. Knibb, P.A. Maksym: "Theory of RHEED by Reconstructed Surfaces." Direct comparisons of theory and experiment with the theoretical calculations treated as experiments in modeling

5.4d S.Y. Tong, T.C. Zhao, H.C. Poon: "Accurate Dynamical Theory of RHEED Rocking-curve Intensity Spectra." Calculations for Ag(0 0 1) explained and compared with experiment, demonstrating the importance of considering evanescent beams

5.4e A.L. Bleloch, A. Howie, R.H. Milne, M.G. Walls: Inelastic Scattering Effects in RHEED and Reflection Imaging." When a complex potential is used as required by inelastic scattering all waves are in some sense evanescent. Inelastic scattering probes to depths of the order of 4 nm

5.4f G. Meyer-Ehmsen: "Resonance Effects in RHEED." Direct comparison of experiments for Pt with 9 beam calculations. If the writer could have reproduced the simple delta function model given here, he would have used it as the basis of this chapter

5.4g H. Marten: "Inelastic Scattering and Secondary Electron Emission Under Resonance Conditions in RHEED from Pt(1 1 1)." The confinement of the internal beam to the near-surface region on the conditions of Bragg reflection accentuates the yield from inelastic events

5.4h J.C.H. Spence, Y. Kim: "Adatom Site Determination Using Channeling Effects in RHEED on X-ray and Auger Electron Production." The discussion of dynamical theory of channeling is illustrated by a calculated picture of the wave field at resonance

5.4i J. Gjønnes: "A Note on the Bloch Wave and Integral Formulation of RHEED theory." A 4-beam calculation is used to bring out the essential features of dynamical theory

5.4j M.G. Lagally, D.E. Savage, M.C. Tringides: "Diffraction from Disordered Surfaces: An Overview." To be read as a supplement to Sect. 2 of this chapter. At least the figures and their captions are required reading

5.4k W. Moritz: "Theory of Electron Scattering from Defect: Steps on Surfaces with Non-Equivalent Terraces." Reflection from Si steps differs from metals because alternate terraces scatter differently

5.4l M. Henzler: "Diffraction from Stepped Surfaces". Spot profile LEED is used to show that oscillations present at the out-of-phase condition are not seen in LEED for the in-phase condition. Oscillations develop after two layers of deposition

5.4m B. Bölger, P.K. Larsen, G. Meyer-Ehmsen: "RHEED and Disordered Surfaces" The interaction of Kikuchi lines and reflected intensities is considered

5.4n M. Albrecht, G. Meyer-Ehmsen, Temperature Diffuse Scattering RHEED. The phonons change the lattice on a time scale long compared to the interference of each electron with itself

5.4o J.M. Cowley: "Reflection Electron Microscopy in TEM and STEM Instruments. "The discussion of resonance and channeling is directly applicable to RHEED

5.4p R.H. Milne: "Reflection Microscopy in a Scanning Transmission Electron Microscope." Fig. 4(a) shows a RHEED pattern for the rod splitting for a copper (0 1 8) surface with regular steps. Fig. 4(b) shows the *chevron* RHEED pattern which results on faceting from the oxidation of Cu

5.4q P.R. Pukite, P.I. Cohen, S. Batra: "The Contribution of Atomic Steps to Reflection High Energy Electron Diffraction from Semiconductor Surfaces." Many of the observed effects of surface structure are interpreted within the framework of the kinematic column approximation

5.4r J. Aarts and P.K. Larsen: "RHEED studies of Growing Ge and Si surfaces." The details o Si and Ge growths are clearly given, showing the complications not seen in metals

5.4s D.E. Savage, M.G. Lagally: "Quantitative Studies of the Growth of Metals on GaAS(1 1 0) Using RHEED." RHEED is used to study systems that do not grow layer-by-layer

5.4t G. Lilienkamp, C. Koziol E. Bauer: "RHEED Intensity Oscillations in Metal Epitaxy." Oscillations on metals were discovered independently in several laboratories. This is one of the early reports

5.5 Thin Film Growth Techniques for Low-Dimensional Structures, ed. by R.F.C. Farrow, S.S.P. Parkin, P.J. Dobson, J.H. Neave A.S. Arrott (Plenum Press, 1987). Selected chapters are listed below:

References

5.5a B.A. Joyce, J.H. Neave, J. Zhang, P.J. Dobson, P. Dawson, K.J. Moore, C.T. Foxon: "Dynamic RHEED Techniques and Interface Quality in MBE-Grown GaAs/(Al,Ga)As structures"

5.5b P.I. Cohen, P.R. Pukite: "Diffraction Studies of Epitaxy: Elastic, Inelastic and Dynamic Contributions of RHEED"

5.5c P.A. Maksym: "Some Aspects of RHEED Theory"

5.5d T. Sakamoto, K. Sakamoto, S. Nagao, G. Hashiguchi, K. Kuniyoshi, Y. Bando; "RHEED Intensity Oscillations-an Effective Tool of Si and Ge_xSi_{1-x} Molecular Beam Epitaxy."

5.5e L.A. Kolodziejski, R.I. Gunshor, A.V. Nurmikko, N. Otsuka: "RHEED Intensity Oscillations and the Epitaxial Growth of Quasi-2D Magnetic Semiconductors"

5.5f G.A. Prinz: "Growth and Characterization of Magnetic Transition Metal Overlayers on GaAs Substrates."

5.6 *Kinetics of Ordering and Growth at Surfaces*, ed. by M.G. Lagally (Plenum Press, New York, 1990). Selected chapters are listed below:

5.6a M.G. Lagally, Y.-W. Mo, R. Kariotis, B.S. Swartzentruber, M.B. Webb: "Microscopic aspects of the intial stages of epitaxial growth: A scanning tunneling microscopy study of Si on Si(0 0 1)."

5.6b P.I. Cohen, G.S. Petrich, A.M. Dabiran, P.R. Pukite: "From Thermodynamics to Quantum Wires: a Review of RHEED"

5.6c T. Sakamoto, K. Sakamoto, K. Miki, H. Okumura, S. Yoshida, H. Tokumoto: "Silicon Molecular Beam Epitaxy: Si-on-Si Homoepitaxy"

5.6d D.D. Vvedensky, S. Clarke, K.J. Hugill, A.K. Meyers-Beaghton, M.R. Wilby: "Growth Kinetics on Vicinal (0 0 1) Surfaces: The solid-on-solid model of Molecular Beam Epitaxy"

5.6e A.S. Arrott, B. Heinrich, S.T. Purcell: "RHEED intensities and Oscillations During Growth of Iron on Iron Whiskers"

5.6f J.A. Venables, T. Doust, J.S. Drucker, M. Krishnamurthy: "Studies of Surface Diffusion and Crystal Growth by SEM and STEM"

5.6g J.J. de Miguel, A. Cebollada, J.M. Gallego, R. Miranda "On the Magnetic Properties of Ultrathin Epitaxial Cobalt Films and Superlattices"

5.7a *Electron Diffraction 1927–1977*, Inst. Phys. Conf. Ser. 41, ed. by P.J. Dobson, J.B. Pendry C.J. Humphreys, (Inst. Phys. London, 1978), preface by M. Blackett

5.7b H.A. Bethe, Ann. Phys. (Leipzig) **87**, 55–129 (1928)

5.7c E. Bauer: "Reflection Electron Diffraction (RED)," in *Techniques in Metals Research* Vol. II, part 2 (ed. by R.F. Bunshah (Interscience, New York, 1969) pp. 501–558

5.8a G.W. Simmons, D.E. Mitchell, K.R. Lawless: "LEED and HEED Studies of the Interaction of Oxygen with Single Crystal Surfaces of Copper," Surf. Sci. **8**, 130–164 (1967)

5.8b A.Y. Cho, J.R. Arthur, Jr.: Molecular Beam Epitaxy, in *Progress in Solid State Chemistry*, ed. by G. Somorjiai, J. MicCaldin (Pergamon Press, New York, 1975) pp. 157–191, with extensive bibliography to 1975

5.8c K. Matysik: "Diffuse Scattering from Chemisorption Induced Surface Step Distortions: Co on W (1 1 0), "Surf. Sci. **46**, 457 (1974); J.Appl. Phys. **47**, 3826 (1976)

5.8d K. Ploog, A. Fischer: "In situ Characterization of MBE Grown GaAs and $Al_xGa_{1-x}As$ Films Using RHEED, SIMS and AES Techniques," Appl. Phys. **13**, 111–121 (1977)

5.9a A.A. van Gorkum, M.R.T. Smits, P.K. Larsen, R. Raue: "High-brightness and high-resolution RHEED system," Rev. Sci. Instrum. **60**, 2940 (1989)

5.9b D. Barlett, C.W. Snyder, B.G. Orr, Roy Clarke: "CCD-based RHEED diffraction detection and analysis system," Rev. Sci. Instrum. **62**, 1263–1269 (1991)

5.9c J.S. Resh, K.D. Jamison, J. Strozier, A. Ignatiev: "Multiple reflection high-energy electron diffraction beam intensity measurement system" Rev. Sci. Instrum. **61**, 771 (1990)

5.9d C.J. Sa, H.H. Wieder: "RHEED intensity oscillation recorder for molecular-beam epitaxy systems" Rev. Sci. Instrum. **61**, 917 (1990)

5.9e B. Bölger, P.K. Larsen: "Video system for quantitative measurements of RHEED patterns", Rev. Sci. Instrum. **57**, 1363 (1986)

5.9f   J.C.H. Spence, J.M. Zue: "Large dynamic range, parallel detection system for electron Difraction and imaging," Rev. Sci. Instrum, **59**, 2102 (1989)

5.10a J.F. Menadue: "S(1 1 1) Surface Structures by Glancing Incidence High-Energy Electron Diffraction," Acta. Cryst. **A28**, 1–11 (1972)

5.10b R. Collela: "n-Beam Dynamical Diffraction of High-Energy Electrons at Glancing Incidence, General Theory and Computational Methods," Acta. Cryst. **A28**, 11–15 (1972)

5.10c R. Collela, J.F. Menadue: "Comparison of Experimental and N-Beam Calculated Intensities for Glancing Incidence High-Energy Electron Diffraction" Acta. Cryst. **A28**, 16 (1972)

5.11a J.H. Neave, B.A. Joyce, P.J. Dobson, N. Norton: "Dynamics of Film Growth of GaAs by MBE from RHEED Observations," Appl. Phys. **A31**, 1–8 (1983)

5.11b J.H. Neave, P.K. Larsen, B.A. Joyce, J.P. Goweres J.F. van der Veen: "Some observations on Ge: GaAs (0 0 1) and GaAs: Ge(0 0 1) interfaces and films," J. Vac. Sci. Technol. **B1**, 668 (1983)

5.11c B.A. Joyce, J.H. Neave, P.J. Dobson, P.K. Larsen, "Analysis of reflection high-energy electron-diffraction data from reconstructed semiconductor surfaces," Phys. Rev. B, **29**, 814 (1984)

5.11d J.H. Neave, P.J. Dobson, B.A. Joyce, Jing Zhang: "RHEED oscillations from vicinal surfaces-a new approach to surface diffusion measurements," Appl. Phys. Lett. **47** 100–3 (1985)

5.11e J. Aarts, W.M. Gerits, P.K. Larsen: "Observations on intensity oscillations in RHEED during epitaxial growth of Si(0 0 1) and Ge(0 0 1)" Appl. Phys. Lett. **48**, 931–6 (1986)

5.11f J. Zhang, J.H. Neave, P.J. Dobson, B.A. Joyce: "Effects of Diffraction Conditions and Processes on RHEED Intensity Oscillations During the MBE Growth of GaAs," Appl. Phys. **A42** 317–326 (1987). The article reproduces color enhanced intensity maps of RHEED screens.

5.11g J. Aarts, P.K. Larsen: "Monolayer and Bilayer Growth on Ge (1 1 1) and Si(1 1 1)," Surf. Sci. **188** 391 (1987)

5.11h A. Yoshinaga, M. Fahy, S. Dosanjh, J. Zhang, J.H. Neave, B.A. Joyce: " Relaxation kineitics of MBE grown GaAs(0 0 1) surfaces." Surf. Sci. Lett. **264**, L157–L161 (1992)

5.12a J.M. Van Hove, P. Pukite, P.I. Cohen, C.S. Lent: "RHEED streaks and instrument response," J. Vac. Sci. Technol, **A1** 609-13 (1983)

5.12b J.M. Van Hove, C.S. Lent, P.R.Pukite, P.I. Choen: "Damped Oscillations in RHEED during GaAs MBE" J. Vac. Sci. Technol. **B1** 741 (1983)

5.12c C.S. Lent, P.I. Cohen: "Diffraction from Stepped Surfaces I, reversible surfaces," Surf. Sci. **139**," 121–54 (1984)

5.12d C.S. Lent, P.I. Cohen: "Quantitative analysis of streaks in RHEED: GaAs and AlAs deposited on GaAs(0 0 1)," Phys. Rev. **B33**, 8329–35 (1984)

5.12e P.R. Pukite, J.M. Van Hove, P.I. Cohen: "Sensitive RHEED measurements on the local misorientation of vicinal GaAS surfaces," Appl. Phys. Lett. **44**, 456 (1984)

5.12f P.I. Cohen, P.R. Pukite, J.M. Van Hove, C.S. Lent: "RHEED studies of epitaxial growth on semiconductor surfaces," J. Vac. Sci. Technol. **A4**, 1251–7 (1986)

5.12g P.R. Pukite, C.S. Lent, P.I. Cohen: "Diffraction from Stepped Surfaces II, arbitrary terrace distribution," Surf. Sci. **161** 39 (1985)

5.12h J.M. Van Hove, P.I. Cohen: "RHEED measurements of surface diffusion during the growth of GaAs by MBE," J. Crystal Growth **81**, 13–18 (1987)

5.12i P.R. Pukite, G.S. Petrich, G.J. Whaley, P.I. Cohen: "RHEED Studies of Diffusion and Cluster Formation During Molecular Beam Epitaxy," in *Diffusion at Interfaces: Microscopic Concepts*, ed. by M. Grunze, H.J. Kreuser, J.J. Weimer (Springer Berlin, Heidelberg 1988) pp 19–36 Fourier transform of a checker board pattern of defects gives RHEED arcs

5.13a A. Ichimya: "Surface Resonance State by RHEED," Phys. Jpn, **49** 684–88 (1980)

5.13b A. Ichimya: "Many Beam Calculation of Reflection High Energy Electron Diffraction (RHEED) intensities by the Multi-Slice Method," Jpn. J. of Appl. Phys. **22**, 176–180 (1983), Correction: Jpn. J. Appl. Phys. **24**, 1365 (1985)

References

5.13c T. Kawamura, A. Ichimiya, P.A. Maksym: "Comparison of RHEED Dynamical Calculation Methods," **27** 1098–1099 (1988)

5.13d A. Ichimiya: "Bethe's Correction Method for Dynamical Calculations of RHEED Intensities from General Surfaces," Acta Cryst. **A44** 1042–44 (1988)

5.13e A. Ichimiya: "Numerical convergence of dynamical calculations of reflection high-energy electron diffraction intensities" Surf. Sci. 235 75–83 (1990)

5.13f A. Ichimiya: "Structural Analysis of Adsorption Processes on Silicon by RHEED," in *Advances in Surface and Thin Film Diffraction*, Mat. Res. Soc. Symp. Proc. Vol. 208, 3–10 (1991)

5.13g G. Lehmpfuhl, A. Ichimiya, H. Nakahara: "Interpretation of RHEED oscillations during MBE growth," Surf. Sci. Lett. **245** L159–L162 (1991)

5.14a S.Y. Tong, H.C. Poon: "Two-dimensional boundary conditions and finite-size effects in angle-resolved photoelectron emission spectroscopy, low-energy electron diffraction, and high-resolution electron-energy-loss spectroscopy, "Phys. Rev. **B37**, 2884–91 (1988) (Many of the points made are of direct importance to RHEED)

5.14b T.C. Zhao, H.C. Poon, S.Y. Tong: "Invariant-embedding R-matrix scheme for RHEED," Phys. Rev. B **38** 1172 (1988)

5.14c S.Y. Tong: "Exploring Surface Structure," Physics Today, **37**(8), 50 (1984)

5.15 J. Lekner: *Theory of Reflection* (Martinus Nijhoff Publishers, Dordrech 1987)

5.16 R. Petit: "A Tutorial Introduction," in *Electromagnetic Theory of Gratings*, (ed. by R. Petit, Springer Berlin, Heidelberg 1980) pp 1–40

5.17 C.J. Humphreys: "The scattering of fast electrons by crystals," Rept. Progr. Phys. **42**, 1825–87 (1979) (sect 2.2)

5.18 It has been suggested that the terraces should have dimensions of at least 100 atoms across to be considered singular. This is of the order of the sideways coherence of the RHEED beam (J.L. Beeby: Surf. Sci. **80**, 56 (1979)). The vicinal angle then should be less than 5 mrad.

5.19 P. Pukite: "Reflection High Energy Electron Diffraction Studies of Interface Formation," Ph.D. Thesis, University of Minneota, (1988) (Graduate students have valued this thesis highly as an introduction of RHEED.)

5.20 G.H. Rao, T. Hibma, "Determination of the distribution of steps on a surface from LEED spot profiles, Surf. Sci. **250**, 207–219 (1991). This provides an up-to-date reference list. Read the discussion first and then work backward. They show how to analyse data to deduce the mean terrace length, the width of the distribution and a measure of the skewness of the distribution.

5.21a M. Henzler: Atomic Steps on Single Crystals: Experimental Methods and Properties, "Appl. Phys. **9**, 11–17 (1976)

5.21b M. Henzler: "LEED Studies of Surface Imperfections" Appl. Surf. Sci **11/12**, 450–469 (1982)

5.22 For a discussion of the importance of the Mathieu equation in solid state physics see J.C. Slater: "Interaction of Waves in Crystals," Rev. Mod. Phys. (1959). For solutions, see G. Blanch in Handbook of Mathematical Functions, ed. by M. Abramowitz, I.A. Stegun (National Bureau of Standards, U.S. Printing Office, 1964) pp 721–759

5.23 Z. Mitura, M. Jalochowski: "RHEED intensity calculations for Au (0 0 1) ultrathin films on an Ag(0 0 1) substrate and for Au/Ag (0 0 1) superlattice," Surf. Sci. **222**, 247–258 (1989)

5.24 L.-M. Peng: "A Note on the General Bloch Wave Theory and Boundary Problems in RHEED and REM," Surf, Sci. **222**, 296–312, (1989)

5.25a N. Masud, J.B. Pendry: "Theory of RHEED," J. Phys. C. **9**, 1833–44 (1976)

5.25b P. Maksym, J.L. Beeby: "A Theory of RHEED," Surf. Sci. **110**, 423–438 (1981)

5.25c S. Holloway, J.L. Beeby: "The origins of streaked intensity distribution in RHEED ," J. Phys. C**11**, L247–L251 (1977)

5.25d S. Holloway: "Calculations of RHEED Scattering Intensities," Sur. Sci. **80**, 62–68 (1979)

5.25e P.A. Maksym, J.L. Beeby: "Calculation of MEED Intensities in the 5–10 keV Electron Energy Range," Surf. Sci. **140**, 77–84 (1984)

5.25f P.A. Maksym: "Analysis of Intensity Data for RHEED by the MgO(0 0 1) Surface," Surf. Sci. **149**, 157–174 (1985)

5.25g M.G. Knibb, P.A. Maksym: "The effect of reconstruction on RHEED intensities for the GaAs(0 0 1) 2 × 4 surface," Surf. Sci. **195**, 475–498 (1987)

5.25h M. Stock, G. Meyer-Ehmsen, "An Efficient Multiparameter Evaluation of Measured RHEED Rocking Curves Applied to Pt(1 1 1)," Surf. Sci. Lett. **226**, L59–L62 (1990)

5.26a T. Kawamura, P.A. Maksym, T. Iijima: "Calculation of RHEED intensities from Stepped Surfaces," Surf. Sci. **148** L671–6 (1984), Surf. Sci. **161** 12–24 (1985)

5.26b M.G. Knibb: "Computation of the time dependence of dynamic and kinematic RHEED intensities for growing surfaces," Surf. Sci. **257**, 389–401 (1991). An assessment of the current understanding of RHEED.

5.27a L.M. Peng, J.M. Cowley: "Dynamical calculations for RHEED and REM," Acta Cryst. **A42** 545 (1986)

5.27b J.M. Cowley: "Reflection Electron Microscopy, in Surface Interface Characterization by Electron Optical Methods," ed. by A. Howie, U. Valdre (Plenum Press, New York 1988) pp. 127–157

5.27c Z.L. Wang: "Dynamical Calculations of RHEED and REM Including the Plasmon Inelastic Scattering," Surf. Sci. **215**, 201–216, 217–231 (1988); and references therein;

5.27d Z.L. Wang: "A multislice Theory of Electron Inelastic Scattering in a solid," Acta Cryst. **A45**, 636–644 (1989)

5.27e Z.L. Wang: "Studies of surface resonance waves in RHEED," Phil. Mag. **B60**, 617–26 (1989)

5.27f Z.L. Wang, J. Liu, P. Lu, J.M. Cowley: "Electron Resonance Reflections form Perfect Crystal Surfaces and Surfaces with Steps," Ultramicroscopy **27**, 101–112 (1989)

5.28 G. Meyer-Ehmsen: "Direct Calculation of the Dynamical Reflectivity Matrix for RHEED," Surf. Sci. **219** 177–188 (1989)

5.29 P.A. Bennett, X. Tong, J.R.Butler: "Thin-film crystallography using reflection high-energy electron diffraction "rod intensity profiles": Ni/Si(1 1 1)," J. Vac. Sci. Technol. **B6**, 1336–40, (1988)

5.30a M. Doyama, R. Yamamoto, T. Kaneko, M. Imafuku, C. Kokubu, T. Izumiya, T. Hanamure: "The first observation of RHEED intensity oscillations during the growth of a metal-metal multilayered film by MBE and the electrical resistivity measurement of Mo/Al multilayered films grown by rf sputtering," Vacuum **36**, 909–911 (1986)

5.30b S.T. Purcell, B. Heinrich, A.S. Arrott: "Intensity oscillations for electron beams reflected during epitaxial growth of metals," Phys. Rev. **B35**, 6458–60 (1987)

5.30c S.T. Purcell, A.S. Arrott, B. Heinrich: "Reflection High-energy Electron Diffraction Oscillations During Growth of Metallic Overlayers on Ideal and Non-ideal Metallic Substrates," J. Vac. Sci. Technol. **B6**, 794–797 (1988)

5.30d M. Jalochowski, E. Bauer: "RHEED Intensity and Electrical Resistivity Oscillations During Metallic Film Growth," Surf. Sci. 213, 556–563 (1989)

5.31a G.S. Petrich, P.R. Pukite, A.M. Wowchak, G.J. Whaley, P.I. Cohen, A.S. Arrott, "On the origin of RHEED intensity oscillations," J. Cryst. Growth **95**, 23–27 (1989)

5.31b P.I. Cohen, G.S. Petrich, P.R. Pukite, G.J. Whaley, A.S. Arrott, "Birth-Death Models of Epitaxy," Surf. Sci. 216, 222–48 (1989)

5.32a P.J. Dobson, B.A. Joyce, J.H. Neave, J. Zhang, "Current understanding and applications of the RHEED intensity oscillation technique," J. Cryst. Growth **81** 1–8, (1987)

5.32b J. Zhang, J.H. Neave, B.A. Joyce, P.J. Dobson, P.N. Fawcett, "On the RHEED Specular Beam and its Intensity Oscillation During MBE Growth of GaAs," Surf. Sci. **231**, 379–388 (1990)

5.32c D. Kashchiev, Yu.O. Kanter, "Oscillations of Specular Beam Intensity in Reflection Diffraction from the Surface of Growing Epitaxial Film," Phys. Stat. Sol. a **110**, 61–76 (1988)

5.32d S. Clarke, M.R. Wilby, D.D. Vvedensky, "Theory of homoepitaxy on Si(0 0 1)," Surf. Sci. **255** 91–110 (1991)

5.33 Geometric patterns are formed by one electron at a time interfering with itself as it leaves the crystal starting from an atom below the surface. The pattern is essentially the same whether that electron was emited as the result of a photoelectric process, and Auger Process or the inelastic scattering of an electron from the incoming beam. In the latter case it produces the Kikuchi pattern. This is explicity shown by Z.-L. Han, S. Hardcastle, G.R. Harp, H. Li, X.-D Wang, J. Zhang and B.P. Tonner, Structural effects in single-crystal photoelectron, Auger-electron, and Kikuchi-electron angular diffraction patterns. Surf. Sci. **258**, 313–327 (1991)

5.34 Z. Celinski, B. Heinrich: "Ferromagnetic resonance linewidth of Fe ultrathin films grown on a bcc Cu substrate.," J. Appl. Phys. **70**, 5935–7 (1991)

5.35 D.E. Sanders, A.E. DePristo, "Metal/metal homo-epitaxy on fcc (0 0 1) surfaces: Is there transient mobility of adsorbed atoms?," Surf. Science **254** 341–353 (1991). They answer the question in the negative.

5.36a J.W. Evans: "Modelling of epitaxial thin-film growth on fcc(1 0 0) substrates at low temperatures," Vacuum **41**, 479–481, (1990).

5.36b J.W. Evans, D.E. Sanders, P.A. Thiel, A.E. DePristo: "Low-temperature epitaxial growth of thin metal films," Phys. Rev. B **41**, 5410 (1990)

5.36c H.C. Kang, D.K. Flynn-Sanders, P.A. Thiel, J.W. Evans: "Diffraction profile analysis for epitaxial growth on fcc(1 0 0) substrates: diffusionless models," Surf. Sci. **256**, 205–215 (1991)

5.37 J.C. Slonczewski: "Fluctuation Mechanism for Biquadratic Exchange Coupling in Magnetic Multilayers," Phys. Rev. Lett. **67**, 3172–5, (1991)

5.38 B. Heinrich, Z. Celinski, J.F. Cochran, A.S. Arrott, K. Mrytle, S.T. Purcell: Bilinear and biquadratic exchange coupling in bcc Fe/Cu/Fe trilayers: Ferromagnetic resonance and surface magneto-optical Kerr effect studies. Phys. Rev. B**47**, 5077–89, (1993)

5.39a A. Schmid: "Growth of Epitaxial Co Thin Films on Cu(1 0 0) Studied by Scanning Tunneling Microscopy," Doctoral Dissertation, Free University, Berlin (1992)

5.39b R. Wiesendanger, I.V. Shvets, D. Bürgler, G. Tarrach, H.J. Güntherodt, J.M.D. Coey, S. Gräser: "Topographic and Magnetic-Sensitive STM Study of Magnetite," Science, Dec. 1991

Section 5.2

5.40 G. Margaritondo J.E. Rowe: J. Vac. Sci. Technol. **17**, 561 (1980)
5.41 S.Y. Tong, C.H. Li: CRC Crit. Rev. Solid State Mat. Sci. **9**, 209 (1981)
5.42 S.Y. Tong, C.H. Li: in *Chemistry and Physics of Solid Surfaces*, Vol. 3, ed. by R. Vanselow, W. England (CRC Press, Boca Raton, 1982) p. 287
5.43 D.A. Shirley: CRC Crit. Rev. Solid State Mat. Sci. **10**, 373 (1982)
5.44 D.P. Woodruff: Le Vide **38**, 189 (1983)
5.45 C.S. Fadley: Prog. Surf. Sci. 16, 275 (1984)
5.46 Y. Margoninski: Contemp. Phys. **27**, 203 (1986)
5.47 C.S. Fadley: in *X-Rays in Materials Analysis: Novel Applications and Recent Developments*, ed. by T.W. Rusch, ed. by (Springer Berlin, Heidelberg 1987)
5.48 C.S. Fadley: Phys. Scr. T **17**, 39 (1987)
5.49 C.S. Fadley: in *Core Level Spectroscopy in Condensed Systems*, ed. by J. Kanamori, A. Kotani, (Springer, Berlin, Heidelberg, 1988) p. 236
5.50 G. Grenet, Y. Jugnet, S. Holmberg, H.C. Poon, T.M. Duc: Surf. Interf. Anal, 14, 367 (1989).
5.51 M. Sagurton, E.L. Bullock, C.S. Fadley: Surf. Sci. **182**, 287 (1987)
5.52 C.S. Fadley: In *Synchrotron Radiation Research: Advances in Surface Science* ch. 11, ed. by R.Z. Bachrach (Plenum, New York, 1990)
5.53 S.A. Chambers: Adv. Phys. **40**, 357 (1991)
5.54 W.F. Egelhoff: Jr., CRC Crit. Rev. Solid State & Mat. Sci. **16**, 213 (1990)
5.55 J.J. Barton: Phys. Rev. Lett. **61**, 1356 (1988)
5.56 G.R. Harp, D.K. Saldin, B.P. Tonner: Phys. Rev. Lett. **65**, 1012 (1990)
5.57 G.R. Harp, D.K. Saldin, B.P. Tonner: Phys. Rev. B **42**, 9199 (1990)
5.58 C.M. Wei, T.C. Zhao, S.Y. Tong; Phys. Rev. Lett. **65**, 2278 (1990)
5.59 C.M. Wei, T.C. Zhao, S.Y. Tong; Phys. Rev. B **43**, 6354 (1990)
5.60 B.P. Tonner, Z.L. Han, G.R. Harp, D.K. Saldin: Phys. Rev. B **43**, 14423 (1991)

5.61   B.P. Tonner: Sync. Rad. News **4**, 27 (1991)
5.62   Z.-L. Han, S. Hardcastle, G.R. Harp, Hong Li, X.-D. Wang, J. Zhang, B.P. Tonner: Surf. Sci. **258**, 313 (1991)
5.63   D.K. Saldin, G.R. Harp, B.L. Chen, B.P. Tonner: Phys. Rev. B **44**, 2480 (1991)
5.64   S.Y. Tong, C.M. Wei, T.C. Zhao, H. Huang, Hua Li: Phys. Rev. Lett. **66**, 60 (1991)
5.65   S. Hardcastle, Z.-L. Han, G.R. Harp, J. Zhang, B.L. Chen, D.K. Saldin, B.P. Tonner: Surf. Sci. Lett. **245**, L190 (1991)
5.66   S. Thevuthasen, G.S. Herman, A.P. Kaduwela, R.S. Saiki, Y.J. Kim, C.S. Fadley: Phys. Rev. Lett. **67**, 469 (1991)
5.67   J.J. Barton, J. Electron Spectrosc. **51**, 37 (1990)
5.68   J.J. Barton, L.J. Terminello: In *Structure of Surfaces III*, ed. by S.Y. Tong, M.A. Van Hove, X. Xide, and K. Takayanagi, (Springer, Berlin, Heidelberg 1991) p. 107
5.69   S.Y. Tong, H. Huang, Hua Li: M.R.S. Symp. Proc. **208**, 13 (1991)
5.70   H. Huang, Hua Li, S.Y. Tong: Phys. Rev. B **44**, 3240 (1991)
5.71   J. Osterwalder, T. Greber, A. Stuck, L. Schlapbach: Phys. Rev. B **44**, 13768 (1991)
5.72   L.H. Germer, C.C. Chang: Surf. Sci. **4**, 498 (1966)
5.73   G.S. Herman, S. Thevuthasen, T.T. Tran, Y.J. Kim, C.S. Fadley: Phys. Rev. Lett. **68**. 650 (1992)
5.74   J.J. Barton: Phys. Rev. Lett. **67**, 3106 (1991)
5.75   D.K. Saldin, G.R. Harp, B.P. Tonner: Phys. Rev. B **45**, 9629 (1992)
5.76   P. Hu, D.A. King: Nature **353**, 831 (1992)
5.77   A. Stuck, K. Naumović, H.A. Aebischer, T. Greber, J. Osterwalder, L. Schlapbach: Surf. Sci. **264**, 380 (1992)
5.78   G.R. Harp, D.K. Saldin, X. Chen, Z.-L. Han, B.P. Tonner: J. Elect. Spect. Rel. Phenom. **57**, 331 (1991)
5.79   K. Siegbahn, U. Gelius, H. Siegbahn, E. Olson: Phys. Lett. **32A**, 221 (1970)
5.80   K. Siegbahn, U. Gelius, H. Siegbahn, E. Olson: Phys. Scr. **1**, 272 (1970)
5.81   Photographs of Kikuchi bands may be found in almost any textbook on electron microscopy.
5.82   D.M. Zehner, J.R. Noonan, L.H. Jenkins: Phys. Lett. **62A**, 267 (1977)
5.83   R.J. Baird, C.S. Fadley, L.F. Wagner: Phys. Rev. **B15**, 666 (1977)
5.84   C.S. Fadley, S.Å.L. Bergström: In *Electron Spectroscopy*, ed. by D.A. Shirley, (North-Holland, Amsterdam, 1972) 233
5.85   C.S. Fadley, R.J. Baird, W. Siekhaus, T. Novakov, S.Å.L. Bergström, J.: Elect. Spect. Rel. Phenom. **4**, 93 (1974).
5.86   T.A. Carlson, *Photoelectron and Auger Spectroscopy*, (Plenum, New York, 1975) p. 266
5.87   J.M. Hill, D.G. Royce, C.S. Fadley, L.F. Wagner, F.J. Grunthaner: Chem. Phys. Lett. **44**, 225 (1976)
5.88   C.S. Fadley: Prog. Solid State Chem. **11**, 265 (1976)
5.89   N.E. Erickson: Phys. Scr. 16, 462 (1977)
5.90   C.C. Chang: Appl. Phys. Lett. 31, 304 (1977)
5.91   C.S. Fadley: in *Electron Spectroscopy*, ed. by C.S. Brundle, A.D. Baker, Vol. 2, (Academic, London, 1977) p. 132
5.92   T.A. Carlson: *X-Ray Photoelectron Spectroscopy* (Dowden, Hutchinson, and Ross, Stroudsburg, PA, 1978) pp. 184, 202
5.93   M. Kudo, M. Owari, Y. Nihei, Y. Gshshi, H. Kamada: Jpn. J. Appl. Phys. **17**, Suppl. 17-2, 275 (1978)
5.94   M. Aono, C. Oshima, T. Tanaka, E. Bannai, S. Kawai: J. Appl. Phys. **49**, 2761 (1978)
5.95   D. Briggs, R.A. Marbrow, R.M. Lambert: Solid State Comm. **26**, 1 (1978)
5.96   T. Matsudiara, M. Onchi: Surf. Sci. 74, 684 (1978)
5.97   V.V. Nemoshkalenko, V.G. Aleshin: *Electron Spectroscopy of Crystals*, (Plenum, New York, 1979), pp. 311–312
5.98   H. Hilferink, E. Lang, K. Heinz: Surf. Sci. **93**, 398 (1980)
5.99   N. Koshizaki, M. Kudo, M. Owari, Y. Nihei, H. Kamada: Jpn. J. Appl. Phys. **19**, L349 (1980)

5.100   S. Kono, S.M. Goldberg, N.F.T. Hall, C.S. Fadley: Phys. Rev. **B 22**, 6085 (1980)
5.101   S.M. Goldberg, R.J. Baird, S. Kono, N.F.T. Hall, C.S. Fadley: J. Elect. Spectr. Rel. Phenom. **21**, 1 (1980)
5.102   R. Weissmann, K. Müller: Surf. Sci. Rep. 1,251 (1981)
5.103   Y. Nihei, M. Owari, M. Kudo, H. Kamada: Jpn. J. Appl. Phys. **20**, L420 (1981)
5.104   M. Owari, M. Kudo, Y. Nihei, H. Kamada: J. Elect. Spectr. Rel. Phenom. 22, 131 (1981)
5.105   D. Briggs, J.C. Rivière: in *Practical Surface Analysis* ed. by D. Briggs, M.P. Seah, (Wiley, Chichester, England, 1983) p. 136
5.106   M. Schärli, H.P. Waldvogel: J. Phys. C **17**, 2427 (1984)
5.107   H.P. Waldvogel, M. Schärli: J. Elect. Spectr. Rel. Phenom. **34**, 115 (1984)
5.108   M. Thompson, M.D. Baker, A. Christie, J.F. Tyson: *Auger Electron Spectroscopy*, (Wiley, New York, 1985) p. 280
5.109   M. Owari, M. Kudo, Y. Nihei, H. Kamada: Jpn. J. Appl. Phys. **24**, L394 (1985)
5.110   D.P. Woodruff, T.A. Delchar: *Modern Techniques in Surface Science*, (Cambridge Univ. Press, New York, 1986) p. 118
5.111   W.F. Egelhoff, Jr: Phys. Rev. B **30**, 1052 (1984)
5.112   W.F. Egelhoff, Jr: J. Vac. Sci. Technol. A **2**, 350 (1984)
5.113   W.F. Egelhoff, Jr: J. Vac. Sci. Technol. A **3**, 1511 (1985)
5.114   W.F. Egelhoff, Jr: Mat. Res. Soc. Symp. Proc. **37**, 443 (1985)
5.115   W.F. Egelhoff, Jr: in *Structure of Surfaces*, ed. by M.A. Van Hove, S.Y. Tong, (Springer, Berlin, Heidelberg 1985) p. 199
5.116   R.A. Armstrong, W.F. Egelhoff, Jr: Surf. Sci. **154**, L225 (1985)
5.117   W.F. Egelhoff, Jr: J. Vac. Sci. Tecnhol. A **4**, 758 (1986)
5.118   W.F. Egelhoff, Jr: J. Vac. Sci. Tecnhol. A **5**, 1684 (1987)
5.119   W.F. Egelhoff, Jr: Mat. Res. Soc. Symp. Proc. **83**, 189 (1987)
5.120   W.F. Egelhoff, Jr: Phys. Rev. Lett. **59**, 559 (1987)
5.121   The use of XPS diffraction data as a fingerprint technique for determining crystal structure was developed earlier without resorting to a physical interpretation of the data. See Refs. [5.122–124]
5.122   S. Evans, J.M. Adams, J.M. Thomas: Phil. Trans. Roy. Soc. London **292**, 563 (1979)
5.123   S. Evans, E. Raftery, J.M. Thomas: Surf. Sci. 89, 64 (1979)
5.124   S. Evans, M.D. Scott: Surf. Interf. Anal. **3**, 269 (1981)
5.125   For early suggestions that forward scattering can play a role in the angular distribution of XPS peak intensities from crystal lattices, see [5.100, 126, 127]
5.126   S. Takahashi, S. Kono, H. Sakurai, T. Sagawa: J. Phys. Soc. Jpn. 51, 3296 (1982)
5.127   S.A. Chambers, L.W. Swanson, Surf. Sci. **131**, 385 (1983)
5.128   R. Trehan, J. Osterwalder, C.S. Fadley: J. Elect. Spectr. Rel. Phenom. 42, 187 (1987)
5.129   L.-G. Petersson, S. Kono, N.F.T. Hall, C.S. Fadley, Phys. Rev. Lett. **42**, 1545 (1979)
5.130   C.S. Fadley, S. Kono, L.-G. Petersson, S.M. Goldberg, N.F.T. Hall, J.T. Lloyd, Z. Hussain: Surf. Sci. **89**, 52 (1979)
5.131   L.-G. Petersson, S. Kono, N.F.T. Hall, S.M. Goldberg, J.T. Lloyd, C.S. Fadley, J.B. Pendry: Mat. Sci. Eng. **42**, 111 (1980)
5.132   S. Kono, C.S. Fadley: Nucl. Inst. Meth. **177**, 207 (1980)
5.133   P.J. Orders, S. Kono, C.S. Fadley, R. Trehan, J.T. Lloyd: Surf. Sci. **119**, 371 (1982)
5.134   S. Kono, H. Sakurai, K. Higashıyama, T. Sagawa: Surf. Sci. 130, L299 (1983)
5.135   W.F. Egelhoff, Jr: Surf. Sci. 141, L324 (1984)
5.136   B. Sinković, P.J. Orders, C.S. Fadley, R. Trehan, Z. Hussain, J. Lecante: Phys.Rev. **B30**, 1833 (1984)
5.137   P.J. Orders, B. Sinković, C.S. Fadley, R. Trehan, Z. Hussain, J. Lecante: Phys. Rev. **B30**, 1838 (1984)
5.138   K.A. Thompson, C.S. Fadley, Surf. Sci. 146, 281 (1984)
5.139   K.A. Thompson, C.S. Fadley: J. Elect. Spectr. Rel. Phenom. **33**, 29 (1984)
5.140   K.C. Prince, E. Holub-Krappe, K. Horn, D.P. Woodruff: Phys. Rev. B **32**, 4249 (1985)
5.141   E. Holub-Krappe, K.C. Prince, K. Horn, D.P. Woodruff: Surf. Sci. **173**, 176 (1986)

5.142   A. Chassé, P. Rennert: Phys. Stat. Sol. (b) **138**, 53 (1986)
5.143   D.A. Wesner, F.P. Coenen, H.P. Bonzel: Phys. Rev. **B33**, 8837 (1986)
5.144   D.A. Wesner, G. Pirug, F.P. Coenen, H.P. Bonzel, Surf. Sci. **178**, 608 (1986)
5.145   D.A. Wesner, F.P. Coenen, H.P. Bonzel: J. Vac. Sci. Technol. A **5**, 927 (1987)
5.146   H.P. Bonzel, G. Pirug, J.E. Müller; Phys. Rev. Lett. **58**, 2138 (1987)
5.147   D.A. Wesner, F.P. Coenen, H.P. Bonzel: Phys. Rev. Lett. **60**, 1045 (1988)
5.148   D.A. Wesner, F.P. Coenen, H.P. Bonzel: Surf. Sci. **199**, L419 (1988)
5.149   D.A. Wesner, F.P. Coenen, H.P. Bonzel: Phys. Rev. B **39**, 10770 (1989)
5.150   D.A. Wesner: Vacuum **41**, 85 (1990)
5.151   R.S. Saiki, G.S. Herman, M. Yamada, J. Osterwalder, C.S. Fadley: Phys. Rev. Lett. **63**, 283 (1989)
5.152   A. Nilsson, H. Tillborg, N. Mårtensson: Phys. Rev. Lett. **67**, 1015 (1991)
5.153   J.B. Pendry: J. Phys. **C8**, 2413 (1975)
5.154   R.N. Lindsay, C.G. Kinniburgh, J.B. Pendry: J. Elect. Spectr. Rel. Phenom. **15**, 157 (1979)
5.155   H.C. Poon, S.Y. Tong: Phys. Rev. **B30**, 6211 (1984)
5.156   M. Owari, M. Kudo, Y. Nihei, H. Kamada: J. Elect. Spectr. Rel. Phenom. **34**, 215 (1984)
5.157   E.L. Bullock, C.S. Fadley: Phys. Rev. B **31**, 1212 (1985)
5.158   S.Y. Tong, H.C. Poon, D.R. Snider: Phys. Rev. B **32**, 2096 (1985)
5.159   S.A. Chambers, S.B. Anderson, J.H. Weaver: Phys. Rev. B **32**, 581 (1985)
5.160   S.A. Chambers, S.B. Anderson, J.H. Weaver: Phys. Rev. B **32**, 4872 (1985)
5.161   K. Tamura, M. Owari, M. Kudo, Y.Nihei: Bull. Chem. Soc. Jpn. **58**, 1873 (1985)
5.162   H.C. Poon, D.R. Snider, S.Y. Tong: Phys. Rev. B **33**, 2198 (1986)
5.163   S.Y. Tong, M.W. Puga, H.C. Poon, M.L. Xu: in *Chemistry and Physics of Solid Surfaces VI*, ed. by R. Vanselow, R. Howe, (Springer, Berlin, Heidelberg, 1986) p. 509
5.164   S.A. Chambers, H.W. Chen, S.B. Anderson, J.H. Weaver: Phys. Rev. B **34**, 3055 (1986)
5.165   D. Jousset, J.P. Langeron: J. Vac. Sci. Technol. A **5**, 989 (1987)
5.166   M. Sagurton, E.L. Bullock, R. Saiki, A. Kaduwela, C.R. Brundle, C.S. Fadley, J.J. Rehr: Phys. Rev. B **33**, 2207 (1986)
5.167   J. Osterwalder, E.A. Stewart, D. Cyr, C.S. Fadley, I. Mustre de Leon, J.J. Rehr: Phys. Rev. **B35**, 9859 (1987)
5.168   A. Chassé, P. Rennert: Phys. Stat. Sol. (b) **141**, K147 (1987)
5.169   P. Rennert, A. Chassé: Expt. Tech. Phys. **35**, 27 (1987)
5.170   S.A. Chambers, I.M. Vitomirov, J.H. Weaver: Phys. Rev. B **36**, 3007 (1987)
5.171   S.A. Chambers, I.M. Vitomirov, S.B. Anderson, H.W. Chen, T.J. Wagener, J.H. Weaver: Superlat. Microst. **3**, 563 (1987)
5.172   Y. Jugnet, G. Grenet, N.S. Prakash, Tran Minh Duc, H.C. Poon: Phys. Rev. B **38**, 5281 (1988)
5.173   M.L. Xu, J.J. Barton, M.A. Van Hove: J. Vac. Sci. Technol. **A6**, 2093 (1988)
5.174   J.J. Barton, M.L. Xu, M.A. Van Hove: Phys. Rev. B **37**, 10475 (1988)
5.175   M.L. Xu, M.A. Van Hove: Surf. Sci. 207, 215 (1989)
5.176   H. Cronacher, K. Heinz, K. Müller, M.-L. Xu, and M.A. Van Hove, Surf. Sci. **209**, 387 (1989)
5.177   M.L. Xu, J.J. Barton, M.A. Van Hove: Phys. Rev. B **39**, 8275 (1989)
5.178   D.E. Parry: J. Elect. Spectr. Rel. Phenom. **49**, 23 (1989)
5.179   R.N. Lindsay, C.G. Kinniburgh, Surf. Sci. 63, 162 (1977)
5.180   D.J. Friedman, C.S. Fadley: J. Elect. Spect. Rel. Phenom **51**, 689 (1990)
5.181   J. Osterwalder, A. Stuck, D.J. Friedman, A. Kaduwela, C.S. Fadley, J. Mustre de Leon, J.J. Rher: Phys. Scr. 41, 990 (1990)
5.182   T. Fujikawa, M. Hosoya: J. Phys. Soc. Jpn. **59**, 3750 (1990)
5.183   H.A. Aebischer, T. Greber, J. Osterwalder, A.P. Kaduwela, D.J. Friedman, G.S. Herman, C.S. Fadley: Surf. Sci. **239**, 261 (1990)
5.184   J. Osterwalder, T. Greber, S. Hüfner, L. Schlapbach: Phys. Rev. **B41**, 12495 (1990)
5.185   H.P. Bonzel, U. Breuer, O. Knauff: Surf. Sci. 237, L398 (1990)
5.186   S. Hüfner, J. Osterwalder, T. Greber, L. Schlapbach: Phys. Rev. **42**, 7350 (1990)
5.187   G.S. Herman, C.S. Fadley: Phys. Rev. B **43**, 6792 (1991)

5.188 T. Greber, J. Osterwalder, S. Hüfner, L. Schlapbach: Phys. Rev. **44**, 8958 (1991)
5.189 G.S. Herman, A.P. Kaduwela, T.T. Tran, Y.J. Kim, S. Lewis, C.S. Fadley: In [5.68] p. 85
5.190 J. Osterwalder, T. Greber, S. Hüfner, H.A. Aebischer, L. Schlapbach: In [5.68] P. 91
5.191 A.P. Kaduwela, D.J. Friedman, Y.J. Kim, T.T. Tran, G.S. Herman, C.S. Fadley, J.J. Rehr, J. Osterwalder, H.A. Aebischer, A. Stuck: In [5.68] p. 77
5.192 X.-D. Wang, Y.Chen, Z.-L. Han, S.Y. Tong, B.P. Tonner: In [5.68] p.65
5.193 A.P. Kaduwela, D.J. Friedman, and C.S. Fadley: J. Elect. Spect. Rel. Phen. **57**, 223 (1991)
5.194 R.D. Rydgren, H.F. Helbig, C.A. Moyer: Surf. Sci. **244**, 81 (1991)
5.195 S.A. Chambers, T.R. Greenlee, C.P. Smith, J.H. Weaver: Phys. Rev. B **32**, 4245 (1985)
5.196 B. Sinković, B. Hermsmeier, C.S. Fadley: Phys. Rev. Lett. **55**, 1227 (1985)
5.197 B. Sinković, C.S. Fadley: Phys. Rev. B **31**, 4665 (1985)
5.198 B. Hermsmeier, B. Sinković, J. Osterwalder, C.S. Fadley: J. Vac. Sci. Technol. A**5**, 1082 (1986)
5.199 B. Sinkovic, B. Hermsmeier, C.S. Fadley: Magn. Magn. Mater. **54–57**, 975 (1986)
5.200 S.A. Chambers, H.W. Chen, I.M. Vitomirov, S.B. Anderson, J.H. Weaver: Phys. Rev. B **33**, 8810 (1986)
5.201 S.A. Chambers, F. Xu, H.W. Chen, I.M. Vitomirov, S.B. Anderson, J.H. Weaver: Phys. Rev. B **34**, 6605 (1986)
5.202 S.A. Chambers, S.B. Anderson, H.W. Chen, J.H. Weaver: Phys. Rev. B **34**, 913 (1986)
5.203 F. Xu, J.J. Joyce, M.W. Ruckman, H.W. Chen, F. Boscherini, D.M. Hill, S.A. Chambers, J.H. Weaver: Phys. Rev. B **35**, 2375 (1987)
5.204 S.A. Chambers, S.B. Anderson, H.W. Chen, J.H. Weaver: Phys. Rev. B 35, 2592 (1987)
5.205 S.A. Chambers, I.M. Vitomirov, S.B. Anderson, J.H. Weaver: Phys. Rev. B **35**, 2490 (1987)
5.206 S.A. Chambers, T.J. Wagener, J.H. Weaver: Phys. Rev. B **36**, 8992 (1987)
5.207 P. Alnot, J. Olivier, F. Wyczisk, C.S. Fadley: J. Elect. Spectr. Rel. Phenom. **43**, 263 (1987)
5.208 R. Saiki, A. Kaduwela, J. Osterwalder, M. Sagurton, C.S. Fadley, C.R. Brundle: J. Vac. Sci. Technol. A **5**, 932 (1987)
5.209 J. Osterwalder, E. Stewart, R. Saiki, D. Cyr, C.S. Fadley: J. Vac. Sci. Technol. A **5**, 661 (1987)
5.210 R.W. Judd, M.A. Reichelt, E.G. Scott, R.M. Lambert: Surf. Sci. **185**, 529 (1987)
5.211 Y. Jugnet, N.S. Prakash, G. Grenet, Tran Minh Duc, H.C. Poon: Surf. Sci. **189/190**, 649 (1987)
5,212 V. Fritzsche, P. Rennert: Phys. Stat. Sol. (b) **135**, 49 (1986)
5.213 V. Fritzsche, P. Rennert: Phys. Stat. Sol. (b) **142**, 15 (1987)
5.214 V. Fritzsche: Phys. Stat. Sol. (b) **147**, 485 (1988)
5.215 V. Fritzsche: Surf. Sci. **213**, 648 (1989)
5.216 V. Fritzsche: J. Phys. Cond. Mat. **2**, 9735 (1990)
5.217 V. Fritzsche, A. Chassé, A. Mróz: Surf. Sci. 231, 59 (1990)
5.218 T. Jing-Chang: In *Surface Physics*, ed. by X. Xide, (World Scientific, Singapore, 1987) p. 466
5.219 S.A. Chambers, T.J. Irwin: Phys. Rev. B **38**, 7484 (1988)
5.220 S.A. Chambers, T.J. Irwin: Phys. Rev. B **38**, 7858 (1988)
5.221 S.A. Chambers, H.W. Chen, T.J. Wagener, J.H. Weaver: J. Vac. Sci. Technol. A **6**, 1994 (1988)
5.222 T. Abukawa, C.Y. Park, S. Kono: Surf. Sci. **201**, L513 (1988)
5.223 Hong Li, B.P. Tonner, Phys. Rev. B **37**, 3959 (1988)
5.224 W.F. Egelhoff, Jr: J. Vac. Sci. Technol. A **6**, 730 (1988)
5.225 W.F. Egelhoff, Jr., D.A. Steigerwald, J.E. Rowe, T.D. Bussing: J. Vac. Sci. Technol. A **6**, 1495 (1988)
5.226 D.A. Steigerwald, W.F. Egelhoff, Jr: J. Vac. Sci. Technol. A **6**, 1995 (1988)
5.227 D.A. Steigerwald, I. Jacob, W.F. Egelhoff, Jr: Surf. Sci. **202**, 472 (1988)
5.228 D.A. Steigerwald, W.F. Egelhoff, Jr: Phys. Rev. Lett. **60**, 2558 (1988)
5.229 A.D. Johnson, J.A.C. Bland, C. Norris, H. Lauter: J. Phys. C **21**, L899 (1988)
5.230 J. Kawai, K. Tamura, M. Owari, Y. Nihei: In 6*th Int. Conf. Surf. Coll. Sci.*, Chem. Soc. Jpn., 1988, p. 475.

5.231 A.D. Johnson: in *Physics, Fabrication, and Applications of Multilayers*, ed. by P. Dhez and C. Weisbuch., (Plenum, New York, 1988) p. 343
5.232 T. Abukawa, S. Kono, Phys. Rev. B **37**, 9097 (1988)
5.233 K. Tamura, U. Bardi, M. Owari, Y. Nihei: In *The Structure of Surfaces II*, ed. by J.F. van der Veen, M.A. Van Hove (Springer, Berlin, Heidelberg 1988) p. 404
5.234 S.A. Chambers: J. Vac. Sci. Technol. A **7**, 2461 (1989)
5.235 W.F. Egelhoff, Jr: J. Vac. Sci. Technol. A **7**, 2060 (1989)
5.236 W.F. Egelhoff, Jr., D.A. Steigerwald: J. Vac. Sci. Technol. A **7**, 2167 (1989)
5.237 A. Vuoristo, M. Valden, C.J. Barnes: to be published.
5.238 J. Olivier, P. Alnot: Semicond. Sci. Technol. **4**, 63 (1989)
5.239 P. Alnot, J. Olivier, C.S. Fadley: J. Elect. Spectr. Rel. Phenom., in press
5.240 P. Alnot, J. Olivier, F. Wyczisk, R. Joubard: J. Electrochem. Soc., in press
5.241 Y.U. Idzerda, D.M. Lind, G.A. Prinz: J. Vac. Sci. Technol. A **7**, 1341 (1989)
5.242 S.A. Chambers: Phys. Rev. B **39**, 12664 (1989)
5.243 S.A. Chambers: J. Vac. Sci. Technol. B **7**, 737 (1989)
5.244 T. Abukawa, S. Kono: Surf. Sci. **214**, 141 (1989)
5.245 R. Saiki, A. Kaduwela, J. Osterwalder, C.S. Fadley, C.R. Brundle: Phys. Rev. B **40**, 1586 (1989)
5.246 K. Tamura, U. Bardi, Y. Nihei: Surf. Sci. **216**, 209 (1989)
5.247 G. Lilienkamp, C. Koziol, E. Bauer: Surf. Sci. **226**, 358 (1990)
5.248 E.L. Bullock, G.S. Herman, M. Yamada, D.J. Friedman, C.S. Fadley: Phys. Rev. B **41**, 1703 (1990)
5.249 A. Stuck, J. Osterwalder, S. Hüfner, L. Schlapbach: Phys. Rev. Lett. **65**, 3029 (1990)
5.250 S.A. Chambers, V.A. Loebs: Phys. Rev. B **42**, 5109 (1990)
5.251 W.F. Egelhoff, Jr., I. Jacob, J.M. Rudd, J.F. Cochran, B. Heinrich: J. Vac. Sci. Technol. A **8**, 1582 (1990)
5.252 S.A. Chambers: Phys. Rev. B **42**, 10865 (1990)
5.253 K. Higashiyama, C.Y. Park, S. Kono: In *The Structure of Surfaces II*, ed. by J.F. van der Veen, M.A. Van Hove, (Springer, Berlin, Heidelberg, 1988) p. 348
5.254 S.A. Chambers: Langmuir **6**, 1427 (1990)
5.255 S.A. Chambers: Science **248**, 1129 (1990)
5.256 W.F. Egelhoff, Jr., J.W. Gadzuk, C.J. Powell, M.A. Van Hove: Science **248**, 1129 (1990)
5.257 X.D. Wang, Z.-L. Han, B.P. Tonner, Y. Chen, S.Y. Tong: Science **248**, 1129 (1990)
5.258 D.P. Woodruff: Science **248**, 1129 (1990)
5.259 Hong Li, B.P. Tonner: Phys. Rev. B **40**, 10241 (1989)
5.260 Hong Li, B.P. Tonner: Surf. Sci. **237**, 141 (1990)
5.261 U. Breuer, O. Knauff, H.P. Bonzel: Phys. Rev. B **41**, 10848 (1990)
5.262 U. Breuer, O. Knauff, H.P. Bonzel: J. Vac. Sci. Technol. A **8**, 2489 (1990)
5.263 D.A. Wesner, U. Breuer, O. Knauff, H.P. Bonzel: Phys. Scr. T**31**, 247 (1990)
5.264 E.A. Murphy, H.E. Elsayed-Ali, K.T. Park, J. Cao, Y. Gao: Phys. Rev. B **43**, 12615 (1991)
5.265 B.D. Hermsmeier, R.F.C. Farrow, C.H. Lee, E.E. Marinero, C.J. Lin, R.F. Marks, C.J. Chien: J. Appl. Phys. **69**, 5646 (1991)
5.266 B.T. Jonker, G.A. Prinz, J. Appl. Phys. **69**, 2938 (1991)
5.267 B.T. Jonker, G.A. Prinz, Y.U. Ydzerda: J. Vac. Sci. Technol. B**9**, 2437 (1991)
5.268 C. Krembel, M.C. Hanf, J.C. Peruchetti, D. Bolmont, G. Gewinner: Phys. Rev. B **44**, 8407 (1991)
5.269 M. Seelmann-Eggebert, H.J. Richter: Phys. Rev. B **43**, 9578 (1991)
5.270 M. Seelmann-Eggebert, H.J. Richter, J. Vac. Sci. Technol. B **9**, 1861 (1991)
5.271 L.E. Cox, W.P. Ellis, Solid State Commune. **78**, 1033 (1991)
5.272 L. Kubler, F. Lutz, J.L. Bischoff, D. Bolmont: Surf. Sci. **251**, 305 (1991)
5.273 A. Stuck, J. Osterwalder, L. Schlapbach, H.C. Poon: Surf. Sci. **251**, 670 (1991)
5.274 S.A. Chambers, V.S. Sundaram: J. Vac. Sci. Technol. B **9**, 2256 (1991)
5.275 T. Abukawa, T. Okane, S. Kono: Surf. Sci. **256**, 370 (1991)

5.276 G.S. Herman, D.J. Friedman, T.T. Tran, C.S. Fadley, G. Granozzi, G.A. Rizzi, J. Osterwalder, S. Bernardi: J. Vac. Sci. Technol. B **9**, 1870 (1991)
5.277 R. Duszak, S. Tatarenko, J. Cibert, K. Saminadayar, C. Deshayes: J. Vac. Sci. Technol. **9**, 3025 (1991)
5.278 L.J. Terminello, J.J. Barton, Science **251**, 1218 (1991)
5.279 H. Daimon, Y. Tezuka, N. Kanada, A. Otaka, S.K. Lee, S. Ino, H. Namba, H. Kuroda: In Ref. [5.68] p. 96
5.280 G.S. Herman, A.P. Kaduwela, D.J. Friedman, Y. Yamada, E.L. Bullock, C.S. Fadley, Th. Lindener, D. Ricken, A.W. Robinson, A.M. Bradshaw: In Ref. [5.68] p. 600
5.281 S.A. Chambers: In Ref. [5.68] p. 72
5.282 Y.U. Idzerda, D.E. Ramaker: in press
5.283 J.J. Barton, L.J. Terminello: in press
5.284 T.T. Tran, D.J. Friedman, Y.J. Kim, G.A. Rizzi, C.S. Fadley: In Ref. [5.68] p. 522
5.285 C. Akita, T. Tomioka, M. Owari, A. Mizuike, Y. Nihei: Japn. J. Appl. Phys., in press
5.286 G.S. Herman, T.T. Tran, K. Higashiyama, C.S. Fadley: Phys. Rev. Lett. **68**, 1204 (1992)
5.287 M.A. Newton, S.M. Francis, M. Bowker: J. Phys. Cond. Mat. 3, S139 (1991)
5.288 T. Greber, J. Osterwalder, S. Hüfner, L. Schlapbach: Phys. Rev. B **45**, 4540 (1992)
5.289 A.B. Yang, F.C. Brown, J.J. Rehr, M.G. Mason, Y.T. Tan: Phys. Rev. B **45**, 6188 (1992)
5.290 S.A. Chambers, V.A. Loebs: Appl. Phys. Lett. 60, 38 (1992)
5.291 E. Gürer, K. Klier: Phys. Rev.B **46**, 4884 (1992)
5.292 H. Ascolani, M.M. Guraya, G. Zampieri: Phys. Rev. B **43**, 5135 (1991)
5.293 H. Ascolani, R.O. Barrachina, M.M. Guraya, G. Zampieri: Phys. Rev. B **46**, 4836 (1992) press
5.294 A. Hoffman, G.L. Nyberg, J. Liesgang: Phys. Rev. B **45**, 5679 (1992)
5.295 P.A.M. Dirac:*The Principles of Quantum Mechanics*, 4th Ed., Oxford, 1958, p. 121
5.296 K.W. Ford, J.A. Wheeler: Ann. Phys. 7, 287 (1959) and **7**, 295 (1959)
5.297 R.B. Bernstein: Adv. Chem. Phys. **10**, 75 (1966)
5.298 M.V. Berry, K.E. Mount: Rep. Prog. Phys. **35**, 315 (1972)
5.299 E.J. Heller: J. Chem. Phys. **62**, 1544 (1975)
5.300 E.J. Heller: J. Chem. Phys. **75**, 2933 (1981)
5.301 L.I. Schiff: *Quantum Mechanics*, 2nd Ed., (McGraw -Hill, New York, 1955) p. 187
5.302 V.L. Moruzzi, J.F. Janak, A.R. Williams: *Calculated Electronic Properties of Metals* (Pergamon, New York, 1978)
5.303 W.F. Egelhoff, Jr., D. Barak: to be published
5.304 C. Cohen-Tannoudji, B. Diu, and F. Laloë, *Quantum Mechanics*, Vol. 1, (Wiley, New York, 1977) p. 64
5.305 M. Fink, A.C. Yates: Atomic Data, **1**, 385 (1970)
5.306 M. Fink, J. Ingram, Atomic Data, **4**, 1 (1972)
5.307 D. Gregory, M. Fink: Atomic Data, **14**, 39 (1974)
5.308 M.B. Webb, M.G. Lagally: Solid State Phys. **28**, 301 (1973)
5.309 W.F. Egelhoff, Jr: results using WKB methods following Ref. [5.310] to be published
5.310 A.P. Jauho, J.W. Wilkins, M. Cohen, R.P. Merrill, in *Determination of Surface Structure by LEED*, ed. by P.M. Marcus, F. Jona, Eds., (Plenum, New York, 1985) p. 129
5.311 J.J. Barton, S.W. Robey, C.C. Bahr, D.A. Shirley: in *Structure of Surfaces*, ed. by M.A. Van Hove, S.Y. Tong, (Springer Berlin, Heidelberg 1985) p. 191
5.312 Specialists may find it interesting to see how the generalized Ramsauer–Townsend minima come about. Note first that, in general, the phase shift upon scattering increases continuously with scattering angle. This increase occurs because the larger the scattering angle, the deeper into the atom the electron must go, picking up kinetic energy as it falls through the potential, and thus incorporating more wavelengths in its path. This principle applies also to the trajectories that loop around the atom having scattering angles, as it were, larger than 180°. In this sense, two trajectories, one entering with its impact parameter on, say, the left side of the atom and scattering through 178° and one entering through the right side of the

atom looping around it and scattering through 182° would coincide and exit together at 178° off to the right side. However, their phase shifts would differ due to the 4° difference in total scattering angle. A 4° difference is not large enough to produce a significant difference in phase shifts, however, for larger values the difference in phase shifts eventually reaches π, whereupon destructive interference produces a minimum in the scattering amplitude. [5.309] Heavier atoms such as Au can exhibit several maxima and minima as a function of angle as this interference goes alternatively constructive and destructive. A characteristic of this behavior is a sudden change in the combined phase shift see (5.68) as a function of observed angle. Examples of these sudden changes may be found in [5.53, 305–307]. They are easy to understand if visualized as the sum of two vectors (in the complex plane) with similar magnitude and a difference in phase angles of about π. With the phase angle of one vector is increasing and the other decreasing, then just as the difference goes through π the phase angle of their sum will jump suddenly, by about π, just as the magnitude of their sum goes through a minimum [5.309]

5.313 See, for example, pp. 120–126 and Table IV in Chapter X of [5.309] and p. 111 in Ref. [5.314]. Note that the semiclassical method for calculating the partial-wave phase shifts termed "Jeffreys's" is known today as the WKB method and that termed "Langer's is a first order correction to the WKB method

5.314 N.F. Mott, H.S.W. Massey: *The Theory of Atomic Collisions*, 2nd ed., (Oxford, University Press London, 1949)

5.315 H.S.W. Massey, E.H.S. Burhop, *Electronic and Ionic Impact Phenomena*, (Oxford University Press., London, 1952)

5.316 M. Gryzinski: Phys Rev. **107**, 1471 (1957)

5.317 M. Gryzinski: Phys Rev. **115**, 374 (1959)

5.318 M. Gryzinski: Phys Rev. **138**, A 305 (1965)

5.319 M. Gryzinski: Phys Rev. **138**, A 336 (1965)

5.320 B.-G. Englert: *Semiclassical Theory of Atoms* (Springer, Berlin, Heidelberg, 1988)

5.321 R.S. Judson, D.B. McGarrah, O.A. Sharafeddin, D.J. Kouri, D.K. Hoffman: J. Chem. Phys: **94**, 3577 (1991)

5.322 D. Lu, J.J. Rehr: J. de Phys., Coll. C8, Supp. 12, T 47, C8–67 (1986)

5.323 R.H. Ritchie, A. Howie: Phil. Mag. A **58**, 753 (1988)

5.324 F. Fujimoto, S. Takagi, K. Komaki, H. Koike, Y. Uchida: Radiation Effects **12**, 153 (1972)

5.325 There exists an extensive literature on the modelling of electron transport in solids using Monte Carlo methods to incorporate elastic and inelastic scattering event in single path (viz. classical) trajectory calculations. Reviews of this literature may be found in [5.326–328]

5.326 D.F Kyser: in *Electron Beam Interactions with Solids*, ed. by D.F. Kyser, D.E. Newbury, H. Niedrig, R. Shimizu, (SEM, Inc., AMF, O'Hare, IL, 1984) p. 119

5.327 J.I. Goldstein: in *Practical Scanning Electron Microscopy*, ed. by. J.I. Goldstein, H. Yakowitz (Plenum, New York, 1975) p. 49

5.328 D.E. Newbury, D.C. Joy, P. Echlin, C.E. Fiori, J.I. Goldstein: *Advanced Scanning Electron Microscopy and X-Ray Microanalysis* (Plenum, New York, 1986) pp. 19–43

5.329 D.G. Frank, N. Batina, J.W. McCarger, A.T. Hubbard: Langmuir **5**, 1141 (1989)

5.330 D.G. Frank, N. Batina, T. Golden, F. Lu, A.T. Hubbard: Science **247**, 182 (1990)

5.331 D.G. Frank, N. Batina, T. Golden, F. Lu, A.T. Hubbard, MRS Bulletin **15**, 19 (1990)

5.332 C. Shannon, D.G. Frank, and A.T. Hubbard: Annu. Rev. Phys. Chem. **42**, 393 (1991)

5.333 D.G. Frank, O.M.R. Chyan, T. Golden, A.T. Hubbard: J. Vac. Sci. Technol. A **10**, 158 (1992)

5.334 N. Batina, O.M.R. Chyan, D.G. Frank, T. Golden, A.T. Hubbard: Naturwissenshaft **77**, 557 (1990)

5.335 W.F. Egelhoff, Jr: to be published.

5.336 P.N. Tomlinson, A. Howie: Phys. Lett. **27A**, 491 (1968)

5.337 H. Raether: *Excitation of Plasmons and Interband Transitions by Electrons* (Springer, Berlin, Heidelberg, 1980) p. 50

5.338 W.F. Egelhoff, Jr: unpublished results (to be published)

5.339 This effect may be the reason that Barton and Terminello (239, [5.278, 283] 244) report that an enhance intensity is observed at 0° for Cu 3p photoelectrons of 56.6 eV emitted by a Cu(1 0·0) crystal whereas $M_{2,3} M_{4,5} M_{4,5}$ Auger electrons of the same energy show a reduced intensity (much like the same Auger electrons from Ni(1 0 0) in Fig 5.26). The Auger electron in this case is known to be $L = 3$ while the 3p photoelectron is dominated by $L = 0$; see [5.179, 340–342]

5.340 S.M. Goldberg, C.S. Fadley, S. Kono: J. Elect. Spectr. Rel. Phenom. **21**, 285 (1981)
5.341 P.J. Feibelman, E.J. McGuire: Phys. Rev. B **15**, 3575 (1977)
5.342 H.L. Davis: in *Proc. 7th Int. Vac. Cong.* (Vienna) eds., title, North-Holland, Amsterdam (1977) p. 2281
5.343 M.J. Seaton, G. Peach: Proc. Phys. Soc. London, **79**, 1296 (1962)
5.344 S.T. Manson, J.W. Cooper: Phys. Rev., **165**, 126 (1968)
5.345 J.W. Cooper, R.N. Zare: J. Chem. Phys. **48**, 942 (1968)
5.346 J.W. Gadzuk: Phys. Rev. B **12**, 5608 (1975)
5.347 R.F. Reilman, A. Msezane, S.T. Manson: J. Elect. Spect. Rel. Phenom. **8**, 389 (1976)
5.348 J.W. Gadzuk, Surf. Sci. **60**, 76 (1976)
5.349 T.N. Rhodin, J.W. Gadzuk: in *The Nature of the Surface Chemical Bond*, ed. by T.N. Rhodin, G. Ertl, (North-Holland, Amsterdam, 1979) p. 115
5.350 J.J. Barton, D.A. Shirley: Phys. Rev. A **32**, 1019 (1985)
5.351 J.J. Barton, D.A. Shirley: Phys. Rev. B **32**, 1892 (1985)
5.352 J.J. Barton, D.A. Shirley: Phys. Rev. B **32**, 1906 (1985)
5.353 J.J. Barton, S.W. Robey, D.A. Shirley: Phys. Rev. B **34**, 778 (1986)
5.354 J.J. Rehr, R.C. Albers, C.R. Natoli, E.A. Stern: Phys. Rev. B **34**, 4350 (1986)
5.355a J. Mustre de Leon, J.J. Rehr, C.R. Natoli, C.S. Fadley, J. Osterwalder: Phys. Rev. B **39**, 5632 (1989)
5.355b G. Tréglia, M.C. Desjonquères, D. Spanjaard, D. Sébilleau, C. Guillot, D. Chauveau, J. Lecante: J. Phys. Cond. Mat. **1**, 1879 (1989)
5.355c D. Sébilleau, G. Tréglia, M.C. Desjonquères, D. Spanjaard, C. Guillot, D. Chauveau, J. Lecante: J. Physique **49**, 227 (1988)
5.356 Sometimes these effects are quite small (see [5.164, 170, 171]), but sometimes they can be significant (see [5.357–378])
5.357 B.D. Grachev, A.P. Komar, Yu. S. Korobochko, V.I. Mineev: JETP Lett. **4**, 163 (1966)
5.358 A.R. Scul'man, V.V. Korablev, Yu. A. Morozov: Sov. Phys. Solid State **10**, 1246 (1968)
5.359 A.P. Komar, Yu. S. Korobochko, B.D. Grachev, V.I. Mineev: Sov. Phys. Solid State **10**, 1222 (1968)
5.360 A.R. Scul'man, V.V. Korablev, Yu. A. Morozov: Sov. Phys. Solid State **10**, 1512 (1968)
5.361 B.D. Grachev, A.P. Komar, Yu.S. Korobochko, V.I. Mineev: Sov. Phys. Solid State **10**, 1894 (1968)
5.362 A. Gervais, R.M. Stern, M. Menes: Acta Cryst. A **24**, 191 (1968)
5.363 A.M. Baró, M. Salmerón: Phys. Stat. Sol. (b) **49**, K135 (1972)
5.364 T.W. Rusch, J.P. Bertino, W.P. Ellis: Appl. Phys. Lett. **23**, 359 (1973)
5.365 H.E. Bishop, B. Chornik, C. LeGressus, A. LeMoel: Surf. Interf. Anal. **6**, 116 (1984)
5.366 F.E. Doern, L. Kover, N.S. McIntyre: Surf. Interf. Anal. **6**, 282 (1984)
5.367 B. Bennett, H. Viefhaus: Surf. Interf. Anal. **8**, 127 (1986)
5.368 Y. Sakai, A. Mogami: J. Vac. Sci. Technol. A5, 1222 (1987)
5.369 W.O. Barnard, H.C. Snyman, F.D. Auret: S. Afr. J. Phys. **10**, 153 (1987)
5.370 S. Mróz, A. Mróz: Surf. Sci. **224**, 235 (1989)
5.371 L.H. Tjeng, R.W. Bartstra, G.A. Sawatzky: Surf. Sci. **211/212**, 187 (1989)
5.372a H.E. Bishop: Surf. Inter. Anal. **15**, 37 (1990)
5.372b H.E. Bishop: Surf. Inter. Anal. **16**, 118 (1990)
5.373 F. Peeters, E.R. Puckrin, A.J. Slavin, J. Vac. Sci. Technol. A **8**, 797 (1990)
5.374 Y. Gao, K.T. Park, Phys. Rev. B, in press
5.375 E.E. Gorodnichev, S.L. Dudarev, D.B. Rogozkin, M.I. Ryazanov, Phys. Lett. A **144**, 411 (1990)
5.376 M.V. Gomoyunova, S.L. Dudarev, I.I. Pronin, Surface Sci. **235**, 156 (1990)

5.377a B.L. Maschhoff, J. Pan, T.E. Madey, Surface Sci. **259**, 190 (1991)
5.377b H. Ascolani, M.M. Guraya, G. Zampieri: In *Surface Science*, ed. by F.A. Ponce, M. Cardona (Springer, Berlin, Heidelberg 1992) p. 163
5.377c Y. Gao and K.T. Park, Phys. Rev. **B 46**, 1743 (1992)
5.378   A. Jablonski, H.S. Hansen, C. Jansson, S. Tougaard, Phys. Rev. **B 45**, 3694 (1992)
5.379   For typical data see Refs. [5.58, 62, 75, 77, 83, 98, 102, 109, 192, 380–385]
5.380   D.C. Johnson, A.U. MacRae: J. Appl. Phys. **37**, 1945 (1966)
5.381   R. Baudoing, R.M. Stern, H. Taub, Surf. Sci. **11**, 255 (1968)
5.382   R.M. Stern, H. Taub, Phys. Rev. Lett. **20**, 1340 (1968)
5.383   G. Allié, E. Blanc, D. Dufayard, R.M. Stern: Surf. Sci. **46**, 188 (1974)
5.384   A. Mosser, Ch. Burggraf, S. Goldsztaub, Y.H. Ohtsuki: Surf. Sci. **54**, 580 (1976)
5.385   M.V. Gomoyunova, I.I. Pronin, I.A. Shmulevitch: Surf. Sci. **139**, 443 (1984)
5.386   For example: J.M. Cowley: *Diffraction Physics*, (North-Holland, Amsterdam, 1975)
5.387   K.J. Strandburg, ed., *Bond-Orientational Order in Condensed Matter Systems*, (Springer, Berlin, Heidelberg 1992)
5.388   W.H. Gries: Surf. Inter. Anal. **17**, 803 (1991)
5.389   J.M. Bloch, M. Sagurton, I. Jacob, C. Binns: to be published (determining bond angles in Langmuir-Blodgett films)
5.390   E. Bauer: Z. Kristallogr., **110**, 372 (1958)
5.391   E. Bauer: Z. Kristallogr., **110**, 395 (1958)
5.392   J.G. Gay, J.R. Smith, R. Richter, F.J. Arlinghaus, R.H. Wagoner: J. Vac. Sci. Technol. A **2**, 931 (1984)
5.393   J.R. Smith, A. Banerjea: Phys. Rev. Lett. **598**, 2451 (1987)
5.394   Although [5.392, 393] contain some of the most important values for magnetic thin films, no complete list exists. Approximate values for the surface free energies of other metals can be estimated by scaling these values in proportion to cohesive energies.
5.395   F.R. deBoer, R. Boom, W.C.M. Mattens, A.R. Miedema, A.K. Niessen: *Cohesion in Metals: Transition Metal Alloys*, Elsevier, Amsterdam, 1988)
5.396   This conclusion was briefly disputed on the basis of Auger "breakpoints" [5.397, 398], but then it was found that the first breakpoint came at 2 ML [5.399–401]. It may be noteworthy that Bauer [5.402] recently showed that the coverage at which the breakpoints occur can vary strongly depending on the detection angle for the Auger electrons. For a review of Auger breakpoints: [5.403]
5.397   D. Pescia, M. Stampanoni, G.L. Bona, A. Vaterians, R.F. Willis, F. Meier: Phys. Rev. Lett. **60**, 2559 (1988)
5.398   R. Germar, W. Dürr, J.W. Krewer, D. Pescia, W. Gudat: Appl. Phys. A **47**, 393 (1988)
5.399   H. Glatzel, Th. Fauster, B.M.U. Scherzer, V. Dose: Surf. Scii. **254**, 58 (1991)
5.400   H. Landskron, G. Schmidt, K. Heinz, K. Müller, C. Stuhlmann, U. Beckers, M. Wuttig, H. Ibach: Surf. Sci. 256, 115 (1991)
5.401a J. Thomassen, B. Feldmann, M. Wuttig: Surf. Sci. **264**, 406 (1992)
5.401b D.D. Chambliss, R.J. Wilson, S. Ching: J. Vac. Sci. Technol. B 10, 1993 (1992)
5.402   E. Bauer, C. Koziol, G. Lilienkamp: to be published
5.403   C. Argile, G.E. Rhead: Surf. Sci. Rep. **10**, 277 (1989)
5.404   Cu core-level shifts induced by titration of the surface with carbon monoxide [5.227]
5.405   M.T. Kief, W.F. Egelhoff, Jr: Phys. Rev. B **47**, 10785 (1993)
5.406   For example, B. Heinrich, A.S. Arrott, C. Liu, S.T. Purcell: J. Vac. Sci. Technol. A **5**, 1935 (1987), and references therein
5.407   For a review of this literature: [5.405]
5.408   For example, D.P. Pappas, C.R. Brundle, H. Hopster: Phys. Rev. B 45, 8169 (1992), and references therein.
5.409   For example, in the present book, Chap. 1 and Sects. 5.1 and 5.3
5.410   B.T. Jonker, J.J. Krebs, G.A. Prinz: Phys. Rev. B **39**, 1399 (1989)
5.411   M.T. Kief, G.J. Mankey, R.F. Willis: J. Appl. Phys. **69**, 5000 (1991)
5.412   G.J. Mankey, M.T. Kief, R.F. Willis: J. Vac. Sci. Technol. A **9**, 1595 (1991)

5.413   M.T. Kief, G.J. Mankey, R.F. Willis: In [5.68] p. 422
5.414   G.J. Mankey, M.T. Kief, R.F. Willis: J. Elect. Spect. Rel. Phenom. **54**, 510 (1990)
5.415   C.M. Schneider, J.J. de Miguel, P. Bressler, J. Garbe, S. Ferrer, R. Miranda, J. Kirschner: J. de Phys. Coll. **C8**-1657, Tome 49 (1988)
5.416   C.M. Schneider, P. Bressler, P. Schuster, J. Kirschner, J.J. de Miguel, R. Miranda, S. Ferrer: Vacuum **41**, 503 (1990)
5.417   C.M. Schneider, P. Bressler, P. Schuster, J. Kirschner, J.J. de Miguel, R. Miranda: Phys. Rev. Lett. **64**, 1059 (1990)
5.418   J.J. de Miguel, A. Cebollada, J.M. Gallego, R. Miranda, in *Kinetics of Ordering and Growth at Surfaces*, ed. by M.G. Lagally, (Plenum, New York, 1990) p. 483
5.419a J.J. de Miguel, A. Cebollada, J.M. Gallego, R. Miranda, C.M. Schneider, P. Schuster, J. Kirschner: J. Mag. Mag. Mat. **93**, 1 (1991)
5.419b J. Ferrón, J.M. Gallego, A. Cebollada, J.J. de Miguel, S. Ferrer: Surf. Sci. 211/212, 797 (1989)
5.420   D.W. Bassett, P.R. Webber, Surf. Sci. **70**, 520 (1978)
5.421   J.D. Wrigley, G. Ehrlich, Phys. Rev. Lett. **44**, 661 (1980)
5.422   T.J. Raeker, D.E. Sanders, A.E. DePristo: J. Vac. Sci. Technol. **A8**, 3531 (1990)
5.423   C. Chen, T.T. Tsong: Phys. Rev. Lett. **64**, 3147 (1990)
5.424   P.J. Feibelman: Phys. Rev. Lett. **65**, 729 (1990)
5.425   G.L. Kellogg, P.J. Feibelman: Phys. Rev. Lett. **64**, 3143 (1990)
5.426   G.L. Kellogg, A.F. Wright, M.S. Daw: J. Vac. Sci. Technol. **A 9**, 1757 (1991)
5.427   G.L. Kellogg: Phys. Rev. Lett. **67**, 216 (1991)
5.428   S.C. Wang, G. Ehrlich: J. Chem. Phys. **94**, 4071 (1991)
5.429   S.C. Wang, G. Ehrlich: Phys. Rev. Lett. **67**, 2509 (1991)
5.430   W.D. Lutke, U. Landman: Phys. Rev. B **44**, 5970 (1991)
5.431   T.T. Tsong, C.-I. Chen, Phys. Rev. B **43**, 2007 (1991)
5.432   C.L. Liu, J.M. Cohen, J.B. Adams, A.F. Voter: Surf. Sci. **253**, 334 (1991)
5.433   W.F. Egelhoff, Jr: Proc. Mat. Res. Soc. Proc. **229**, 27 (1991)
5.434   B. Heinrich, S.T. Purcell, A.S. Arrott, J.F. Cochran: Phys. Rev. B **38**, 12879 (1988)
5.435   B. Heinrich, J.F. Cochran, A.S. Arrott, S.T. Purcell, K.B. Urquhart, J.R. Dutcher, W.F. Egelhoff, Jr: Appl. Phys. A **49**, 473 (1989)
5.436   B. Heinrich, J.F. Cochran, A.S. Arrott, K.B. Urquhart, K. Myrtle, Z. Celinski, Q.-M. Zhong: Mat. Res. Soc. Symp. Proc. **151**, 177 (1989)
5.437   B. Heinrich, and A.S. Arrott, to be published
5.438   See [5.434–436] and references therein
5.439   D.D. Chambliss, R.J. Wilson, S. Chiang: J. Vac. Sci. Technol. **A10**, 1993 (1992)
5.440   S.C. Wang, G. Ehrlich: Phys. Rev. Lett. **62**, 2297 (1989)
5.441   S.C. Wang, G. Ehrlich, Surf. Sci. **217**, L397 (1989)
5.442   K. Meinel, M. Klaua, H. Bethge: Phys. Stat. Sol. (a) **110**, 189 (1988)
5.443   H.A.M. de Gronckel, C.H.W. Swüste, K. Kopinga, W.J.M. de Jonge: Appl. Phys. A **49**, 467 (1989)
5.444   H.A.M. de Gronckel, J.A.M. Bienert, F.J.A. den Broeder, W.J.M. de Jonge, J. Magn. Magn. Mater. **93**, 457 (1991)
5.445   F.J. Lamelas, C.H. Lee, H. He, W. Vavra, R. Clarke: Phys. Rev. B **40**, 5837 (1989)
5.446   K. Le Dang. P. Veillet, H. He, F.J. Lamelas, C.H. Lee, R. Clarke: Phys. Rev. B **41**, 12902 (1990)
5.447   C.H. Lee, H. He, F.J. Lamelas, W. Vavra, C. Uher, R. Clarke: Phys. Rev. B **42**, 1066 (1990)
5.448   H.A.M. de Gronckel, K. Kopinaga, W.J.M. de Jonge, P. Panissod, J.P. Schillé, F.J.A. den Broeder: Phys. Rev. B **44**, 9100 (1991)
5.449   M. Copel, M.C. Reuter, E. Kaxiras, R.M. Tromp: Phys. Rev. Lett. **63**, 632 (1989)
5.450   M. Copel, M.C. Reuter, M. Horn von Hoegen, R.M. Tromp: Phys. Rev. B **42**, 11682 (1990)
5.451   K. Fuktani, H. Daimon, S. Ino, in S.Y. Tong, et al., op cit. (Ref. 29), p. 615
5.452   J.M.C. Thornton, A.A. Williams, J.E. Macdonald, R.G. van Silfhout, J.F. van der Veen, M. Finney, C. Norris: J. Vac. Sci. Technol. B **9**, 2146 (1991)
5.453   F.K. LeGoues, M. Horn von Hoegen, M. Copel, R.M. Tromp: Phys. Rev. B **44**, 12894 (1991)

5.454 M. Horn von Hoegen, F.K. LeGoues, M. Copel, M.C. Reuter, R.M. Tromp: Phys. Rev. Lett. **67**, 1130 (1991)
5.455 H.J. Osten, G. Lippert, J. Klatt: J. Vac. Sci. Technol. B **10** 1151 (1992)
5.456 R.M. Tromp, M.C. Reuter: Phys. Rev. Lett. **68**, 954 (1992)
5.457 R. Cao, X. Yang, J. Terry, P. Pianetta: Phys. Rev. B **45**, 13749 (1992)
5.458 H.A. van der Vegt, H.M. van Pinxteren, M. Lohmeier, E. Vleig, J.M.C. Thornton: Phys. Rev. Lett. **68**, 3335 (1992)
5.459a J.M.C. Thornton, A.A. Williams, J.E. Macdonald, R.G. van Silfhout, M. Finney, C. Norris: Surface Sci., in press
5.459b H.J. Osten, J. Klatt, G. Lippert, B. Dietrich, E. Bugiel: Phys. Rev. Lett. **69**, 450 (1992)
5.459c H.J. Osten, J. Klatt, G. Lippert, E. Bugiel, S. Hinrich: Appl. Phys. Lett. **60**, 2522 (1992)
5.459d H.J. Osten, G. Lippert, J. Klatt: J. Vac. Sci. Technol. B **10**, 1151 (1992)
5.460 F. Jona, P.M. Marcus: Solid State Comm. **64**, 667 (1987)
5.461 K.E. Johnson, D.D. Chambliss, R.J. Wilson, S. Chiang,: J. Vac. Sci. Technol. A22, 1654 (1993)
5.462 J. Chen, M. Drakaki, J.L. Erskine: Phys. Rev. B **45**, 3636 (1992)

Section 5.3

5.463 V.L. Moruzzi, J.F. Janak, A.R. Williams: *Calculated Electronic Properties of Metals* (Pergamon, New York 1978)
5.464 M.N. Baibich, J.M. Broto, A. Fert, F. Nguyen Van Dau, F. Petroff: Phys. Rev. Lett. **61**, 2472 (1988)
5.465 C.H. Lee, H. He, F. Lamelas, W. Vavra, C. Uher, R. Clarke: Phys. Rev. Lett. **62**, 653 (1989)
5.466 F.J. Lamelas, C.H. Lee, Hui He, W. Vavra, Roy Clarke: Phys. Rev. B **40**, 5837 (1989)
5.467 U. Gradmann: J. Magn. Magn. Mat. **6**, 173 (1977)
5.468 C.J. Chien, R.F.C. Farrow, C.H. Lee, C.J. Lin, E.E. Marinero: J. Magn. Magn. Mat. **93**, 47 (1991); S. Chikazumi, *Physics of Magnetism* (Wiley, New York 1964) p. 359
5.469 T.W. Barbee Jr., W.K. Warburton, J.H. Underwood: J. Opt. Soc. Am. B **1**, 691 (1984)
5.470 A. Segmüller, A.E. Blakeslee: J. Appl. Cryst. **6**, 19 (1973)
5.471 D.B. Mc Whan, M. Gurvitch, J.M. Rowell, L.R. Walker: J. Appl. Phys. **54**, 3886 (1983)
5.472 W. Sevenhans, M. Gijs, Y. Bruynseraede, H. Homma, I.K. Schuller: Phys. Rev. B **34**, 5955 (1986)
5.473 B.M. Clemens, J.G. Gay: Phys. Rev. B **35**, 9337 (1987)
5.474 J.-P. Locquet, D. Neerinck, L. Stockman, Y. Bruynseraede, I.K. Schuller: Phys. Rev. B **39**, 13338 (1989)
5.475 F.J. Lamelas, H.D. He, R. Clarke, Phys. Rev. B **43**, 12296 (1991)
5.476 K. Le Dang, P. Veillet, P. Beauvillain, C. Chappert, Hui He, F.J. Lamelas, C.H. Lee, Roy Clarke: Phys. Rev. B **43**, 13 228 (1991); K. Le Dang, P. Veillet, Hui He, F.J. Lamelas, C.H. Lee, Roy Clarke: Phys. Rev. B **41**, 12 902 (1990)
5.477 P.F. Miceli, D.A. Neumann, H. Zabel: App. Phys. Lett. **48**, 24 (1986)
5.478 C.H. Lee, Hui He, F.J. Lamelas, W. Vavra, C. Uher, R. Clarke: Phys. Rev. B **42**, 1066 (1990)
5.479 G.H.O. Daalderop, P.J. Kelly, M.F.H. Schuurmans: Phys. Rev. B **42**, 7270 (1990)
5.480 B.E. Warren: *X-ray Diffraction* (Addison-Wesley, Reading 1969)
5.481 A. Guinier: *X-ray Diffraction* (W.H. Freeman, San Francisco 1963)
5.482 J.Q. Zheng, J.B. Ketterson, G.P. Felcher: J. Appl. Phys. **53**, 3624 (1982)
5.483 Roy Clarke, F. Lamelas, C. Uher, C.P. Flynn, J.E. Cunningham: Phys. Rev. B **34**, 2022 (1986)
5.484 P. Bödeker, A. Abromeit, K. Bröhl, P. Sonntag, N. Metoki, H. Zabel: Phys. Rev. B **47**, 2353 (1993)
5.485 H.A.M. de Gronckel, K. Kopinga, W.J.M. de Jonge, P. Panissod, J.P. Schillé, F.J.A. den Broeder: Phys. Rev. B **44**, 9100 (1991)

References

5.486 C. Mény, P. Panissod, R. Loloee: Phys. Rev. B **45**, 12 269 (1992)
5.487 H. Jagodzinski: Acta. Cryst. **2**, 208 (1949)
5.488 H. Holloway: J. Appl. Phys. **40**, 4313 (1969)
5.489 M.T. Sebastian, P. Krishna: Phys. Status Solidi (a) **101**, 329 (1987)
5.490 M.B. Salamon, S.K. Sinha, J.J. Rhyne, J.E. Cunningham, E. Ross, J. Borchers, C.P. Flynn: Phys. Rev. Lett. **56**, 259 (1986)
5.491 M. Hong, R.M. Fleming, J. Kwo, L.F. Schneemeyer, J.V. Waszczak, J.P. Mannaerts, C.F. Majkrzak, D. Gibbs, J. Bohr: J. Appl. Phys. **61**, 4057 (1987)
5.492 For example, B.K. Tanner, M.J. Hill: Adv. X-ray Anal. **29**, 337 (1986)
5.493 R. Clarke, W. Dos Passos, W. Lowe, B.G. Rodricks, C. Brizard: Phys. Rev. Lett. **66**, 317 (1991)
5.494 W. Lowe, R.A. MacHarrie, J.C. Bean, L. Peticolas, R. Clarke, W. Dos Passos, C. Brizard, B. Rodricks: Phys. Rev. Lett. **67**, 2513 (1991)
5.495 Y. Huai, R.W. Cochrane, M. Sutton: Phys. Rev. B **48**, 2568 (1993)

6. Polarized Neutron Reflection

J.A.C. BLAND

"Science has this great weapon: experiment...."
J. Polkinghorne: 'Physics and Theology', *Cavendish Laboratory,* 1992

At first sight it may seem surprising that neutron scattering has emerged during the last decade as a valuable magnetometric technique for studying ultrathin magnetic films since the interaction between neutrons and matter is particularly weak. Furthermore, a plethora of sensitive magnetometric techniques such as superconducting quantum interference device (SQUID) magnetometry, torque magnetometry and the surface magneto-optic Kerr effect (SMOKE) have been refined over the last decade in order to investigate static magnetic parameters in ultrathin films with high precision. Therefore it is the purpose of this chapter to answer the question: why use polarized neutron reflection (PNR) to study ultrathin magnetic films?

6.1 Introduction

The motivation for using neutrons to study ultrathin magnetic films lies with the theoretical predictions of significantly enhanced magnetic moments in ultrathin transition metal films, described for example by *Gay* and *Richter* in this volume (around 35% for a Fe monolayer on a noble metal substrate [6.1, 2]). In order to unambiguously test these predictions, a technique is needed in the case of supported ultrathin films which will yield the absolute value of the magnetic moments within the ferromagnetic layer with the required accuracy and without the need to correct for a significant magnetic signal due to the supporting non-magnetic substrate. In constrast with conventional magnetometric methods, these requirements are in principle met in neutron scattering [6.3] since the magnetic interaction is well defined and no spin dependent scattering occurs for an unpolarized medium. Moreover the scattering intensity can be accurately calculated and compared with experiment because of the weakness of the overall scattering (i.e. dynamical or even kinematic diffraction theory can be used). Also the non-magnetic part of the effective neutron-solid interaction associated with the nuclear potential is often comparable in magnitude with that of the Zeeman interaction between the neutron spin and the total magnetic field due to the scattering medium [6.4]: this result in a high intensity contrast upon reversing the neutron spin with respect to the sample magnetization. However, for ultrathin magnetic films it is not possible to simply apply conventional scattering techniques in studying the extremely small amounts of magnetic material present in such samples: for example, a film of five monolayers of Fe embedded in a Ag sample of 2 mm thickness corresponds to an Fe 'impurity' level of approximately 4×10^{-5}%. Since the scattered signal is proportional to the amount of material present, the relative contribution of the magnetic layer to the total scattered intensity would be unacceptably small in this case. An enhanced contribution from the magnetic film is needed. One way of doing this is to prepare a multilayer sample and carry out conventional diffraction measurements [6.5], but the magnetic behavior of a multilayer is not necessarily simply the sum of those of the individual films since collective effects occur due to interactions between the layers. Also the constituent layers may not be of exactly the same thickness, thus introducing uncertainty in the data analysis. However in PNR, the intensity reflectivities approach 0.01 in the range of wavevector where the reflectivity is most sensitive to the magnetization of an embedded ultrathin layer (typically around $1-2 \times 10^{-2}$ Å^{-1}): for example for single Fe films of few monolayers thickness the spin dependent intensity contrast (defined as the normalized difference in spin dependent reflectivity) approaches two in this region. Therefore the total magnetic signal associated with the Fe layer is of the order of 1% of the full beam intensity, from which a remarkable degree of sensitivity results.

In this chapter no attempt is made to review neutron diffraction studies of magnetic multilayers nor is a comprehensive survey of recent PNR studies of ultrathin films attempted. The aim of the chapter is to describe the technique and to discuss the unique combination of magnetic and structural information that PNR yields, its advantages and limitations with respect to other techniques, and recent case studies.

6.2 Theory of Polarized Neutron Reflection

The phenomenon of the critical reflection of neutrons was first demonstrated by *Fermi* and *Zinn* in 1944 who observed the specular reflection of thermal neutrons from a smooth solid surface [6.6]. This experiment marked the beginning of the field of neutron optics which has led to many of the optical phenomena associated with coherent elastic scattering such as Fraunhofer and Fresnel diffraction, and multiple beam interference being demonstrated [6.7]. Determinations of the index of refraction have been routinely carried out to accuracy determine the coherent scattering length for many materials [6.7]. The surface quality was known to influence the reflectivity in such measurements but it remained until the 1980's for PNR to be recognized as a useful probe of ultrathin magnetic films [6.8, 9].

In PNR the partially reflected neutron intensity is measured as a function of the incident spin state and incident wave vector [6.4]. Such measurements typically permit the refractive index profile of the solid medium to be determined with a depth resolution in the nm range. Since the strength of the Zeeman interaction is determined by the magnetic induction B only (the neutron moment is known), the total magnetic moment per atom (spin and orbital components) can be accurately obtained for a magnetically saturated thin ferromagnetic film of known density. This contrasts with the situation for X-rays where the magnetic part of the interaction is typically a factor of 10^{-6} smaller than the non-magnetic part. For antiferromagnetic materials, the average magnetization is zero and so no magnetic contrast arises in this case. PNR is in principle a *self-calibrating magnetometric technique* since the spin dependence of the reflectivities yields the total magnetic moment of the layer while the layer thicknesses can be determined independently from the wavevector dependence of the reflectivity. This is because two different optical potentials are presented to the neutrons according to the incident spin state: the potentials differ only in strength as dictated by the magnetic interaction but have *identical* layer thickness. However, in practice only layer thicknesses exceeding approximately 2 nm can be obtained directly with presently available techniques. PNR has the further important advantage over conventional magnetometric techniques that no magnetic signal arises from the non-magnetic substrate in the case of supported ultrathin films. In the case of SMOKE, the magneto-optic signal associated with a non-magnetic substrate is often weak (due to the short skin depth of light in metals) – see the chapter by *Bader* and *Erskine* in Volume II. However SMOKE does not yield the absolute value of the magnetic moment. Since the neutron reflectivity is affected by interface quality, (e.g., substrate roughness) as we discuss in detail in Sect. 6.2.5, PNR also yields valuable information concerning the interface fluctuations of nm amplitude. The magnetic properties (e.g., magnetic moment, magnetic anisotropy) are very sensitively dependent on interface roughness on this lengthscale.

6.2.1 The Optical Potential for a Magnetized Medium

We begin with a quantum mechanical description of the scattering process. The interface is assumed to lie in the (x, z) plane and the interaction of the neutron with the reflecting (magnetic) medium can be represented by a one-dimensional (1D) optical potential $V(y)$ [6.7] if we assume no lateral (i.e. in-plane) variation in scattering density. A structure which varies in scattering density along a direction normal (y axis) to the surface of the film can therefore be described by a sequence of infinitesimally thin layers (i.e. a stratified medium) each with a constant interaction potential. In such a bounded continuous medium, the spatially averaged optical potential V_i, may be approximated by

$$V_i = \frac{2\pi\hbar^2}{m_n} \rho_i b_i - \boldsymbol{\mu}_n \cdot \boldsymbol{B}_i, \qquad (6.1)$$

where m_n is the neutron mass, ρ_i is the atomic density, b_i is the bound coherent neutron scattering length of the material [6.3, 7], μ_n is the neutron magnetic moment, $\boldsymbol{B}_i$ is the total magnetic induction in the medium (arising from the atomic moments) and the suffix i labels the medium. Absorption can be parameterized by introducing a wavelength dependent imaginary part to b_i which is therefore in general a complex quantity [6.3, 7]. The first term on the right hand side of (6.1) which we denote $V_{in} = 2\pi\hbar^2 \rho_i b_i/m_n$ corresponds to an effective potential associated with the short ranged neutron–nucleus interaction which was first introduced by *Fermi* [6.10, 11]. The second term corresponds to the Zeeman interaction and so depends upon the relative orientation of the neutron spin with respect to the magnetic induction within the magnetic material. The neutron magnetic dipole moment is related to the Pauli spin operator σ_n by $\boldsymbol{\mu}_n = \gamma \mu_N \sigma_n$ where the gyromagnetic ratio $\gamma = -1.913$ and μ_N is the nuclear magneton. Initially we will assume that the incident neutron spin is either parallel or antiparallel to the magnetic induction vector.

The Schrödinger equation for neutrons in the ith magnetic medium can be written as

$$\left[-\frac{\hbar^2 \nabla^2}{2m_n} + V_i \right] \Psi(y) = E \Psi(y) \tag{6.2}$$

where Ψ denotes the total neutron wavefunction (spin and spatial parts) and E the total energy of the neutron. A further simplification can be made in the 1D case since we can write $\Psi(y)$ in the form $\psi(y) \exp(i\boldsymbol{k}_{//} \cdot \boldsymbol{r})$ where $\boldsymbol{k}_{//}$ is the in-plane component of the wavevector and $\boldsymbol{r}$ is the position vector.

We now solve for $\psi(y)$ for some simple illustrative cases. Consider first a system composed of two semi-infinite media i, j separated by a planar boundary at $y = y_0$. Consider the neutron wave to be incident at a grazing angle θ_i from medium i as shown schematically in Fig. 6.1(a) with the x, y axes in the scattering plane. This figure defines the geometry for PNR used in subsequent discussions. The incident perpendicular wavevector component q_i is therefore given by $k_i \sin \theta_i$ where k_i is the incident wavevector. The solution (6.2) for the ith medium is given by the sum of a forward and backward travelling wave with coefficients A_i and B_i, respectively, as

$$\psi_i(y) = A_i \exp(iq_i(y - y_0)) + B_i \exp(-iq_i(y - y_0)) \tag{6.3}$$

and a solution of the same form holds for the jth medium. In medium j a single transmitted wave occurs and therefore $A_j = t_{ij}$ and $B_j = 0$ whereas in the ith medium $A_i = 1$ and $B_i = r_{ij}$ where r_{ij}, t_{ij} are the reflection and transmission coefficients respectively. The amplitude reflection coefficients are found by matching wavefunctions and derivatives at the boundary. In exact analogy with the Fresnel reflectivities for transverse electric (TE) polarized light, the results are:

$$r_{ij} = \frac{q_i - q_j}{q_i + q_j} \text{ and } t_{ij} = \frac{2q_i}{q_i + q_j}, \tag{6.4}$$

6.2 Theory of Polarized Neutron Reflection

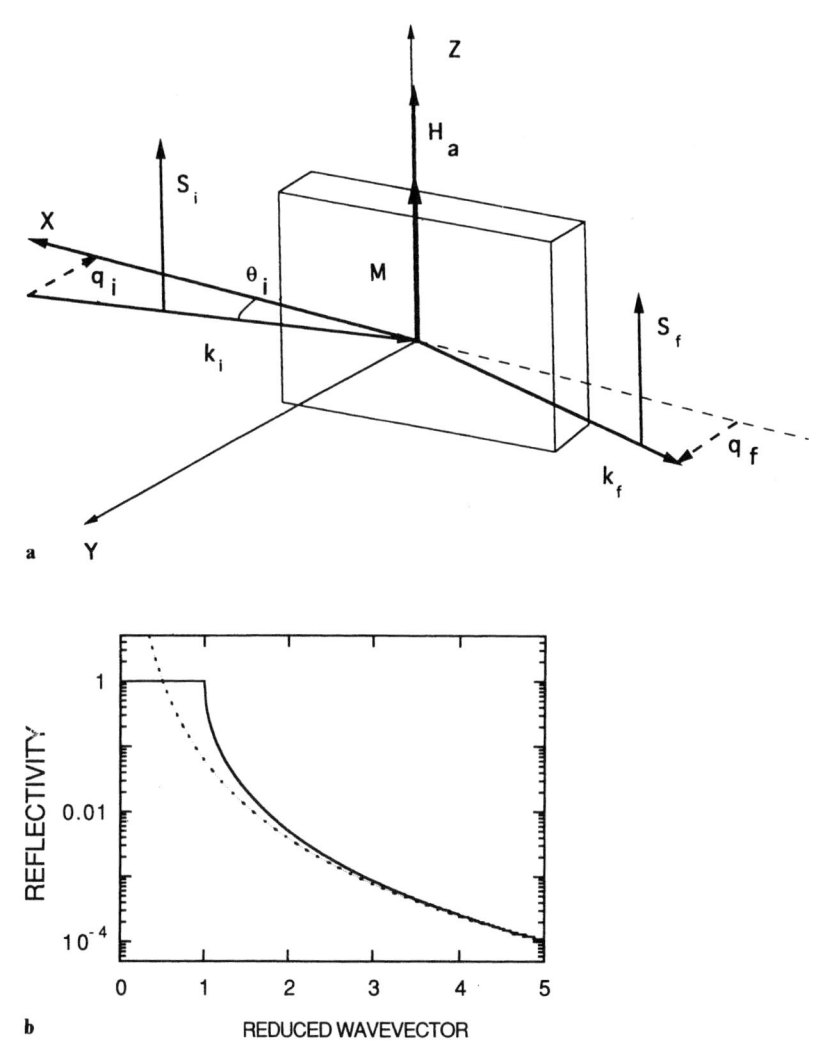

Fig. 6.1. a The geometry for PNR from a single interface. The neutron wave is incident at a grazing angle θ_i on the solid medium (rectangular box), as shown schematically, with x, y the scattering plane (see text). The incident perpendicular wavevector component is given by $q_i = k \sin(\theta_i)$. The reflected beam has a wavevector k_f with a perpendicular component q_f equal in magnitude to the incident perpendicular wavevector component. The incident and final spin states S_i, S_f are parallel (or antiparallel) to the applied field H_a and the sample magnetization M which is assumed to lie parallel to H. Away from the sample the spin polarization is maintained by a weak guide field H_g (not shown) directed along the incident spin direction. **b** The reflectivity for a single interface is shown as a function of reduced perpendicular wavevector q_i/q_c (solid line). At large wavevectors the reflectivity can be approximated by $(q_c/2q)^4$ (shown as a dashed line)

where

$$q_j = \sqrt{q_i^2 + q_{ci}^2 - q_{cj}^2} \tag{6.5}$$

and with the critical wavevector q_{ci} for the ith medium given by

$$q_{ci}^2 = \frac{2m_n}{\hbar^2} V_i. \tag{6.6}$$

Total reflection (for which $|r_{ij}| = 1$) therefore occurs for $q_i^2 < q_{cj}^2 - q_{ci}^2 = k^2 \sin^2 \theta_{cij}$ where θ_{cij} is the critical angle for the i, j interface. In this case the wavevector is real in the ith medium and pure imaginary in the jth medium assuming $q_{cj}^2 > q_{ci}^2$. The critical angle is related to the relative refractive index $n_{ij} = k_i/k_j = n_i/n_j$ (where n_i, n_j are the absolute refractive indices of the media) by

$$\sin^2 \theta_{cij} = 1 - n_{ij}^2. \tag{6.7}$$

Since the refractive index is slightly less than unity the vacuum has the larger refractive index in contrast to the optical case. The weakness of the interaction potential means that reflection will only occur with appreciable intensity at glancing incidence and values of the critical angle are typically $\sim 10^{-2}$ radians. Figure 6.1(b) shows the form of the intensity reflectivity for a single interface.

It is useful to see how the scattering density is related to the refractive index by considering the total coherent scattering of a medium (usually considered to be a slab for computational convenience). Assuming bound scatterers, b_i defines the strength of the s wave scattering by the atoms in the ith medium [6.7], and an element of volume δV within the solid gives rise to a scattered wave of the form:

$$\delta \psi_s(r) = \psi_o \rho_i b_i \frac{\exp(-ikr)}{r} \delta V \tag{6.8}$$

where r defines the radial distance from the scatterer and ψ_o is the incident wave amplitude. The wave emerging from the slab (assumed to lie in the x, y plane) is therefore described by summing the incident wavefield and the total coherently scattered wavefield due to all s wave scatters as given by (6.8). Exactly the same summation applies in physical optics [6.12] and the resulting expression must correspond to a wave of the form:

$$\psi_r = \psi_o \exp(-ik(n_i - 1)y) \tag{6.9}$$

This procedure yields an expression for the refractive index which is the same to the first order in the term $(1 - n_i)$ as that obtained from (6.7), provided we assume an optical potential proportional to $\rho_i b_i$, so justifying the form of the Fermi pseudopotential of (6.1).

Since we are interested in studying the magnetic moment per atom we need to see how the critical angle depends on the magnetisation. Using the relation $\mathbf{B}_i = \mu_o(\mathbf{H} + \mathbf{M}_i)$ for a ferromagnetic medium and recalling that the components

6.2 Theory of Polarized Neutron Reflection

of H parallel to the (x, z) plane and the normal component of B are continuous across the interface, we obtain using (6.1) the well known result [6.13]:

$$\sin^2 \theta_{cij} = \frac{1}{E}(V_{jn} - V_{in} - \mu_o \mathbf{\mu}_n \cdot (\mathbf{M}_{j//} - \mathbf{M}_{i//})) \qquad (6.10)$$

Thus only the in-plane components of the magnetisation $\mathbf{M}_{j//}$, $\mathbf{M}_{i//}$ determine the reflectivity from a ferromagnetic medium. We can make further progress if we write the magnetisation $\mathbf{M}_i$ as an average dipole moment per unit volume, since this scales with the density of scatterers: $\mathbf{M}_i = \rho_i \mathbf{\mu}_i$, where the magnetic dipole moment per atom $\mu_i = g\mu_B s_i$ with $g = 2$, μ_B is the Bohr magneton and s_i is the spin per atom, assuming magnetic saturation of the medium. We can now write the spin-dependent critical angle in terms of a spin-dependent scattering length $b^\pm$:

$$\sin^2 \theta_{cij}^\pm = \frac{4\pi}{k^2}(\rho_j b_j^\pm - \rho_i b_i^\pm)$$

where:

$$b_i^\pm = b_i \pm 2cs_i \qquad (6.11)$$

with the signs corresponding to the neutron spin parallel ($+$) or antiparallel ($-$) to the magnetization where $c = m_n/2\pi\hbar^2 \mu_B \mu_n = 0.2695 \times 10^{-15}$ m, i.e. c converts the moment per atom in Bohr magnetons to an effective length (e.g., in Fe, $2s = 2.22$). The real part of $\sin^2 \theta_{cij}^\pm$ is seen to vary as λ^2 and so dispersion occurs; the imaginary part is found in practice to scale with λ since it relates to a constant absorption cross section which is itself related to the product of the imaginary part of the scattering length and wavelength via the optical theorem [6.14, 15]. This feature does no emerge from the *Fermi* pseudopotential [6.7].

We should note that (6.6) applies to one spin state only, and that via (6.1), the value of r_{ij} (given in (6.4)) will change according to the orientation of the spin with respect to the magnetic induction. This result provides the basis for the use of polarized neutron reflection as a magnetometric technique. The measured quantity is $R^\pm(q) = |r_{ij}^\pm(q)|^2$. In PNR magnetometry it is useful to define a quantity known as the spin asymmetry, S given by

$$S = \frac{R^+ - R^-}{R^+ + R^-} \qquad (6.12)$$

This quantity allows a direct comparison of the difference in the reflectivities due to the spin dependent magnetic interaction, and as we shall see later yields the in-plane magnetization directly. A similar quantity is used to define the beam polarization in spin-polarized electron spectroscopies – see for example Sect. 4.1 by Hopster in this volume and the chapter by *Unguris, Celotta* and *Pierce* in Volume II. The related quantity $F = R^+/R^-$ known as the flipping ratio is often reported since it is the ratio of the directly measured reflectivities. In principle, a measurement of the critical angle is sufficient to determine the coherent scattering length. Two such measurements for parallel and antiparallel spin states allow the magnetization discontinuity and the nuclear scattering length to be determined. In the case of a stratified medium, the entire reflectivity curve has to be measured as a function of wavevector in order to determine the profile corresponding to $V(y)$. An inversion of the reflectivity curve data to yield the profile cannot be performed except in certain special limiting cases [6.16]. In practice an assumed profile is varied iteratively until agreement is obtained between the measured and calculated reflectivity values. The procedure is only reliable if the data extends over a sufficiently large wavevector range (limited in practice by the rapid fall in reflectivity with increasing wavevector).

It is straightforward to extend the above results to a multilayer if we follow an approach frequently used in multilayer optics. Consider a three medium system with the wave incident from the first medium. Defining $A_i = 1$ and $B_i = r_{123}$ we obtain the following expression for r_{123} by summing over all possible reflection processes (as in an etalon):

$$r_{123} = r_{12} + t_{12}r_{23}t_{21}\phi_2(1 + (r_{21}r_{23}\phi_2) + (r_{21}r_{23}\phi_2)^2 + \ldots)$$

where

$$\phi_2 = \exp(2iq_2 t_2), \qquad (6.13)$$

with t_2 the thickness of the second layer. This result is instructive since we see that infinitely high order reflection processes in principle to r_{123}, whereas we expect in practice only the lower order terms to contribute. Beyond the total reflection region, the small reflectivity will lead to a rapid reduction in the contribution of higher order terms, but even in the total reflection region where the reflectivities are both of unit magnitude, the finite coherence length of the beam must mean that sufficiently high order processes are suppressed. However, near the critical region high order processes do indeed occur in PNR experiments on high quality samples, as we shall discuss in Sect. 6.4. Provided we are beyond the critical region, i.e. $|r_{21}r_{23}| < 1$, we can sum the series, obtaining:

$$r_{123} = \frac{r_{12} + r_{23}\phi_2}{1 + r_{12}r_{23}\phi_2}. \qquad (6.14)$$

Having obtained a similar expression for r_{ijk} we can to write down an iterative expression for the reflectivity of a four medium system in terms of r_{12} and r_{234}. This approach can be extended to an arbitrary number of layers provided the spin states are always eigenstates of each of the magnetic media. In this case the reflectivity can be computed separately for each spin state.

6.2.2 Transfer Matrix Methods and the Polarization Dependent Reflectivity

The following section extends the discussion to the case of non-aligned layers and can be omitted if the reader is concerned only with single, spin aligned magnetic films.

6.2 Theory of Polarized Neutron Reflection

It is appropriate at this point to introduce the transfer matrix method often used for calculating the reflection coefficients for structures consisting of an arbitrary number of layers using an approach also widely used in the physical optics of multilayers [6.17]. In the neutron-optical case the wave in the ith medium is in general the sum of a spin-up and spin-down wave, in analogy to the optical case where the wave is in general a sum of two polarization modes. This wave can be represented as a four-component vector given by $\psi_i = (\psi_i^+, \psi_i^-)$ where each of the spin dependent component waves are represented by a two-component vector of the form $\psi_i^\pm = (A_i^\pm, B_i^\pm)$ the components of which are the spin dependent amplitudes of the forward and backward travelling waves in the ith medium [6.3]. The perpendicular wavevector $q_i^\pm$ depends on the spin orientation according to (6.5, 6).

In the case where the magnetization orientation varies in the (x, z) plane between layers, we need to take into account the axis of spin quantization. This is because a spin-up state defined with respect to the external magnetic field projects onto both the up and down components of the spin states defined with respect to the magnetization within the medium. The case of neutron *reflection* from magnetically non-aligned layers has been recently worked out in detail [6.18, 19], although this situation is familiar in the case of diffraction from superlattices or bulk structures. If the direction of the quantization axis changes in the plane of the film by an angle Φ_{ij} between media i and j then the process of refraction at the interface boundary the adjacent media i and j can be described as a transformation of the vector Ψ_i to Ψ_j as [6.19]:

$$D_i \Psi_i = R_{ij} D_j \Psi_j: \tag{6.15}$$

where the matrix D_i is a 4×4 'refraction' matrix and where the spin rotation matrix R_{ij} satisfies the 4π spinor symmetry and is a function of Φ_{ij} only. The matrix D_i is a function of $q_i^\pm$ only. Equation (6.15) describes the continuity of the wavefunction amplitudes and derivatives at the interface. It therefore provides information on how to obtain the amplitudes of the forwards and backwards travelling waves in the medium j, given that we know the form of these waves in the medium i. Once we have these amplitudes we have fully described the wave since we know the value of the wavevector within the medium from (6.5, 6). For the case of semi-infinite media i and j, (6.15) can be used to obtained the Fresnel reflectivity of (6.4) directly. The ith medium extends over a thickness t_i and so we need to introduce a further 4×4 matrix L_i which describes the change in phase associated with the wave propagating from $y - y_i$ to $y = y_i - t_i$: accordingly, the matrix L_i changes the phase of each component due to the propagation of the wave in the medium and is therefore a function $(q^\pm_i t_i)$. The result corresponds to the wave in the same medium i but at the extreme side of the layer in the negative y direction.

We must now consider four possible reflection processes: a spin conserving process and a spin flipping process for each of the two possible initial states. For example, for incident spin-up neutrons, both a spin-down and spin-up reflected state occurs if in any layer the magnetization is inclined with respect to the guide

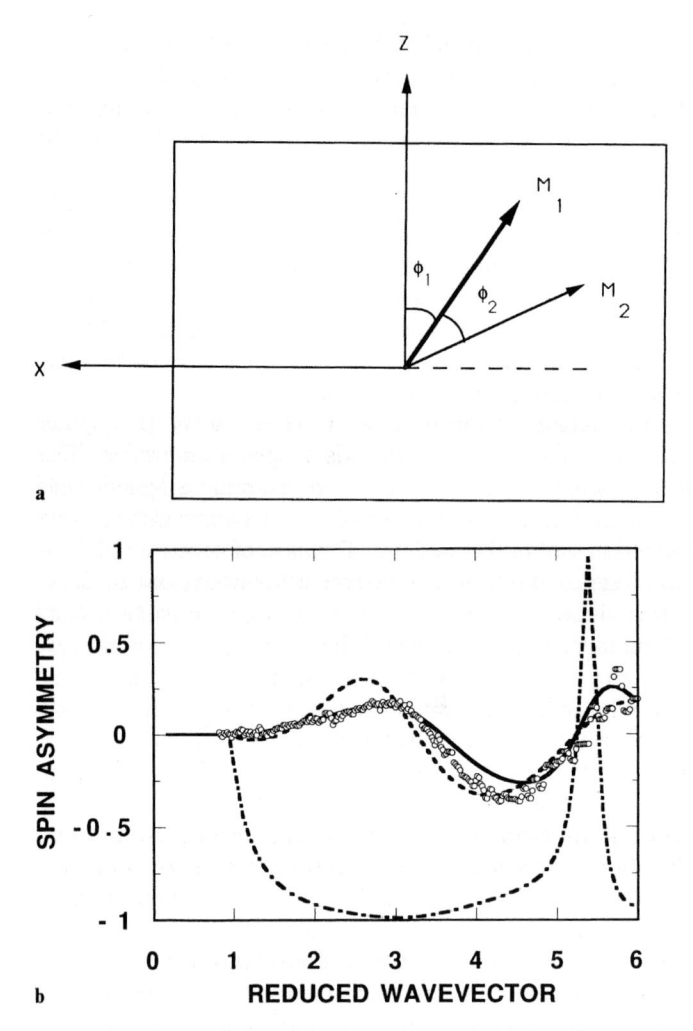

Fig. 6.2. a The sample geometry for a multilayer sample with non-aligned layer magnetizations. The sample surface is contained within the (x, z) plane as in Fig. 6.1a. The magnetization vector M_1 (thick arrow) of the first layer and the magnetization vector M_2 of the second Fe layer are constrained to lie in the film plane. The angle Φ_1 refers to the orientation of M_1 with respect to the applied field direction along z and the angle Φ_2 defines the angular separation of M_1 and M_2 (narrower arrow), with a positive value corresponding to the counter-clockwise sense. b The observed flipping ratio for the 100 Å Cr/50 Å Fe/15 Å Cr/50 Å Fe/Si sputtered sandwich structure at $H_a = 12$ G upon reducing the field from positive saturation (open circles) [6.20]. The solid line shows the ratio calculated for $\Phi_1 = 3\pi/8$, $\Phi_2 = \pi/4$ with $M_1 = M_2 = M_s$. The dot-dashed line shows the ratio calculated for $\Phi_1 = \pi$, $\Phi_2 = \pi$ with $M_1 = M_2 = M_s$

field, with the associated reflectivities given by r^{++} and r^{+-}. Where there is no reorientation of the magnetization direction, then an up-spin incident on the sample is transmitted and reflected in the same spin state. The reflectivity and transmission coefficient of a multilayer can be directly obtained from the matrix relation

$$\Psi_1 = T_{1\ldots N} \Psi_N \tag{6.16}$$

where

$$\Psi_1 = \begin{pmatrix} 1 \\ r^{++} \\ 0 \\ r^{+-} \end{pmatrix} \text{ and } \Psi_N = \begin{pmatrix} t^{++} \\ 0 \\ t^{+-} \\ 0 \end{pmatrix} \tag{6.17}$$

and where the matrix $T_{1\ldots N}$ is given by

$$T_{1\ldots N} = D_1^{-1} R_{1,2,\ldots\ldots} D_{N-1} L_{N-1} D_{N-1}^{-1} R_{N-1,N} D_N. \tag{6.18}$$

The wave Ψ_N in (6.16) is operated on by a sequence of matrices according to (6.18). First the wave in medium N at the $N-1, N$ interface is transformed to that in medium $N-1$ on the positive side of the interface. The wave is then propagated to the other side of the medium $N-1$ and the process continued until the vacuum is reached (medium 1). Here the substrate is the Nth medium and the neutron wave is incident in the first medium. In addition to the reflectivity, the transfer matrix $T_{1\ldots N}$ provides a direct method for calculating the depth dependent intensity of the neutron wave within the solid.

In the conventional PNR magnetometric measurement, the magnetization is saturated along the direction of the incident neutron polarization as shown in Fig. 6.1; however, in the case where the magnetization in the sample is not aligned with the incident spin state the reflected beam does not retain the polarization state of the incident beam. For general orientations of the incident spin (including the case where the incident spin is perpendicular to the magnetization) the reflectivity is in general dependent upon both of the in-plane components of the magnetization vector. In order to measure only the nuclear potential, it is therefore necessary to magnetically saturate the sample along the surface normal. This measurement is useful in determining the non-magnetic part of the optical potential independently of the magnetic part. In general, the reflectivity matrix is a function of both the magnitude and the orientation of the magnetization. In this case polarization analysis of the reflected beam can be used to investigate the magnetization vector orientation and magnitude within the sample [6.18], including the case where the magnetization in each layer is perpendicular to the polarization of the incident beam. The spin asymmetry is then zero independently of the magnetization magnitude.

For multilayer samples with non-aligned magnetizations, the spin asymmetry can be very sensitive to both the orientation and magnitude of the magnetizations, and so conventional PNR can provide an experimental method for investigating the layer-dependent magnetization vector [6.19]. PNR has been used in this way to investigate the field-dependent vector orientation of the individual Fe layer average magnetizations in an antiferromagnetically coupled polycrystalline Cr/Fe/Cr/Fe/Si sandwich structure (Fig. 6.2(a) [6.20]. The magnetically saturated state is found to correspond to a uniform ferromagnetic alignment of the full Fe layer magnetizations. The PNR measurements show that upon reducing the field from the saturation value to almost zero, the magnetization vectors of the two Fe layers do not align purely antiparallel as a single domain state. The observed spin asymmetry is compared in Fig. 6.2(b) with the results of calculations which assume that each layer is uniformly magnetized. A canted orientation almost fits the data. However, no single domain configuration (i.e. with the magnetization in each layer given by the full saturation value) is found to yield an exact fit to the data throughout the entire wavevector range studied experimentally and a significant difference remains between the simulated and observed flipping ratio for all values of the orientation of the magnetization vectors, suggesting that multidomain formation occurs. It is found that a close fit can be obtained by varying both the magnitude and the orientation of the magnetization vectors in each layer. These results illustrate the sensitivity of PNR to the magnetization vector profile in multilayer structures.

6.2.3 PNR Magnetometry of Single Magnetic Films

In this section we see how magnetometric information can be obtained from PNR for simple overlayer systems. We first consider a single uncoated magnetic layer and refer to the results we obtained in Sect. 6.2.2 for a three layer medium which we now choose to correspond to the wave incident in the vacuum (1) upon a magnetic film (2) supported by a non-magnetic substrate (3). From (6.14) we can rewrite r_{123} without approximation in terms of the uncoated substrate reflectivity r_{13} as

$$r_{123} = r_{13} + \Delta \quad \text{where} \quad \Delta = \frac{(\phi_2 - 1)r_{23}(1 - r_{12}r_{23})}{(1 + r_{12}r_{23})(1 + r_{12}r_{23}\phi_2)} \tag{6.19}$$

A number of insights can be gained from this (exact) result. First, the thickness dependence of the perturbation is due only to the phase dependent term ϕ_2. The magnetic overlayer introduces an oscillatory perturbation Δ to the reflectivity, but the reflectivity has a value equal to that of the uncoated substrate when $q_2 t_2 = n\pi$ where n is an integer (corresponding to $\phi_2 = 1$). Notice that this condition is spin dependent (since q_2 is itself spin dependent) and approximately corresponds to the condition for a whole number of half wavelengths to fit in the magnetic film so that the wavefield is perfectly matched onto the substrate. Secondly, the strength and sign of the spin dependent perturbation Δ is dictated by the difference between V_2 and V_3 (i.e. the contrast) and so scales with r_{23}: for example we see that in the case when $V_2 < V_3$ then $R_{123} < R_{13}$. We note that $R^+ > R^-$ since $V^+ > V^-$, i.e. the two spin dependent reflectivities are in antiphase. The minima in R^- can reach zero when the antireflection condition $q_2^2 = q_3 q_1$ is satisfied corresponding to $r_{12} = r_{23}$. This result can readily be seen from (6.19) for $\phi_2 = -1$, and is analagous to the condition for a $\lambda/4$ coating in 'blooming' a camera lens (i.e. the impedance matching condition). The minima approximately occur in all cases when an odd number of quarter wavelengths span the film. These results are illustrated in Fig. 6.3 where we show the reflectivity for a 100 Å Fe/Au overlayer structure as a function of reduced wavevector q/q_{c3}, where q refers to the incident perpendicular wavevector. Throughout this chapter we shall always use the substrate critical wavevector to define the reduced wavevector. As expected, the spin asymmetry first peaks for the quarter wavelength condition corresponding to a minimum in R^-. The value of wavevector at which this occurs can be used to accurately determine the overlayer thickness.

We now need to calculate the spin asymmetry by considering the spin dependence of the perturbation Δ. This can be accomplished exactly using (6.19) but it is insightful to consider the dependence of Δ on the magnetic properties of the second layer. In the case of an ultrathin film satisfying $q_2 t_2 \ll 1$ we can expand the exponential term ϕ_2 in increasing powers of $q_2 t_2$ and obtain to second order the following approximate expression:

$$\phi_2 \sim 1 + 2iq_2 t_2 - 2q_2^2 t_2^2. \tag{6.20}$$

Combining this result with (6.19) we see that the perturbation in the reflectivity Δ due to the magnetic layer results from the *phase change* associated with the neutron wave traversing the magnetic layer. This is quite a remarkable fact since the perpendicular component of the wavelength λ/θ is orders of magnitude larger than the thickness. In the wavevector region for which (6.20) applies, the change in intensity reflectivity due to the magnetic layer therefore scales as the thickness t_2 squared since the part linear in t_2 is out of phase by $\pi/2$ with respect

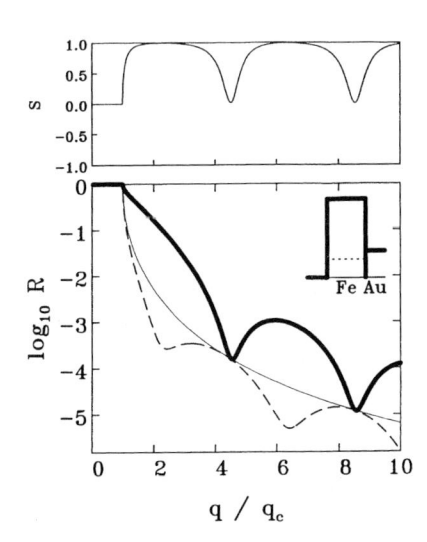

Fig. 6.3. Lower panel: The reflectivity for spin-up (+) neutrons (shown as a solid line) and for spin-down (–) neutrons (shown as a dashed line) for a 100 Å Fe/Au overlayer structure. Upper panel: the corresponding spin asymmetry S. Inset: the optical potential for the two spin states

to the beam reflected from the substrate. In this approximation the following approximatie expression for S is found to apply:

$$S = f(q)\rho_2 s_2 t_2^2, \tag{6.21}$$

where the function f depends only on the wavevector and the relevant nuclear potentials, and can be readily calculated from (6.19). We can summarize this result by stating that for a single uncoated magnetic layer the spin asymmetry scales linearly with the product of the total magnetic moment of the layer m_i and its thickness t_2, where $m_2 = M_2 t_2$ (Sect. 6.2.1). A significant spin asymmetry occurs only at large wavevectors where the film thickness approaches the half wavelength condition. Since the reflectivity falls dramatically with increasing wavevector, falling as q^{-4} at large wavevectors (Fig. 6.7b), a method is required for enhancing the asymmetry in the experimentally accessible wavevector region.

By overcoating the ultrathin ferromagnetic film with a non-magnetic layer of appropriate thickness, the sensitivity of the spin asymmetry to the layer magnetization at low wavevectors is increased. By using the same non-magnetic metal for both the single crystal substrate and the overlayer, both of the magnetic/non-magnetic metal interfaces become equivalent. Extensive spin polarized band structure calculations have been made for this specific case in order to study the role of the interface in determining the magnetic moment [6.1, 2]. The flipping ratio F for such a four medium system is given to first order in the magnetic film thickness t_3 by:

$$F = 1 + \frac{4(cs_3\rho_3)q_1 t_3}{b_2 \rho_2} \sin(2q_2 t_2) \text{ for } q_3 t_3 \ll 1. \tag{6.22}$$

For very thin magnetic layers this expression can be used to approximate the flipping ratio quite well (Fig. 6.4a) where the approximation is compared with the exact calculation for a monolayer of Fe). For thicker layers (Fig. 6.4b) the approximation fails but it is useful in showng that the spin asymmetry is a function only of the effective nuclear scattering densities, the non-magnetic overlayer layer thickness and the total magnetic moment of the layer. We note only that in the region of the enhancement in the spin asymmetry the *form* of S as a function of wavevector is to a first approximation *independent of the ultrathin magnetic layer thickness* for a symmetrically sandwiched ultrathin film structure, as shown by (6.22). The non-magnetic overlayer thickness determines the wavevector value at which the peak asymmetry occurs, while the value of the peak asymmetry yields the moment of the magnetic layer. To a good approximation, the spin asymmetry now scales linearly with the magnetic film thickness, reaching a maximum for an overlayer thickness t_2 corresponding to one eighth of a wavelength, and is therefore enhanced substantially with respect to that obtained for the equivalent uncoated structure at low wavevectors. The magnitude of the relative enhancement in spin asymmetry at low wavevectors with respect to the uncoated case increases with increasing overlayer thickness. The mechanism for the enhancement in the spin asymmetry is associated with

6.2 Theory of Polarized Neutron Reflection

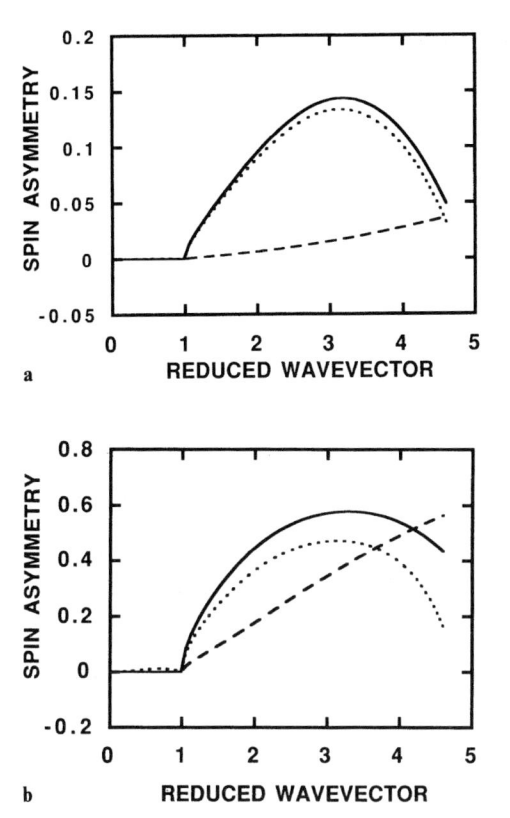

Fig. 6.4. a The spin asymmetry S for a 1 MLFe/Ag overlayer structure (dashed line) and for a 50 Å Ag/1 ML Fe/Ag sandwich structure (solid line) showing the enhancing effect of a non-magnetic overlayer. The dotted line corresponds to the approximation of (6.22). **b** The spin asymmetry, S for a 5.8 ML Fe/Ag overlayer structure (dashed line) and for a 50 Å Ag/5.8 ML Fe/Ag sandwich structure (solid line) showing the effect of a non-magnetic overlayer. Enhancement in this case is only significant at small wavevectors. The dotted line corresponds to the approximation of (6.22) which begins to fail at this thickness

the additional change in the phase of the neutron wave in transversing the overlayer, allowing the term in the expansion of ϕ_3 which is linear in t_3 to contribute.

By comparing the observed spin asymmetry with that calculated exactly for the sandwich structure, it is possible to extract the absolute value of the magnetic moment $m_3 = g\mu_B \rho_3 s_3 t_3$ of the ferromagnetic layer when it is magnetically saturated in-plane. The approach was used in early measurements of the magnetic moment per atom in symmetrically sandwiched ultrathin epitaxial layers [6.21–23]. However, in practice, the magnetic layer thickness has to be known for the magnetization to be extracted from measurements of the size of the peak spin asymmetry at low wavevectors while all other relevant parameters can be obtained by fitting the reflectivity data over a sufficiently large wavevector range. For a very high brightness neutron source it would be possible to determine t_3 from the reflectivity measurements at large wavevectors provided that the background and diffuse scattering is sufficiently low for this region to be accessed. The spin asymmetry calculated as a function of reduced wavevector for both a free and a sandwiched 5.8 ML bcc Fe layer is shown in Fig. 6.4. The Fe atoms are assumed for the purpose of calculation to have a magnetic moment per atom equal to 2.6 Bohr magnetons. The strongly enhanced response obtained for the sandwich structure at low wavevector compared with that obtained for the uncoated sample is clearly demonstrated. The size of the peak asymmetry is similar in both cases (outside the wavevector range of the figure for the uncoated structure) but the effect of the overlayer is to compress the oscillation in q in order to bring the maximum into the observable range. In the absence of ferromagnetic ordering of the Fe layer the asymmetry is unity independently of wavevector. For thicker films for which $q_3 t_3$ approaches unity, the distorted sine curve expression is no longer valid. It is indeed fortunate that chemically sealing an ultrathin layer by incorporating it within a sandwich structure enhances the spin asymmetry since it becomes possible to investigate such structures *ex situ*. In SMOKE, the overlayer attenuates the Kerr signal, but the sensitivity of SMOKE makes it nonetheless possible to investigate submerged ultrathin layers provided the overlayer does not greatly exceed the optical skin depth. In metals this length is typically around 10–20 nm at visible wavelengths and therefore of the same magnitude as the overlayer thickness required for PNR enhancement. It is thus possible to investigate the same sandwich structures with both SMOKE and PNR which provide complementary measurements of the field-dependent relative magnetization M/M_s and M_s respectively.

Since the reflectivity falls rapidly with wavevector, away from the critical angle a single reflection approximation is often valid for reduced wavevectors exceeding two or three, with the exact value depending on the parameters of the media. In this case the denominator in (6.14) can be replaced by unity with percent accuracy, so simplifying the expression of equation (6.14) but retaining the effect of refraction. This approximation is useful in interpreting the oscillatory structure which appears in the reflectivity curve for a multilayer.

6.2.4 The Diffraction Limit

If the incident kinetic energy associated with the perpendicular wavevector is large compared with the optical potentials of each of the layers in a multilayer system then we approach the conditions for kinematical diffraction to occur. Expanding q_j to first order in V_j in (6.6) we can rewrite the reflectivity r_{ij} to first order in V as

$$r_{ij} = \frac{1}{4q^2}(V_j - V_i) \tag{6.23}$$

We can therefore write the reflectivity of a multilayer to first order in V_i, V_j as

$$r_{ij} = \frac{1}{4q^2}\sum_j \exp(2iqy_j)(V_{j+1} - V_j) \quad \text{for } j = 1, \; n-1 \tag{6.24}$$

At large wavevectors this expression corresponds to an approximation which is formally equivalent to the kinematic limit used to describe X-ray diffraction from metal multilayers, as discussed in Sect. 6.5.3. Here the wavevector change Δq upon scattering is given by $2q_i$. This limit is also predicted by (6.14) where the denominator is approximated by unity and the wavevector within the medium is replaced by the incident wavevector. Several approximate results for the reflectivity have also been given *Lekner* for overlayer systems [6.25]. For a medium described by a continuously varying 1D potential $V(y)$ we can rewrite (6.24) as the Fourier transform of the potential gradient $V'(y)$ [6.19, 24]. In the case of ultrathin metal overlayers the kinematic limit is reached at typically around $q_i/q_c \sim 5\text{–}6$, corresponding to the upper end of the wavevector range accessible in PNR. As an illustration of these approximations, we show in Fig. 6.5 the spin asymmetry for a 20 Å Ag/60 Å FeMn/40 Å FeNi/20 Å Cu/40 Å FeNi/Si overlayer structure assuming ferromagnetic alignment of the ferromagnetic FeNi layers and that that FeMn layer is antiferromagnetic. The asymmetry is calculated in (i) the exact case (thick solid line) and the (ii) single reflection (thin solid line) and (iii) diffraction limit (dashed line). It can be seen from this simulation that at high wavevector the spin asymmetry is well described by the diffraction result. At intermediate wavevectors, refraction is seen to remain important and this is retained in the single reflection model. Using (6.24) it can be shown that the spin conserving and spin-flipping reflectivity amplitudes are different functions of the magnetization vector (as is the general case for the region of partial reflection) and so measurements of these two reflectivities can be used to obtain the magnetization vector [6.18].

PNR also provides a means of determining the magnetization profile in single layers [6.26, 27]. In the kinematic limit, the magnetization vector profile can in principle be directly determined from PNR. In practice, the limit on the intensity of the incident beam places an upper limit on the wavevectors which can be accessed. This corresponds to a depth resolution in the nm thickness range for currently available neutron sources. Such a profile can arise from the *intrinsic* magnetic properties of the system investigated: one example is the magnetization profile due to the influence of surface anisotropies; a second example is the surface magnetization profile on the scale of the magnetic

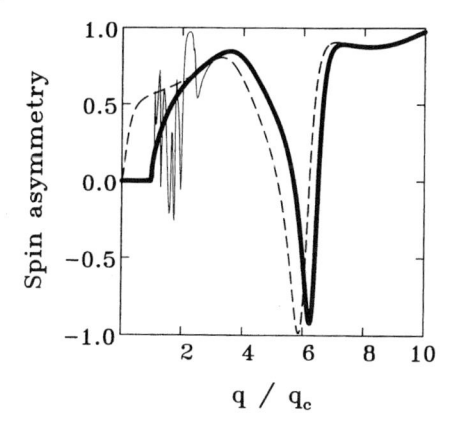

Fig. 6.5. The spin asymmetry S calculated for a ferromagnetically aligned 20 Å Ag/60 Å FeMn/40 Å FeNi/20 Å Cu/40 Å FeNi/Si structure without approximations (solid line), in the single reflection approximation (thin line) and in the diffraction limit (dashed line). The single reflection approximation is seen to be valid for this structure for $q > 3q_c$, since the thin line and thick line coincide in this wavevector range

coherence length which in a ferromagnet diverges at the Curie temperature [6.8]. Unfortunately the bulk signal dominates the magnetic response in the temperature region in which the coherence length approaches the depth resolution of PNR, making the direct observation of the surface magnetization profile very difficult.

6.2.5 Rough Interfaces and Wave Coherence

So far we have assumed the existence of sharp interfaces. In practice defects, steps, interdiffusion, etc. will destroy this condition introducing an effective interface roughness. We shall need to take this into account in quantitatively describing the reflectivities obtained for real samples. In optics, for multiple beam interference to occur in a multilayer stack, the surfaces must be optically smooth such that the variation the phase $\Delta(2q_i t_i) \ll 2\pi$. We see that the condition for optical smoothness in reflection is that interface (thickness) fluctuations Δt_i must satisfy

$$\Delta t_i \ll \frac{\lambda}{2\theta} \tag{6.25}$$

at large wavevectors for which $q \gg q_{ci}$. For light, this implies that the surfaces must be smooth on a lengthscale corresponding to a fraction of a wavelength. For neutrons the situation is different since θ_c (and therefore θ) is small and we require that the fluctuations are small on the scale of the perpendicular wavelength, i.e. $\Delta t \ll 0.02$ μm for cold neutrons ($\lambda \sim 10$ Å). This is fortunate, since PNR studies of real metal ultrathin films would be impossible if the refractive index was as large on the scale of as in the optical case.

It is also appropriate to consider the requirements for the monochromatization and angular collimation of the beam in terns of the coherence of the incident neutron wave. We can write the uncertainty in total wavevector in terms of the wavelength and angular spread of the incident beam $\Delta\lambda$ and $\Delta\theta$ as

$$|\Delta q|^2 = q^2(\alpha^{-2} + \beta^{-2}), \tag{6.26}$$

where the wavelength resolution $\alpha = \lambda/\Delta\lambda$ and the angular resolution $\beta = \theta/\Delta\theta$. We can write $\Delta q = 2\pi/l_c^{\text{trans}}$ where l_c^{trans} is the transverse coherence length which is therefore determined by α and β. Typically these quantities are around 10 in the actual experimental situation from which a value for l_c^{trans} of $\sim 1 \mu$m results. We can conclude that wave interference is possible across a multilayer stack of this thickness and that multireflection processes are important in ultrathin films. The longitudinal coherence length of the neutron beam l_c^{long} for comparison is in the 10–20 nm range for cold neutrons. Furthermore, in order for the sample not to degrade the wavevector spread which can be achieved in PNR experiments, the surface must be flat to better than 10^{-3} radians.

What is the distance in-plane over which the neutron wave is coherent? To estimate this distance, consider the neutron wave undergoing reflection from a surface, as shown schematically in Fig. 6.6a. The coherent part of the incident wave front extends over a width $\sim l_c^{\text{trans}}$ perpendicular to the incident wavevector as defined by the points S_0, S_0'. The coherent part of the beam is incident on the surface at the points A_1, A_1' which correspond to the virtual points S_1 and S_1'. Provided the distance $S_1 S_1' \tan\theta$ does not exceed the longitudinal coherence length, a condition which is well satisfied in most experiments for small θ, the effective in-plane coherence length can be estimated as the distance $A_1 A_1' \sim l_c^{\text{trans}}/\sin\theta$. Typically this length is of the order 100 μm. At large θ the longitudinal coherence length determines the effective in-plane coherence length. In the case of reflection from an overlayer, the source point S_0 on the incident wavefront appears as the virtual sources $S_1, S_2 \ldots$ due to multiple reflection, as shown in Fig. 6.6b. In principle an infinite number of reflections occurs, but the effective number reflections N is determined by the beam coherence according to the condition that the distance $S_1 S_N \sim l_c^{\text{trans}}$.

We now need to take into account the effect of a local variation in scattering density at the interface (equivalent to an interface roughness associated with variations in the position of the interface) upon the reflectivity. *Nevot* and *Croce* [6.28] showed that for X-rays reflected from an interface exhibiting a Gaussian roughness distribution, the specular reflectivity r_{ij} is modified by a roughness factor to become $r_{ij} \exp(-W_{ij})$ where W_{ij} is given by

$$W_{ij} = 2q_i q_j \sigma_{ij} \tag{6.27}$$

and where $\sigma_{ij} = \langle \Delta y_{ij}^2 \rangle$ is the variance of the local fluctuation in interface position Δy_{ij} defined by

$$\Delta y_{ij} = y_{ij} - \langle y_{ij} \rangle. \tag{6.28}$$

The angular brackets indicate averaging over in-plane positions. It should be noted that for small roughness amplitudes, i.e. for $q_i q_j \langle \Delta y_{ij}^2 \rangle \ll 1$, then the exponential term expanded to first order in $\langle \Delta y_{ij}^2 \rangle$ is equivalent, in a first approximation, to the perturbed reflectivity given by (6.19) where the thin overlayer now corresponds to the 'rough' perturbed region with a scattering density reduced from that of the bulk substrate. A reduced specular reflectivity results with a corresponding increase in the transmission coefficient but no diffuse (off-specular) scattering results. A smoothly graded scattering density perpendicular to the interface, e.g., where the optical profile varies as $\tanh(\alpha y)$ provides a better approximation to the Nevot–Croce result. In principle a rough interface with no in-plane correlation of the fluctuations can be modelled by a graded interface. The Navot–Croce result is valid in PNR when the Gaussian roughness amplitude is small as defined above, uncorrelated (i.e. the fluctuations are random with respect to each other) in-plane and the average interface fluctuation is zero. This does not often correspond to the experimental situation for single crystal metal substrates, although for structures supported by optically flat semiconductor substrates the Nevot–Croce approach is appropriate. The general conclusions which can be drawn in considering the effect of roughness upon the neutron reflectivity for a single interface with root mean square

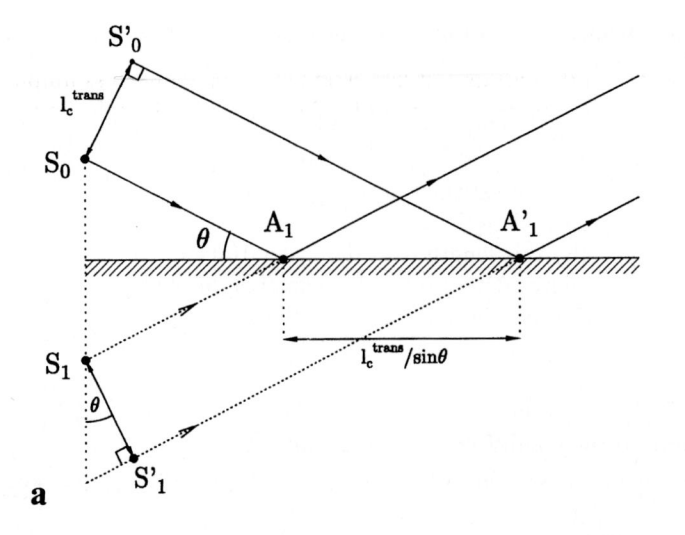

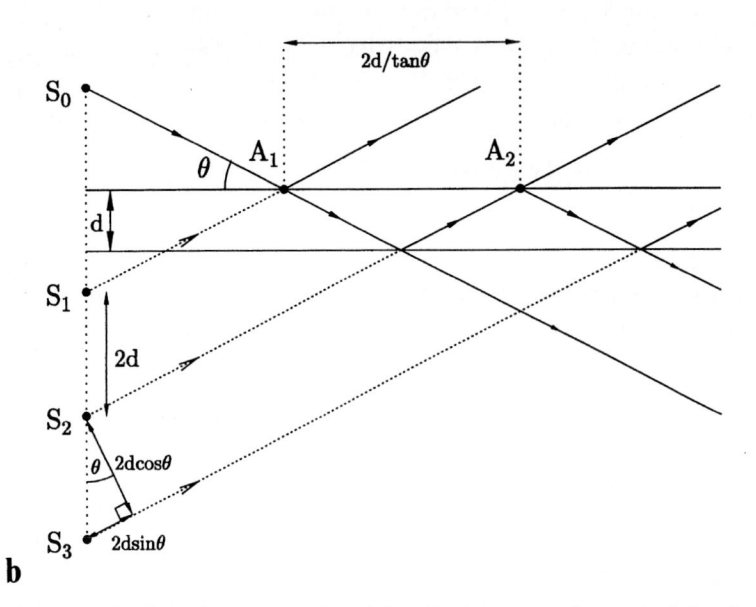

Fig. 6.6. a A schematic representation of the reflection process from a single interface, illustrating the effective in-plane coherence length in PNR. The coherent part of the incident wave front extends over a width l_c perpendicular to the incident wavevector defined by the points S_0, S'_0. The beam is incident on the surface at the points A_1, A'_1 which correspond to the virtual points S_1 and S'_1. The effective in-plane coherence length is given by $A_1 A'_1$. **b** The case of reflection from an overlayer is shown: the source point S_0 on the incident wavefront appears as the virtual sources $S_1, S_2, \ldots$ due to multiple reflection

roughness amplitudes satisfying $\sigma q_i^2 \ll 1$ are: (i) the ratio of the observed to ideal reflectivity curve, $\exp(-2W_{ij})$ yields a direct estimate of the roughness amplitude; (ii) the uncorrelated roughness does not affect the critical angle or the reflectivity in the total reflection region in this approximation; and that (iii) to first order we might therefore expect the effect of uncorrelated roughness not to significantly affect the spin asymmetry as is in fact the case for root mean square roughness amplitudes of 10 Å or less.

In practice roughness correlations do of course occur as a result of steps, terraces and dislocations at the surface or interface and the correlation function $\langle \Delta y_{ij}(r_{//}) \Delta y_{ij}(0) \rangle$ (where the angular brackets indicate averaging over radial-position $r_{//}$ in-plane) is now non-zero. Diffuse reflection occurs and we need to identify how this arises and the conditions for which it is important in PNR, particularly in the region of the enhancement condition used in PNR magnetometry. The diffuse scattering process in reflection has been considered by several authors for light [6.29] and also for X-rays and neutrons [6.30]. The geometry for diffuse scattering is illustrated schematically in Fig. 6.7 for a single interface bounding two semi-infinite media denoted as i and j. The wave incident at a glancing angle θ_i is refracted but is subsequently scattered from an impurity or defect within the vicinity of the interface. This gives rise to an incoherently scattered beam which emerges at a general angle θ_f with respect to the interface. The element of diffusely reflected intensity dI^d for scattering from the scattering from the incident wave state $\mathbf{k}_i$ into a solid angle element $d\Omega$ along the direction of the scattering state $\mathbf{k}_f$ is given in the distorted wave Born approximation by [6.30, 31]:

$$\frac{dI^d}{d\Omega} = \frac{I_0 q_{cs}^4 A_s}{16\pi^2 A_b} |G(\mathbf{q}_f, \mathbf{q}_i)|^2 |E(\mathbf{k}_{//f} - \mathbf{k}_{//i})|^2 \tag{6.29}$$

for $\sigma q^2 \ll 1$ where the function $|G(\mathbf{q}_f, \mathbf{q}_i)|^2 = |t_{ij}(\mathbf{q}_i)|^2 t_{ij}(\mathbf{q}_f)|^2$, I_0 is the total intensity of the beam incident upon the sample (integrated over the range of incident wavevectors), A_s is the sample surface area, A_b is the usable beam cross sectional area and q_{cs} is the critical wavevector of the medium containing the scatterer (in this case q_{cj}). The quantity E is the Fourier transform with respect to the wavevector transfer $\Delta k_{//} = k_{//f} - k_{//i}$ of the correlation function $e(r_{//})$ defined as $\langle \Delta y_{ij}(r_{//}) \Delta y_{ij}(0) \rangle$ where $r_{//}$ is the in-plane position. For comparison, the total specularly reflected intensity is given by $I_0 R(q)$, assuming a perfectly coherent incident beam. The approximation described by (6.29) is usually adequate for interpreting PNR data except at the critical angle. In the kinematic limit $G = 1$ and the result of (6.29) reduces to the Born approximation for scattering. Equation 6.29 describes a diffuse intensity varying as a function of scattering angle within the vicinity of the specular beam and also contributes to the intensity at the specular position itself. The diffuse intensity is proportional within the above approximation to the power spectrum of the roughness fluctuations $|E|^2$ weighted by the product of the transmitted intensity coefficients $|t_{ij}(\mathbf{q}_i)|^2$ and $|t_{ij}(\mathbf{q}_f)|^2$ (corresponding to the incoming and scattered waves, respectively) calculated for the sharp interface. A spin dependence of (6.29) enters

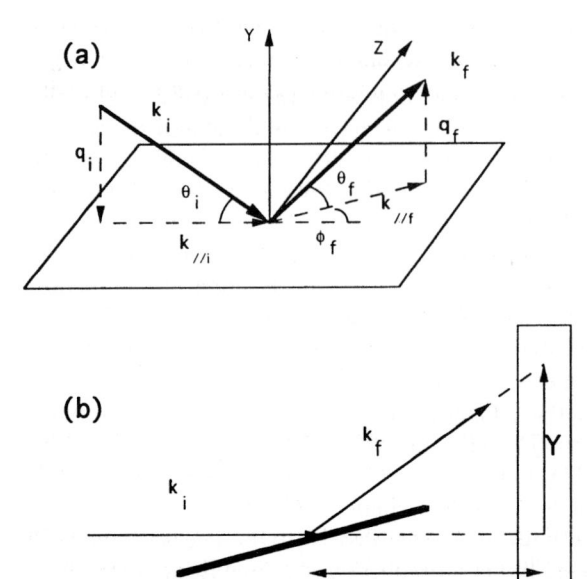

Fig. 6.7. a The diffuse scattering geometry is illustrated schematically. The incident wavevector k_i has components $k_{//i}, q_i$ parallel and a perpendicular to the sample surface respectively. The scattered wavevector k_f has components $k_{//f}, q_f$ parallel and perpendicular to the surface respectively. The incident glancing angle is θ_i and the scattered beam is inclined at an angle θ_f with respect to the surface. The azimuthal angle ϕ_f is also shown for the general case. **b** The scattering geometry for the multidetector is shown schematically for $\phi_f = 0$. The transmitted beam is detected at $Y = 0$ and the specular beam is detected at $Y = Y_r$. Off-specular scattering is detected at general positions Y. The perpendicular wavevector transfer $|q_f - q_i|$ is given to a good approximation by KY/R where R is the detector sample distance

via the spin dependence of these unperturbed transmission coefficients. The existence of a distribution of scatterers within the vicinity of the interface is also described by the function $E(\Delta r_{//})$ if interpreted as the Fourier transform of the local density of impurity scatters $n_s(r_{//})$ multiplied by an appropriate constant. We shall assume, following *Steyerl* [6.31], that the roughness spectrum is Gaussian and short-ranged and therefore can be described in terms of a roughness amplitude $\sigma^{1/2}$ and a correlation length w in-plane as

$$|E(\Delta k_{//})|^2 = \frac{\sigma w^2}{2\pi} \exp\left(-\frac{w^2 \Delta k_{//}^2}{2}\right). \tag{6.30}$$

To treat the multilayer case we now assume the scattering centre to be located in a medium l embedded in an N medium multilayer system. The diffuse scattering can be calculated for this case in the distorted Born wave approximation. At large perpendicular wavevectors (i.e. in the region of the enhancement peak in the asymmetry), to a good approximation the function G scales with the strength of the wave in the medium l. The term $|G|^2$ contains oscillatory structure with scattering angle due to interference terms arising in the wavefield within the medium. The depth dependence of the wave intensity in the solid determines which rough interfaces contribute to the diffuse scattering. As a result, the flipping ratio F can fall below unity in the vicinity of the critical region, as is observed experimentally for sandwiched ultrathin magnetic films prepared on rough substrates [6.22, 32], since the down-spin state illuminates the substrate interface more strongly in this wavevector range. For large wavevectors, all interface scatterers contribute equally. It can be shown for a single interface with short-range Gaussian roughness that the modified specular reflectivity has the same form as that given in (6.27).

For our present purpose the important conclusions to be drawn from (6.25–30) are as follows. (i) The absolute level of the diffuse scattering is maximized close to the critical condition since the transmitted wave intensity within the solid is maximized (for reflection from a semi-infinite medium $1 + r = t$ and $t = 2$ at the critical condition), but the diffuse scattering is a small fraction of the total scattered intensity because the reflectivity is also maximized. (ii) The *total* diffuse scattering intensity I_d integrated over all final scattering wavevectors satisfies

$$I_d \propto \frac{q_{cs}^4 \sigma}{q_i^2} \tag{6.31}$$

beyond the critical condition showing that the effect of a given roughness distribution at an interface will depend upon the scattering density of the material (the absolute reduction in specular reflectivity due to roughness is also proportional to the same quantity in the Nevot–Croce model for large wavevectors). (iii) The diffuse intensity accepted ΔI^d within a solid angle $\Delta \Omega$ at the specular position is independent of q at large wavevectors and satisfies

$$\Delta I^d \propto q_{cs}^4 \sigma w^2 \tag{6.32}$$

and so dominates the specular reflectivity (which falls as $(q_c/2q)^4$) for a sufficiently large wavevector value. The wavevector value at which this occurs can be used to estimate the magnitude of σw^2 and for a sufficiently well-collimated incident beam and sufficiently small detector acceptance angle (typically less than 10^{-6} steradians) diffuse scattering is negligible. (iv) At large wavevector, since $|G| \sim 1$, the diffuse scattering as a function of scattering angle θ_f has the form of a Gaussian centered at the specular position with an angular width $\Delta \theta_f \sim 1.4/kw$ [6.1] and so yields a direct estimate of the correlation length of the roughness fluctuations. (v) Structure in the diffuse scattering intensity as a function of wavevector becomes significant away from the specular condition only when either the incoming or outgoing beam is close to the critical condition, and therefore can usually be neglected. (vi) The effect of diffuse scattering is to reduce the spin asymmetry in practice. This is because the diffuse cross section in the specular direction has a weaker spin dependence than the specular reflectivity.

The approach outlined above often provides an adequate description of the experimentally measured diffuse scattering for realistic values of the parameters σ, w. An exact solution of the diffuse scattering intensity is not justified in general unless the actual roughness variation at a surface or interface is accurately known.

To fit the observed reflected beam profile as a function of θ_f it is necessary to include each of the effects of (i) an incident wavevector spread, (ii) a variation in the macroscopic sample flatness and (iii) diffuse scattering (usually important only at large wavevector). In addition, the background varies with θ_f and also must be included. A distortion of the reflected beam profile occurs in practice for values of q_i not too far from the critical angle and is principally due to the wavevector spread of the incident beam resulting in a significant intensity reflected at wavevectors smaller than the specular value corresponding to the mean incident wavevector. This is because the reflectivity is varying sharply with wavevector over the range defined by Δq_i. A lack of macroscopic surface flatness will further add to the effective incident wavevector spread. The wavevector transfer is determined by fitting the separation between the mean position of the reflected beam (Y) on the detector correlated for the reflectivity distortion of the profile in the intermediate wavevector region and the mean position of the transmitted beam (Fig. 6.7). At larger wavevectors reflectivity distortion of the profile is often less pronounced since the reflectivity varies less sharply with q_i. Nonetheless an asymmetry in the wings of the profile due to the reflectivity distortion is often still visible. However, the FWHM, approaches that of the incident beam with increasing wavevector in the absence of diffuse scattering. In order to fit the profile the scattered intensity $I(q_f)$ is assumed to be of the form:

$$I(q_f) = \int_{q_f - \Delta q_i/2}^{q_f + \Delta q_i/2} I(q_i) R(q_i) \exp(-2W) \, dq_i + \left(\frac{dI^d}{d\Omega}\right) \Delta\Omega \qquad (6.33)$$

where the first and second terms are the roughness modified specularly and diffusely reflected intensities respectively, where the factor $\exp(-2W)$ describes the overall reduction in reflectivity of the entire structure computed for rough interfaces according to equation 6.27 and $\Delta\Omega$ is the solid angle subtended by the detector. The function $I(q_i)$ describes the incident intensity distribution as a function of wavevector and the second term is evaluated using the expression in (6.29) evaluated at the mean incident wavevector since the diffuse scattering is a weakly varying function of q_i except in the vicinity of the critical wavevector. In practice, at large wavevector the lineshape can often be adequately represented as the sum of two components: (i) a 'specular' component with the lineshape of the incident beam and (ii) a 'diffuse component' with a Gaussian shape and a width $(\Delta q)_{\text{diff}}$. The diffuse scattering is fitted by treating the intensities of these two terms as adjustable parameters and the width $(\Delta q)_{\text{diff}}$ also an adjustable parameter.

6.3 Experimental Methods

The following section deals with the details of how the experiments can be performed. The reader who is not concerned with this aspect can move directly on to the discussion of results for ultrathin Fe films in Sect. 6.4

The weakness of the optical potential leads to critical reflection angles of the order of a degree or less for cold neutrons ($\lambda \sim 10$ Å) in most solids, requiring high collimation. This makes the use of short wavelength thermal neutrons ($\lambda < 1$ Å) unsuitable for PNR since the collimation requirements at such short wavelengths are prohibitive. In the PNR experiment a highly collimated long wavelength (typically 12 Å) neutron beam is incident at grazing angle upon the reflecting surface. The neutron spin is arranged to lie parallel or anti-parallel to the in-plane magnetization in the solid and the spin dependent neutron reflectivity is measured as a function of wavevector transfer normal to the surface close to the total reflection condition. Two basic approaches have been chiefly adopted in carrying out PNR experiments on ultrathin films: the first is the rotating sample (RS) method [6.9] in which the incident wavelength is fixed and the incident angle is varied, and the second a time of flight (TOF) method [6.33–35] at fixed incidence angle with variable wavelength. In both methods the reflectivity is determined as a function of variable incident wavevector. In the RS method a modified small-angle diffraction geometry can be used with either a multidetector or a θ–2θ scan with a single detector. The RS method is most appropriate for a continuous neutron source and provides the advantage that the usable beam width increases with incident wavevector. The TOF method has the advantages of the high wavelength resolution intrinsic to TOF techniques and a fixed sample geometry, and it is best suited to a pulsed source. However, it is the angular collimation of the detected beam which determines the level of diffuse scattering accepted and so it is this quantity which must be reduced as far as possible, making an intense collimated source necessary for both methods. The incident flux and background levels determine the signal to noise ratio which can be achieved with a given wavevector resolution. It is clear from (6.26) that it is desirable to match the angular resolution and the wavelength resolution if neither is to dominate the wavevector resolution; however, this consideration is only important when $S(q)$ varies rapidly with wavevector q (as occurs in thick films or multilayers) and as we have seen this is not the case for ultrathin films. We will briefly describe an apparatus of the TOF and RS types and refer the reader to the references of this section for further details.

6.3.1 Time of Flight Methods

A TOF polarized beam reflectometer at the Rutherford-Appelton Laboratory, U.K. has been fully described by *Penfold* [6.33] with a wavelength range of

330   6. Polarized Neutron Reflection

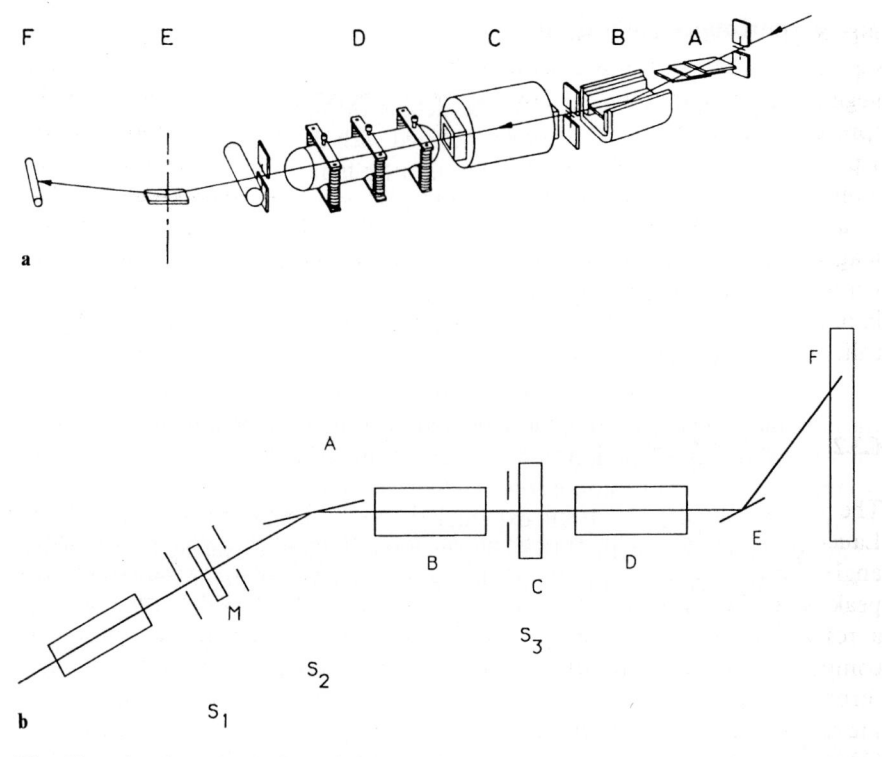

Fig. 6.8. a A schematic outline of the T.O.F. polarized beam reflectometer at the Rutherford–Appleton Laboratory (U.K.) (see text). The principal components are: (A) frame overlap Ni mirrors inclined at 1.3°; (B) a Co-Ti supermirror; (C) a two coil Drabkin flipper; (D) permanent magnet guide fields; (E) the sample position; (F) the detector [6.24]. **b** A schematic outline of a R.S. polarized beam reflectometer (see text) based on the D17 reflectometer at ILL, Grenoble. The principal components are: (A) a Co-Ti supermirror; (B) permanent magnet guide fields; (C) (D) a Miezei spin flipper; (D) permanent magnet guide fields; (E) the sample position; (F) the detector. The beam is monochromatized by a velocity selector and collimated (apertures S_1 and S_2) before illuminating A

2.5–13 Å (shown in Fig. 6.8a). The beam first passes through a 180° aperture disk-chopper rotating at 25 Hz to define the wavelength band. Ni mirrors (A) reflect out long wavelength contamination of the incident beam. The beam is reflected from a horizontal Co-Ti polarizing supermirror (B) [6.4 6] placed in the magnetic field of two Sm-Co permanent magnets placed in a C-shaped soft iron core. The spin state is reversed by a Drabkin two coil non-adiabatic spin flipper [6.37, 38] (C), passes through a uniform guide field region (to preserve the spin polarization) (D) onto the horizontal sample (E) and is detected by a He^3 single detector or multidetector at F placed 1.75 m from the sample behind a 2 mm cadmium sheet slit (S_3) which determines the single detector acceptance. The final beam size and incident collimation is defined by two cadmium sheet slits S_1, S_2 (placed between B and C and between D and E, respectively) separated by 2.63 m and usually of width 2 mm. Since the wavelength spread is negligible, the q resolution is determined principally by the overall angular spread of the beam, but this is determined by the ratio of the width of S_1 to the separation of the first slit and the sample position when the sample is over-illuminated, resulting in a fixed value of $\Delta q/q = 0.10$. Glancing incidence angles of around 0.35–0.6 degrees are typical for metal substrates. The sample is magnetized in-plane along a direction corresponding to the normal to the scattering plane using an electromagnet (H_a up to 1T). This apparatus was used in obtaining the measurements discussed in Sect. 6.2.2. The beam polarization is calibrated using a polarizing supermirror at the sample position [6.34].

6.3.2 Fixed Wavelength Methods

The measurements described below in Sect. 6.4 were carried out at the Institut Laue-Langevin (ILL), Grenbole using the RS method by adapting the D17 small angle diffractometer located on the H17 cold canal. A wavelength close to the peak of the liquid deuterium cold source distribution at 12 Å is selected with a resolution of 10% using a mechanical chopper. The arrangement of the components is schematically depicted in Fig. 6.8b. The beam passes in turn through a collimating slit S_1, an optical mirror M (used for laser alignment of the sample) and a collimating slit S_2 and is reflected from a vertical supermirror (A) in order to polarize the beam. The beam polarization at this point is close to 100%. The beam then passes through a guide field (B) (with $H_g \sim 100$ Oe) onto a Mezei Larmor precession spin flipper (C) [6.4, 36], through a second guide field (D) and is reflected from the sample (E) onto the detector (F). The spin flipper consists of two interpenetrating coils wound from 1 mm diameter Al wire in order to null the stray vertical guide field and produce a resultant horizontal precession field of approximately 5 Oe. A third collimating slit S_3 is inserted between the first guide field and the spin flipper. An evacuated tube with glass windows is placed between S_3 and the sample to reduced air scattering. A BF_3 multidetector is used which can be operated in either the high resolution (128×128 pixels) or low resolution (64×64 pixels). Each pixel is of area 5 mm $\times$ 5 mm and the sample-detector distance D is adjustable in the range 0.8–3 m. The detector is usually operated at a distance D of 2.83 m corresponding to an acceptance per pixel of 3×10^{-6} steradians. A 'box' of 3×3 pixels is normally used to determine the specular reflectivity. For over-illumination of the sample the incident angular resolution is controlled by the ratio of the width of S_2 to its separation from the sample (2.4 m) corresponding to $\Delta \theta_i \sim 10^{-3}$ radians. The overall q resolution is determined by the wavelength spread also. Typically $\Delta q_i = 0.14 q_i$ at the critical angle. In RS reflectometry, the q resolution increases with increasing wavevector. The sample is magnetized in-plane along the scattering plane normal using either an external electromagnet or a purpose built permanent magnet assembly around the sample holder ($H_a = 0.83$ kOe).

The peak flux per wavelength per unit of solid angle from the H17 canal is 2×10^{11} cm^{-2} sec^{-1} Å^{-1} sterad^{-1} resulting in a beam intensity of around 10^6 cm^{-2} sec^{-1} at the sample position for $\Delta\lambda = 1.2$ Å. Since the usable beam area is $L_x L_z \theta$ where L_x, L_z are the sample dimensions in plane, the intensity incident upon the sample is approximately 10^4 neutrons sec^{-1}. The sample vessel is mounted on a stepper motor controlled turntable. The angular orientation is controlled to within 0.01° by a computer controlled stepper motor. The flipping ratio is determined at each position of the sample in order to eliminate positioning errors. The current supply for the spin flipper also computer controlled and the counting time is determined by a count present determined from a monitor intercepting the incident beam. The background intensity profile is measured with sample removed from the beam. The beam polarization at the sample is usually close to 90% and is reduced mainly by the effect of depolarizing stray fields along the beam path and by the inefficiency of the flipper (which is a single wavelength device and so can only be tuned for the peak wavelength). The value of the beam polarization S_0 obtained using a supermirror at the sample position is used to correct the measured spin asymmetry S_m according to the relation:

$$S = \frac{S_m}{S_o}. \tag{6.34}$$

6.4 Experimental Results for Fe Films

6.4.1 Magnetic Moments in Ultrathin Fe Films

Epitaxial growth techniques make it possible to prepare metastable fcc Fe and strained bcc Fe ultrathin films, both of which are predicted to show a sensitive dependence of magnetic moment upon atomic volume [6.39] and thickness [6.40] Until recently the growth mode of fcc Fe and the sensitivity of the magnetic properties to growth conditions has been a controversial subject, although a consensus is beginning to emerge concerning the magnetic properties. For bcc Fe, there is better agreement concerning the growth and magnetic properties. However early PNR experiments on bcc Fe/Ag films did not confirm the predicted enhanced moments whereas later studies of high quality samples were found to show enhanced moments using PNR. The comparison reveals that a high structural quality (in particular substrate flatness) is required for accurate PNR studies. This is because the structural quality not only affects the magnetic properties themselves but because rough surfaces strongly perturb the neutron reflectivity [6.41]. We will now discuss recent results on bcc Fe films obtained by the groups in Cambridge (U.K.) and Simon Fraser (B.C., Canada).

Recently developed first-principles band structure calculations have predicted enhanced magnetic moments for ultrathin Fe films supported by noble metal [6.42–44], Cu [6.45] and Pd [6.42] substrates. The origin of the enhancement is principally associated with the significantly increased density of 3d states for films of less than 5 ML thickness (Chap. 1 of this volume) which is only weakly reduced by sp–d hybridization with the non-magnetic substrate. Pd is of particular interest since a significant magnetic polarization is thought to be induced by Fe on the Pd interface layers [6.42]. Several investigations of epitaxial Fe films have been carried out by PNR to search for the predicted enhancement. [6.41, 46]. In an early PNR experiment on fcc Fe/Cu (0 0 1) films no in-plane ferromagnetism was detected. The existence of an antiferromagnetic phase has been deduced for fcc Fe/Cu (0 0 1) epitaxial films using Mossbauer spectroscopy [6.47] and this result is consistent also with recent theoretical calculations [6.40], although the existence of the antiferromagnetic phase is dependent on growth conditions. The epitaxial growth of well defined bcc phase Fe(0 0 1) layers can be experimentally achieved for Ag(0 0 1) [6.48–49] and Pd(0 0 1) [6.50–51] single crystal substrates, as has been described in the chapter by *Heinrich* in Volume II. Reflection High Energy Electron Diffraction (RHEED) patterns and RHEED intensity oscillations indicate that the interface roughness is limited to one monolayer [6.48–50]. Ultrathin Fe layers therefore provide an ideal model system for systematic experimental study of the role of thickness, interface effects and reduced atomic coordination in determining the absolute value of the magnetic moment per atom, although to date, due to the experimental difficulties involved, conclusive evidence for enhanced magnetism in thin Fe films and of the relative influence of different non-magnetic metal overlayers has been lacking.

The sandwich structures given in Table 6.1 were prepared by molecular beam epitaxy (MBE), as described in [6.50]. Ni, Pd, Cu and Ag epitaxial overlayers were used for Fe layer thicknesses close to 5 ML. The comparison of the moment for 10.9 ML Fe sample with that of thinner samples prepared in the same way provides a crucial test of the theoretical prediction that significantly enhanced moments occur in the vicinity of the film interfaces [6.43] and the complete set of samples permits the effect of the overlayer to be systematically studied. The thicknesses of the films were determined by RHEED intensity oscillations and were found to be reproducible to within 0.2 ML.

In Fig. 6.9 we show as a function of q_f the raw beam profile and background intensity (dotted line) for a Au/Ag/10.9 ML Fe/Ag(001) reference sample for at $q_i = 2.8 q_c$ using the multidetector operated at high resolution. It can be seen that at this value of incident wavevector the background intensity is small but that it constitutes a significant fraction of the spin-up intensity and corresponds to around 8% of the spin-down specular intensity at this wavevector. Since magnetometric measurements are usually made using data at lower wavevectors the background intensity can often be neglected. In the absence of diffuse scattering and surface undulations, the profile for $q_i < q_c$ has the same wavevector width Δq_f as the incident beam FWHM of $\Delta q_i = 0.16 q_c (\sim 6.7 \times 10^{-4} \text{ Å}^{-1})$. The intensity profile obtained at high resolution and integrated over three vertical pixels is shown in Fig. 6.10a for a 20 ML Au/7ML Ag/5.5 ML Fe/Ag

(001) sample as a function of scattered perpendicular wavevector $(q_f - q_i)/q_c$. At $q_i = 0.8 q_c$ the beam is close in shape to that incident upon the sample. This shows that the macroscopic surface undulations correspond to a variation in glancing angle which is less than the angular spread of the incident beam. By carefully fitting the form of the profile using (6.33), the diffuse scattering is found to correspond to a constant correlation length $w = 198 \pm 20$ Å and a roughness amplitude $\sigma^{1/2} = 14 \pm 2$ Å [6.52]. In the range $q_i > q_c$ the relative strength of the diffuse scattering accepted by the detector compared with the specular reflectivity is found to increase with wavevector, as expected from (6.32): as a fraction of the total signal, the diffuse scattering intensity accepted for a detector solid angle $\Delta\Omega = 2.8 \times 10^{-4}$ sr, corresponding to 3×3 pixels is found to be $\sim 8\text{-}10\%$ for $q_i = 1.5 q_c$, reaching $\sim 20\%$ for $q_i = 2 q_c$ and $\sim 80\%$ for $q_i = 3 q_c$ [6.52]. In Fig. 6.10b the contribution of the diffuse scattering to the detected signal is shown. In an earlier PNR study a Ag/8 ML bcc Fe/Ag(0 0 1) structure was prepared on a vicinal Ag(0 0 1) crystal surface [6.32–46]. The diffuse scattering is strong enough in this case to strongly perturb the reflected beam profile. Fitting of the reflectivity curve yields a value for the roughness amplitude $\sigma^{1/2}$ of ~ 50 Å for the Ag substrate, considerably higher than that found for the singular substrates. For a GaAs(0 0 1) water with a high degree of substrate flatness, the profile at large wavevectors remains narrow with a width close to that of the incident distribution.

In fitting the specular neutron reflectivity, the nuclear scattering lengths appropriate to the bulk materials and the experimentally determined values of lattice parameter and layer thickness are assumed. The degree of strain in the layer is sufficiently small for bulk densities to be assumed. The only variable parameters are therefore the layer-dependent magnetization and the roughness amplitude. However the spin asymmetry is insensitive to small roughness amplitudes and so the spin asymmetry can be fitted with the magnetic moment being the only adjustable parameter. In Fig. 6.11a we show the reflectivity and the spin asymmetry corrected for partial incident beam polarization (87%) and background intensity observed for the 20 Au/7 Ag/5.5 Fe/Ag(0 0 1) and the asymmetry calculated assuming uniformly magnetized ferromagnetic layers with a moment per atom of $\mu_{Fe} = 2.6\,\mu_B$ (shown as a solid line) and assuming $\mu_{Fe} = 2.2\,\mu_B$ (shown as a dashed line). We have also assumed that no spin polarization is induced in the Ag substrate. The reflectivity is well-fitted for both spin states in the vicinity of the enhancement peak, although the fit is less good at smaller wavevectors, possibly due to beam distortion. By comparing the observed asymmetry with that calculated for $\mu_{Fe} = 2.2\,\mu_B$ we see the moment is clearly enhanced with respect to the bulk value. The error in thickness determination in this case is small (around 4%). It should also be noted that since the magnetic moment is directly determined in PNR, it is sufficient to know the magnetic layer thickness in ML (as yielded by RHEED) rather than in absolute units. This is because the perpendicular lattice constant is cancelled in the product $\rho_i r_i$. The wavevector spread of the incident beam $\Delta q_i = 0.16\, q_c$ has a negligible effect on the spin asymmetry except at $q_i = q_c$ where a rounding of

6.4 Experimental Results for Fe Films

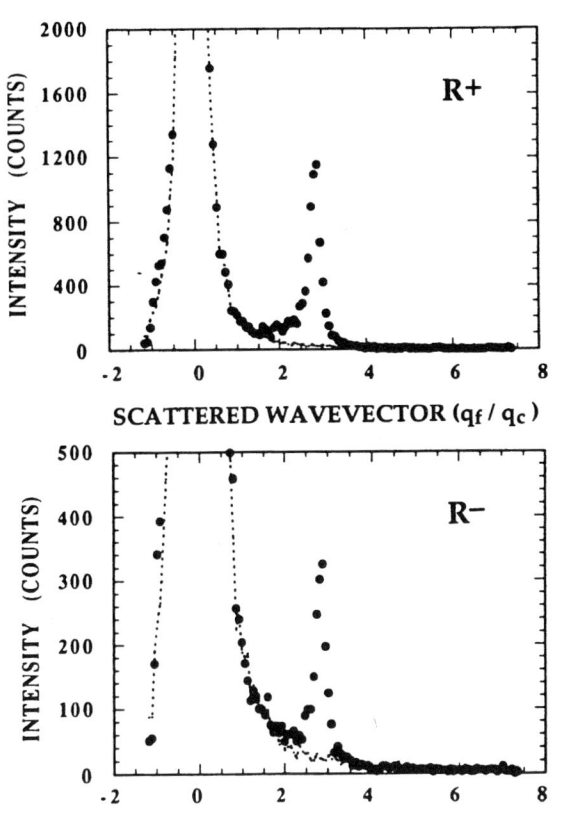

Fig. 6.9. The beam profile (solid points) and background intensity (dashed line) for the Au/10.9 ML Fe/Ag sample referred to in the text measured as a function of scattered perpendicular wavevector at q_f for $q_i = 2.8\, q_c$. The scattering geometry is given in Fig. 6.7. The detector position Y is proportional to q_f. The peak at $q_f = 0$ corresponds to the transmitted beam. The negative values of q_f refer to the fact that the detector extends beyond the forward beam position. The peak at $q_f = 2.8\, q_c$ corresponds to the reflected beam. The reduction in background intensity with increasing wavevector is clearly revealed. This occurs because much of this intensity is small angle scattering associated with the collimating slits for example. The skewing of the reflected beam profile at small wavevectors is due in part to the fact that there is a spread of wavevectors incident upon the sample (due to the finite incident beam collimation) but the smaller wavevectors are reflected more strongly. Additionally diffuse scattering occurs as discussed in the text. Accounting for all of these contributions to the reflected intensity is important in quantitative PNR magnetometry

the reflectivity is observed. By varying the fitted moment we conclude that $\mu_{Fe} = 2.58 \pm 0.09\, \mu_B$ provides the best estimate of the layer averaged moment per atom for the 5.5 ML thickness sample from the χ^2 variation with μ_{Fe}. The calculations of *Ohnishi* et al. [6.2] for a Ag/5.5 ML Fe/Ag (0 0 1) sandwich structure predict a moment of 0.08 μ_B for the first Ag layer which is too small to be determined in the present experiments, and a layer averaged moment per Fe atom of 2.4 μ_B, close to our result. Comparable values of the layer-averaged moment per atom are also expected for Fe layers overcoated with Au [6.3]. The

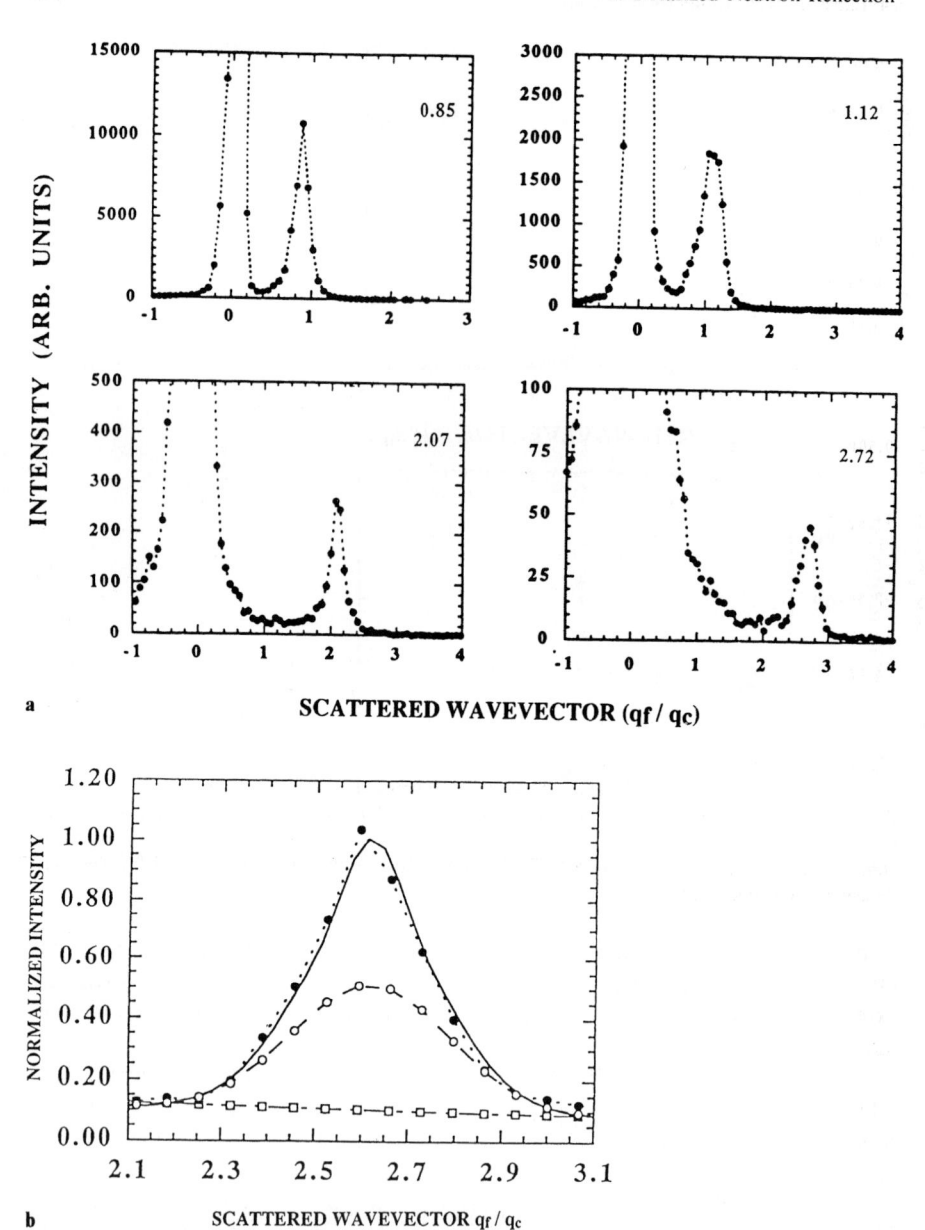

Fig. 6.10. a The measured beam profile for the Au/Ag/5.5 ML Fe/Ag sample (referred to in the text and in Table 1) as a function of scattered perpendicular wavevector q_f for values of the incident wavevector q_i of (a) 0.85 q_c (b) 1.12 q_c (c) 2.07 q_c. and (d) 2.72q_c. At small q_i the beam profile is similar to that of the incident beam, showing that the sample surface is reasonably flat. At larger incident wavevectors, the reflected beam contains an increasing component due to diffuse scattering associated with interface inhomogeneities. **b** The reflected beam profile of the Au/7 ML Ag/5.5 ML Fe/Ag(0 0 1) sample for $q = 2.6 q_c$ as a function of reduced final wavevector q_f/q_c. The measured data (full circles), the background (squares) and the diffuse contribution (open circles) are shown. The fit to the data is shown as a solid line

6.4 Experimental Results for Fe Films

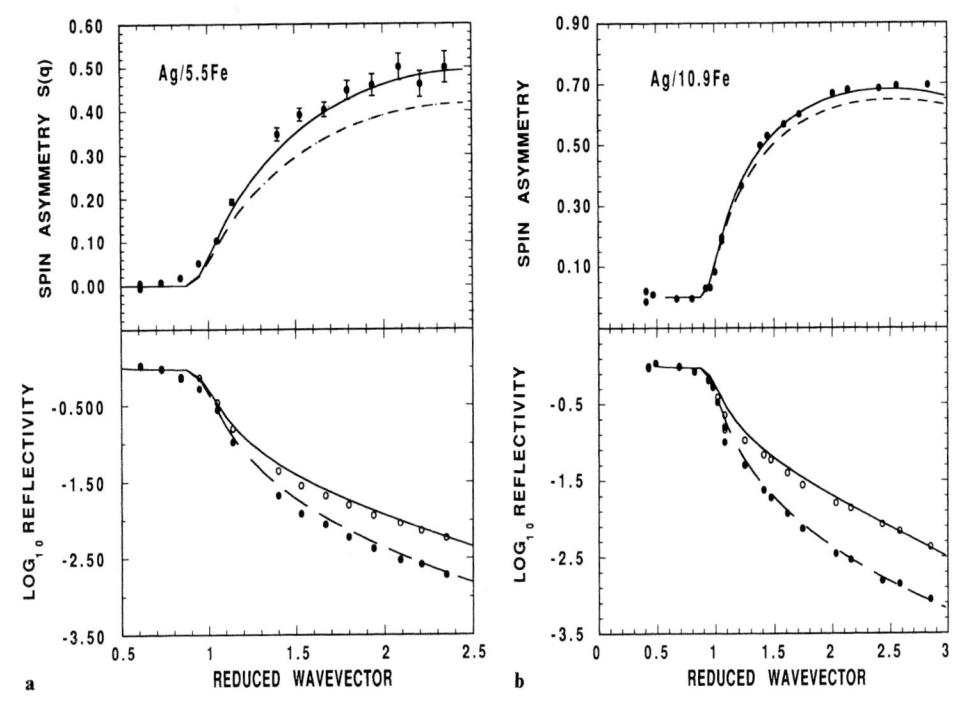

Fig. 6.11. The background corrected reflectivity and spin asymmetry obtained at low temperature for: (a) 20 Au/7 Ag/5.5 Fe/Ag(0 0 1) (b) 20 Au/7 Ag/10.9 Fe/Ag(0 0 1). The dashed and solid lines in the plots of the reflectivity refer to the spin-down and-up states, respectively. The dashed and solid lines in the plots of the spin asymmetry refer to model fits for the bulk moment and assuming an enhanced moment, respectively (see text). The spin asymmetry data are corrected for the spin-dependent diffusive scattering

calculations predict a significantly enhanced moment (around 2.5 μ_B) for the interface Fe layer for Ag and Au sandwiches but our measurements do not extend sufficiently far in wavevector to allow direct observation of the predicted magnetization profile.

The spin asymmetry of the Au/Ag/10.9 ML Fe/Ag structure is shown in Fig. 6.11b. The dashed line shows the asymmetry calculated for $\mu_{Fe} = 2.2\ \mu_B$ and the solid line corresponds to $\mu_{Fe} = 2.3\ \mu_B$ and we estimate the layer-averaged moment per atom to be $\mu_{Fe} = 2.33 \pm 0.05\ \mu_B$ from the variation of χ^2. The reflectivity is well-fitted for both spin states throughout the wavevector range. The result for the layer averaged magnetic moment per Fe atom is consistent with the enhanced interface Fe atom moments predicted by *Ohnishi* et al. [6.2]. The ratio of the magnetizations of these two samples was determined to within 2% by ferromagnetic resonance (FMR) (Table 6.1). The relative increase of the moment of the 5.5 ML sample with respect to the 10.9 ML reference sample provides conclusive evidence for an increase of the moment with reduced thickness while the PNR measurements show that in both cases the absolute value of the moment is enhanced with respect to the bulk value, yielding a ratio in agreement with that deduced by FMR within experimental error. FMR measurements of the relative magnetization were reproducible to 2% accuracy for all of the samples shown in Table 6.1. It should be noted that the PNR results provide a more accurate value for the magnetic moment than for the ratio of moments between samples: an overall accuracy of $\pm 3\%$ is obtained for the absolute moment of the 5.5 ML sample. FMR does not yield the absolute value of the moment of the sample.

In order to assess the reproducibility of the PNR measurements and the question of the structures best suited to PNR, two sets of samples were prepared and measured in a separate experimental investigation at low detector resolution. For the second versions of each sample a total overlayer thickness closer to 50 ML was used (Table 6.1). In these versions the degree of surface flatness is poorer in several cases, as judged by the form of the reflectivity as a function of wavevector in the vicinity of the critical angle. A large error is obtained for the second measurements which is associated with the increased diffuse scattering and poorer surface flatness, but within experimental error, the estimates of the absolute value of the magnetic moment are in agreement for the two investigations. The question of the structures best suited to PNR magnetometry is an important one: for thin overlayers the wavevector at which the asymmetry first peaks is large and hence a background subtraction needs to be carefully made. Also the diffuse scattering needs to be substracted at large wavevectors, making thicker overlayers desirable for a given substrate roughness amplitude. However there is clearly an upper limit in the usable overlayer thickness since the enhancement peak is shifted closer to the critical wavevector with increasing thickness where beam distortion (due to the incident wavevector spread and a lack of surface flatness) and wavevector resolution effects are important. Our study shows the importance of preparing high quality samples on sufficiently flat substrates since the most accurate estimates of the magnetic moment were achieved for samples showing the best flatness but with thinner overlayers. For all of the samples prepared on singular substrates, good fits to the reflectivity data can be obtained up to intermediate wavevectors using only an expression for the roughness modified reflectivity of the Nevot–Croce form with roughness amplitudes around 1 nm. This is perhaps surprising since it would appear to suggest that the film interface is ill defined. Clearly interference of the neutron wave within the structure would not occur if this were the case (see the discussion in Sect. 6.2.3). The value for the roughness amplitude corresponds to an effective value associated with the variations in the surface of the substrate over the entire sample area sampled by the neutron beam but on a local scale (as probed by short coherence length probes such as LEED or RHEED) the interface is indeed atomically sharp.

6.4.2 Comparison of the Experimentally Determined Moment with Theory

The combined results obtained for the Cu, Au, Pd and Ni coated bcc Fe samples are summarized in Table 6.1. The PNR and FMR measurements of the magnet-

6.4 Experimental Results for Fe Films

Table 6.1. Parameters of the bcc Fe/Ag(001) samples

Sample. [Thickness in ML]	μ_{Fe} [μ_B]	$\frac{\mu_{Fe}}{\mu_{bulk}}$	$\frac{M_{FMR}}{M(5.7)}$	$\frac{M_{PNR}}{M(5.7)}$	$\frac{2K_1}{M_s}$ [kOe]	$4\pi M_{eff}$ [kG]
20 Au/7 Ag/5.5 Fe/Ag(001)	2.58 ± 0.09	1.16 ± 0.04	0.99 ± 0.01	1.03 ± 0.05	0.218	1.21
20 Au/7 Ag/10.9 Fe/Ag(001)	2.33 ± 0.05	1.05 ± 0.02	0.93 ± 0.01	0.93 ± 0.05	0.544	7.06
52 Au/5.7 Fe/Ag(001)	2.5 ± 0.1	1.13 ± 0.05	1.00	1.00	0.255	7.44
20 Au/9 Fe/Ag(001)	2.3 ± 0.2	1.03 ± 0.09			0.479	9.702
20 Au/7 Cu/5.8 Fe/Ag(001)	2.48 ± 0.08	1.12 ± 0.04	1.02 ± 0.01	0.99 ± 0.05	0.325	0.94
42 Au/8 Cu/5.7 Fe/Ag(001)	2.5 ± 0.1	1.13 ± 0.05		1.0 ± 0.06	0.378	0.91
20 Au/7 Pd/5.6 Fe/Ag(001)	2.66 ± 0.05	1.20 ± 0.02	1.03 ± 0.01	1.06 ± 0.04	0.223	10.91
42 Au/8 Pd/5.7 Fe/Ag(001)	2.6 ± 0.2	1.17 ± 0.09	1.044 ± 0.01	1.04 ± 0.09	0.225	9.87
24 Au/3 Ni/5 Fe/Ag(001)	2.6 ± 0.1*	1.17 ± 0.05		1.04 ± 0.06	0.212	8.30

* This value for the Fe moment assumes that the Ni moment is close to the bulk value of 0.6 Bohr magnetons.

The magnetic and structural parameters of the bcc Fe/Ag(001) sandwich structures are defined in column 1. The second column gives the value of the Fe layer magnetic moment used in fitting the spin asymmetry data. The third column gives the ratio of the layer averaged moment per Fe atom estimated by PNR to the bulk moment (2.22 μ_B). The fourth column lists the ratio of the moment per atomic layer for the sample scaled by that of a 20 Au/5.7 ML Fe/Ag(001) reference sample as determined by FMR at 77 K. The fifth column gives the corresponding value estimated from the PNR measurements at low temperature. The sixth column lists the in-plane anisctropy strength determined by FMR. The seventh column lists the value of $4\pi M_{eff}$ as measured by FMR

ization of the samples relative to that of a reference sample are found to agree within experimental error. In all cases the Fe moment per atom is found to be significantly enhanced with respect to the bulk value [6.53]. These combined results show that the magnetic moment at the Fe/Cu interface is not significantly decreased compared to that at the Fe/Ag interface. The reported decrease of the magnetic moment in Cu/Fe interfaces [6.45] therefore seems to be incorrect. The result for Fe/Pd is in reasonable agreement with FMR measurements of the magnetization normalized to that of a Au/5.7 ML Fe/Au reference sample (see Table 1), which yield a relative enhancement of $3 \pm 1\%$ of the Fe/Pd sample magnetization, assuming no Pd polarization. *Blugel* et al. [6.42] have predicted an induced magnetic moment of 0.32 μ_B and 0.17 μ_B on the first and second Pd layers respectively. If this were to occur in our sample then the average magnetic moment per Fe atom deduced from the PNR data would be reduced by approximately 0.1 μ_B. The accuracy of the combined measurements is therefore insufficient to determine whether such an interface polarization occurs or not. However, we can conclude that the Fe layer magnetization is enhanced in either case with respect to the bulk, but we cannot distinguish between the degree of enhancement observed for Fe/Pd and that observed for the other interfaces. We can also exclude the possibility of ferromagnetism occuring in the entire Pd film [6.53]. The results for the Ni/Fe sample also fit well with the observed thickness-dependent trends. No significant differences are found between the degree of enhancement for Ag/Fe and Au/Fe interfaces, in agreement with the theoretical predictions. The combined results are compared with the theoretical predictions for the Ag/Fe interface in Fig. 6.12. It is seen that the experimentally determined values of the moment are in general higher than the predicted values for the 5–6 ML thickness films but that very good agreement is obtained for the 10.9 ML film. The former result is surprising since in general experimental factors tend to reduce the spin asymmetry (diffuse scattering in particular). It is possible that the trend can be explained by roughness on a scale too small to

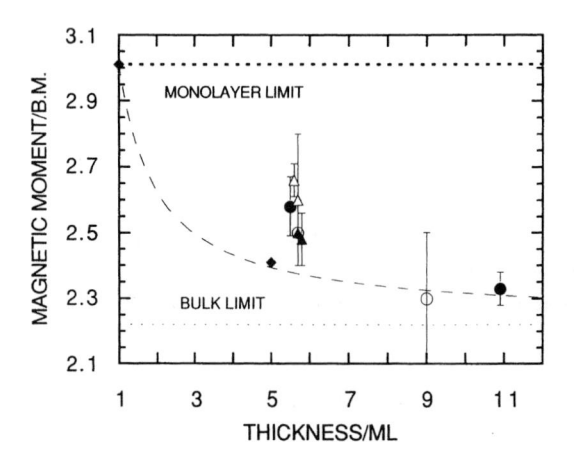

Fig. 6.12. The values of the layer averaged moment per Fe atom deduced from PNR measurements for samples referred to in Table 6.1 and compared with the predictions for the layer averaged moment for 1 ML Fe/Ag(0 0 1) [6.42] and Ag/5 ML Fe/Ag [6.43] shown as solid diamonds. The dashed line is a guide to the eye only. The Ag/Fe data is shown as solid circles, the Au/Fe data as open circles, the Cu/Fe data as solid triangles and the Pd/Fe data as open triangles

significantly perturb the neutron reflectivity but large enough to reduce the effective coordination of a significant number of Fe atoms. These experimental results should play a role in stimulating further theoretical work on real interface systems for which roughness is incorporated into the calculations.

6.4.3 Conclusions

We have shown that PNR provides a useful magnetometric technique which has been successfully used in determining the absolute value of the magnetic moment in embedded ultrathin Fe films of known thickness. In favorable cases, the magnetic moment can be determined to within 5% accuracy, as a test of theoretical predictions of enhanced moments. No attempt has been made to review the use of PNR in this field: a brief review by *Felcher* [6.54] outlines some of the recent work on magnetic films. The particular value of PNR is due to the unique combination of magnetic and structural information that it provides and its applicability to embedded layers. The magnetic moment and layer thickness can be independently determined for supported ferromagnetic layers in the few nm thickness range. The total magnetic moment of the magnetic layer is yielded directly while no magnetic signal due to the non-magnetic substrate occurs. We have discussed the regions of wavevector in which multiple reflection, single reflection and kinematic diffraction approximations are valid. We demonstrate that it is the long transverse coherence length which can be achieved experimentally that makes it possible to carry out PNR measurements on samples with roughness amplitudes in the nm range, but that the surfaces need to be flat typically to within 10^{-3} radians. The angular profile of the reflected beam permits the macroscopic flatness to be determined and the roughness correlation length can also be estimated in certain cases, while the specular reflectivity curve versus wavevector can be used to yield the effective roughness amplitude. For thicker films PNR provides a means of probing the magnetization profile on nm lengthscales and, in the case of magnetically coupled films, can be used to probe the magnetization vector profile. PNR also provides a means of determining the temperature-dependent magnetization of ultrathin films [6.55]. In the future, more intense neutron sources may permit measurements at sufficiently large wavevectors that the thicknesses of ultrathin layers can be directly determined, thereby removing the need to calibrate the thickness of the films using *in-situ* analysis. However this will require significantly flatter samples than it is currently possible to achieve on single crystal substrates since the upper limit in wavevector is often determined by the diffuse scattering due to interface roughness. More detailed information concerning the vector character of the magnetization in single films and multilayers can be obtained by using polarization analysis of the reflected beam [6.18]. The PNR technique therefore has considerable future potential as a structural and spatially resolved magnetometric probe of ultrathin films and multilayers.

Acknowledgements. I would like to thank my research students and also my colleagues, especially Prof. B. Heinrich, Drs. Z. Celinski, J. Penfold, V. Speriosu, B. Gurney and Dr. H.P. Hughes, for their part in the PNR experiments described in this chapter and for valuable discussions. The support of the staff of the Institut Laue-Langevin and the Rutherford-Appleton Laboratory is gratefully acknowledged.

References

6.1   R. Richter, J.G. Gay, J.R. Smith: Phys. Rev. Lett. **54**, 2704 (1985)
6.2   C.L. Fu, A.J. Freeman, T. Oguchi: Phys. Ref. Lett. **54**, 2700 (1985)
6.3   W. Marshall, S.W. Lovesey: *Theory of Thermal Neutron Scattering*, Oxford University Press, Oxford 1971)
6.4   W.G. Williams: *Polarised Neutrons*, Oxford University Press, Oxford 1988)
6.5   N. Hosoito, K. Mibu, S. Araki, T. Shinjo, S. Itoh, Y. Endoh: J. Phys. Soc. Jpn. **61** 300 (1992)
6.6   E. Fermi, W. Zinn: Phys. Rev. **70**, 103 (1946)
6.7   V.F. Sears: *Neutron Optics*, (Oxford University Press, New York 1989)
6.8   G.P. Felcher: Phys. Ref. **B24**, 1595 (1981)
6.9   J.A.C. Bland, D. Pescia, R.F. Willis, O. Schaerpf: Physica Scripta **35**, 528 (1987)
6.10 E. Fermi: Ric. Sci. **7**, 13 (1936)
6.11 E. Fermi, L. Marshall: Phys. Rev. **71**, 666 (1947)
6.12 S.G. Lipson, H. Lipson: *Optical Physics* (Cambridge University Press, Cambridge 1969)
6.13 M. Lax: Rev. Mod. Phys. **23**, 287 (1951)
6.14 E. Feenberg: Phys. Rev. **40**, 40 (1932)
6.15 R.G. Newton: Am. J. Phys. **44**, 639 (1976)
6.16 S. Dietrich, R. Schack: Phys. Rev. Lett. **58**, 140 (1987)
6.17 M. Born, E. Wolf: *Principles of Optics* (Pergamon, Oxford 1970)
6.18 J.A.C. Bland, H.P. Hughes, S.J. Blundell, N.F. Johnson: J. Magn. Magn. Mat., in press (1993)
6.19 S.J. Blundell, J.A.C. Bland: Phys. Rev. **B46**, 3391 (1992)
6.20 J.A.C. Bland, R.D. Bateson, N.F. Johnson, S.J. Blundell, V.S. Speriosu, S. Metin, B.A. Gurney: J. Magn. Magn. Mat. **123**, 320 (1993)
6.21 J.A.C. Bland, D. Pescia, R.F. Willis: Phys. Rev. Lett. **58**, 1244 (1987)
6.22 J.A.C. Bland, D. Pescia, R.F. Willis: Physica Scripta **T19**, 413 (1987)
6.23 D. Pescia, R.F. Willis, J.A.C. Bland: Surf. Sci. **189/190**, 724 (1987)
6.24 J. Penfold, R.K. Thomas: J. Phys. CM **2**, 1369 (1990)
6.25 J. Lekner: *Theory of Reflection* (Martinus Nijhof, Dordrecht 1987)
6.26 S.S. Parkin, V.R. Deline, R.O. Hilleke, G.P. Flecher: Phys. Rev. **B 42**, 10583 (1990)
6.27 J.A.C. Bland, R.D. Bateson, P.C. Riedi, R.G. Graham, H.J. Lauter, C. Shackleton, J. Penfold: J. Appl. Phys. **69**, 4989 (1991)
6.28 L. Nevot, P. Croce: Revue Phys. Appl. **15**, 761 (1980)
6.29 J.A. De Santo, R.J. Wombell: Waves in Random Media **1**, S41 (1991)
6.30 S.K. Sinha, E.B. Sirota, S. Garoff, H.B. Stanley: Phys. Rev. **B38**, 2297 (1988)
6.31 A. Steyerl: Z. Physik **254**, 169 (1972)
6.32 J.A.C. Bland, A.D. Johnson, R.D. Bateson, S.J. Blundell, H.J.L. Lauter, C. Shackleton, J. Penfold: J. Magn. Magn. Mat. **93**, 513 (1991)
6.33 J. Penfold: Physica B **173**, 1 (1991)
6.34 R. Felici, J. Penfold, R.C. Ward, W.G. William: J. Appl. Phys. A **45**, 169 (1988)
6.35 G.P. Felcher, R.O. Hilleke, R.K. Crawford, J. Haumann, R. Kleb, G. Ostrowski: Rev. Sci. Instrum. **58**, 609 (1987)
6.36 P.A. Degleish, J.B. Hayter, F. Mezei: *Neutron Spin Echo, Lecture Notes in Physics No. 128*, ed. by F. Mezei (Springer, Berlin, Heidelberg 1980)

6.37 G.M. Drabkin, E.I. Zabidarov, Y.A. Kasman, A.I. Okorokov: Sov. Phys. JETP **29**, 261 (1969)
6.38 D.J. Hughes and M.T. Burgy, Phys. Rev. **81**, 498 (1951)
6.39 D. Bagayoko, J. Callalway: Phys. Rev. **B 28**, 5419 (1983)
6.40 A.J. Freeman, C.J. Fu, S. Ohnishi, M. Weinert: *Polarised Electrons in Surface Physics*, ed. by R. Feder, Advanced Series in Surface Physics (World Scientific, Singapore 1985)
6.41 J.A.C. Bland, R.D. Bateson, A.D. Johnson, B. Heinrich, Z. Celinski, H.J. Lauter: J. Magn. and Magn. Mat. **93**, 331 (1991)
6.42 S. Blugel, B. Drittler, R. Zeller, P.H. Dederichs: Appl. Phys. A**49**, 547 (1989)
6.43 S. Ohnishi, M. Weinert, A.J. Freeman: Phys. Rev. B**30**, 36 (1984)
6.44 M.E. McHenry, J.M. Maclaren, M.E. Eberhart, S. Crampin: J. Magn. Magn. Mat. **88**, 134 (1990)
6.45 C.L. Fu, A.J. Freeman: Phys. Rev. **B35**, 925 (1987)
6.46 J.A.C. Bland, A.D. Johnson, C. Norris, H.J. Lauter: J. Appl. Phys. **67**, 5397 (1990)
6.47 W.A.A. Macedo, W. Keune: Phys. Rev. Lett. **61**, 475 (1988)
6.48 B Heinrich, S.T. Purcell, J.R. Dutcher, K.B. Urquhart, J.F. Cochran, A.S. Arrott: Phys. Rev. B. **38**, 12879 (1988)
6.49 B. Heinrich, Z. Celinski, J.F. Cochran, W.B. Muir, J. Rudd, Q.M. Zhong, A.S. Arrott, K. Myrtle, J. Krischner: Phys. Rev. Lett. **64**, 673 (1990)
6.50 Z. Celinski, B. Heinrich, J.F. Cochran, W.B. Muir, A.S. Arrott, J. Kirschner: Phys. Rev. Lett. **65**, 1156 (1990)
6.51 H. Chen, N.E. Brener, J. Callaway: Phys. Rev. **B40**, 1443 (1990)
6.52 R.D. Bateson, G.W. Ford, J.A.C. Bland, H.J. Lauter, B. Heinrich, Z. Celinski: J. Magn. Magn. Mat. **121**, 189 (1993)
6.53 J.A.C. Bland, R.D. Bateson, B. Heinrich, Z. Celinski, H.J. Lauter: J. Magn. Magn. Mat. **104–107**, 1909 (1992)
6.54 G.P. Felcher: Physica B, in press (1993)
6.55 J.A.C. Bland, G.A. Gehring, B. Kaplan, C. Daboo: J. Magn. Magn. Mat. **113**, 173 (1992)

Subject Index

ABCABC stacking sequence 52, 57
absolute refractive index 309
action integral 226
adsorbates as surfactants 261
adsorption 308
Ag/Fe/Ag(001) 333, 335, 339
Ag (100) 232
agglomeration 251, 255
  agglomeration of Fe on Cu(100) 263
alternating gradient magnetometry
    (AGM) 74
angular distributions of scattered
    electrons 248
angular momentum 245
angular resolution 329
anisotropic demagnetizing field 69
anisotropy
  surface 154, 164
  space 158
  spin 169
anisotropy energy as a function of band
    filling 45, 46
anisotropy energy density 40
Anisotropy field
  mean 169, 170
  transferred exchange 162
annealing 217
anti-Bragg condition 195, 217
anti-reflection condition 317
asperities 179
atomic layers 177–220
atomic multiplet structure 24
atomic sphere approximation (ASA) 42
atomic volume 253
Au/Fe/Ag(001) 339
Auger electron forward scattering 221
Auger electron spectroscopy 180
Auger emission 246
  Auger emission from Cu 224, 243
Auger forward scattering 222
average dipole moment 310

average magnetic anisotropy K 66, 67
band structure
  majority, minority spin 3d bands 43, 44
basic spin Hamiltonian
  anisotropy, dipolar, exchange & Zeeman
      terms 94
bcc cobalt 276
bcc Fe/Ag 332
beam coherence 191
beam polarization 332
BF_3 multidetector 331
binding site 242
biquadratic exchange 219
blocking temperature 158
Born approximation 195
Bose Einstein function 102, 112
Bragg condition 195, 200, 209, 210, 212
Brillouin function 165, 168
Brillouin light scattering (BLS) 73, 107

channeling 180, 217
classical electron channeling 249
classic model of Auger angular
    distributions 231
Co-Au superlattice 274
Co-Cu superlattice 277
coherence length 311
Co on Cu(001) 165, 252
Co-Ti polarizing supermirror 330
constructive interference peaks 240
continuous neutron source 329
contraction in layer spacing 254
core level 149
correlation length and roughness amplitude
    for a rough surface 326, 334
Coulomb limit 227
Cr on Ag(100) 252
Cr, sputtered spacer layer 160
critical exponents in magnetism 138
critical reflection of neutrons 306
critical wavevector 308

cross section for large-angle scattering of electrons 234
Cu(100) 248, 252
Cu(111) 261
Cu $2p_{3/2}$ core level 224
Cu/Fe/Ag(001) 339
Cu grown on Ni(100) 236, 251
Cu Auger lines 224, 243
Cu monolayer on Ni(100) 240
Cu scattering potential 224
Curie temperature 97, 114, 163

de Broglie wavelength 223
Debye-Waller effect 248
defects at surfaces 189, 193, 214
defocussing in forward electron scattering 238
demagnetizing energy 4
demagnetizing field 69
demagnetizing tensor 69
depolarization energy of ferromagnetic thin films 37, 38
depolarizing stray fields 332
depth resolution in neutron reflection 321
detector acceptance in neutron reflection 327
differential scattering cross section 227, 246
diffraction grating 182–199
diffuse cross section in the specular direction in neutron reflection 328
  diffuse intensity 327
  diffuse reflection 325
diffusion 216, 218, 219
dipole lattice sums 101
  at zero wave vector 101
dipole selection rules 243
Dirac equation 22, 24
disk-chopper rotating at 25 Hz 330
dispersion surface 202, 203, 208
domain structure
  domain size 76
  experiment of Allenspach et al. 108
  experiments reported by Pappas & co-workers 109
  linear 108
  wall energy 76
Drabkin two coil non-adiabatic spin flipper 330
dynamic theory 189, 192, 200–215
dynamical diffraction 249

effective anisotropy 66
effective electrostatic potential 25
effective induction 25
effective range of the potential 225

effective vector potential 25
elastic backscattered distribution 247, 249
elastic modulus 71
elastic scattering 178, 180, 191, 200
electron capture spectroscopy 138
electron density 189–191, 203
electron-electron interaction 22
enhanced moment for interface Fe atoms 337
enhancement in the spin asymmetry 318
epitaxial growth 216–219
epitaxial structure 278
etalon 311
evanescence 179
Ewald sphere 183, 184, 191, 197, 201, 217, 229
EXAFS 221
exchange asymmetries 136
exchange correlation potential 25
exchange diffusion 258, 259
exchange energy 4
  coefficient of exchange interaction 5
exchange field 25
exchange interaction 93
exchange length 5, 67, 99
exchange scattering 140
exchange splitting 142
exchange stiffness constant 94, 101, 160, 166

Faraday balance 74
fcc Cobalt 281
fcc Fe and strained bcc Fe ultrathin films 332
fcc Fe/Cu(001) 333
Fe and Co deposited on Cu(100) 255
Fe/Cr/Fe/Si 315
FeMn layer 321
FeNi 321
Fe on Ag(100) 241, 258
Fe on Cu(001) 252, 259, 261
Fe on Cu(110) 252
Fe on Cu(111) 260
Fe sputter deposited on Ta 167
Fe: growth on Cu 261
Fermi potential 308
ferromagnetic alignment 97
ferromagnetic resonance (FMR) 73
fine structure constant α^2 23, 25, 26
first principle calculations of MAE 37–39, 40, 45–51, 53–64
first principles electronic structure calculations 26
fit the observed reflected beam profile 328
fluxgate magnetometry 73
"focussed" 247

Subject index

'force theorem' 30
"forward focussing" 223
forward scattering 234, 248
forward scattering peaks 249
Fourier analysis 181, 185, 187–191, 205, 213, 214
Fresnel reflectivities 308
full-potential-linear-augmented-plane-wave (FLAPW) method 32, 52
funneling down model 218
FWHM 328

GaAs(001) 335
GaAs source 131
GaAs substrates 278
Gd 171
geometry for PNR 308
giant magnetic molecules 5
goniometer 216
Green's function 117
growth islands 196, 197, 217, 219
growth modes
   layer by layer 78
   Volumer-Weber 78
   Stranski-Krastanov 78
gyromagnetic ratio 308

Heisenberg model 93
He^3 single detector or multidetector 330
hysteresis loops 74, 159
   regular and inverted 161

impact parameter 225, 233, 234, 247
induced orbital moment 27
inelastic scattering 180, 181, 216
information depth 126
inner potential 236
in-plane strain 253
interdiffusion 271
interface anisotropy 40
interface quality 82
   annealing treatment 82
   diffuse interfaces 82
   sharpening of interfaces 82
interface roughness 322
interface states 148
inverse t dependence 72, 82
inversion of the reflectivity 311
iron 183, 193, 195, 200, 216, 217
islands 158
isotropic magnetostriction 70

K_2CuF_4 166
Kikuchi bands 222

Kikuchi lines 178, 180, 217
Kikuchi patterns 249
kinematical diffraction 320
kinematic theory 181–199
Kohn-Sham
   eigenvalues 41
   equations 24, 25, 32, 41
   orbital 25

L states 243
Lagrangian 226
Langmuire-Blodgett films 93
Large-angle scattering 234
lattice expansion of Fe or Co films 253
lattice mismatch 71
lattice registry
   coherent regime 71
   coherent-incoherent transition 81
   incoherent regime 71
   misfit dislocations 71
lattice strain 280
layer-by-layer growth 179, 216–219, 255
layer projected density of states 33
LEED 181, 182, 197, 213, 215, 216, 220, 229
LEED I–V 229
LEED pattern 248
linear muffin thin orbital (LMTO) 41
local density approximation (LDA) 24, 25, 41
local-spin-density approximation (LSDA) 41
long wavelength low frequency thermal excitations 103
long-range ferromagnetic order 92, 103
longitudinal coherence length 322
Low Energy Electron Diffraction (LEED) 86
low kinetic energies 235

MAE-convergence of the BZ integral 53
macroscopic sample flatness in neutron reflection 328
macroscopic surface undulations 334
magnetic anisotropies in ultrathin films (experiment)
   Co(111) wedge 82–84
   Co(111), Co(100), Co(110) 80
   Co/Ni multilayers 84, 85
   Fe(001), Fe(110), Fe(111) 79
   Ni(100), Ni(111) 79
Magnetic anisotropy (MA), energy (MAE) 4, 65, 281
   angular dependence of the magnetization 76
   "area method" 76
   easy plane 75

Magnetic anisotropy (*Contd.*)
  field dependent measurements 74
  perpendicular easy 75
  quartic inplane 107
  saturation fields 76
magnetic dipolar interaction 97
  in ellipsoidal ferromagnet 69
  shape dependent contribution 66
  strength 98
magnetic moment of bulk ferromagnets (band calculations)
  Co 27
  Fe 27
  Ni 28
magnetic moments of ferromagnetic thin films (band calculations)
  fcc Fe on Cu (100) 35
  Fe/Ag (100) 35
  Ni(001) 33
  W/Fe/Ag 35
magnetic moments of ultrathin ferromagnetic films (neutron reflections experiments) 319, 332–339
  Ag/Fe/Ag(001) 334, 335, 339
  Au/Fe/Ag(001) 337, 339
  Cu/Fe/Ag(001) 339, 340
  Pd/Fe/Ag(001) 339, 340
magnetically non-aligned layers 312
magnetization
  critical exponent 155
  orbital 154
  spontaneous 153
magnetization at non-zero temperature 103
  beyond the low temperature regime 111
  parameter $\Delta(T)$ 106, 107, 111
magnetization orientation 312
magnetization profile 321
magneto-optic Kerr effect (MOKE) 73, 77
  polar Kerr effect measurements 77–78
MA of bulk ferromagnets (band calculations)
  Co 27
  Fe 27
  Ni 28
Ma of ultrathin films (band calculations)
  Co multilayer 45
  Co/Ni multilayers 57–60
  Co/Pd (111) multilayers 53–56
  Co/Pd (001) multilayers 56, 57
  Fe monolayer on Ag(001), Cu(001), Au(001) and Pd(001) 39
magnetoelastic energy, anisotropy 54, 66, 81, 82
magnetometric information in neutron reflection 315

magnetostriction constant 54, 71
magnetostrictive anisotropy 66
Mathieu's equation 205
matrix methods 213–215
mean free path 126
Mermin-Wagner Theorem 102, 114
misfit dislocations 71, 279
MnAg(100) 254
Mn on Ag(100) and Cu(100) 252, 254
Mn films on Cu(100) 255
molecular beam epitaxy (MBE) 2, 182, 216, 276
Monochromatization and angular collimation in neutron reflection 322
Mossbauer spectroscopy
  hyperfine field measurements 106
Mott detector 109, 129
multidomain formation 315
multilayer 311
multiple forward scattering 236
multiple-scattering calculations 223, 238

Néel's model of surface anisotropy 60, 70
neutron magnetic dipole moment 308
neutron optics 306
Nevot-Croce 323, 327
next-nearest-neighbour atom 234
next-nearest-neighbour axes 231
next-nearest-neighbour forward scattering 232
Ni(100) 235
Ni(110) 258
Ni/Cu(100) 241, 261
Ni/5Fe/Ag(001) 339
Ni over the Cu monolayer 257
$Ni_{78} Fe_{22}$ (permalloy) 159, 163
non-aligned layers 311
non-flip scattering 141
nuclear magneton 308
nuclear potential 313
Nuclear Magnetic Resonance (NMR) measurements 82, 274

(1D) optical potential 307
optical potentials 307
orbital angular momentum 26, 94
overcoating (the ultrathin ferromagnetic film) 318
oxygen 263
oxygen in the Fe on Ag(100) system 263

pair correlation function 190, 196, 217
pair interaction energy, model 40, 70
Pauli equation 23
Pauli spin operator 308

Pd(111)   232
Pd/Fe/Ag(001)   339
pendulum magnetometry   74
perpendicular, in-plane orientation   48
perpendicular magnetic energy (PMA)   40, 65
perturbation of neutron reflectivity   315
phase change in neutron reflection from a
    magnetic layer   317
phase shifts due to scattering   234, 244, 245
photoelectron holography   231
photoelectrons, the angular distribution   246
plane-wave phase   246
plasmons   187
Poisson's ratio   254
polarisation analysis in neutron reflection   313
polarised neutron reflection (PNR)   305
product Kt versus t   67, 82
Pt   258
Pt(100)   258
pulsed source   329

quantum well (resonance) states   6
quasi-two-dimensional ferromagnets,
    films   92, 105, 110
quenching of forward scattering   244

Ramsauer-Townsend minima   227
random phase approximation (RPA)   118
    self energy diagrams (Hartree, exchange)   120
reciprocal lattice point (rlp)   179, 190–199, 212
reciprocity   247
reconstruction at a surface   185, 216
reflection and transmission   308
Reflection High Energy Electron Diffraction
    (RHEED)   2, 82
4 × 4 refraction matrix   312
refraction   236
relative refractive index   309
relativistic calculations   27, 29
renormalization group method   104, 110
    by Pescia and Pokrovsky   111
    quantum dominated systems   111
    temperature renormalization of the
        exchange constant   113
renormalized effective anisotropy   110
RHEED attenuation   192, 200–209
RHEED intensity   178–220
RHEED oscillations   179, 203, 216–220
rotating sample (RS) method   329
roughness amplitude   326

roughness correlations   325
RS   331

sandwich structures   320
scattering angle   233
Schrödinger equation   180, 181, 187, 189, 200–205, 210, 308
segregation   257
selfconsistent local-orbital (SCLO)   32
shape anisotropy   38, 82
    manipulation of the effective shape
        anisotropy   84
Sherman function   129
short-range order   249
single reflection approximation   320
single uncoated magnetic layer   318
small angle scattering, x-ray   275
Sm-Co permanent magnets   330
specular reflection   179, 184–187, 194–199
spherical-wave correction   246
spin asymmetry   153, 310, 318
spin blocks   158
spin conserving process   312
spin-dependent critical angle   310
spin flip scattering   140
spin flipping process   312
spin moment   27, 94
    raising and lowering operators   117
spinor symmetry   312
spin-orbit coupling matrix   42
spin-orbit interaction, coupling, term   26, 42, 44, 66, 94
    degenerate energy levels   47
    parameter, strength   42, 94
spin-orbit operator, computational difficulty
    arising from   29
spin-other orbit-interaction   24, 26
spin polarization
    definition   153
    detectors   128
    remanent   164
    saturation   164
    spontaneous   164
spin polarized electron source   131
spin-polarized Hamiltonian   41
spin-polarized spectroscopies   123
    Auger electrons   125
    electron diffraction   135
    electron energy loss   139
    field emission   152
    inverse photoemission   151
    photoemission   145
    secondary electrons   132
    tunnelling   152

spin quantization   312
spin rotation matrix R_{ij} 312
spin waves   93, 100, 165, 170
   "acoustical, optical"   105, 106
   dispersion   100
   fluctuations   165, 170
   "soft mode"   107, 108, 109
stacking faults   260, 282
stacking sequence of the spheres in Muffin in calculations   42
step densities   192, 217
Stoner exchange integral   43
Stoner excitations   94, 142
strain   70
   compressive   71
   in-plane   71
   residual   72
   tensile   71
   thermal   71
Stranski-Krastanov growth   251
stratified medium   311
stress   71
superconducting quantum interference device (SQUID)   73
superlattice   5–6, 271
surface alloying   255
surface and interface free energies   251
surface enhanced magnetic order   138
surface (interface) anisotropy energy   4, 40, 66, 70
surface relaxation   213
surface resonance   181
surface segregation   255, 256
symmetry breaking elements   65
superparamagnetism   159, 164

Ta, sputtered spacer layer   166, 170
temperature dependence of the magnetization M(T)   101
   phase transition to paramagnetic phase   104
   transition to quasi-three-dimensional behaviour   115
terraces   191–196, 213

theorist's view of film   91, 92
three layer medium   315
time of flight (TOF)   329
titration of CO or H_2 256
torque magnetometry measurements   73
torsion oscillating magnetometry (TOM) measurements   73
total coherent scattering   309
transfer matrix method   312
Transmission Electron Microscopy (TEM) measurements   82
transverse coherence length   322
two-beam approximation   249

ultrathin film   4, 91, 99
uncorrelated fluctuations   323
uncorrelated roughness   325
uniaxial anisotropy   66, 102
   strength   100

vector orientation   313
vertical atomic layer   253
vibrating sample magnetometry (VSM) measurements   73
vibrations in adsorbed diatomic molecules   238
vicinal surface   192–199, 216
Volmer–Weber growth   251
volume or bulk magnetic anisotropy energy   40, 66

wavelength resolution   329
wedged-shaped magnetic layers   82, 86
Wentzel-Kramers-Brillouin (WKB) approximation   229
whiskers   183, 200, 216, 217

XPS   221, 222
XPS and Auger angular distributions   241
x-ray diffraction experiments   82
x-ray scattering   264
"xy" model   95, 166

Zeeman interaction   305

Printing: Krips bv, Meppel
Binding: Litges & Dopf, Heppenheim